AF361856

Lignocellulosic Biomass

Lignocellulosic Biomass

Editors

Alejandro Rodríguez Pascual
Eduardo Espinosa Víctor

MDPI • Basel • Beijing • Wuhan • Barcelona • Belgrade • Manchester • Tokyo • Cluj • Tianjin

Editors
Alejandro Rodríguez Pascual
Chemical Engineering
Department
Universidad de Córdoba
Córdoba
Spain

Eduardo Espinosa Víctor
Inorganic Chemistry and
Chemical Engineering
Department
Universidad de Córdoba
Córdoba
Spain

Editorial Office
MDPI
St. Alban-Anlage 66
4052 Basel, Switzerland

This is a reprint of articles from the Special Issue published online in the open access journal *Molecules* (ISSN 1420-3049) (available at: www.mdpi.com/journal/molecules/special_issues/lignocellulosic).

For citation purposes, cite each article independently as indicated on the article page online and as indicated below:

LastName, A.A.; LastName, B.B.; LastName, C.C. Article Title. *Journal Name* **Year**, *Volume Number*, Page Range.

ISBN 978-3-0365-2475-7 (Hbk)
ISBN 978-3-0365-2474-0 (PDF)

Contents

About the Editors

Alejandro Rodríguez Pascual

Dr. Alejandro Rodríguez Pascual is a Full Professor of the Chemical Engineering Department. He studied Chemistry and Food and Science Technology before obtaining his doctorate in Chemical Sciences from the University of Córdoba in 2002. He worked as "Profesor Ayudante and Ayudante Doctor" at Pablo de Olavide University, Seville. In January 2007, he joined the Chem. Eng. Area of the University of Córdoba with a "Ramón y Cajal" Contract (call 2006). Initially, his research line was focused on the use of agricultural residues to obtain cellulosic pulps. As a result of the RyC contract as well as the management of two national projects, the applicant began to develop a new research line focused on the biorefinery of the lignocellulosic residues from agri-food activity. He has 112 indexed articles (70% Q1), >80 International Congresses. He co-authored three books and five book chapters and has edited one. He has been the supervisor of nine doctoral theses, currently supervising five (international mention).

Eduardo Espinosa Víctor

Dr. Eduardo Espinosa Víctor is Assistant Professor of the Chemical Engineering Department in Universidad de Córdoba (Spain). He graduated in Environmental Sciences in 2014 from the University of Cordoba, attained a Master's degree in Molecular, Cellular and Genetic Biotechnology to get onto a PhD in the area of Chemical Engineering at the University of Cordoba. He started his PhD in October 2015 after receiving a grant for the training of university faculty (FPU) in 2015 and finished it in September 2019 with the qualification of Outstanding Cum Laude and with International Doctorate Mention. He participated as part of a working team in a Spanish national project (CTQ2016-78729-R), as well as in eight other research projects as a research team. His research activity is focused on the biorefinery of lignocellulosic materials, especially agri-food waste, focused on the separation of the different fractions (cellulose, hemicelluloses and lignin) and their characterization and application.

Preface to "Lignocellulosic Biomass"

In the last century, the production and energy sectors have been highly dependent on fossil resources. Only in the last few years has the increase in oil prices and the environmental impact caused by some processes of obtaining and using petroleum products motivated a change in the thinking of the industrial sector and public organizations. Currently, more sustainable technologies are gradually being incorporated to produce clean energy and bio-based products that serve as raw materials for transformation or final use by consumers. This paradigm shift focuses on the use of available natural resources for the implementation of more economical and environmentally friendly production systems.

The origin of the concept of sustainable development is not up-to-date, as it was first defined in the Brundtland report published in 1987 (*"Our Common Future"*, Brundtland, 1987) and has served to nurture future policies such as the Montreal and Kyoto protocols, Agenda 21, the Millennium Development Goals (MDGs), and more recently, the Sustainable Development Goals (SDGs). These policies not only highlight the problem of energy supply and climate change, but the concept of sustainability encompasses much broader implications, such as food, production processes, agriculture, fisheries, and quality of life.

In this sense, lignocellulosic biomass is a valuable, renewable and undervalued source of chemicals for use in the processing industry and can be used directly or indirectly for the production of platform molecules or bioproducts through chemical, physical, microbial, or enzymatic treatments and can also be used in sectors such as food, health, medicine, energy, materials, and the chemical industry. In maintaining this process, the scientific community plays a very important role in generating the basic knowledge that gives rise to technology and allows developments in the laboratory to be transferred to society. The integral valorization of lignocellulosic biomass is a fundamental pillar of sustainable development. Given the origin of this biomass, as well as its composition, lignocellulosic biomass is a vast resource.

In creating this Special Issue, focused on the valorization of lignocellulosic biomass, we aim to establish an invaluable source of information that will serve as a reference for other researchers.

Alejandro Rodríguez Pascual, Eduardo Espinosa Víctor
Editors

Editorial

Special Issue "Lignocellulosic Biomass"

Alejandro Rodríguez * and **Eduardo Espinosa**

BioPrEn Group, Chemical Engineering Department, Universidad de Córdoba, 14014 Córdoba, Spain;
a02esvie@uco.es
* Correspondence: a.rodriguez@uco.es; Tel.: +34-957-21-22-74

Citation: Rodríguez, A.; Espinosa, E. Special Issue "Lignocellulosic Biomass". *Molecules* **2021**, *26*, 1483. https://doi.org/10.3390/molecules26051483

Received: 5 March 2021
Accepted: 8 March 2021
Published: 9 March 2021

Publisher's Note: MDPI stays neutral with regard to jurisdictional claims in published maps and institutional affiliations.

The use of lignocellulosic biomass as potential raw material for fractionation and transformation into high value-added products or energy is gathering the attention of scientists worldwide in seeking to achieve a green transition in our production systems. In this context, the aim of compiling recent advances in this field has motivated this Special Issue.

Lignocellulosic biomass is a valuable renewable and undervalued source of chemicals for use in the processing industry and can be used directly or indirectly for the production of platform molecules or bioproducts through chemical, physical, microbial, or enzymatic treatments and be used in sectors such as food, health, medicine, energy, materials, and the chemical industry. In maintaining this process, the scientific community plays a very important role in generating the basic knowledge that gives rise to technology and allows developments in the laboratory to be transferred to society. The integral valorization of lignocellulosic biomass is a fundamental pillar of sustainable development. Given the origin of this biomass, as well as its composition, lignocellulosic biomass is a vast resource.

In this Special Issue, twelve original research articles and three research reviews covering some of the most recent advances in lignocellulosic biomass fractionation and conversion for different applications are reported. Three articles deal with the production of biomass-derived biocomposites to replace oil-based products in different applications. Ehman et al. [1] reported on the use of bio-polyethylene (BioPE), sugarcane bagasse pulp, and two compatibilizers (fossil and bio-based) to produce biocomposite filaments for 3D printing. They show the influence of bagasse fiber fraction on the mechanical properties of the biocomposites as well as the reduction in CO_2-equivalent emissions from replacing fossil compatibilizers with a bio-based compatibilizers in a cradle-to-gate life cycle analysis. Serra-Parareda et al. [2] studied the feasibility of incorporating barley straw fiber as reinforcement in a bio-based polyethylene to develop a fully bio-based and 100% recyclable material. They analyze the efficiency of barley fibers by the addition of anhydride maleic polyethylene as coupling agent to determine the flexural behavior of the material to determine the suitability of the material for several applications. Karagiannidis et al. [3] evaluated the use of micro-fibrillated cellulose (MFC) in waterborne adhesive systems applied in the manufacture of composite wood-based panels. They test the potential of improving the performance of wood-based panel types such as particleboard, waferboard, or randomly-oriented strand board and plywood, by the application of MFC and the substitution of conventional and non-renewable chemical compounds.

Five articles explore and develop the use of biomass-derived products as new industrial alternatives for different applications. Ortiz et al. [4] designed and prepared fully bio-based epoxy resins by combining epoxidized linseed oil, lignin, and bio-based diamine derived from fatty acid dimers. They showed that as the lignin content in the resin increases, the glass transition, the Young's Modulus and the onset of thermal degradation increases. This correlation is non-linear, and the higher the percentage of lignin, the more pronounced the effect. All the components of the epoxy resin being commodity chemicals, the present system provides a realistic opportunity for the preparation of fully biorenewable resins at an industrial scale. Bascón-Villegas et al. [5] explored the use of horticultural plant residues (tomato, pepper, and eggplant) as new sources for lignocellulose nanofiber (LCNF) isolation using two different pretreatments, mechanical and TEMPO (2,2,6,6-Tetramethyl-piperidin-1-oxyl)-mediated oxidation, and followed by high-pressure homogenization. LCNF were added as reinforcing agent in recycled paperboard and compared with conventional mechanical beating. It showed that the addition of LCNF is a viable alternative to mechanical beating, achieving greater reinforcing effect and increasing the products' life cycles. Szufa et al. [6] investigated the effect of biogas plant waste on the physiological activity, growth, and yield of Jerusalem artichoke and the energetic usefulness of the biomass obtained in this way after the torrefaction process. They achieved an increase in the calorific value process from 15.82 to 22.12 MJ kg^{-1} by this process. Nikolic et al. [7] investigated the potential of kraft lignin as a support material for the removal of hydrogen sulfide (H_2S) from gaseous streams, such as biogas. The removal of H_2S was enabled by copper ions that were previously adsorbed on kraft lignin. They reported a removal capacity of 2 mg H_2S per gram of kraft lignin, and further studies in this technology being necessary to increase the viability of this technology. Michel et al. [8] reported the modification with β-cyclodextrin (βCD) of TEMPO-oxidized cellulose nanofibers (toCNF) samples with different carboxyl contents for biomedical applications. They reported covalent esterification binding between β-cyclodextrin and toCNF under acidic pH by freeze-drying, showing an interesting impact on the mechanical properties in the swollen state. This study is a step towards the production of mechanically tailored cryogels containing cyclodextrin, making them promising materials for the sustained delivery of active principal ingredients.

Three articles report novel advances in the use of biomass for the production of platform molecules and their conversion into high value-added products. Padilla-Rascón et al. [9] evaluated the conditions under which furfural concentration is maximized from a synthetic, single-phase, homogeneous xylose medium. They performed the experiments in a microwave reactor using $FeCl_3$ and sulfuric acid as catalysts, showing the best operational conditions for a 57% furfural yield production at 210 °C, 0.5 min, and 0.05 M $FeCl_3$. Oliva-Taravilla et al. [10] investigated the use of four biosurfactants, namely, horse-chestnut escin, *Pseudomonas aeruginosa* rhamnolipid, and saponins from red and white quinoa varieties, on the enzymatic saccharification of steam-pretreated spruce. They reported that the use of biosurfactants improved hydrolysis up to 24%, showing the potential of biosurfactants for enhancing the enzymatic hydrolysis of stem-pretreated softwood. Jiménez-Quero et al. [11] explored the optimization of solid-state fermentation (SSF) with lignocellulosic biomasses using *Aspergillus terreus* and *Aspergillus oryzae* to produced itaconic and fumaric acids. *A. oryzae* on corn cobs at specific conditions showed the best yield in acid production, achieving 0.05 mg of itaconic acid and 0.16 mg of fumaric acid per gram of biomass after 48 h in a large-scale fermentation process.

One of the articles present a novel protocol for the modification of dialdehyde cellulose (DAC) by silanization processing. Lucia et al. [12] presented a straightforward protocol, which meets common green chemistry principles, for the direct silanization of DAC, one of the most recently developed and studied cellulose derivatives, describing the direct silanization of DAC with (3-aminopropyl)triethoxysilane (APTES), through thermal treatment and freeze-drying.

Three review papers complete this Special Issue. The first one summarizes the evolution, year by year, of the development of the "lignin first" biorefinery approach [13]. A compact summary of achievements, future prospects, and remaining challenges is provided in this review. The second one outlines the possibility of applying known biorefinery processes to banana agro-industrial residues to generate high-value products from this residual biomass source [14]. It details information on the Central and Latin American context of this residue and the advantages of its use as raw material for the production of different biofuels, nanocellulose fibers, different bioplastics, and other high value products. The last review addresses the current knowledge on the electrochemical conversion of bio-based chemicals, particularly saccharides, to commodity chemicals [15]. Both oxidation and reduction pathways are shown with the most recent examples. Further recommendations are reported about the research needs, choice of electrocatalyst and electrolyte, as well as upscaling the technology.

Given the diversity of the contributions, it is evident that a multidisciplinary approach is needed to continue advancing the development of technologies and processes for the valorization of lignocellulosic biomass. There are still significant barriers to be overcome in order to achieve a complete transition of our production systems, from a petroleum-based economy to a bio-based economy. It is therefore expected that this field will be of particular relevance in the coming years. Finally, the guest editors would like to sincerely thank all the authors for their valuable contributions.

Funding: This research received no external funding.

Conflicts of Interest: The authors declare no conflict of interest.

References

1. Ehman, N.V.; Ita-Nagy, D.; Felissia, F.E.; Vallejos, M.E.; Quispe, I.; Area, M.C.; Chinga-Carrasco, G. Biocomposites of Bio-Polyethylene Reinforced with a Hydrothermal-Alkaline Sugarcane Bagasse Pulp and Coupled with a Bio-Based Compatibilizer. *Molecules* **2020**, *25*, 2158. [CrossRef] [PubMed]
2. Serra-Parareda, F.; Julián, F.; Espinosa, E.; Rodríguez, A.; Espinach, F.X.; Vilaseca, F. Feasibility of Barley Straw Fibers as Reinforcement in Fully Biobased Polyethylene Composites: Macro and Micro Mechanics of the Flexural Strength. *Molecules* **2020**, *25*, 2242. [CrossRef] [PubMed]
3. Karagiannidis, E.; Markessini, C.; Athanassiadou, E. Micro-Fibrillated Cellulose in Adhesive Systems for the Production of Wood-Based Panels. *Molecules* **2020**, *25*, 4846. [CrossRef] [PubMed]
4. Ortiz, P.; Vendamme, R.; Eevers, W. Fully Biobased Epoxy Resins from Fatty Acids and Lignin. *Molecules* **2020**, *25*, 1158. [CrossRef] [PubMed]
5. Bascón-Villegas, I.; Espinosa, E.; Sánchez, R.; Tarrés, Q.; Pérez-Rodríguez, F.; Rodríguez, A. Horticultural Plant Residues as New Source for Lignocellulose Nanofibers Isolation: Application on the Recycling Paperboard Process. *Molecules* **2020**, *25*, 3275. [CrossRef] [PubMed]
6. Szufa, S.; Piersa, P.; Adrian, Ł.; Sielski, J.; Grzesik, M.; Romanowska-Duda, Z.; Piotrowski, K.; Lewandowska, W. Acquisition of Torrefied Biomass from Jerusalem Artichoke Grown in a Closed Circular System Using Biogas Plant Waste. *Molecules* **2020**, *25*, 3862. [CrossRef] [PubMed]
7. Nikolic, M.; Najarro, M.C.; Johannsen, I.; Iruthayaraj, J.; Ceccato, M.; Feilberg, A. Copper Adsorption on Lignin for the Removal of Hydrogen Sulfide. *Molecules* **2020**, *25*, 5577. [CrossRef] [PubMed]
8. Michel, B.; Bras, J.; Dufresne, A.; Heggset, E.B.; Syverud, K. Production and Mechanical Characterisation of TEMPO-Oxidised Cellulose Nanofibrils/β-Cyclodextrin Films and Cryogels. *Molecules* **2020**, *25*, 2381. [CrossRef] [PubMed]
9. Padilla-Rascón, C.; Romero-García, J.M.; Ruiz, E.; Castro, E. Optimization with Response Surface Methodology of Microwave-Assisted Conversion of Xylose to Furfural. *Molecules* **2020**, *25*, 3574. [CrossRef] [PubMed]
10. Oliva-Taravilla, A.; Carrasco, C.; Jönsson, L.J.; Martín, C. Effects of Biosurfactants on Enzymatic Saccharification and Fermentation of Pretreated Softwood. *Molecules* **2020**, *25*, 3559. [CrossRef] [PubMed]
11. Jiménez-Quero, A.; Pollet, E.; Avérous, L.; Phalip, V. Optimized Bioproduction of Itaconic and Fumaric Acids Based on Solid-State Fermentation of Lignocellulosic Biomass. *Molecules* **2020**, *25*, 1070. [CrossRef] [PubMed]
12. Lucia, A.; Bacher, M.; Van Herwijnen, H.W.G.; Rosenau, T. A Direct Silanization Protocol for Dialdehyde Cellulose. *Molecules* **2020**, *25*, 2458. [CrossRef] [PubMed]
13. Korányi, T.I.; Fridrich, B.; Pineda, A.; Barta, K. Development of 'Lignin-First' Approaches for the Valorization of Lignocellulosic Biomass. *Molecules* **2020**, *25*, 2815. [CrossRef] [PubMed]
14. Redondo-Gómez, C.; Quesada, M.R.; Astúa, S.V.; Zamora, J.P.M.; Lopretti, M.; Vega-Baudrit, J.R. Biorefinery of Biomass of Agro-Industrial Banana Waste to Obtain High-Value Biopolymers. *Molecules* **2020**, *25*, 3829. [CrossRef] [PubMed]
15. Vedovato, V.; Vanbroekhoven, K.; Pant, D.; Helsen, J. Electrosynthesis of Biobased Chemicals Using Carbohydrates as a Feedstock. *Molecules* **2020**, *25*, 3712. [CrossRef] [PubMed]

 *molecules*

Review

Development of 'Lignin-First' Approaches for the Valorization of Lignocellulosic Biomass

Tamás I. Korányi [1,*] , Bálint Fridrich [2], Antonio Pineda [3] and Katalin Barta [2,4]

1 Surface Chemistry and Catalysis Department, Centre for Energy Research, Konkoly Thege M. u. 29-33, 1121 Budapest, Hungary
2 Stratingh Institute for Chemistry, University of Groningen, Nijenborgh 4, 9747 AG Groningen, The Netherlands; b.fridrich@rug.nl (B.F.); k.barta@rug.nl (K.B.)
3 Department of Organic Chemistry, University of Cordoba, Ed. Marie Curie (C 3), Campus of Rabanales, Ctra Nnal IV-A, Km 396, E14014 Cordoba, Spain; q82pipia@uco.es
4 Department of Chemistry, Organic and Bioorganic Chemistry, University of Graz, Heinrichstrasse 28/II, 8010 Graz, Austria
* Correspondence: koranyi.tamas@energia.mta.hu

Academic Editors: Alejandro Rodríguez and Eduardo Espinosa
Received: 27 May 2020; Accepted: 14 June 2020; Published: 18 June 2020

Abstract: Currently, valorization of lignocellulosic biomass almost exclusively focuses on the production of pulp, paper, and bioethanol from its holocellulose constituent, while the remaining lignin part that comprises the highest carbon content, is burned and treated as waste. Lignin has a complex structure built up from propylphenolic subunits; therefore, its valorization to value-added products (aromatics, phenolics, biogasoline, etc.) is highly desirable. However, during the pulping processes, the original structure of native lignin changes to technical lignin. Due to this extensive structural modification, involving the cleavage of the β-O-4 moieties and the formation of recalcitrant C-C bonds, its catalytic depolymerization requires harsh reaction conditions. In order to apply mild conditions and to gain fewer and uniform products, a new strategy has emerged in the past few years, named 'lignin-first' or 'reductive catalytic fractionation' (RCF). This signifies lignin disassembly prior to carbohydrate valorization. The aim of the present work is to follow historically, year-by-year, the development of 'lignin-first' approach. A compact summary of reached achievements, future perspectives and remaining challenges is also given at the end of the review.

Keywords: lignocellulose valorization; 'lignin-first'; reductive catalytic fractionation

1. Introduction

Plant cells' composite material is lignocellulose, which mainly consists of cellulose, hemicellulose, and lignin and in total accounts for ca. 90% of dry matter of land-based biomass. Depending on its origin, lignocellulose can be divided into three main categories—i.e., softwood, hardwood, and grass. In fact, the lignin content is highest in softwood, followed by hardwood and lowest in grasses [1,2].

Cellulose and hemicellulose are both polysaccharides, differing in building units, degree of polymerization and morphology. Lignin is a complex aromatic biopolymer built up from three monolignols: p-coumaryl alcohol, coniferyl alcohol, sinapyl alcohol (Figure 1). These monolignols differ in the number of methoxy groups (none, one, and two) attached to the aromatic ring and make up the three key lignin units (H (hydroxyphenyl), G (guaiacyl), and S (syringyl), respectively). G units constitute approximately 90–95% of softwood lignin, whereas 25–50% of G and 50–75% of S units are typically found in hardwood lignin. Because the coupling of the monolignols is a process involving radicals, there are many possible linkages between the sub-units, involving different C-C and C-O bonds with certain linkages being more prevalent than others. A typical linkage in both softwood and hardwood lignins is the β-O-4 ether bond (Figure 1), approximately reaching half of the lignin in softwood and more than 60% in hardwood. Hardwood lignin contains less C-C linkages than softwood, because the additional methoxy groups on the aromatic rings, mainly in the S units, prevent their formation [1,2].

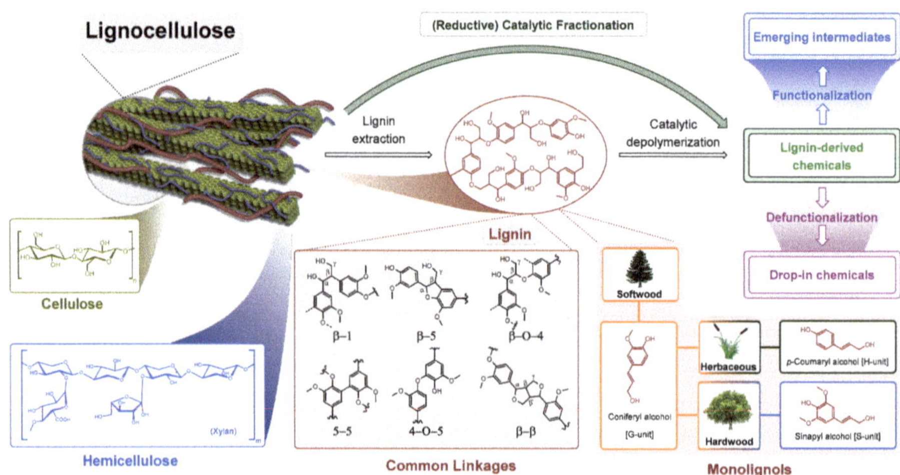

Figure 1. General strategies for lignocellulose valorization and application of lignin-derived platform chemicals with representative lignin structure displaying typical lignin subunits and linkages.

Currently, only the cellulosic part of lignocellulosic biomass is used effectively as feedstock for the pulp and paper industry and as precursor of second-generation bioethanol. Traditionally, lignin is isolated from lignocellulosic biomass by fractionation in the pulp and paper industry (route *a* in Figure 2) and by fermentation in biorefineries producing cellulosic ethanol (route *b* in Figure 2). Pulping methods result in structurally heavily modified lignins (route *a* in Figure 2), while enzymatic lignin displays mild structural modification (route *b* in Figure 2) [3]. The chemical structure of native lignin is altered during conventional lignocellulose fractionation methods: ether bonds (β-O-4 and 4-O-5 in Figure 1) are cleaved and new stable C-C linkages are formed, resulting in more condensed and unreactive technical lignins [4]. Therefore, essentially under pulping conditions lignin undergoes structural rearrangement leading to the formation of unnatural C-C bonds [5]. High-yield lignin depolymerization methods are limited by the presence of these linkages formed during lignin extraction as well as the interunit carbon–carbon bonds within native lignin [6].

Several methods have investigated the catalytic conversion of recalcitrant lignins obtained from the pulp and paper industry, generally requiring harsher reaction conditions in order to achieve feasible product yields [2,3,5,7–9]. On the other hand, lignins obtained upon milder enzymatic digestion would lead to higher aromatic monomer yields, after catalytic depolymerization [1,10,11]. Many studies focusing on catalyst development for lignin depolymerization have used organosolv lignins prepared in the respective laboratories, using different fractionation procedures, which in terms of severity would fall between the industrial pulping conditions and enzymatic digestion [2,3,5,7]. Lignin valorization leads to value-added products such as biofuels, macromolecules (carbon fiber, polyurethane) and aromatics (BTX, monophenolic compounds) [12]. Functionalized lignin monomers can be regarded as perspective lignin-derived platform chemicals, from which emerging intermediates can be produced leading to pharmaceuticals, fine chemicals, polymers, and fuels. The other groups are drop-in chemicals with an existing market platform leading to bulk chemicals (Figure 1) [13].

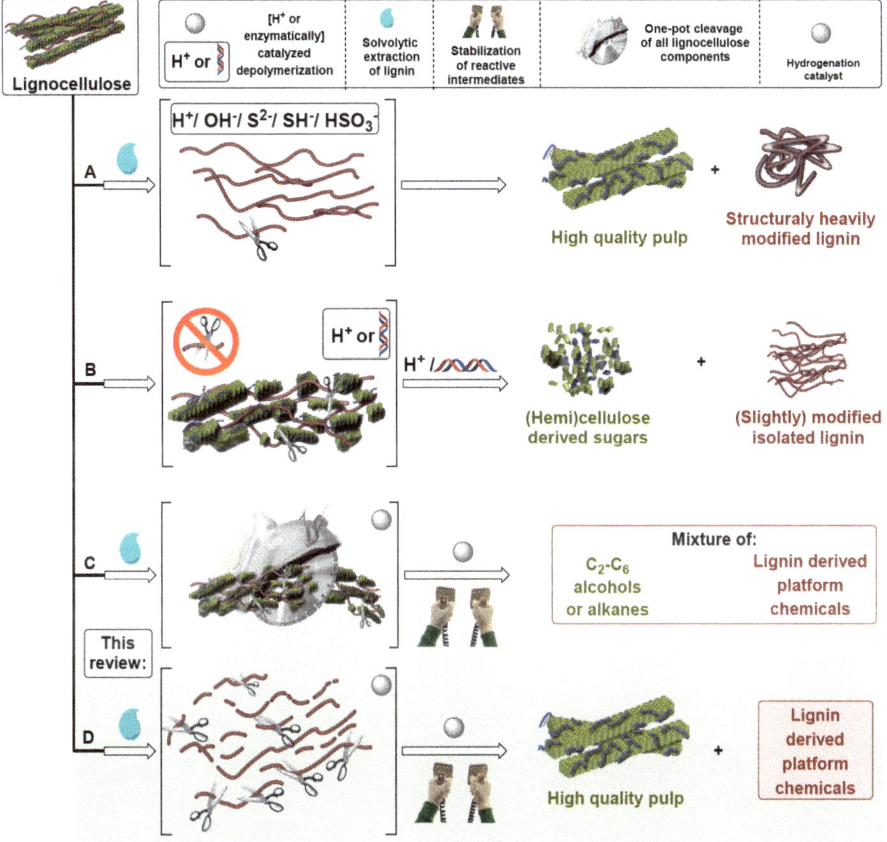

Figure 2. Types of lignocellulose valorization processes. (**A**) Harsh (pulping) fractionation. (**B**) Mild (enzymatic) fractionation. (**C**) One-step reductive catalytic processing. (**D**) Reductive catalytic fractionation (RCF).

Other approaches that emerged for the valorization of the whole lignocellulose are methods, which allow for the catalytic conversion of all constituents of lignocellulose, simultaneously. During these processes, product mixtures of cellulose as well as lignin are generated. For example, complete conversion of all lignocellulose components to lignin monomers and C_2–C_6 alcohols and/or alkanes is also possible by one-step reductive catalytic processing (route *c* in Figure 2) [3,13].

A method that emerged as alternative to lignin valorization is the so-called 'lignin-first' process. Already, several excellent reviews have discussed this powerful strategy [3,5,14–20]. Dubbed 'lignin-first', this approach considers the catalytic conversion of lignin during biomass fractionation, in other words solvolytic extraction of lignin accompanied by instant lignin depolymerization and predominantly reductive stabilization of reactive intermediates (route *d* in Figure 2) [4]. This approach is also called 'early-stage catalytic conversion of lignin' (ECCL) [14] or, when using a metal catalyst under reductive atmosphere, as 'reductive catalytic fractionation' (RCF) [15]. The extraction and immediate catalytic conversion of lignin to monomers by these methods from lignocellulose directly in the presence of a catalyst, usually under reductive conditions, results in higher yield of aromatic monomers due to the higher presence of cleavable C-O linkages and less C-C linkages [3]. RCF is also a two-step process, extracting lignin from whole biomass with a polar-protic solvent, and then selectively cleaving C-O ether bonds using a hydrogen donor and a heterogeneous catalyst. As a result, the process yields a depolymerized lignin oil rich in phenolic monomers, dimers, and oligomers, next to a solid carbohydrate pulp, which is amenable to further valorization. The most common solvents are alcohols (mainly methanol) and water/organic solvent mixtures such as water/dioxane and water/ethanol. The hydrogen donor can be pressurized hydrogen gas or can originate from the solvent or from the lignocellulose itself [16]. Generally, nickel or noble metal catalysts are used.

Different pathways and mechanisms have been proposed for the RCF process. The key steps in all cases are solvolysis as well as hydrogenolysis of ether bonds, removal of benzylic OH-groups (OH_α), and possible removal of OH_γ-groups. These primary reactions lead to the formation of substituted methoxyphenols and small oligomeric fragments. Additional hydrogenation of alkenyl and carbonyl groups, as well as hydrogenolysis, can take place (secondary reactions) [14,16,21].

A detailed overview of RCF processes is presented in Figure 3 and discussed fully in the paper of Rinaldi et al. [21]. The lignin fragments present in the liquor are prone to several types of reaction. Under pulping conditions lignin fragments undergo recondensation, into technical lignins containing strong C-C bonds [14]. The main difference between pulping (route *a* in Figure 2) and these processes (routes *c* and *d* in Figure 2), is that in the latter two, catalytic processing of lignin happens in its native form in conjunction with fractionation, or parallel to cellulose processing. Under these conditions, reactive fragments or intermediates that originate during pulping immediately undergo stabilization (typically by hydrogenation see Figure 3) to form more stable molecules. Possible lignin-derived monomers obtained thus far from routes *c* and *d* combined (Figure 2) are collected in Figure 4.

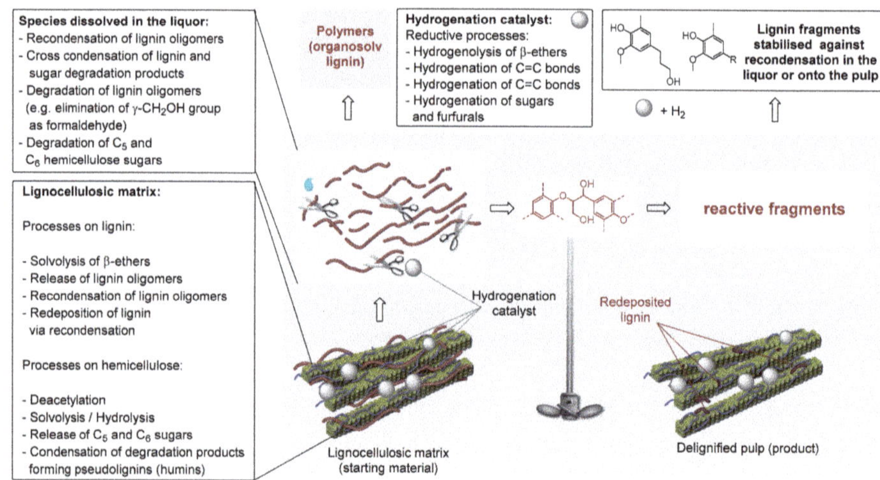

Figure 3. Schematic representation of chemical processes involved in the 'lignin-first' biorefining. For clarity, hemicellulose sugars and their degradation products were omitted.

The aim of this review is to follow historically the development of 'lignin-first' approach with inclusion of one-pot reductive catalytic depolymerization of all lignocellulosic components in the scientific literature. The number of these publications has increased significantly in the past years, new ideas emerged, and some novel methods were developed.

Table 1. 'Lignin-first' or reductive catalytic fractionation (RCF) papers in the literature.

Entry	Year	Feedstock	Catalyst	Solvent	Reaction Conditions			Yield of Main Products wt %	Lignin Monomer Yield (wt %)	Reference
					T (°C)	p (bar)	t(h)			
1	1940	maple, spruce	$Cu_2Cr_2O_5$	dioxane	250	333 (H_2)	12	4-propylcyclohexanol, 4-propylcyclohexanediol		[22]
2	1941	maple, spruce	$Cu_2Cr_2O_5$	dioxane	280	H_2	10	4-propylcyclohexanol	36	[23]
3	1943	maple	$Cu_2Cr_2O_5$	dioxane	280	H_2	20	4-propylcyclohexanol, 3-cyclohexyl-1-propanol	40	[24]
4	1948	maple	Raney-Ni + NaOH	dioxane/H_2O (1:1)	173	210 (H_2)	6	15% 4-ethylsyringol, 6% 4-ethanolsyringol	27	[25]
5	1963	aspen	Raney-Ni	dioxane/H_2O (1:1)	220	35 (H_2)	5	29% 4-propylsyringol, 13% 4-propanolsyringol	59	[26]
6	1966	spruce	Pd/C	dioxane/H_2O (1:1)	195	35 (H_2)	10	17% 4-propanolguaiacol	17	[27]
7	1969	spruce	Rh/C	dioxane/H_2O (1:1)	195	35 (H_2)	5	11% 4-propanolguaiacol	34	[28]
8	1970	spruce	Raney-Ni	dioxane/H_2O (1:1)	195	35 (H_2)	5	17% 4-propanol-cyclohexanol	17	[29]
9	1978	aspen	Rh/C	dioxane/H_2O (1:1)	195	34(H_2)	5	26% 4-propanolsyringol, 13% 4-propylsyringol	50	[30]
10	1978	spruce	Rh/C	dioxane/H_2O (1:1)	195	35(H_2)	5	21% 4-propanolguaiacol	21	[31]
11	1986	aspen poplar	Rh/C	dioxane/H_2O (1:1)	195	35(H_2)	5	21% 4-propanolguaiacol	21	[32]
12	1993	rice husks	polyvalent metals	dioxane	250	50(H_2)	2	33% 4-propylsyringol	33	[33]
13	2008	birch	H_3PO_4 + Pt/C	dioxane/H_2O (1:1)	200	40(H_2)	4	21% 4-propylsyringol, 15% 4-propanolsyringol	46	[34]
14	2011	pine	Pd/C	dioxane/H_2O (1:1)	195	35(H_2)	24	21% 4-propanolguaiacol	22	[35]
15	2012	birch	Ni-W_2C/C	ethylene glycol	235	60(H_2)	4	18% 4-propylsyringol, 10% 4-propanolsyringol	47	[36]
16	2013	birch	Ni/C	MeOH	200	1(Ar)	6	36% 4-propylsyringol, 12% 4-propylguaiacol	54	[37]
17	2014	birch	Pd/C	EtOH/H_2O (1:1)	195	4(Ar)	1	49% 4-propenylsyringol	49	[38]
18	2014	poplar	Raney-Ni	2-PrOH/H_2O (7:3)	180	autogenous	3	4-propanolsyringol	25	[39]
19	2015	poplar	Zn-Pd/C	MeOH	225	35(H_2)	12	30% 4-propylsyringol, 24% 4-propylguaiacol	54	[40]
20	2015	birch	Ru/C	MeOH	250	30(H_2)	6	31% 4-propylsyringol, 10% 4-propylguaiacol	52	[41]
21	2015	lignocellulose	Ru/C + $H_4SiW_{12}O_{40}$	org.phase/H_2O	140–300	50(H_2)	< 24	n-hexane, cyclohexane, methylcyclopentane	82 *	[42]
22	2015	corn stalk	Ru/C + $LiTaMoO_6$	H_3PO_4	230	60(H_2)	24	6% 4-ethylphenol	24	[43]
23	2015	birch	Pd/C	MeOH	250	30(H_2)	3	35% propanolsyringol, 10% propanolguaiacol	49	[44]
24	2015	birch	Pd/C	ethylene glycol	250	30(H_2)	3	35% propanolsyringol, 12% propanolguaiacol	50	[45]
25	2015	birch	Ni/C	MeOH	200	2(N_2)	6	18% 4-propylsyringol, 10% 4-propylguaiacol	32	[46]
26	2015	cedar	H_2SO_4	toluene/MeOH	170	1(air)	0.1	5% homovanillyl aldehyde dimethyl acetal	10	[47]
27	2016	poplar	Ru/C	MeOH	250	40(H_2)	15	49% propylsyringol, 22% propanolsyringol	78 *	[6]
28	2016	birch	Pd/C	EtOH/H_2O (1:1)	210	1(Ar)	15	20% 4-propylsyringol, 11% 4-propenylsyringol	36	[48]
29	2016	poplar	Zn-Pd/C	MeOH	225	35(H_2)	12	28% 4-propylsyringol, 14% 4-propylguaiacol	43	[49]
30	2016	miscanthus	Ni/C	MeOH	225	60(H_2)	12	19% 4-propylsyringol, 21% 4-propylguaiacol	68	[50]
31	2016	poplar	$(Rh(cod)Cl)_2$ + $Sc(OTf)_3$	Dioxane/H_2O	175		2	3% 4-propenylsyringol, 2% 4-methylguaiacol	10	[51]
32	2016	birch	Pd/C + $Yb(OTf)_3$	MeOH	200	30(H_2)	2	4-propanolsyringol, 4-propanolguaiacol	48	[52]
33	2016	beech	Ni/C	MeOH/H_2O (3:2)	200	60(H_2)	5	29% 4-propanolsyringol, 10% 4-propanolguaiacol	51	[53]
34	2016	corn stover	H_3PO_4 + Ni/C	MeOH	200	30(H_2)	6	15% methyl coumarate, 15% methyl ferulate	38	[54]
35	2016	poplar	H_3PO_4 + Pd/C	MeOH	200	20(H_2)	3	21% 4-propanolsyringol, 14% 4-propanolguaiacol	42	[55]
36	2016	poplar	Pd/C	MeOH/H_2O (7:3)	200	20(H_2)	3	23% 4-propanolsyringol, 13% 4-propanolguaiacol	44	[56]
37	2017	birch	Pd/C + $Al(OTf)_3$	MeOH	180	30(H_2)	2	34% methoxypropylsyringol, 8% methoxypropylguaiacol	55	[57]
38	2017	oak	Pd/C + $Al(OTf)_3$	MeOH	180	30(H_2)	2	12% methoxypropylsyringol, 10% propanolsyringol	46	[58]
39	2017	birch	NiFe/C	MeOH	200	20(H_2)	6	24% 4-propylsyringol, 11% 4-propylguaiacol	40	[59]
40	2017	*pinus radiata*	Pd/C	Dioxane/H_2O (1:1)	195	34(H_2)	24	22% propanolguaiacol, 3% propylguaiacol	78 +	[60]
41	2017	birch	Ni/Al_2O_3	MeOH	250	30(H_2)	3	21% propanolsyringol, 5% propylsyringol	36	[4]
42	2017	birch	H_3PO_4 + Pd/C	MeOH/H_2O (7:3)	180	30(H_2)	3	18% propanolsyringol, 11% propylsyringol	37	[61]
43	2017	poplar	Ni/C	MeOH	190	60(H_2)	3	12% propylguaiacol and propylsyringol	17	[62]

Table 1. *Cont.*

Entry	Year	Feedstock	Catalyst	Solvent	Reaction Conditions T (°C)	p (bar)	t(h)	Yield of Main Products wt %	Lignin Monomer Yield (wt %)	Reference
44	2018	poplar	Ni/C	MeOH	200	30(H_2)	1	8% propylsyringol, 5% propylguaiacol	48	[63]
45	2018	pine	Cu_{20}PMO	MeOH	220	40(H_2)	18	8% propanolguaiacol, 4% propylguaiacol	13	[64]
46	2018	birch	Ru/C + $H_4SiW_{12}O_{40}$	petrol/H_2O	220	50(H_2)	5	21% propanolsyringol, 5% propylsyringol	39	[65]
47	2018	poplar, spruce	Raney-Ni, Ni_2P/SiO_2	2-PrOH/H_2O (7:3)	180	autogenous	3	50–60% phenolic species	20–25	[66]
48	2018	poplar, spruce	Raney-Ni	2-PrOH/H_2O (7:3)	200	autogenous	6			[67]
49	2018	sorghum	Fenton	EtOH	250	autogenous	12	phenolic oil	76	[68]
50	2018	oak	Al(OTf)$_3$ + Pd/C	MeOH	160	autogenous	2	9% propylsyringol, 5% propylguaiacol	25	[69]
51	2018	cornstalk	Ru/C	H_2O	200	30(H_2)	8	67% ethylcyclohexane, 16% propylcyclohexane	97 *	[70]
52	2018	birch	Pd/C+Yb(OTf)$_3$	MeOH	250	20(H_2)	20	lignin-oil	83 *	[71]
53	2018	eucalyptus	Ru/C	BuOH/H_2O (1:1)	200	30(H_2)	2	41% propanolsyringol and propanolguaiacol	49	[72]
54	2018	bark	Pd/C	MeOH/H_2O (2:1)	200		2	3% ethylguaiacol	42	[73]
55	2018	vanilla seeds	Ni/C	MeOH	250	30(H_2)	3	18% propylcatechol, 3% propenylcatechol	21	[74]
56	2018	corncob	$ZnMoO_4$/MCM-41	MeOH	220	30(H_2)	4	16% methyl coumarate, 13% methyl ferulate	38	[75]
57	2018	cork	Rh/C	2-methyl tetrahydrofuran	200	40(H_2)	4	bio-oil	43	[76]
58	2018	apple wood	Ru/SiC	MeOH	250	10(H_2)	3	propylsyringol and ethylsyringol	48	[77]
59	2018	corn stover	MoS_2	MeOH	20	30(H_2)	2	4% propylguaiacol, 3% ethylphenol	18	[78]
60	2018	birch	CdS	MeOH/H_2O	r.t.	1(N_2)	8	14% propanonesyringol, 7% propanoneguaiacol	27	[79]
61	2019	birch	Co-phen/C	EtOH/H_2O (1:1)	200	autogenous	4	10% 4-propylsyringol, 9% 4-propenylsyringol	34	[80]
62	2019	oak	Ni-Al/AC	HCOOH/EtOH/H_2O	190	autogenous	3	9% propylsyringol, 5% propylguaiacol	23 ˣ	[81]
63	2019	birch	Pt/Al_2O_3	MeOH/H_2O (1:2)	230	autogenous	3	40% propylsyringol, 6% propylguaiacol	49	[82]
64	2019	poplar	Raney-Ni	2-PrOH/H_2O (7:3)	200	autogenous	3	11% propanolsyringol, 10% propanolguaiacol	34	[83]
65	2019	poplar, spruce	Raney-Ni	2-PrOH/H_2O (7:3)	220	autogenous	3	10% ethylsyringol, 7% propylsyringol	36	[84]
66	2019	poplar	Ni/C	MeOH	225	35(H_2)	12	propylsyringol, propylguaiacol	90	[85]
67	2019	apple wood	Mo_xC/CNT	MeOH	250	10(H_2)	3	propylsyringol, propylguaiacol	42	[86]
68	2019	eucalyptus	Ni@ZIF-8	MeOH	260	30(H_2)	8	24% propylsyringol + propylguaiacol	44	[87]
69	2019	basswood	binuclear Rh complex	NaOH/H_2O	110	1(Ar)	24	2% propanonesyringol, 1.6% ethanonesyringol	5	[88]
70	2020	poplar	binuclear Rh complex	NaOH/H_2O	110	1(Ar)	12	9% ethanonesyringol, 6% ethanoneguaiacol	17	[89]
71	2020	birch	Ru/C	MeOH	235	30(H_2)	3	20% phenol, 9% propylene	29	[90]
72	2020	eucalyptus sawdust	Ru/C	BuOH/H_2O (1:1)	200	30(H_2)	2	propanol-substituted phenolics	49	[91]
73	2020	pine	H_2SO_4	dimethyl carbonate	140	autogenous	0.67	8% G-C2-acetal	9	[92]
74	2020	eucalyptus	Pd/C	MeOH	240	30(H_2)	4	32% propanolsyringol, 13% propanolguaiacol	50	[93]
75	2020	beech	NiMo/Al_2O_3	EtOH	260	26(H_2)	3	11% propylsyringol, 6% propylguaiacol	20	[94]
76	2020	pine	Pt/C	MeOH/H_2O	230	30(H_2)		10 mol% phenol	15	[95]
77	2020	birch	H-BEA	EtOH/H_2O	220		2		20	[96]
78	2020	poplar			250	autogenous	0.33			[97]
79	2020	birch		H_2O	195					[98]
80	2020	poplar	emimAce	emimAce	110	1(air)	4	38% lignin		[99]
81	2020	maple	Zr-KIT-5	γ-valerolactone	250	30(H_2)	18	3.5% 2-phenylpropan-2yl acetate, 1.6% 3,4-dimethoxyphenol	7	[100]

* mol%, ⁺ oil yield in wt %, ˣ C%.

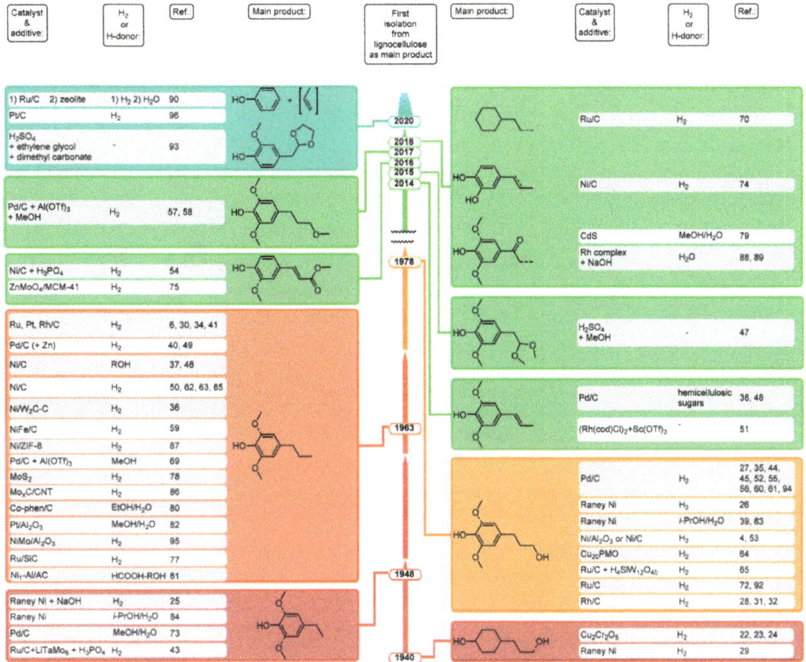

Figure 4. Main lignin monomeric products with functionalized sidechains, the date of their first isolation from lignocellulose as main product, typical catalysts, and additives with added hydrogen or hydrogen donors applied in the process, and literature references. For simplicity, guaiacols are not shown if the same syringol derivatives exist as main product. The detailed process characteristics and monomer yields of each individual study can be found in Table 1.

2. Chronological Overview

2.1. From 1940 to 2014

Originally, high pressure hydrogenation and hydrogenolysis of wood dates back to the 1940s with the aim of clarifying lignin structure [19]. Various woods (maple, aspen, spruce) were hydrogenated using copper-chromite, Raney-Ni, Pd/C, or Rh/C catalysts in dioxane(/water) solvent (mixture) under various conditions (173–280 °C, 35–333 bar H_2, 5–20 h reaction time) resulted in 17–59 wt % monomer yields with 4-propylcyclohexanol (1940) [22], 4-ethylsyringol (1948) [25], 4-propylsyringol (1963) [26], 4-propanolguaiacol (1966) [27], and 4-propanolsyringol (1978) [30] as main products (entries 1–11 in Table 1 and Figure 4) [22–32]. Under harsh reaction conditions (250–280 °C, 240–333 bar H_2 pressure) not only saturation of the benzene ring occurred, but also hydrogenation and hydrogenolysis of the holocellulose part took place [5,22–24]. The 4-n-propylphenol nature of the lignin monomers was confirmed in these early studies [5]. The efficiency of various (Ni, Pd, Rh, and Ru) catalysts was compared for the hydrogenolysis of spruce wood under mild reaction conditions (195 °C, 35 bar H_2), the Pd/C catalyzed reaction gave the highest (16%) 4-propanolguaiacol yield, and the highest monomeric yield (34%) was obtained when Rh/C was used [5,28]. An increased monomers yield was demonstrated when softwood was replaced by hardwood [30], and hydrogen pressure did not affect the yield [31]. The increased monomers yield of hardwood can be explained by its higher S to G ratio compared to softwood, and therefore less C-C bonds, as was explained in the introduction.

Later, (1993–2011) new feedstocks (rice husks [33], birch [34], pine [35]) and catalysts (polyvalent metals [33], H_3PO_4 + Pt/C [34]) were tried in the dioxane(/water) solvent (mixture) under similar conditions and similar main products (4-propylsyringol and 4-propanolguaiacol) with comparable monomer yields (22–46 wt %) were gained (entries 12–14 in Table 1 and Figure 4). Four catalysts (Ru/C, Pd/C, Rh/C, and Pt/C) were tested for birch treatment and the monomer yields depended on the applied conditions: the total yield of monomers was 34% over the Pt/C catalyst, it was improved to 38% with addition of acid (H_3PO_4), the addition of dioxane further improved the yield to 46% [34]. The effect of additives was demonstrated in this work. Beside dimeric and oligomeric products, 4-propanolguaiacol was produced as a monomeric product almost exclusively in the Pd/C catalyzed hydrogenolysis of *Pinus radiata* wood [35]. Birch was converted in new solvents (ethylene glycol [36], methanol [37], ethanol/water mixture [38]) over new catalysts (Ni-W_2C/C [36], Ni/C [37]) with high (47–54 wt %) monomer yields. Two processes [37,38] did not require addition of external hydrogen, alcohol solvents provided the active hydrogen species. One major monomeric product was generated in some processes (entries 12, 14, and 17 in Table 1) and a new one (4-propenylsyringol) appeared in 2014 (Figure 4) [33,35,38].

In 2012, Zhang and co-workers [36] described the direct conversion of birch over a Ni-W_2C/C catalyst in a one-pot one-step reductive catalytic depolymerization process: the carbohydrate fraction was converted to ethylene glycol and other diols with a total yield of 76%, while the lignin component was transformed selectively into monophenols with a yield of 47% (entry 15 in Table 1 and 18% 4-propylsyringol in Figure 4). Different feedstocks, solvents and catalysts were tested. Hardwood, compared to softwood, led to a better conversion of both lignin and carbohydrates as expected. Replacing the original water solvent to methanol and later to ethylene glycol resulted increasing monophenol yields. Using a Pd/C catalyst led to the highest selectivity (56%) towards 4-propanolsyringol. This method uses high pressure (60 bar) hydrogen gas due to the complete conversion of all lignocellulose components (route *c* in Figure 2) [36].

In 2014, Ferrini and Rinaldi [39] realized that lignin was released solvolytically from the plant cell by simply "cooking" wood in the presence of Raney-Ni in 2-propanol (2-PrOH)/H_2O) and partially depolymerized lignin, a non-pyrolytic lignin bio-oil was produced in addition to pulps that are amenable to enzymatic hydrolysis. The suspension of wood pellets, Raney Ni catalyst and aqueous solution of 2-propanol was heated under mechanical stirring (e.g., at 180 °C for 3 h) and 25 wt % lignin oil and 71 wt % pulp were produced (entry 18 in Table 1). The lignocellulosic feed was processed in the absence of molecular hydrogen and acetone generated by the hydrogen transfer can be hydrogenated to 2-PrOH. The holocellulose fraction or pulp (i.e., cellulose and hemicellulose) was isolated by filtration and washed with the 2-PrOH/water solution. Raney Ni was removed from the suspension with a magnet. Finally, the non-pyrolytic lignin bio-oil was isolated by solvent removal from the extracting liquor. This lignin oil is readily susceptible to further hydroprocessing (hydrodeoxygenation) under low-severity conditions. The complexity of the low-molecular weight lignin product mixture is a disadvantage, but the autogenous hydrogen usage and lignin-only conversion are advantages of this method. The pulp (holocellulose) is suitable for the production of glucose and xylose through enzymatic hydrolysis [39].

2.2. 2015

Lignin-first biorefineries were described in two works in 2015. Abu-Omar et al. [40] presented a selective hydrogenolysis of poplar wood with bimetallic Zn-Pd/C in methanol with external H_2, focusing on the lignin monomers (30% 4-propylsyringol and 24% 4-propylguaiacol) and the enzymatic conversion of the retained pulp to glucose in 95% yield (entry 19 in Table 1). Sels et al. [41] presented reductive lignocellulose fractionation of birch sawdust through simultaneous solvolysis and catalytic hydrogenolysis in the presence of Ru/C in methanol under hydrogen at 250 °C resulting in carbohydrate pulp and lignin oil. The thermal and solvolytic disassembly of lignin (delignification) was immediately followed by the reductive stabilization of lignin's most reactive intermediates into soluble and stable low-molecular-weight phenolic products. This fractionation strategy was denominated as a 'lignin-first' biorefinery, as the valorization of lignin to chemicals was performed before carbohydrate processing.

The lignin oil yields above 50% of phenolic monomers (mainly 4-propylguaiacol and 4-propylsyringol) and about 20% of a set of phenolic dimers, relative to the original lignin content, next to phenolic oligomers. The separated carbohydrate pulp contains up to 92% of the initial polysaccharides (entry 20 in Table 1).

In a patent in 2015, Sels et al. described an interesting biorefinery concept for the direct production of light naphtha (hexane, pentane, methyl cyclopentane, cyclohexane, etc.) for converting lignocellulose in the presence of an acidic reactive aqueous phase and a redox catalyst (Ru/C + $H_4SiW_{12}O_{40}$) in the organic extracting/reaction phase (entry 21 in Table 1) [42]. The products are useful as feedstock for steam and catalytic cracking, as precursors for the synthesis of bioaromatics, and as fuel additives. Another one-pot lignocellulose conversion into gasoline alkanes and monophenols was published by Ma et al. [43]. Raw biomass feedstocks (pine, corn, wheat, rice, etc.) were processed with Ru/C + $LiTaMoO_6$ catalysts in phosphoric acid and pentanes and hexanes were produced from the carbohydrates with up to 82% total yield and monophenols, related alcohols and hydrocarbons from the lignin fraction. Partial hydrocracking of the monophenol fraction was suggested (entry 22 in Table 1).

Propylphenolics and propanolphenolics are the most abundant main lignin monomeric products using Ru/C + H_2 and Pd/C + H_2 catalytic systems, respectively, as shown in Figure 4. Accordingly, changing the catalyst from Ru/C to Pd/C drastically increased the OH-content of the phenolic monomers (entry 23 in Table 1) [44], and the solvent choice (methanol and ethylene glycol) has an impact on both pulp retention and delignification efficiency [45]. Reductive catalytic fractionation (RCF), as a new expression, was used first (2015) in the latter paper [45]. The effect of substrate and catalyst loading was studied over Ni/C catalyst in methanol: birch resulted higher monomer yields than poplar and eucalyptus, while higher catalysts loading caused higher monomer yields due to the presence of more hydrogen produced from methanol reforming [46].

As an example of metal-free catalytic systems (entry 26 in Table 1), the acid-catalyzed degradation of cedar and eucalyptus wood samples in a toluene-methanol solvent mixture resulted in selective production of lignin monomers, homovanillyl aldehyde dimethyl acetal and homosyringaldehyde dimethyl acetal (by 2015 in Figure 4), due to the trapping of enol intermediates with alcohol [47].

2.3. 2016

Luterbacher et al. reported that adding formaldehyde during biomass pretreatment followed by reductive depolymerization of this stabilized lignin over Ru/C as catalyst in methanol, produced guaiacyl and syringyl monomers at near theoretical yields (78 mol% for poplar) during hydrogenolysis (entry 27 in Table 1). These yields were three to seven times higher than those obtained without formaldehyde, which prevented lignin condensation by forming 1,3-dioxane structures with lignin side-chain hydroxyl groups [6].

In 2016, the utilization of hemicellulose as a hydrogen donor for the reductive lignin transformations and the separation of biomass into three main components: solid carbohydrate residue (mainly cellulose), liquid bio-oil (mainly lignin monomers and oligomers), and water-solubilized sugars (originating mainly from hemicellulose) emerged as a new idea [48]. Usage of Zn^{II} as a co-catalyst beside Pd/C increased the selectivity toward 4-propylsyringol and 4-propylguaiacol production (entry 29 in Table 1) through removal of the hydroxyl group at the $C\gamma$ position of the β-O-4 ether linkage [49]. All three major components of Miscanthus biomass (lignin, cellulose, and hemicellulose) were effectively (with 55% overall conversion) utilized into high value chemicals with mass balance of 98% using a Ni/C catalyst over 68% yield into four phenolic products from lignin (entry 30 in Table 1) [50].

Bruijnincx et al. described a tandem catalysis process for ether linkage cleavage within lignin, involving ether hydrolysis by water-tolerant Lewis acids (metal triflates) followed by aldehyde decarbonylation by a Rh complex (entry 31 in Table 1) [51]. In situ decarbonylation of the reactive aldehydes limited loss of monomers by recondensation, and surprisingly 4-(1-propenyl)phenols (4-propenylsyringol for poplar) were the main monomeric products (Figure 4). Hensen et al., also used metal triflate (Yb(OTf)$_3$) catalysts for rapid cleavage of the chemical bonds between lignin and carbohydrates combined with Pd/C hydrogenolysis catalysts and 36–48% aromatic monomer yields (4-propanol derivatives, entry 32 in Table 1) were reached from different woods in the tandem process [52]. Xu et al. efficiently hydrogenated beech to natural phenolic alcohols (4-propanolsyringol and 4-propanolguaiacol) with 51% total yield using Ni/C catalyst in a methanol–water co-solvent (entry 33 in Table 1) [53]. Breaking the intramolecular hydrogen bonds in lignin β-O-4 motifs accelerated the C$_\beta$-O cleavage, maintaining the original structure of lignin [53].

Román-Leshkov et al. investigated the RCF of corn stover using Ru/C and Ni/C catalysts and H$_3$PO$_4$ cocatalyst in methanol at 200 and 250 °C [54]. The monomer yields increased up to 38% as a function of time, with the addition of acid cocatalyst, and methyl coumarate/ferulate (2016 in Figure 4) were the main products (entry 34 in Table 1). Clear trade-offs existed between the levels of lignin extraction, monomer yields, and carbohydrate retention in the residual solids [54].

The influence of acidic and alkaline additives was studied on the Pd/C-catalyzed reductive processing of poplar wood in methanol: under acidic (H$_3$PO$_4$) conditions both delignification (to 4-propanolsyringol/guaiacol as main monomeric products, entry 35 in Table 1) and alcoholysis of hemicellulose are promoted, leaving behind a cellulose-rich pulp, alkaline (NaOH) conditions also enhanced delignification, but other lignin products (C$_2$-substituted phenolics with loss of hydroxyl groups) were formed, lignin depolymerization was hampered, and cellulose loss was found in the pulp [55]. Synergetic effects of alcohol/water mixing were studied under similar conditions but without acid/alkaline addition in another paper. Low (30 vol %) water concentrations enhanced the removal of lignin from the biomass, while the majority of the carbohydrates were left untouched, high (70 vol %) water concentrations favored the solubilization of both hemicellulose and lignin, resulting in a cellulosic residue of higher purity [56].

2.4. 2017

The tandem metal triflate and Pd/C catalysis was further investigated by the Hensen group in 2017. Metal triflates were involved in cleaving not only ester and ether linkages between lignin and the carbohydrates, but also β-O-4 ether linkages within the aromatic lignin structure. Pd/C is required for cleaving α-O-4, 4-O-5 and β–β linkages. Synergy was revealed between Pd/C and metal triflates: under optimized conditions, 55 wt % mono-aromatics (entry 37 in Table 1)—mainly alkylmethoxyphenols (2017 in Figure 4)—were obtained from the lignin fraction (24 wt %) of birch wood [57]. Instead of metal triflates the effect of possible alternative acid co-catalysts (HCl, H$_2$SO$_4$, H$_3$PO$_4$, and CH$_3$COOH) was studied for the tandem RCF process in another publication [58]. Al(OTf)$_3$ and HCl, respectively, afforded 46 wt % (entry 38 in Table 1) and 44 wt % lignin monomers from oak wood sawdust in tandem catalytic systems with Pd/C at 180 °C in 2 h, therefore HCl is a promising alternative to the metal triflates [58].

Birch was effectively depolymerized using Ni-Fe/C catalyst with alloy structure in methanol reaching 40% monomer yield (entry 39 in Table 1) with 88% selectivity to 4-propylsyringol and 4-propylguaiacol [59]. *Pinus radiata* wood was depolymerized by mild hydrogenolysis in dioxane-water mixture by Pd/C catalyst to give an oil product, from which new biobased epoxy resins were prepared [60].

The role of Ni/Al$_2$O$_3$ catalyst was elucidated in the solubilization, depolymerization, and stabilization of lignin from birch in methanol in the 2017 work of Sels et al. [4]: the solvent is responsible for the first two processes, while the catalyst is for the stabilization through hydrogenation of reactive intermediates. Recuperation and reuse of the Ni/Al$_2$O$_3$ pellets was facilitated using a catalyst basket. This catalytic reduction also prevents undesirable repolymerization reactions.

Continuous systems are vital for realistic scale-up because time-resolved product distributions and yields can be obtained from these experiments. The first two papers using flow-through systems instead of batch reactors for RCF were published by Samec [61] and Román-Leshkov [62] in 2017. During RCF in a flow-through system, separate reactors are used for pulping and delignification processes (Figure 5). A percolation reactor filled with birch and consecutively a fixed catalytic bed reactor filled with Pd/C catalyst was used in the first paper [61]. A methanol–water solution of phosphoric acid was percolated through the system at 30 bar pressure (left reactor system in Figure 5); under optimized conditions 37% yield of monophenolic compounds (18% propanolsyringol and 11% propylsyringol) was reached (entry 42 in Table 1). It was concluded that organosolv pulping and transfer hydrogenolysis should be performed under different conditions; and depolymerized lignin can be obtained without the palladium catalyzed step [61]. Two flow-through systems were used in the second paper [62]: a single-bed reactor with a biomass bed located upstream from a catalyst bed and a dual-bed reactor featuring switchable biomass beds physically separated from the catalyst in a separate upstream reactor (right reactor system in Figure 5). RCF of poplar with Ni/C catalyst in methanol solvent was studied in the latter paper, 17% monomer yield (mainly propylguaiacol and propylsyringol) was reached (entry 43 in Table 1). It was concluded that flow-through studies allowed the observation of biomass extraction intermediates, decoupling of solvolysis and hydrogenolysis, simple catalyst recovery and recyclability, and elucidated catalyst deactivation mechanisms [62].

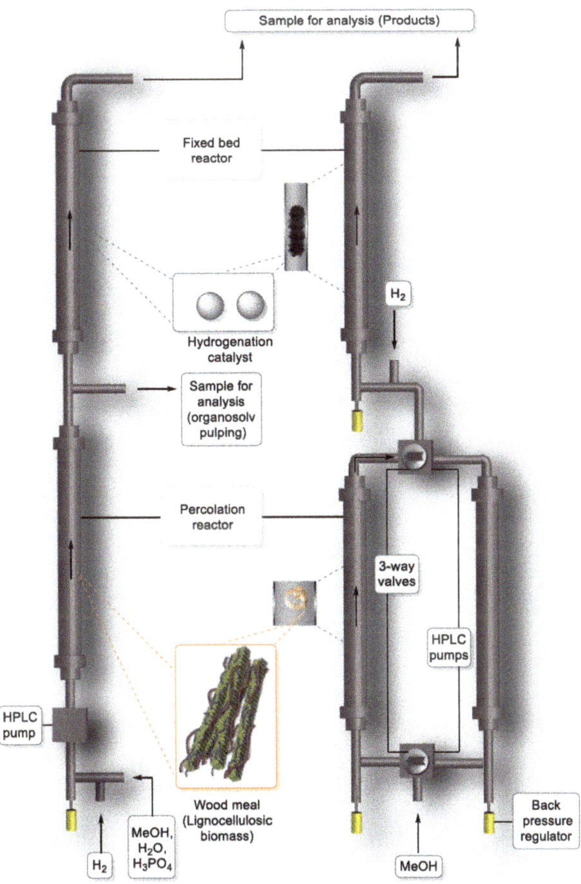

Figure 5. Schematic representations of semi-continuous flow-through reactor systems. (**Left**) Combination of a percolation reactor filled with woody biomass and a fixed catalytic bed reactor filled with Pd/C catalyst (Adapted from [61]). (**Right**) Schematic of the semi-continuous dual-bed flow reactor. Biomass in lower beds, catalyst in upper bed. (Adapted from [62,63]).

2.5. 2018

Continuous systems initiated by Román-Leshkov et al. [62] were developed further in 2018 [63]. Kinetic studies of RCF in flow-through reactors revealed decoupling of the two limiting mechanistic steps, lignin solvolysis and reduction, which can be independently controlled. The difference of activation barriers between flow and batch reactors indicated that lignin extraction under typical RCF conditions was mass-transfer limited [63].

Complete lignocellulose conversion yielding valuable aromatics and fuels was introduced by Barta et al. [64]. In the first, mild depolymerization step following the principles of RCF using $Cu_{20}PMO$ (porous metal oxide), propanolguaiacol was obtained as main monomeric product (entry 45 in Table 1) that could be converted to plethora of value-added aromatic building blocks, also including amines. In the second step, the cellulose and unreacted lignin residues that were mixed with the heterogeneous catalyst were converted in supercritical methanol, resulting in aliphatic alcohols. Hydrothermal conditions suitable for full conversion of the residues allows for efficient catalyst recycling. Value-added products were produced in further catalytic pathways.

A comprehensive strategy for the smooth integration of an RCF-based biorefinery process into current petrorefinery schemes was carried out by Sels and coworkers [65]. Birch wood processed by RCF provided nearly theoretical amounts of phenolic monomers (entry 46 in Table 1 and Figure 4) and a solid carbohydrate pulp with 83% C_5 and 93% C_6 sugar retention in the presence of Ni/Al_2O_3 using methanol solvent [4]. This pulp can be converted either into bioethanol by fermentation [4] or to alkanes using liquid phase cellulose-to-naphtha (LPCtoN) technology with petrol as the organic solvent [65]. Bio-enriched gasoline was produced from the (hemi)cellulose pulp using a two-phase (water/fossil naphtha) catalytic slurry process followed by isomerization [65].

Rinaldi et al. [66] introduced a deep converting 'lignin-first' biorefinery concept in 2018, which means production of gasoline and kerosene/diesel drop-in fuels in two steps. Poplar and spruce were deconstructed over Raney Ni catalyst in isopropanol–water solvent mixture yielding lignin oils along with cellulosic pulps in the first step, next the lignin oils were catalytically upgraded to aliphatics or aromatics by simply changing hydrogen pressure and temperature in the presence of a Ni_2P/SiO_2 catalyst (entry 47 in Table 1). The self-sufficiency in hydrogen was achieved through the gasification of the delignified holocellulose. The role of Raney-Ni catalyst in the process was clarified in another paper by the Rinaldi group: it suppresses formic acid formation via sugar hydrogenation [67].

Alternatively, the use of hydrogen could also be avoided through the utilization of Fenton's reagent (Fe^{3+}, H_2O_2), that combined with enzymatic hydrolysis transformed sweet sorghum bagasse into phenolic monomers and sugars [68]. Initially, the feedstock's molecular structure was modified through iron chelation and free radical oxidation via Fenton's reagent. The lignin component of the modified feedstock was then selectively depolymerized in supercritical ethanol (250 °C, 6.5 MPa) under nitrogen to produce a phenolic oil (entry 49 in Table 1). Thus, Fenton's reagent seems to provide beneficial effect in lignin depolymerization through the modification of lignin structure by hydroxylation and demethoxylation reactions of lignin substituents in the aromatic rings as well as by the formation of an iron-lignin complex. These two modifications are considered to make the ß-O-4 bond cleavage energetically more favorable. Fenton modification not only increased the yields of phenolic monomers, particularly ethyl-p-coumarate and ethyl-ferulate, but also enhanced enzymatic hydrolysis.

The metal triflate and Pd/C catalyst system was developed further by the Hensen group in 2018 [69]. Beside $Al(OTf)_3$ other homogeneous acid catalysts were tried in the first fractionation step of a two-step process, where oak was converted to lignin-oil and cellulose pulp in the first step, then to phenolic monomers with up to 25 wt % yield over the Pd/C catalysts in the second step (entry 50 in Table 1). Phosphoric acid proved to be the most suitable catalyst because it minimized the repolymerization back to lignin in the first step.

A complete transformation of lignocellulose into valuable platform chemicals was reached by Wang et al. [70,71]. They transferred cornstalk into liquid alkylcyclohexanes (2018 1st row in Figure 4, from the lignin fraction) and polyols (from cellulose and hemicellulose components) over Ru/C catalysts in aqueous phase in one step (entry 51 in Table 1) [70], and various biomasses (birch, beech, cornstalk, and pine) over Pd/C + Yb(OTf)$_3$ catalysts in methanol solvent to bio-oil and carbohydrates in the first step (entry 52 in Table 1), then the lignin-oil to arenes over a Ru/Nb$_2$O$_5$ catalyst in isopropanol, and the carbohydrates phase to 5-hydroxymethylfurfural, and furfural in tetrahydrofuran/seawater in the subsequent steps [71] in high overall yields (entries 51 and 52 in Table 1).

Another RCF process was also published by the Sels group in 2018 [72]. They converted eucalyptus into lignin-derived (mono)phenolics, hemicellulose-derived polyols, and a cellulose pulp in butanol–water 1:1 mixture over Ru/C catalysts and 30 bar hydrogen with 49 wt % lignin monomer yields with propanolsyringol and propanolguaiacol as main lignin products (entry 53 in Table 1). Phase separation of n-butanol and water upon cooling offered a facile and effective strategy to isolate lignin-derived phenolics (n-butanol phase) from polyols (aqueous phase).

Carbohydrates served as an inherent hydrogen donor in a bark RCF process producing hydrocarbon bio-oil in gasoline and diesel ranges and 4-ethylguaiacol [73]. RCF of vanilla seeds was used to investigate the depolymerization of naturally occurring C-lignin, which consists solely of caffeyl alcohol units; only two products (propyl- and propenyl catechol, 2018 second row in Figure 4) were gained with 21 wt % lignin monomer yield (entry 55 in Table 1) [74]. Selective fragmentation into hydroxycinnamic esters (methyl coumarate and methyl ferulate (Figure 4)) was reached by RCF of corncob using MCM-41 supported ZnMoO$_4$ catalyst in methanol [75]. The effect of support, added base, and solvent was studied in the catalytic depolymerization of cork over Rh/C catalyst: the highest bio-oil yield (43 wt %) was reached by a 2-methyl tetrahydrofuran/water 'green' solvent mixture (entry 57 in Table 1) [76]. The reusability of Ru/SiC compared to Ru/C catalyst was emphasized in the RCF of apple wood to lignin-oil further converted to jet fuel aromatics and polyurethane [77]. Unsupported MoS$_2$ catalyst was used in the RCF of corn stover and 18 wt % phenolic monomers yield was reached (entry 59 in Table 1) [78].

Full utilization of biomass by means of solar energy was reached in 2018 by CdS quantum dots catalyzed cleavage of the β-O-4 bonds in birch into functionalized aromatics (2018 third row in Figure 4), xylose, and glucose under visible light at room temperature. The β-O-4 bond in lignin is cleaved by an electron–hole coupled photoredox mechanism based on a C_α radical intermediate, in which both photogenerated electrons and holes participate in the reaction. Due to the colloidal character of the catalyst it could be easily separated and recycled [79].

2.6. 2019

A new 'lignin-first' paper in 2019 discusses RCF of birch wood with high (up to 34 wt %) yields to monophenolic compounds (10% propylsyringol and 9% propenylsyringol) using Co-phen/C catalyst and formic acid or formate as a hydrogen donor (entry 61 in Table 1). The high yield was explained by transfer hydrogenolysis reactions of lignin fragments targeting the β-O-4' bond and stabilizing reactive intermediates due to the cobalt catalyst [80]. Formic acid was also used as hydrogen source and as co-catalysts beside Ni-Al/C in another paper and a positive correlation was suggested between spillover hydrogen on the catalysts and lignin-derived phenolic monomer yields [81]. Pt/Al$_2$O$_3$ not only converted birch into phenolic monomers but also catalyzed methanol reforming in methanol-water mixtures to supply hydrogen. Increased lignin monomer yield (49 wt %, entry 63 in Table 1) was attributed to the stabilization of reactive lignin intermediates by hydrogenation of reactive bonds due to the higher hydrogen yield [82].

A proof-of-concept membrane filtration was demonstrated by Rinaldi and co-workers in 2019, for the separation and concentration of the monophenol-rich fraction (entry 64 in Table 1) from the lignin liquors (poplar, Raney Ni catalyst, isopropanol/water solvent/H-donor mixture) [83]. In a further paper, the impact of process severity (reaction temperature) was studied by the same group, using the same feedstock, catalysts, and solvent mixture in the second one [84]. Higher process temperatures led to improving overall delignification yields (up to 87%), producing low molar mass fragments and preferential cleavage of hydroxyl groups in monolignol sidechains via hydrodeoxygenation, yielding oils with lower oxygen content [84].

Ni/C-catalyzed delignification of poplar resulted recalcitrance to enzymatic digestion of cellulose. Subsequent gelatinization in trifluoroacetic acid greatly enhanced rates of enzymatic digestion or maleic acid-AlCl$_3$ catalyzed conversion to hydroxymethylfurfural (HMF) and levulinic acid (LA). These results informed a 'no carbon left behind' strategy to convert total woody biomass into lignin, cellulose, and hemicellulose value streams for the future biorefinery [85].

Two task-specific catalysts were developed for RCF: Mo$_x$C/CNT for hardwood, and Ru/CMK-3 for softwood and grass. Using Mo$_x$C/CNT for apple wood led to a carbohydrate (both cellulose and hemicellulose) retention degree in solid product close to theoretical maximum and a delignification degree as high as 98.1% with 42% lignin monomers yield (entry 67 in Table 1) [86].

Chemodivergent hydrogenolysis of eucalyptus sawdust was carried out with Ni@ZIF-8 catalyst: phenolic compounds having either a propyl or propanol end-chain were produced under different reaction conditions. Propanol-substituted phenols (10% propanolsyringol, 5% propanolguaiacol) at 220 °C and 30 bar H$_2$ within 4 h were identified as the major depolymerized products, while propyl-substituted phenols (24% propylsyringol and propylguaiacol) where those at 260 °C and 30 bar H$_2$ within 8h (entry 68 in Table 1) [87].

A homogeneous catalytic system (binuclear Rh complex) for a "lignin-first" biorefinery in water was applied for basswood producing aromatic ketones (2% propanonesyringol, and 1.6% ethanonesyringol (Figure 4)) with almost complete deconstruction of lignin component under mild conditions (110 °C, 1 bar Ar, 24 h, entry 69 in Table 1) [88].

2.7. 2020

The same rhodium terpyridine complexes as homogeneous catalysts were used by the same group as in 2019 [88] for redox-neutral depolymerization of poplar wood under mild conditions affording aromatic ketones as the major monomer products (entry 70 in Table 1) [89].

The integrated biorefinery concept originated from Sels et al. [42,65] was developed further by the same group in 2020 [90]. The process model integrated three catalytic steps: RCF of wood, hydroprocessing of crude monomers extract, and dealkylation of crude alkylphenol product stream. Ru/C catalyst in methanol solvent converted birch in the first RCF step into a carbohydrate pulp amenable to bioethanol production and a lignin oil. Lignin monomers were cost-efficiently extracted from lignin oil with fossil n-hexane and were catalytically funneled into phenol and propylene (entry 71 in Table 1, and 2020, first row in Figure 4). A 78 wt % measure of birch was converted into xylochemicals [90]. An integrated techno-economic assessment of the biorefinery process that directly integrates the results of lab studies with economic costs and benefits was also developed [101]. They found that the scale of the plant, the feedstock-specific output quantities, and output prices highly determine the economic feasibility. The Sels group patented the chemocatalytic biorefinery concept [91]. Accordingly, three separate product fractions are produced in this biorefinery: (i) a lignin oil enriched with high contents of lignin-derived (mono)phenolics, (ii) essentially humin (furanic oligomers)-free hemicellulose-derived polyols, and (iii) a cellulose pulp. An example for eucalyptus sawdust is given as Entry 72 in Table 1.

Softwood lignocellulose was effectively (77–98%) depolymerized in a mild lignin-first acidolysis process (140 °C, 40 min, entry 73 in Table 1) using dimethyl carbonate and ethylene glycol solvents/stabilization agent producing high yield (9 wt %) of aromatic monophenols (2020, second row in Figure 4) and preserving cellulose as evidenced by a 85% glucose yield after enzymatic digestion [92]. The total utilization of lignin and carbohydrates in eucalyptus towards phenolics, levulinic acid, and furfural was emphasized in another paper in 2020: Pd/C catalyst in methanol solvent was used in the hydrogenolysis step, 50 wt % maximum phenolic monomers yield was achieved (entry 74 in Table 1) [93]. Beech wood was directly converted into lignin derived monomers (20 wt % yield) and dimers and holocellulose derived light hydrocarbons in the presence of a sulfided NiMo/Al$_2$O$_3$ catalyst in ethanol solvent at 260 °C (entry 75 in Table 1) [94].

Phenol was produced in a three-step process from pinewood with a 10 mol % overall yield [95]. In the first step pinewood was transformed into monomeric alkylmethoxyphenols using Pt/C catalyst in a methanol/water mixture as solvent at 230 °C and 30 bar H$_2$ pressure by the selective cleavage of β-O-4 lignin bonds (entry 76 in Table 1). Subsequently, the methoxy groups were removed by combination of MoP/SiO$_2$ and H-ZSM-5 catalysts leading to the formation of 4-alkylphenols—including 4-propylguaiacol, ethylguaiacol, and methylguaiacol—that were eventually dealkylated to phenol using H-ZSM-5 catalyst in the third step.

Zeolite-assisted fractionation of lignocellulose by preventing the recondensation reactions of aldehydes and allylic alcohols was achieved by Samec, Corma, and coworkers [96]. This prevention effect was attributed to the shape/size selectivity of protonic Beta zeolites, whose pore size limits undesired side reactions such as bimolecular condensations. In addition, mechanistic studies have pointed out that the reductive dehydration of allylic alcohols is carried out in the pores of metal-free zeolites. The highest lignin monomers yield from the organosolv pulping of birch wood was 20 wt %, using an ethanol/water mixture at 220 °C for two hours (entry 77 in Table 1). In parallel, furfural and ethylfurfural have been obtained as result of cellulose and hemicellulose fractions depolymerization over zeolitic acid sites.

Based on their previous kinetic studies of RCF [62,63], Román-Leshkov et al. [97] developed detailed mesoscale reaction-diffusion models for lignin-first fractionation. The models predict that mass transfer plays a governing role in solvolytic lignin extraction at the mesoscale. Lignin fragment diffusion competes with mass transfer resistance, which seems to be dominant effect when biomass particle size is over 2 nm. It is advisable to perform such tests for catalysts evaluation when the particle size of biomass is smaller than 2 nm, when the kinetics of the reaction is controlled by the diffusion of lignin fractions.

New concepts for lignin-first processes were introduced in 2020. Lignin-first integrated hydrothermal treatment [98], using an ionic liquid for lignin-first fractionation [99], and fractionation of wood with a γ-valerolactone consisting solvent system [100], was suggested. We do not consider the first two procedures as strictly regarded lignin-first processes as either a catalyst was not used [98], or reductive conditions were not applied [98,99]. The last procedure is remarkable, as a continuous operation was used to depolymerize maple wood lignin in a stirred reactor, which means continuous feeding of thermally pretreated lignin solution into the reactor and the products were collected at the outlet in a sample vial at 30 min intervals [100].

3. Summary

Early studies (1940–1986) on wood digestion or conversion using a catalyst were mainly focused on the elucidation of the chemical structure of lignin. The 4-n-propylphenol nature of the lignin monomers was confirmed [5]. With environmental concerns and increasing need to shift away from the dependence on fossil resources, interest in biomass as a sustainable feedstock has been resurgent, and lignocellulose has been identified as an important feedstock because it does not compete with the food supply. Innovative approaches, mainly related to efficient valorization of the cellulose platform have been introduced, together with the definition of top value-added platform chemicals. However, research in lignin has lagged behind due to its recalcitrant structure and it was only predominantly after 2010 that lignin conversion gained momentum and many fundamental works have been published, wherein significant progress has been made.

Notable advances in the one-pot full conversion of lignocellulose include the use of supercritical methanol as hydrogen source for the highly efficient conversion of lignocellulose to aliphatic small molecules in 2010 and 2011. The importance of supressing char formation has been recognized here. Later, interesting works to provide highly useful molecules in a one-pot, one-step process were the direct production of light naphtha (n-hexane, n-pentane, cyclohexane, and methylcyclopentane) by the Sels group in 2015 [42], which was developed further in 2018 [65] and 2020 [83] producing phenol and propylene. Liquid alkylcyclohexanes and polyols over Ru/C catalysts [70], and propylphenols, C_5–C_6 ketones, and furans over sulfided NiMo/Al$_2$O$_3$ catalysts [94] were formed in other one-pot one-step processes.

The concept of stabilization of reactive intermediates emerged in many different aspects in the field. One of the elegant examples of stabilization was introduced by Luterbacher et al. who described the role of formaldehyde and other carbonyl compounds in the protection of the native β-O-4 moiety during extraction, and thereby achieved much higher yields of desired monoaromatic products [6]. Trapping reactive intermediates originating from acid treatment after depolymerization in the form of acetals has also resulted in the suppression of recondensation and increased monomer yield from lignin. This approach has shown success on lignocellulose both in toluene/methanol mixtures [47], or more recently in dimethyl carbonate as solvent [84].

Undoubtedly, reductive catalytic fractionation (RCF) introduced during 2014–2015 by three research groups [39–41], has emerged as highly efficient method (also relying on stabilization) for lignocellulose valorization. Much research has been done regarding catalyst development, the role of additives, and types of hydrogen donors as well as reaction setup. For example using Zn-Pd/C bimetallic and Ru/C catalysts resulted in higher than 50% lignin monomers yield [40,41], isopropanol solvent ensured the source of hydrogen [39], and the direct production of light naphtha Ref. [42] became possible. This method has matured over the years towards achieving integrated biorefinery approaches in 2018, reaching the complete valorization of all lignocellulose constituents. For example, value-added products (propanolguaiacol and aliphatic alcohols) were produced in a model biorefinery with complete lignocellulose conversion. Importantly, these were further converted to a plethora of value-added building blocks with focus on amines [64]. The cellulose fraction was fully converted, in supercritical methanol, allowing for catalyst recycling. The aliphatic alcohols obtained in this step were coupled with cyclopentanone and subsequently converted to hydrocarbons, with target of jet-fuel range cyclic alkanes. Furthermore, an elegant liquid phase cellulose-to-naphtha technology was developed [65]. A deep converting 'lignin-first' biorefinery concept meaning gasoline and kerosene/diesel production was introduced [66]. Liquid alkylcyclohexanes, arenes, polyols, and furfural derivatives were produced [70,71]. The latest novel results were carried out using new catalysts [80,87,88] and by developing task-specific catalysts for hardwood and softwood [86], chemodivergent hydrogenolysis (for propyl- or propanol-methoxyphenol production) [87], membrane filtration [83], and applying new solvents/stabilization agent (dimethyl carbonate) [92]. The first integrated techno-economic assessment of a biorefinery process revealed that using only waste wood as feedstock can make the investment profitable [101].

In all the RCF systems, where the lignin fraction is valorized 'first', the celluloses will remain mixed with the heterogeneous catalyst, which means that the catalyst recycling issue needs to be addressed and many creative solutions have been already found. Possible solutions for catalyst separation were developed using a magnetic catalyst [39], membrane filtration [83], or embedding the metal function in a cage [4]. Adding a second catalytic step that converts all process residues was also developed, which liberated the catalyst for re-use [64].

4. Future Perspectives and Challenges

Several achievements have already been reached in 'lignin-first' process technology in recent years: noble-metal containing catalysts were replaced by more sustainable metals, or other catalysts; hydrogen was produced self-sufficiently from the pulp or solvent, and a good level of uniformity and chemodivergency of the products using mild conditions and appropriate (task-specific) catalysts has been reached, pointing toward exciting possibilities for ultimately converting total woody biomass into value-added products (as per the 'no carbon left behind' strategy). Notably, there were examples for the integration of biorefinery into petrorefinery processes, and the direct usage of lignin oil as a sulphur-free diesel-soluble liquid fuel. Semi-continuous flow-through systems have been established with good efficiency.

Future work will undoubtedly focus on several directions such as replacing batch reactors with really continuous flow-through systems and development of RCF processes applicable to crude biomass and lignocellulosic waste streams (e.g., bark). Effective removal of the catalyst from the pulp to enable subsequent enzymatic or catalytic treatment of the (hemi)cellulose fraction and improving recyclability of the catalysts will be essential to move toward real upscaling efforts, where in addition, the choices of solvent will be very important.

RCF enables to derive more value from lignin by increasing the yield and selectivity of desired aromatic monomers, which enables the more efficient production of well-defined products from lignin, thereby influencing the overall economic feasibility of lignocellulosic biorefineries. In the future, focus will also shift toward establishing downstream processing strategies for all lignocellulose constituents, and the diversification of the product portfolio accessible from lignocellulosic biorefineries.

Author Contributions: T.I.K. wrote the manuscript, B.F. produced the figures, A.P. included relevant patents, K.B. contributed to figures and text, all authors commented the manuscript. All authors have read and agreed to the published version of the manuscript.

Funding: This research received no external funding.

Acknowledgments: The authors acknowledge the networking support by the COST Action CA17128-Establishment of a Pan-European Network on the Sustainable Valorization of Lignin.

Conflicts of Interest: The authors declare no conflict of interest.

References

1. Ragauskas, A.J.; Beckham, G.T.; Biddy, M.J.; Chandra, R.; Chen, F.; Davis, M.F.; Davison, B.F.; Dixon, R.A.; Gilna, P.; Keller, M.; et al. Lignin valorization: Improving lignin processing in the biorefinery. *Science* **2014**, *344*, 1246843. [CrossRef] [PubMed]
2. Zakzeski, J.; Bruijnincx, P.C.A.; Jongerius, A.L.; Weckhuysen, B.M. The catalytic valorization of lignin for the production of renewable chemicals. *Chem. Rev.* **2010**, *110*, 3552–3599. [CrossRef] [PubMed]
3. Sun, Z.; Fridrich, B.; de Santi, A.; Elangovan, S.; Barta, K. Bright side of lignin depolymerization: Toward new platform chemicals. *Chem. Rev.* **2018**, *118*, 614–678. [CrossRef] [PubMed]
4. Van den Bosch, S.; Renders, T.; Kennis, S.; Koelewijn, S.-F.; Van den Bossche, G.; Vangeel, T.; Deneyer, A.; Depuydt, D.; Courtin, C.M.; Thevelein, J.M.; et al. Integrating lignin valorization and bio-ethanol production: On the role of Ni-Al$_2$O$_3$ catalyst pellets during lignin-first fractionation. *Green Chem.* **2017**, *19*, 3313–3326. [CrossRef]
5. Galkin, M.V.; Samec, J.S.M. Lignin valorization through catalytic lignocellulose fractionation: A fundamental platform for the future biorefinery. *ChemSusChem* **2016**, *9*, 1544–1558. [CrossRef] [PubMed]
6. Shuai, L.; Amiri, M.T.; Questell-Santiago, Y.M.; Héroguel, F.; Li, Y.; Kim, H.; Meilan, R.; Chapple, C.; Ralph, J.; Luterbacher, J.S. Formaldehyde stabilization facilitates lignin monomer production during biomass depolymerization. *Science* **2016**, *354*, 329–333. [CrossRef]
7. Xu, C.; Arancon, R.A.D.; Labidi, J.; Luque, R. Lignin depolymerisation strategies: Towards valuable chemicals and fuels. *Chem. Soc. Rev.* **2014**, *43*, 7485–7500. [CrossRef]
8. Kumar, C.R.; Anand, N.; Kloekhorst, A.; Cannilla, C.; Bonura, G.; Frusteri, F.; Barta, K.; Heeres, H.J. Solvent free depolymerization of kraft lignin to alkyl-phenolics using supported NiMo and CoMo catalysts. *Green Chem.* **2015**, *17*, 4921–4930. [CrossRef]
9. Narani, A.; Chowdari, R.K.; Cannilla, C.; Bonura, G.; Frusteri, F.; Heeres, H.J.; Barta, K. Efficient catalytic hydrotreatment of Kraft lignin to alkylphenolics using supported NiW and NiMo catalysts in supercritical methanol. *Green Chem.* **2015**, *17*, 5046–5057. [CrossRef]
10. Rahimi, A.; Ulbrich, A.; Coon, J.J.; Stahl, S.S. Formic-acid-induced depolymerization of oxidized lignin to aromatics. *Nature* **2014**, *515*, 249–252. [CrossRef]
11. Zhao, C.; Xie, S.; Pu, Y.; Zhang, R.; Huang, F.; Ragauskas, A.J.; Yuan, J.S. Synergistic enzymatic and microbial lignin conversion. *Green Chem.* **2016**, *18*, 1306–1312. [CrossRef]

12. Luo, H.; Abu-Omar, M.M. Chemicals from lignin. In *Encyclopedia of Sustainable Technologies*; Abraham, M.A., Ed.; Elsevier: Amsterdam, The Netherlands, 2017; pp. 573–585. ISBN 9780128046777.

13. Van den Bosch, S.; Koelewijn, S.-F.; Renders, T.; Van den Bossche, G.; Vangeel, T.; Schutyser, W.; Sels, B.F. Catalytic strategies towards lignin-derived chemicals. *Top. Curr. Chem.* **2018**, *376*, 36. [CrossRef] [PubMed]

14. Rinaldi, R.; Jastrzebski, R.; Clough, M.T.; Ralph, J.; Kennema, M.; Bruijnincx, P.C.A.; Weckhuysen, B.M. Paving the way for lignin valorisation: Recent advances in bioengineering, biorefining and catalysis. *Angew. Chem. Int. Ed.* **2016**, *55*, 8164–8215. [CrossRef] [PubMed]

15. Renders, T.; Van den Bosch, S.; Koelewijn, S.-F.; Schutyser, W.; Sels, B.F. Lignin-first biomass fractionation: The advent of active stabilisation strategies. *Energy Environ. Sci.* **2017**, *10*, 1551–1557. [CrossRef]

16. Schutyser, W.; Renders, T.; Van den Bosch, S.; Koelewijn, S.-F.; Beckham, G.T.; Sels, B.F. Chemicals from lignin: An interplay of lignocellulose fractionation, depolymerisation, and upgrading. *Chem. Soc. Rev.* **2018**, *47*, 852–908. [CrossRef]

17. Renders, T.; Van den Bossche, G.; Vangeel, T.; Van Aelst, K.; Sels, B. Reductive catalytic fractionation: State of the art of the lignin-first biorefinery. *Curr. Opin. Biotechnol.* **2019**, *56*, 193–201. [CrossRef]

18. Ha, J.-M.; Hwang, K.-R.; Kim, Y.-M.; Jae, J.; Kim, K.H.; Lee, H.W.; Kim, J.-Y.; Park, Y.-K. Recent progress in the thermal and catalytic conversion of lignin. *Renew. Sustain. Energy Rev.* **2019**, *111*, 422–441. [CrossRef]

19. Song, Y. Lignin valorization via reductive depolymerization. In *Chemical Catalysts For Biomass Upgrading*; Crocker, M., Santillan-Jimenez, E., Eds.; Wiley: Weinheim, Germany, 2020; pp. 395–438. [CrossRef]

20. Paone, E.; Tabanelli, T.; Mauriello, F. The rise of lignin biorefinery. *Curr. Opin. Green Sustain. Chem.* **2020**, *24*, 1–6. [CrossRef]

21. Ferrini, P.; Rezende, C.A.; Rinaldi, R. Catalytic upstream biorefining through hydrogen transfer reactions: Understanding the process from the pulp perspective. *ChemSusChem* **2016**, *9*, 3171–3180. [CrossRef]

22. Godard, H.P.; McCarthy, J.L.; Hibbert, H. Hydrogenation of wood. *J. Am. Chem. Soc.* **1940**, *62*, 988. [CrossRef]

23. Godard, H.P.; McCarthy, J.L.; Hibbert, H. Studies on lignin and related compounds. LXII. high pressure hydrogenation of wood using copper chromite catalyst (part 1). *J. Am. Chem. Soc.* **1941**, *63*, 3061–3066. [CrossRef]

24. Bower, J.R.; Cooke, L.M.; Hibbert, H. Studies on lignin and related compounds. LXX. hydrogenolysis and hydrogenation of maple wood. *J. Am. Chem. Soc.* **1943**, *65*, 1192–1195. [CrossRef]

25. Pepper, J.M.; Hibbert, H. Studies on lignin and related compounds. LXXXVII. high pressure hydrogenation of maple wood. *J. Am. Chem. Soc.* **1948**, *70*, 67–71. [CrossRef] [PubMed]

26. Pepper, J.M.; Steck, W. The effect of time and temperature on the hydrogenation of aspen lignin. *Can. J. Chem.* **1963**, *41*, 2867–2875. [CrossRef]

27. Pepper, J.M.; Steck, W.F.; Swoboda, R.; Karapally, J.C. Hydrogenation of lignin using nickel and palladium catalysts. *Lignin Struct. React.* **1966**, 238–248. [CrossRef]

28. Pepper, J.M.; Lee, Y.W. Lignin and related compounds. I. A comparative study of catalysts for lignin hydrogenolysis. *Can. J. Chem.* **1969**, *47*, 723–727. [CrossRef]

29. Pepper, J.M.; Lee, Y.W. Lignin and related compounds. II. studies using ruthenium and raney nickel as catalysts for lignin hydrogenolysis. *Can. J. Chem.* **1970**, *48*, 477–479. [CrossRef]

30. Pepper, J.M.; Fleming, R.W. Lignin and related compounds. V. The hydrogenolysis of aspen wood lignin using rhodium-on-charcoal as catalyst. *Can. J. Chem.* **1978**, *56*, 896–898. [CrossRef]

31. Pepper, J.M.; Supathna, P. Lignin and related compounds. VI. A study of variables affecting the hydrogenolysis of spruce wood lignin using a rhodium-on-charcoal catalyst. *Can. J. Chem.* **1978**, *56*, 899–902. [CrossRef]

32. Pepper, J.M.; Rahman, M.D. Lignin and related compounds. XI: Selective degradation of aspen poplar lignin by catalytic hydrogenolysis. *Cellul. Chem. Technol.* **1987**, *21*, 233–239.

33. Kaplunova, T.S.; Abduazimov, K.A.; Pulatov, B.K.; Ikramutdinova, M.T.; Abidova, M.F. Hydrogenolysis of Rice Husk Lignin. *Chem. Nat. Compd.* **1993**, *29*, 530–532. [CrossRef]

34. Yan, N.; Zhao, C.; Dyson, P.J.; Wang, C.; Liu, L.T.; Kou, Y. Selective degradation of wood lignin over noble-metal catalysts in a two-step process. *ChemSusChem* **2008**, *1*, 626–629. [CrossRef] [PubMed]

35. Torr, K.M.; van de Pas, D.J.; Cazeils, E.; Suckling, I.D. Mild hydrogenolysis of in-situ and isolated *pinus radiata* lignins. *Bioresour. Technol.* **2011**, *102*, 7608–7611. [CrossRef]

36. Li, C.; Zheng, M.; Wang, A.; Zhang, T. One-pot catalytic hydrocracking of raw woody biomass into chemicals over supported carbide catalysts: Simultaneous conversion of cellulose, hemicellulose and lignin. *Energy Environ. Sci.* **2012**, *5*, 6383–6390. [CrossRef]

37. Song, Q.; Wang, F.; Cai, J.; Wang, Y.; Zhang, J.; Yu, W.; Xu, J. Lignin depolymerization (LDP) in alcohol over nickel-based catalysts via a fragmentation–hydrogenolysis process. *Energy Environ. Sci.* **2013**, *6*, 994–1007. [CrossRef]
38. Galkin, M.V.; Samec, J.S.M. Selective route to 2-propenyl aryls directly from wood by a tandem organosolv and palladium-catalysed transfer hydrogenolysis. *ChemSusChem* **2014**, *7*, 2154–2158. [CrossRef] [PubMed]
39. Ferrini, P.; Rinaldi, R. Catalytic biorefining of plant biomass to non-pyrolytic lignin bio-oil and carbohydrates through hydrogen transfer reactions. *Angew. Chem. Int. Ed.* **2014**, *53*, 8634–8639. [CrossRef]
40. Parsell, T.; Yohe, S.; Degenstein, J.; Jarrell, T.; Klein, I.; Gencer, E.; Hewetson, B.; Hurt, M.; Kim, J.I.; Choudhari, H.; et al. A synergistic biorefinery based on catalytic conversion of lignin prior to cellulose starting from lignocellulosic biomass. *Green Chem.* **2015**, *17*, 1492–1499. [CrossRef]
41. Van den Bosch, S.; Schutyser, W.; Vanholme, R.; Driessen, T.; Koelewijn, S.-F.; Renders, T.; De Meester, B.; Huijgen, W.J.J.; Dehaen, W.; Courtin, C.M.; et al. Reductive lignocellulose fractionation into soluble lignin-derived phenolic monomers and dimers and processable carbohydrate pulps. *Energy Environ. Sci.* **2015**, *8*, 1748–1763. [CrossRef]
42. Dusselier, M.J.; Op De Beeck, B.; Sels, B.F. Biphasic Solvent Catalytic Process for the Direct Production of Light Naphta from Carbohydrate-Containing Feedstock. Patent 10,407,623, 10 September 2019.
43. Liu, Y.; Chen, L.; Wang, T.; Zhang, Q.; Wang, C.; Yan, J.; Ma, L. One-pot catalytic conversion of raw lignocellulosic biomass into gasoline alkanes and chemicals over $LiTaMoO_6$ and Ru/C in aqueous phosphoric acid. *ACS Sustain. Chem. Eng.* **2015**, *3*, 1745–1755. [CrossRef]
44. Van den Bosch, S.; Schutyser, W.; Koelewijn, S.-F.; Renders, T.; Courtin, C.M.; Sels, B.F. Tuning the lignin oil OH-content with Ru and Pd catalysts during lignin hydrogenolysis on birch wood. *Chem. Commun.* **2015**, *51*, 13158–13161. [CrossRef] [PubMed]
45. Schutyser, W.; Van den Bosch, S.; Renders, T.; De Boe, T.; Koelewijn, S.-F.; Dewaele, A.; Ennaert, T.; Verkinderen, O.; Goderis, B.; Courtin, C.M.; et al. Influence of bio-based solvents on the catalytic reductive fractionation of birch wood. *Green Chem.* **2015**, *17*, 5035–5045. [CrossRef]
46. Klein, I.; Saha, B.; Abu-Omar, M.M. Lignin depolymerization over Ni/C catalyst in methanol, a continuation: Effect of substrate and catalyst loading. *Catal. Sci. Technol.* **2015**, *5*, 3242–3245. [CrossRef]
47. Kaiho, A.; Kogo, M.; Sakai, R.; Saito, K.; Watanabe, T. In situ trapping of enol intermediates with alcohol during acid-catalysed de-polymerisation of lignin in a nonpolar solvent. *Green Chem.* **2015**, *17*, 2780–2783. [CrossRef]
48. Galkin, M.V.; Smit, A.T.; Subbotina, E.; Artemenko, K.A.; Bergquist, J.; Huijgen, W.J.J.; Samec, J.S.M. Hydrogen-free catalytic fractionation of woody biomass. *ChemSusChem* **2016**, *9*, 3280–3287. [CrossRef]
49. Klein, I.; Marcum, C.; Kenttämaa, H.; Abu-Omar, M.M. Mechanistic investigation of the Zn/Pd/C catalyzed cleavage and hydrodeoxygenation of lignin. *Green Chem.* **2016**, *18*, 2399–2405. [CrossRef]
50. Luo, H.; Klein, I.M.; Jiang, Y.; Zhu, H.; Liu, B.; Kenttämaa, H.I.; Abu-Omar, M.M. Total utilization of miscanthus biomass, lignin and carbohydrates, using earth abundant nickel catalyst. *ACS Sustain. Chem. Eng.* **2016**, *4*, 2316–2322. [CrossRef]
51. Jastrzebski, R.; Constant, S.; Lancefield, C.S.; Westwood, N.J.; Weckhuysen, B.M.; Bruijnincx, P.C.A. Tandem catalytic depolymerization of lignin by water-tolerant lewis acids and rhodium complexes. *ChemSusChem* **2016**, *9*, 2074–2079. [CrossRef]
52. Huang, X.; Zhu, J.; Korányi, T.I.; Boot, M.D.; Hensen, E.J.M. Effective release of lignin fragments from lignocellulose by lewis acid metal triflates in the lignin-first approach. *ChemSusChem* **2016**, *9*, 3262–3267. [CrossRef]
53. Chen, J.; Lu, F.; Si, X.; Nie, X.; Chen, J.; Lu, R.; Xu, J. High yield production of natural phenolic alcohols from woody biomass using a nickel-based catalyst. *ChemSusChem* **2016**, *9*, 3353–3360. [CrossRef]
54. Anderson, E.M.; Katahira, R.; Reed, M.; Resch, M.G.; Karp, E.M.; Beckham, G.T.; Román-Leshkov, Y. Reductive catalytic fractionation of corn stover lignin. *ACS Sustain. Chem. Eng.* **2016**, *4*, 6940–6950. [CrossRef]
55. Renders, T.; Schutyser, W.; Van Den Bosch, S.; Koelewijn, S.F.; Vangeel, T.; Courtin, C.M.; Sels, B.F. Influence of acidic (H_3PO_4) and alkaline (NaOH) additives on the catalytic reductive fractionation of lignocellulose. *ACS Catal.* **2016**, *6*, 2055–2066. [CrossRef]
56. Renders, T.; Van den Bosch, S.; Vangeel, T.; Ennaert, T.; Koelewijn, S.-F.; Van den Bossche, G.; Courtin, C.M.; Schutyser, W.; Sels, B.F. Synergetic effects of alcohol/water mixing on the catalytic reductive fractionation of poplar wood. *ACS Sustain. Chem. Eng.* **2016**, *4*, 6894–6904. [CrossRef]

57. Huang, X.; Morales Gonzalez, O.M.; Zhu, J.; Korányi, T.I.; Boot, M.D.; Hensen, E.J.M. Reductive fractionation of woody biomass into lignin monomers and cellulose by tandem metal triflate and Pd/C catalysis. *Green Chem.* **2017**, *19*, 175–187. [CrossRef]

58. Huang, X.; Ouyang, X.; Hendriks, B.; Gonzalez, O.M.M.; Zhu, J.; Korányi, T.I.; Boot, M.; Hensen, E.J.M. Selective production of mono-aromatics from lignocellulose over Pd/C catalyst: The influence of acid co-catalysts. *Faraday Discuss.* **2017**, *202*, 141–156. [CrossRef] [PubMed]

59. Zhai, Y.; Li, C.; Xu, G.; Ma, Y.; Liu, X.; Zhang, Y. Depolymerization of lignin via a non-precious Ni–Fe alloy catalyst supported on activated carbon. *Green Chem.* **2017**, *19*, 1895–1903. [CrossRef]

60. van de Pas, D.J.; Torr, K.M. Biobased epoxy resins from deconstructed native softwood lignin. *Biomacromolecules* **2017**, *18*, 2640–2648. [CrossRef]

61. Kumaniaev, I.; Subbotina, E.; Sävmarker, J.; Larhed, M.; Galkin, M.V.; Samec, J. Lignin depolymerization to monophenolic compounds in a flow-through system. *Green Chem.* **2017**, *19*, 5767–5771. [CrossRef]

62. Anderson, E.M.; Stone, M.L.; Katahira, R.; Reed, M.; Beckham, G.T.; Román-Leshkov, Y. Flowthrough reductive catalytic fractionation of biomass. *Joule.* **2017**, *1*, 613–622. [CrossRef]

63. Anderson, E.M.; Stone, M.L.; Hülsey, M.J.; Beckham, G.T.; Román-Leshkov, Y. Kinetic studies of lignin solvolysis and reduction by reductive catalytic fractionation decoupled in flow-through reactors. *ACS Sustain. Chem. Eng.* **2018**, *6*, 7951–7959. [CrossRef]

64. Sun, Z.; Bottari, G.; Afanasenko, A.; Stuart, M.C.A.; Deuss, P.J.; Fridrich, B.; Barta, K. Complete lignocellulose conversion with integrated catalyst recycling yielding valuable aromatics and fuels. *Nat. Catal.* **2018**, *1*, 82–92. [CrossRef]

65. Deneyer, A.; Peeters, E.; Renders, T.; Van den Bosch, S.; Van Oeckel, N.; Ennaert, T.; Szarvas, T.; Korányi, T.I.; Dusselier, M.; Sels, B.F. Direct upstream integration of biogasoline production into current light straight run naphtha petrorefinery processes. *Nat. Energy* **2018**, *3*, 969–977. [CrossRef]

66. Cao, Z.; Dierks, M.; Clough, M.T.; de Castro, I.B.D.; Rinaldi, R. A convergent approach for a deep converting lignin-first biorefinery rendering high-energy-density drop-in fuels. *Joule.* **2018**, *2*, 1118–1133. [CrossRef] [PubMed]

67. Graça, I.; Woodward, R.T.; Kennema, M.; Rinaldi, R. Formation and fate of carboxylic acids in the lignin-first biorefining of lignocellulose via h-transfer catalyzed by raney Ni. *ACS Sustain. Chem. Eng.* **2018**, *6*, 13408–13419. [CrossRef]

68. Sagues, W.J.; Bao, H.; Nemenyi, J.L.; Tong, Z. A lignin-first approach to biorefining: Utilizing fenton's reagent and supercritical ethanol for the production of phenolics and sugars. *ACS Sustain. Chem. Eng.* **2018**, *6*, 4958–4965. [CrossRef]

69. Ouyang, X.; Huang, X.; Hendriks, B.M.S.; Boot, M.D.; Hensen, E.J.M. Coupling organosolv fractionation and reductive depolymerization of woody biomass in a two-step catalytic process. *Green Chem.* **2018**, *20*, 2308–2319. [CrossRef]

70. Li, X.; Guo, T.; Xia, Q.; Liu, X.; Wang, Y. One-pot catalytic transformation of lignocellulosic biomass into alkylcyclohexanes and polyols. *ACS Sustain. Chem. Eng.* **2018**, *6*, 4390–4399. [CrossRef]

71. Guo, T.; Li, X.; Liu, X.; Guo, Y.; Wang, Y. Catalytic transformation of lignocellulosic biomass into arenes, 5-hydroxymethylfurfural, and furfural. *ChemSusChem* **2018**, *11*, 2758–2765. [CrossRef]

72. Renders, T.; Cooreman, E.; Van den Bosch, S.; Schutyser, W.; Koelewijn, S.-F.; Vangeel, T.; Deneyer, A.; Van den Bossche, G.; Courtin, C.M.; Sels, B.F. Catalytic lignocellulose biorefining in n-butanol/water: A one-pot approach toward phenolics, polyols, and cellulose. *Green Chem.* **2018**, *20*, 4607–4619. [CrossRef]

73. Kumaniaev, I.; Samec, J.S.M. Valorization of quercus suber bark toward hydrocarbon bio-oil and 4-ethylguaiacol. *ACS Sustain. Chem. Eng.* **2018**, *6*, 5737–5742. [CrossRef]

74. Stone, M.L.; Anderson, E.M.; Meek, K.M.; Reed, M.; Katahira, R.; Chen, F.; Dixon, R.A.; Beckham, G.T.; Román-Leshkov, Y. Reductive catalytic fractionation of c-lignin. *ACS Sustain. Chem. Eng.* **2018**, *6*, 11211–11218. [CrossRef]

75. Wang, S.; Gao, W.; Li, H.; Xiao, L.-P.; Sun, R.-C.; Song, G. Selective fragmentation of biorefinery corncob lignin into p-hydroxycinnamic esters with a supported $ZnMoO_4$ catalyst. *ChemSusChem* **2018**, *11*, 2114–2123. [CrossRef] [PubMed]

76. McCallum, C.S.; Strachan, N.; Bennett, S.C.; Forsythe, W.G.; Garrett, M.D.; Hardacre, C.; Morgan, K.; Sheldrake, G.N. Catalytic depolymerisation of suberin rich biomass with precious metal catalysts. *Green Chem.* **2018**, *20*, 2702–2705. [CrossRef]

77. Huang, Y.; Duan, Y.; Qiu, S.; Wang, M.; Ju, C.; Cao, H.; Fang, Y.; Tan, T. Lignin-first biorefinery: A reusable catalyst for lignin depolymerization and application of lignin oil to jet fuel aromatics and polyurethane feedstock. *Sustain. Energy Fuels* **2018**, *2*, 637–647. [CrossRef]

78. Li, S.; Li, W.; Zhang, Q.; Shu, R.; Wang, H.; Xin, H.; Ma, L. Lignin-first depolymerization of native corn stover with an unsupported MoS$_2$ catalyst. *Rsc Adv.* **2018**, *8*, 1361–1370. [CrossRef]

79. Wu, X.; Fan, X.; Xie, S.; Lin, J.; Cheng, J.; Zhang, Q.; Chen, L.; Wang, Y. Solar energy-driven lignin-first approach to full utilization of lignocellulosic biomass under mild conditions. *Nat. Catal.* **2018**, *1*, 772–780. [CrossRef]

80. Rautiainen, S.; Di Francesco, D.; Katea, S.; Westin, G.; Tungasita, D.; Samec, J. Lignin valorization by cobalt-catalyzed fractionation of lignocellulose to yield monophenolic compounds. *ChemSusChem* **2019**, *12*, 404–408. [CrossRef]

81. Park, J.; Riaz, A.; Verma, D.; Lee, H.J.; Woo, H.M.; Kim, J. Fractionation of lignocellulosic biomass over core-shell Ni-alumina catalysts with formic acid as a co-catalyst and hydrogen source. *ChemSusChem* **2019**, *12*, 1743–1762. [CrossRef]

82. Ouyang, X.; Huang, X.; Zhu, J.; Boot, M.D.; Hensen, E.J.M. Catalytic conversion of lignin in woody biomass into phenolic monomers in methanol/water mixtures without external hydrogen. *ACS Sustain. Chem. Eng.* **2019**, *7*, 13764–13773. [CrossRef]

83. Sultan, Z.; Graça, I.; Li, Y.; Lima, S.; Peeva, L.G.; Kim, D.; Ebrahim, M.A.A.; Rinaldi, R.; Livingston, A. Membrane fractionation of liquors from lignin-first biorefining. *ChemSusChem* **2019**, *12*, 1203–1212. [CrossRef]

84. Rinaldi, R.; Woodward, R.T.; Ferrini, P.; Riverac, H.J.E. Lignin-first biorefining of lignocellulose: The impact of process severity on the uniformity of lignin oil composition. *J. Braz. Chem. Soc.* **2019**, *30*, 479–491. [CrossRef]

85. Yang, H.; Zhang, X.; Luo, H.; Liu, B.; Shiga, T.M.; Li, X.; Kim, J.I.; Rubinelli, P.; Overton, J.C.; Subramanyam, V.; et al. Overcoming cellulose recalcitrance in woody biomass for the lignin-first biorefinery. *Biotechnol. Biofuels* **2019**, *12*, 171. [CrossRef] [PubMed]

86. Qiu, S.; Guo, X.; Huang, Y.; Fang, Y.; Tan, T. Task-specific catalyst development for lignin-first biorefinery toward hemicellulose retention or feedstock extension. *ChemSusChem* **2019**, *12*, 944–954. [CrossRef]

87. Liu, X.; Li, H.; Xiao, L.-P.; Sun, R.-C.; Song, G. Chemodivergent hydrogenolysis of eucalyptus lignin with Ni@ZIF-8 catalyst. *Green Chem.* **2019**, *21*, 1498–1504. [CrossRef]

88. Liu, Y.; Li, C.; Miao, W.; Tang, W.; Xue, D.; Li, C.; Zhang, B.; Xiao, J.; Wang, A.; Zhang, T.; et al. Mild redox-neutral depolymerization of lignin with a binuclear rh complex in water. *ACS Catal.* **2019**, *9*, 4441–4447. [CrossRef]

89. Liu, Y.; Li, C.; Miao, W.; Tang, W.; Xue, D.; Xiao, J.; Zhang, T.; Wang, C. Rhodium-terpyridine catalyzed redox-neutral depolymerization of lignin in water. *Green Chem.* **2020**, *22*, 33–38. [CrossRef]

90. Liao, Y.; Koelewijn, S.-F.; Van den Bossche, G.; Van Aelst, J.; Van den Bosch, S.; Renders, T.; Navare, K.; Nicolaï, T.; Van Aelst, K.; Maesen, M.; et al. A sustainable wood biorefinery for low–carbon footprint chemicals production. *Science* **2020**, *367*, 1385–1390. [CrossRef]

91. Sels, B.; Renders, T.; Cooreman, E.; Van den Bosch, S. Fractionation and depolymerisation of lignocellulosic material. International Application No. PCT/EP2019/066887, 2 January 2020.

92. De Santi, A.; Galkin, M.V.; Lahive, C.W.; Deuss, P.J.; Barta, K. Lignin-first fractionation of softwood lignocellulose using a mild dimethyl carbonate and ethylene glycol organosolv process. *ChemSusChem* **2020**, in press. [CrossRef]

93. Chen, X.; Zhang, K.; Xiao, L.-P.; Sun, R.-C.; Song, G. Total utilization of lignin and carbohydrates in eucalyptus grandis: An integrated biorefinery strategy towards phenolics, levulinic acid, and furfural. *Biotechnol. Biofuels* **2020**, *13*, 1–10. [CrossRef]

94. Parto, S.G.; Jørgensen, E.K.; Christensen, J.M.; Pedersen, L.S.; Larsen, D.B.; Duus, J.Ø.; Jensen, A.D. Solvent assisted catalytic conversion of beech wood and organosolv lignin over NiMo/γ-Al$_2$O$_3$. *Sustain. Energy Fuels* **2020**, *4*, 1844–1854. [CrossRef]

95. Ouyang, X.; Huang, X.; Boot, M.D.; Hensen, E.J.M. Efficient conversion of pine wood lignin to phenol. *Chemsuschem* **2020**, *13*, 1705–1709. [CrossRef] [PubMed]

96. Subbotina, E.; Velty, A.; Samec, J.S.M.; Corma, A. Zeolite-assisted lignin-first fractionation of lignocellulose: Overcoming lignin recondensation via shape-selective catalysis. *ChemSusChem* **2020**, in press. [CrossRef] [PubMed]

97. Thornburg, N.E.; Pecha, M.B.; Brandner, D.G.; Reed, M.L.; Vermaas, J.V.; Michener, W.E.; Katahira, R.; Vinzant, T.B.; Foust, T.D.; Donohoe, B.S.; et al. Mesoscale reaction-diffusion phenomena governing lignin-first biomass fractionation. *ChemSusChem* **2020**, in press. [CrossRef]

98. Lourencon, T.V.; Greca, L.G.; Tarasov, D.; Borrega, M.; Tamminen, T.; Rojas, O.J.; Balakshin, M.Y. Lignin-first integrated hydrothermal treatment (HTT) and synthesis of low-cost biorefinery particles. *ACS Sustain. Chem. Eng.* **2020**, *8*, 1230–1239. [CrossRef]

99. Xu, J.; Dai, L.; Gui, Y.; Yuan, L.; Zhang, C.; Lei, Y. Synergistic benefits from a lignin-first biorefinery of poplar via coupling acesulfamate ionic liquid followed by mild alkaline extraction. *Bioresour. Technol.* **2020**, *303*, 122888. [CrossRef] [PubMed]

100. Nandiwale, K.Y.; Danby, A.M.; Ramanathan, A.; Chaudhari, R.V.; Motagamwala, A.H.; Dumesic, J.A.; Subramaniam, B. Enhanced acid-catalyzed lignin depolymerization in a continuous reactor with stable activity. *ACS Sustain. Chem. Eng.* **2020**, *8*, 4096–4106. [CrossRef]

101. Tschulkow, M.; Compernolle, T.; Van den Bosch, S.; Van Aelst, J.; Storms, I.; Van Dael, M.; Van den Bossche, G.; Sels, B.; Van Passel, S. Integrated techno-economic assessment of a biorefinery process: The high-end valorization of the lignocellulosic fraction in wood streams. *J. Clean. Prod.* **2020**, *266*, 122022. [CrossRef]

Review

Biorefinery of Biomass of Agro-Industrial Banana Waste to Obtain High-Value Biopolymers

Carlos Redondo-Gómez [1], Maricruz Rodríguez Quesada [2,†], Silvia Vallejo Astúa [2,†], José Pablo Murillo Zamora [2,†], Mary Lopretti [3] and José Roberto Vega-Baudrit [1,2,*]

[1] National Laboratory of Nanotechnology LANOTEC, 1174-1200 Pavas, San José, Costa Rica; redox06@gmail.com
[2] School of Chemistry, National University of Costa Rica (UNA), 86-3000 Heredia, Costa Rica; marirquesada@gmail.com (M.R.Q.); vallejo.sil.15@gmail.com (S.V.A.); josepablomu@gmail.com (J.P.M.Z.)
[3] UDELAR University, cp1140 Montevideo, Uruguay; mlopretti@gmail.com
* Correspondence: jvegab@gmail.com; Tel.: +506-2519-5835
† These authors contributed equally.

Academic Editors: Alejandro Rodríguez and Eduardo Espinosa
Received: 16 June 2020; Accepted: 31 July 2020; Published: 23 August 2020

Abstract: On a worldwide scale, food demand is increasing as a consequence of global population growth. This makes companies push their food supply chains' limits with a consequent increase in generation of large amounts of untreated waste that are considered of no value to them. Biorefinery technologies offer a suitable alternative for obtaining high-value products by using unconventional raw materials, such as agro-industrial waste. Currently, most biorefineries aim to take advantage of specific residues (by either chemical, biotechnological, or physical treatments) provided by agro-industry in order to develop high-value products for either in-house use or for sale purposes. This article reviews the currently explored possibilities to apply biorefinery-known processes to banana agro-industrial waste in order to generate high-value products out of this residual biomass source. Firstly, the Central and Latin American context regarding biomass and banana residues is presented, followed by advantages of using banana residues as raw materials for the production of distinct biofuels, nanocellulose fibers, different bioplastics, and other high-value products Lastly, additional uses of banana biomass residues are presented, including energy generation and water treatment.

Keywords: biorefinery; residue; agro-industry; high-value products; banana

1. Introduction

The rising development of industries all over the world has brought a consequential increase in their residue generation, especially in the field of agro-industry. This waste can be denominated as "food supply chain waste" (FSCW) and can be defined as "organic material produced for human consumption lost or degraded primarily at the manufacturing and retail stages" [1]. This concept has emerged in the context of the current vast inefficiency of the food supply chain business. For instance, the Food and Agriculture Organization (FAO) revealed in 2011 that up to a third of the food aimed at human consumption is wasted every year globally [2]. The environmental and economic impacts of this worrying situation have driven the development of technologies pursuing not only conventional waste management and disposal, but also the extraction of as much value as possible out of any given agro-industrial waste.

1.1. Agro-Industry Residues as Biomass Sources

Agro-industries have slowly come to realize that valorization of biomass residues (either by using them as raw materials for the development of high-value products, or investing in recirculating processes that make use of these residues to obtain income in the long run) is not only beneficial from an environmental perspective, but can also help to minimize economic losses or even raise the net value of companies.

Inadequate treatment of these biomass residues has a negative impact on the environment, mainly generating greenhouse gases, contaminating water sources, and causing ecological problems [3,4]. Biorefinery technologies rise as suitable alternatives to mitigate these impacts as they can assist on reducing waste volume [5], but also on producing high-value goods out of revalorized biomass waste for circular economy purposes.

Waste valorization converts polymeric substrates into useful products such as chemicals, materials, and fuels, often by extraction, chemical conversion, or degradation. Historically, the utilization of complex biomasses led the way in pulp and paper production, or biotechnological production of furfural, ethanol, or short chain organic acids since the 19th century [6]. However, it was not until the 1990s when the term "biorefinery" became widespread in the industry, once biomass started to be used as a source of higher-value products [7,8].

1.2. The Concept of Biorefinery

Out of the many published definitions for the term "biorefinery", perhaps the one from the American National Renewable Energy Laboratory (NREL) seems to fit best the approach of this review: "A biorefinery is a facility that integrates biomass conversion processes and equipment to produce fuels, power, and chemicals from biomass" [6]. Definitions such as this one comprise the conversion of biomasses not only into biofuels, biopolymers, high-value products, and fine chemicals, but also include the generation of power (heat and electricity) analogous to today's petroleum-based refineries [7].

In addition to lignocellulosic biomass-based industries, more sectors have shown interest in applying biorefinery approaches to organic waste, for instance: food waste [9], nonedible oils [10] sewage sludge [11], and municipal solid waste [12] just to name a few. Biorefinery technologies enable more efficient use of agricultural resources and sustainable food production [9]. Taking advantage of biorefinery technologies represents a valuable strategy for agro-industries and governments; hence, they can easily navigate the challenges of the green economy era.

2. Applications of Agro-Industrial Residues

Agro-industrial waste is mainly composed of lignocellulose biomass. Lignocellulose waste has been gaining increasing attention due to its mechanical and thermal properties, renewability, wide availability, non-toxicity, low cost, and biodegradability [13,14]. The vast range of lignans and celluloses comprised in agro-industrial residues grants them tremendous potential as feedstocks for chemical and biotechnological conversion processes. For instance, enzymatic breakdown of cellulose and hemicelluloses into glucose and xylose allows further fermentation of these monosaccharides into ethanol by fermentative microorganisms. Furthermore, pyrolysis and anaerobic digestion of lignocellulose biomasses can yield combustion gases such as H_2 and CH_4 [13,15,16].

Currently, organic and agro-industrial residues take up a large portion of overall global waste (Figure 1a), which is one of the reasons to make good use of it. The abundance of biomass feedstocks gives a positive prospect for their future utilization in biorefinery technologies [17]. Estimates on biomass crop residue flows in Latin America show that most of lignocellulose containing biomasses are mostly made of maize, soybean, and sugarcane residues [18]; banana residues are not found within the main agro-industrial residues of developing countries to be used as biorefinery biomass sources (Figure 1b), though in many locations banana waste treatment remains a problem that needs to be addressed [19], as we discuss in the following section.

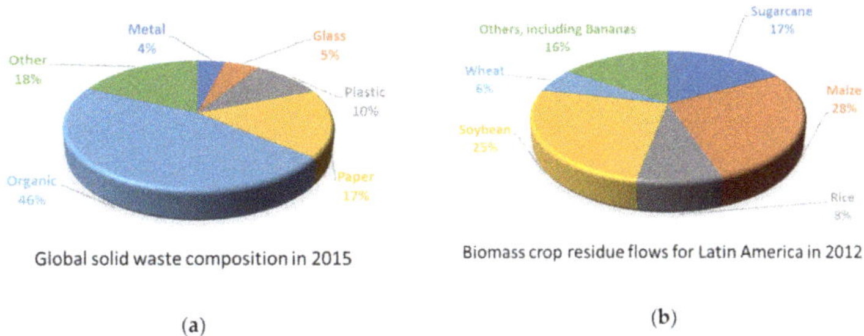

(**a**)   (**b**)

Figure 1. (**a**) Composition of global solid waste in 2015 (adapted from [17]). (**b**) Estimated biomass crop residue flows for Latin America in 2012 (adapted from Table 3.1 in [18]).

2.1. Generation of Banana Residues

A banana plant is a tall and sturdy herbaceous plant, with a succulent and very juicy tubular stem, composed of leaf-petiole sheaths consisting of long and strongly overlapping fibers called pseudostem. Each pseudostem bears fruit only once, before dying and being replaced by a new one; this pseudostem consists of concentric layers of a leaf sheath and a crown of large leaves [15].

Banana biomass mainly consists of four elements, namely: pseudostems, leaves, rachis, and skins. Additionally, a significant number of rejected bananas provide starchy feedstock to feed biorefineries. Feedstock derived from rejected bananas can reach up to 30 wt% of the total production (remaining unsold overripe fruits also fall in this category) [15]. All these biomass residues are normally dumped in rivers, oceans, landfills, and unregulated dumping grounds, creating huge decaying deposits that can lead to the spread of diseases, contamination of water sources, generation of foul odors, and attraction of rodents, insects, and scavengers. Some of the possible ways that enable the utilization of banana waste include compost production and food wrapping. However, these solutions do not always prevent the material from reaching the wasteland after serving its purpose. Recently, a craft type paper of good strength has been made from crushed, washed, and dried banana pseudostems [15].

2.2. Banana Residues in Central and Latin America

Banana is one of the most cultivated fruit crops worldwide (~106.7 million tons of production in 2013). Many industries take advantage of banana pulp, but discard lignocellulosic biomass, including pseudostems, stalks, leaves (normally found in the field), and rachis of the fruit bunches (gathered usually in the packing plants). Leaves, the pseudostem, stalk, and peel generate a huge amount of waste [20]; for instance, banana peels account for more than 41.3 million tons per year, therefore serving as a potential biomass feedstock [21].

The estimated amount of agricultural residue available in Central America in 2011 was about 192 Petajoules (PJ); the countries with the highest energy potentials are Guatemala with 79 PJ and Honduras with 29 PJ, followed by Costa Rica with 22 PJ. Banana residues represent an important fraction of these wastes, as the Central American region provides excellent environmental conditions for optimal development for the banana plant. In this fashion, banana crops rank in the top six agriculture residues in countries such as Belize, Costa Rica, Guatemala, Honduras, and Panama [21].

Nevertheless, Central American producers are far more focused on commercializing the crop itself than valorizing the corresponding waste. In 2011, about 2.9 million tons of banana residue (wet basis) were produced in Central America [21]. However, it is worth mentioning that these residues cannot be fully recovered, as part of them must be left in situ to avoid soil degradation (i.e., reduction of carbon stock in the soil), while other residues have found uses as fertilizers, fodder, and domestic fuel [21]. Nonetheless, there is still a large proportion that can find applications as biorefinery feedstocks.

Other Latin American nations face similar realities when it comes to banana production (Figure 1b). For instance, Brazil produces around 82.8 million tons of bananas annually; each produced ton leaves behind 100 kg of rejected fruit and some 4 tons of lignocellulosic waste (3 tons pseudostems, 480 kg leaves, 160 kg rachis, and 440 kg skins) [22].

Nations such as Ecuador have come up with a series of initiatives regarding bioethanol production using lignocellulosic biomass from banana crops. For instance, Guerrero and collaborators developed a process of production of bioethanol from banana rachis with a positive energy balance [23,24]. Figure 2 presents schematics of the production and use of second-generation ethanol from banana waste and its further blending with regular gasoline, this study employed a Well-to-Wheel (WtW) perspective and concludes that this strategy has great potential to reduce greenhouse gas emissions and fossil depletion, as a consequence of an overall positive energy balance for the process [23]. Bioethanol production from banana wastes is further discussed in Section 3.1 of this review.

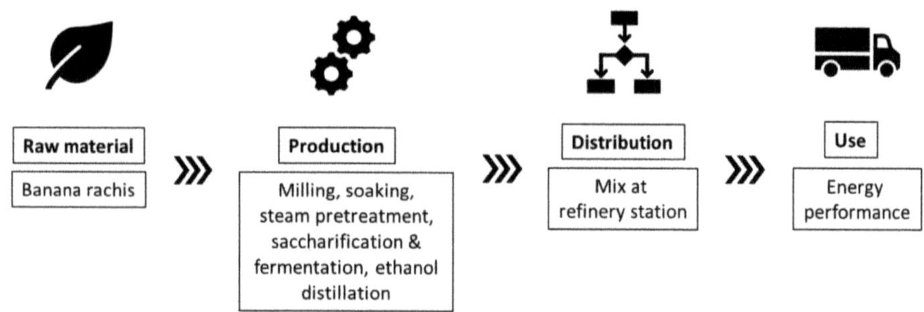

Figure 2. Life cycle system of second-generation ethanol production from banana rachis [23].

2.3. Potential Biorefinery Use of Different Banana Residues

Before addressing the possible ways to convert banana residues into high-value products through biorefinery, it is important to describe their physicochemical properties, as this information allows for their maximum exploitation as raw materials and would help in developing more eco-friendly approaches too. Banana peel and rachis waste are composed mainly of biopolymers such as lignin, pectin, cellulose, hemicellulose [25], fiber, proteins, and some low-molecular-weight compounds such as chlorophylls, phenolic compounds, water-soluble sugars, and minerals.

Table 1 shows the composition of banana peel and rachis on a dry matter basis. It is worth mentioning that the high moisture content of banana residues promotes their biodegradability before processing, thus affecting their handling, transportation, storage, and further uses in biorefinery technologies [20]. The composition of fruit peel residues varies according to species, seasonal variations, geographic location, variety, and stage of maturation [20]. Lignin contents are greater in banana rachis [23], and banana leaves are rich in holocellulose, hemicellulose, and lignin, all promising compounds for biorefinery processing. Lignin is particularly valuable, accounting for 25 wt% of banana leaves, which is higher than other important agro-industry residues such as cotton or straw [26]. Banana rachis and pseudostem residues can be used as biomass feedstocks, both biomasses have a high content of carbohydrates such as hemicellulose, starch [23], and lignin [19]. Banana stem residues also contain lignins, glucans, and most abundantly xylans and ashes [27].

Though mechanical [28] and hydrothermal pretreatments of these residues are still energy and time consuming [13], they have proved necessary for further steps in biorefinery operations. For instance, steam explosion pretreatment increases cellulose content compared to raw materials (from 20.1 to 54.4 wt% going from raw to pretreated pseudostem, and 26.1 to 57.1 wt% going from raw to pretreated rachis); pretreatment also increases free glucose content for further biotransformations [29].

Table 1. Composition analysis of banana peel and rachis on a dry matter basis [20,23].

Parameter	Value (wt%) [1]	
	Peel	Rachis
Cellulose	12.17 ± 0.21	23.0 ± 1.1
Hemicellulose	10.19 ± 0.12	11.2 ± 2.2
(Acid-Detergent) Lignin	2.88 ± 0.05	10.8 ± 0.5
Sucrose	15.58 ± 0.45	-
Glucose	7.45 ± 0.56	-
Fructose	6.2 ± 0.4	-
Protein	5.13 ± 0.14	-
Pectin	15.9 ± 0.3	-
Ashes	9.81 ± 0.42	29.9 ± 0.9

[1] Polyphenolics, fat, and other extractives add up as the remainder of the composition.

3. High-Value Products Obtained from Banana Residues

As shown above, a large number of parts of banana residues can be used as biomass sources; chemical composition of the raw material in question will determine its further suitability as a biomass source. In this section we describe a series of high-value products obtained from banana residues via biorefinery technologies.

3.1. Biofuels

Chemical composition of banana stems provides an indicator of their feasibility for production of fermentable sugars as a function of moisture content, as well as cellulose and hemicellulose contents. In fact, saccharification and further fermentation of banana lignocellulosic content for ethanol production has been extensively investigated [28–30]. Research by Duque and collaborators shows that the potential of ethanol production is 0.259 kg of ethanol per kg of banana stem raw material [31]. Research by Guerrero and co-workers has found high solid loading, low enzyme dosage and a short period conversion process as optimal conditions for bioethanol production from banana pseudostem and rachis, yielding ethanol solutions of 4.0 *v/v* % (87% yield) and 4.8 *v/v* % (74% yield), respectively [29].

Another study by Ingale and co-workers also employed banana stem waste and found that alkali treated banana pseudostems followed by enzymatic saccharification yield higher contents of reducing sugars than those alkali treated only, the corresponding increase in ethanol production was observed (Figure 3) [32]. Not only has lignocellulosic waste from banana production been transformed into ethanol, but fermentation of banana pulp and fruit can yield comparable biofuel efficiencies as corn [28].

Even though further technological improvements are still required, bioethanol yield from banana waste presents a promising alternative for the production of this biofuel. Further research on improved enzymatic cocktail formulations, more robust microorganism strains, as well as optimization of industrial conditions, such as reaction time, water content, and ethanol separation technologies [30], are still required [28].

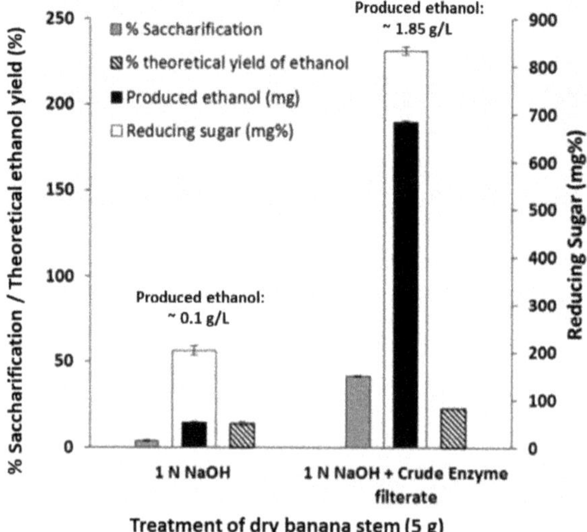

Figure 3. Saccharification percentage and ethanol yield from "alkali treated" and "alkali + enzymatic treatment" banana pseudostems (modified from Reference [32]).

3.2. Fibers for Mechanical Reinforcement

Banana fiber has traditionally found a place in a number of manufactured products such as paper, ropes, table mats, and handbags [15]. Though these materials are inexpensive, biodegradable, and produced from renewable sources, biorefinery approaches offer the possibility to generate higher value outcomes from their raw material. In this fashion, lignocellulosic micro/nanofibers (LCMNF) can be produced from banana leaf residues; Tarrés and co-workers' results show that banana leaves can yield up to 82.44% LCMNF, a significantly high value compared to other agricultural wastes that typically yield around 15% [26].

LCMNFs from banana leaf residue have been used to restore mechanical properties of recycled fluting paper. A study found that incorporation of only 1.5 wt% of LCMNF can restore the original properties of fluting paper, with a low impact in pulp drainability, while increasing the life span of the resulting recycled products [33].

3.3. Nanocellulose Fibers

Banana residues are a great source of cellulosic materials (see Table 1); cellulose provides stiffness and strength to the plants' structure, and approximately a third of the plant's anatomy is composed of this polysaccharide [34–37]. Since cellulose is greatly present in banana peel and rachis (Table 1), those represent biomass sources suitable to obtain nanocellulose fibers (NCFs).

NCFs exhibit many attractive physicochemical properties such as a high bending strength (~10 GPa), a Young's modulus of approximately 150 GPa, a high aspect ratio, and a high specific surface area. Therefore, NCFs have been used as reinforcements for polymer matrixes [38], and as additives for papermaking. Suspensions of NCFs improve the mechanical strength and density of paper while reducing its porosity [15]. The surface of NCFs is decorated with polar hydroxyl groups, which confer high moisture adsorption capacity and surface reactivity [39]. NCFs present a strong potential as oil-water suspension stabilizers in the food industry [35]; as mechanical reinforcement [40] in drug delivery, enzyme supports, biosensors [41], and scaffolds for tissue engineering applications [39,40,42].

Acid hydrolysis of cellulose is the most common process for obtaining NCFs [43], as fractions containing amorphous material can be hydrolyzed with HCl and sulfuric acid, while those containing crystalline cellulose are typically recovered by centrifugation [44,45]. NCFs have been isolated from banana peel using different processes, involving alkaline treatment and bleaching, followed by acid hydrolysis with sulfuric acid and high-pressure homogenization [46]. Transmission electron microscopy (TEM) and atomic force microscopy (AFM) investigations have revealed the clearance of large amounts of amorphous materials to afford highly crystalline NCFs, these NCFs showed good cytocompatibility with human epithelial colorectal adenocarcinoma cells (Caco-2 cell line) in concentrations below 1000 mg/mL, these NCFs exhibit promising features as reinforcement material in composites too (Figure 4) [46].

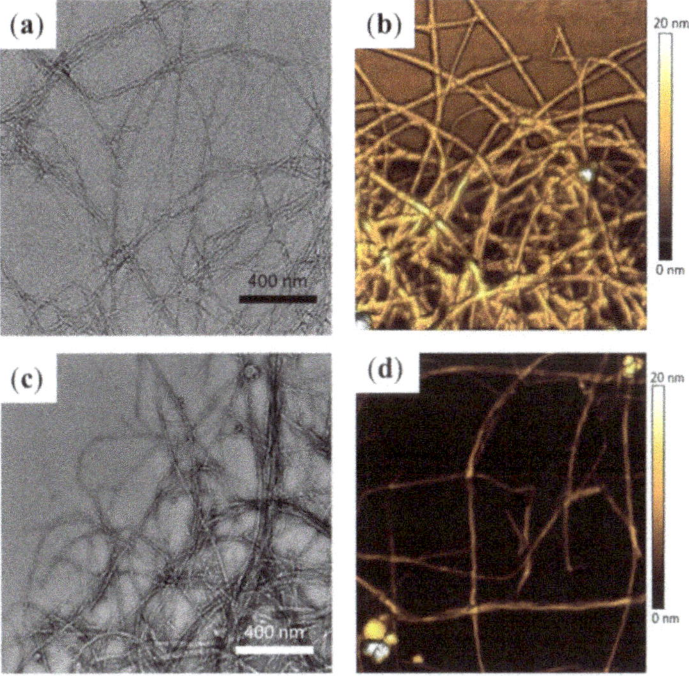

Figure 4. Banana peel-derived nanocellulose fibers (NCFs) produced by chemical hydrolysis without (panels (**a**) and (**b**)) or with mechanical treatment (high-pressure homogenization, panels (**c**) and (**d**)). (**a**) Transmission electron microscopy (TEM) and (**b**) atomic force microscopy (AFM) images of NCFs produced by chemical hydrolysis without mechanical treatment (**c**) TEM and (**d**) AFM images of NCFs produced with mechanical treatment (modified from Reference [46]).

3.4. Bioplastics

Poly-hydroxybutyrates (PHBs) are value-added biocompatible, biodegradable, thermoplastic biopolymers that can be synthesized by microorganisms from diverse carbon sources. Polysaccharides in banana peel can be either chemically or microbiologically transformed into PHBs [8], these biopolymers are hydrophobic, and bear similar mechanical properties to polypropylene or polyethylene. Getachew and collaborators have reported on a series of strains of *Bacillus* sp. able to yield up to 27 *w/w* % of PHB content after the fermentation of hydrolyzed banana peel residues [47]. PHB production from banana waste is still not affordable on an industrial scale, though efforts to couple this production as part of a multiproduct biorefinery are moving the field in this direction; for instance, Naranjo and coworkers reported on how such a kind of integration might save energetic costs and water waste via the fermentation of banana peel hydrolysates using *Burkholderia sacchari* IPT101 [8].

Poly-(*l*-lactic acid) (PLA) is a biodegradable and renewable polyester with many industrial and biomedical applications, including drug delivery systems, bioabsorbable fixation devices, bone regeneration, and tissue engineering scaffolds [48–50]. PLA has been obtained through fermentation of banana (and also pineapple) waste hydrolisates using *Lactobacillus casei* (subspecies rhamnosus), and further microwave-assisted polymerization, as reported by Jiménez-Bonilla and collaborators [48]. This direct melt polycondensation method afforded PLA oligomers with low oxidation losses, better stereopurity and lower energetic cost than conventional heating methods (Figure 5) [48].

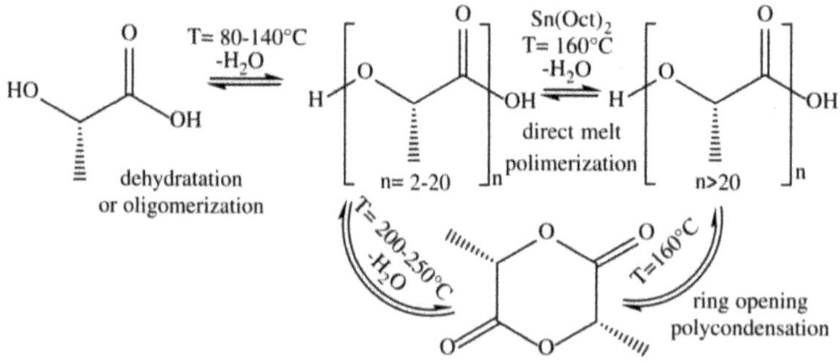

Figure 5. Synthetic path to produce Poly-(*l*-lactic acid) (PLA) from *l*-lactic acid [48].

3.5. Enzymes and Food Additives

Banana stalk residues have been valorized in the bio assisted production of enzymes like laccase, different oxidases, and endoglucanases too [13]. For instance, Reddy and co-workers investigated the use of *Pleurotus ostreatus* and *P. sajor-caju*, to produce lignolytic and cellulolytic enzymes such as laccase, lignin peroxidase, xylanase, endo-1,4-β-D-glucanase (CMCase), and exo-1,4-β-D-glucanase using banana wastes as solid substrate fermentation. Both microorganisms originated comparable levels of enzyme activities and patterns of production.

Leaf biomass was found to be an appropriate substrate (compared to pseudostems) for enzyme production [51]. Banana peel extracts have been studied as antioxidants in fresh orange juices, finding that free radical scavenging capacity increased by adding banana peel extracts to juice formulations. In addition, remarkable increases in antioxidant capacity using 2,2′-azino-bis-(3-ethylbenzothiazoline)-6-sulfonic acid (ABTS) radicals were observed when equal or greater than 5 mg of banana peel extract per ml of freshly squeezed juice was added, though no clear effects were observed in its ability to reduce the extent of lipid peroxidation [52].

4. Additional Uses of Banana Residues

4.1. Energy Generation

There seems to be a worldwide agreement on shifting towards green energy production and lessening the current dependence on fossil fuels. Agro-industrial biomass sources are a natural choice when it comes to exploring environmentally friendly ways to produce energy, and banana wastes are no exception to this. Banana residues have made it into the energy production sector only in recent times and their potential as energetic biomass is becoming evident; for instance, energy generation from dry banana peel can yield up to 18.89 MJ/kg [20]. The reader is invited to consult reference [19] for an extensive assessment of banana biomass as an energy source in the Central American region.

Currently, there are two approaches for the conversion of banana biomass into energy: thermal and biological conversion [20]. The former involves direct combustion and gasification, while the latter involves anaerobic digestion as shown in Figure 6. Compared to other types of waste substrates (such as human sewage, piggery, or feedlot waste), banana residues produce a very clean form of biogas (mostly made of methane and carbon dioxide, with little noxious odors) [15]. Pisutpaisal and collaborators have demonstrated that size reduction of banana peel raw material and its fungal pretreatment might improve methane yield [16].

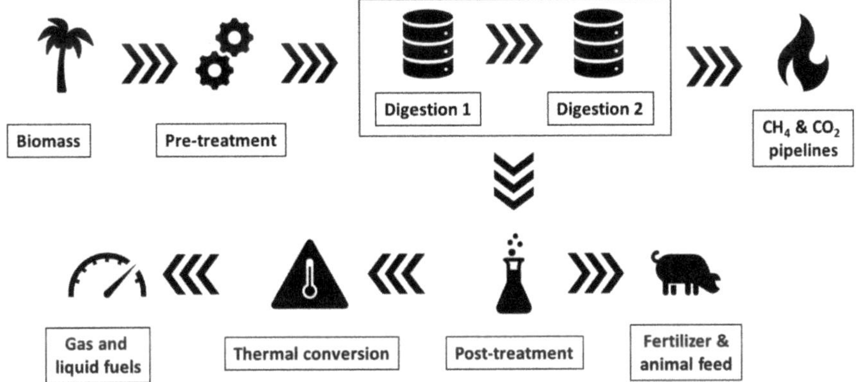

Figure 6. Schematics showing the process of banana waste anaerobic digestion to produce gas and liquid fuels, as well as fertilizer and animal feed (Modified from Figure 1 in Reference [15]).

Theoretical estimates for potential power generation based on both banana waste and banana peels in Malaysia (by Tock and coworkers for the period 2003–2008) suggest that banana biomass is a suitable renewable energy source in this one and other similar tropical nations. This study calculated potential power assuming that 1 PJ can be converted into 46 MW of electrical energy with 21% electrical conversion efficiency. This study estimates that using whole banana residues might generate some 80–95 MW yearly, while banana peels only would generate 12–25 MW per year (15% and 25% of energy production using the whole residue) as detailed in Table 2 [15].

Table 2. Estimated potential power generation based on whole banana residues and banana peel in Malaysia during 2003–2008.

	Whole Residue-Based Estimates [a]				Peel-Based Estimates [b]		
Year	Yield (kt/Year)	Energy from Biomass Residue (MJ/kg)	Energy Potential (PJ)	Potential Power Generation (MW)	Energy from Peel Residue (MJ/kg)	Energy Potential (PJ)	Potential Power Generation (MW)
2003	274	659	8.63	83.35	69	1.30	12.52
2004	317	761	9.97	96.31	79	1.50	14.47
2005	262	629	8.24	79.65	66	1.24	11.96
2006	258	620	8.13	78.50	65	1.22	11.79
2007	265	636	8.34	80.52	66	1.25	12.10
2008	270	649	8.5	82.13	68	1.28	12.34

[a] Residue: Product Ratio = 2.4, [b] Peel: Product Ratio = 0.25, modified from Reference [15].

4.2. Water Treatment

Banana peel has been reported to be used as a bio adsorbent for the removal of contaminants such as heavy metals, dyes, and organic pollutants from wastewaters [53]. A study by Pathak and co-workers reports on the adsorptive removal of benzoic acid (BA) and salicylic acid (SA) using banana peel (Figure 7); the authors report on removal efficiencies between 60–80% for the removal efficiency of these contaminants, with the advantage of possible reuse of the banana peel adsorbent in gasification for power generation [54] (though adsorbing contaminants that may be detrimental for biogas quality must be avoided). Banana pith has been used to produce activated carbons to be employed in divalent heavy metal cations and dye removal from aqueous solutions with satisfactory results [55].

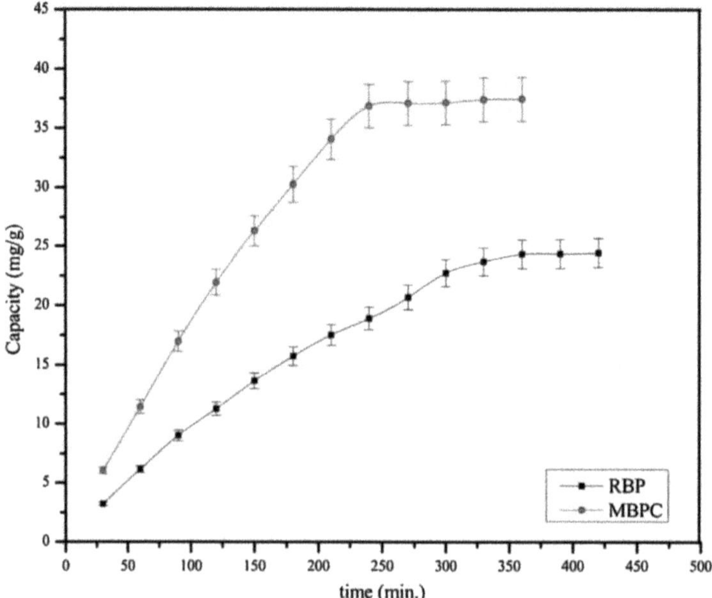

Figure 7. Removal efficiency of benzoic acid (BA) and salicylic acid (SA) from water samples using banana peel as adsorbent (C_0 = 100 mg/L, t = 15 h, T = 303 K, modified from Reference [54]).

5. Conclusions

Banana plants exhibit high growth rates and carbon neutrality; therefore, their agro-industrial residues are a promising alternative to be used as feedstock for biorefinery technologies, though challenges regarding composition variations in the wastes must be addressed, i.e., geographic location, plant variety, ripening stage, among others that difficult standardization for biorefinery processing [20].

Current technologies require further developments in order to extract as much value as possible from banana waste, for instance, by achieving positive energy balances by full integration of different biorefinery processes. Efforts involving the production of bioethanol and biogas are currently moving in this direction [23,24]. Although efforts are being made on the production of enzymes and food additives derived from different components of banana waste, the most promising potential for these residues rely on high-value biopolymers, i.e., micro and nanofibrillar mechanical reinforcements such as nanocellulose fibers, as well as the biotechnologically and chemically assisted production of bioplastics such as PHBs and PLA. More research on biorefinery approaches will be required if banana residues' biomass potential is to be increased. As a result of this, small communities from developing countries, as well as agro-industrial, chemical, and pharmaceutical industries, are most likely to benefit.

Author Contributions: Conceptualization, C.R.-G., J.R.V.-B. and M.L.; methodology and investigation, M.R.Q., S.V.A., J.P.M.Z., C.R.-G, M.L., J.R.V.-B; writing—original draft preparation, M.R.Q., S.V.A., J.P.M.Z., M.L., J.R.V.-B; writing—review and editing, C.R.-G, M.L., J.R.V.-B; supervision, C.R.-G, M.L., J.R.V.-B; project administration, C.R.-G, J.R.V.-B. All authors have read and agreed to the published version of the manuscript.

Funding: This research received no external funding.

Acknowledgments: The authors thank Anna Majkowska for her kind help during the preparation of the manuscript.

Conflicts of Interest: The authors declare no conflict of interest.

References

1. Lin, C.S.K.; Pfaltzgraff, L.A.; Herrero-Davila, L.; Mubofu, E.B.; Abderrahim, S.; Clark, J.H.; Koutinas, A.A.; Kopsahelis, N.; Stamatelatou, K.; Dickson, F.; et al. Food waste as a valuable resource for the production of chemicals, materials and fuels. Current situation and global perspective. *Energy Env. Sci.* **2013**, *6*, 426. [CrossRef]
2. Pfaltzgraff, L.A.; De bruyn, M.; Cooper, E.C.; Budarin, V.; Clark, J.H. Food waste biomass: A resource for high-value chemicals. *Green Chem.* **2013**, *15*, 307. [CrossRef]
3. Siles López, J.Á.; Li, Q.; Thompson, I.P. Biorefinery of waste orange peel. *Crit. Rev. Biotechnol.* **2010**, *30*, 63–69. [CrossRef] [PubMed]
4. Martínez-Ruano, J.A.; Caballero-Galván, A.S.; Restrepo-Serna, D.L.; Cardona, C.A. Techno-economic and environmental assessment of biogas production from banana peel (Musa paradisiaca) in a biorefinery concept. *Env. Sci. Pollut. Res.* **2018**, *25*, 35971–35980. [CrossRef] [PubMed]
5. Gumisiriza, R.; Hawumba, J.F.; Okure, M.; Hensel, O. Biomass waste-to-energy valorisation technologies: A review case for banana processing in Uganda. *Biotechnol. Biofuels* **2017**, *10*, 11. [CrossRef] [PubMed]
6. Kamm, B.; Kamm, M.; Gruber, P.R.; Kromus, S. *Biorefineries-Industrial Processes and Products*; Kamm, B., Gruber, P.R., Kamm, M., Eds.; Wiley: Hoboken, NJ, USA, 2005; ISBN 9783527310272.
7. De Corato, U.; De Bari, I.; Viola, E.; Pugliese, M. Assessing the main opportunities of integrated biorefining from agro-bioenergy co/by-products and agroindustrial residues into high-value added products associated to some emerging markets: A review. *Renew. Sustain. Energy Rev.* **2018**, *88*, 326–346. [CrossRef]
8. Naranjo, J.M.; Cardona, C.A.; Higuita, J.C. Use of residual banana for polyhydroxybutyrate (PHB) production: Case of study in an integrated biorefinery. *Waste Manag.* **2014**, *34*, 2634–2640. [CrossRef]
9. Fava, F.; Totaro, G.; Diels, L.; Reis, M.; Duarte, J.; Carioca, O.B.; Poggi-Varaldo, H.M.; Ferreira, B.S. Biowaste biorefinery in Europe: Opportunities and research & development needs. *N. Biotechnol.* **2015**, *32*, 100–108. [CrossRef]
10. Rojas Alfaro, J.J.; Fernández Araya, L.M.; Redondo-Gómez, C.; Vega-Baudrit, J. Bio Refinery of Oily Wastes. *Org. Med. Chem. Int. J.* **2018**, *6*, 1–7. [CrossRef]
11. Banerjee, A.; Bhaskar, T.; Ghosh, D. A biorefinery approach for sewage sludge. In *Waste Biorefinery*; Elsevier: Amsterdam, The Netherlands, 2020; pp. 293–421.
12. Pérez, V.; Pascual, A.; Rodrigo, A.; García Torreiro, M.; Latorre-Sánchez, M.; Coll Lozano, C.; David-Moreno, A.; Oliva-Dominguez, J.M.; Serna-Maza, A.; Herrero García, N.; et al. Integrated innovative biorefinery for the transformation of municipal solid waste into biobased products. In *Waste Biorefinery*; Elsevier: Amsterdam, The Netherlands, 2020; pp. 41–80.
13. Anwar, Z.; Gulfraz, M.; Irshad, M. Agro-industrial lignocellulosic biomass a key to unlock the future bio-energy: A brief review. *J. Radiat. Res. Appl. Sci.* **2014**, *7*, 163–173. [CrossRef]
14. Cardoso Araújo, D.J.; Machado, A.V.; Lobo Guerra Vilarinho, M.C. Availability and Suitability of Agroindustrial Residues as Feedstock for Cellulose-Based Materials: Brazil Case Study. *Waste Biomass Valorization* **2019**, *10*, 2863–2878. [CrossRef]
15. Tock, J.Y.; Lai, C.L.; Lee, K.T.; Tan, K.T.; Bhatia, S. Banana biomass as potential renewable energy resource: A Malaysian case study. *Renew. Sustain. Energy Rev.* **2010**, *14*, 798–805. [CrossRef]
16. Pisutpaisal, N.; Boonyawanich, S.; Saowaluck, H. Feasibility of Biomethane Production from Banana Peel. *Energy Procedia* **2014**, *50*, 782–788. [CrossRef]
17. Yang, X.; Choi, H.S.; Park, C.; Kim, S.W. Current states and prospects of organic waste utilization for biorefineries. *Renew. Sustain. Energy Rev.* **2015**, *49*, 335–349. [CrossRef]
18. Sommer, S.G.; Hamelin, L.; Olesen, J.E.; Montes, F.; Jia, W.; Chen, Q.; Triolo, J.M. Agricultural waste biomass. In *Supply Chain Management for Sustainable Food Networks*; John Wiley & Sons, Ltd.: Chichester, UK, 2016; pp. 67–106.
19. Padam, B.S.; Tin, H.S.; Chye, F.Y.; Abdullah, M.I. Banana by-products: An under-utilized renewable food biomass with great potential. *J. Food Sci. Technol.* **2014**, *51*, 3527–3545. [CrossRef]
20. Pathak, P.D.; Mandavgane, S.A.; Kulkarni, B.D. Valorization of banana peel: A biorefinery approach. *Rev. Chem. Eng.* **2016**, *32*. [CrossRef]
21. Cutz, L.; Haro, P.; Santana, D.; Johnsson, F. Assessment of biomass energy sources and technologies: The case of Central America. *Renew. Sustain. Energy Rev.* **2016**, *58*, 1411–1431. [CrossRef]

22. Fernandes, E.R.K.; Marangoni, C.; Souza, O.; Sellin, N. Thermochemical characterization of banana leaves as a potential energy source. *Energy Convers. Manag.* **2013**, *75*, 603–608. [CrossRef]

23. Guerrero, A.B.; Muñoz, E. Life cycle assessment of second generation ethanol derived from banana agricultural waste: Environmental impacts and energy balance. *J. Clean. Prod.* **2018**, *174*, 710–717. [CrossRef]

24. Guerrero, A.B.; Aguado, P.L.; Sánchez, J.; Curt, M.D. GIS-Based Assessment of Banana Residual Biomass Potential for Ethanol Production and Power Generation: A Case Study. *Waste Biomass Valorization* **2016**, *7*, 405–415. [CrossRef]

25. Lavoine, N.; Desloges, I.; Dufresne, A.; Bras, J. Microfibrillated cellulose—Its barrier properties and applications in cellulosic materials: A review. *Carbohydr. Polym.* **2012**, *90*, 735–764. [CrossRef] [PubMed]

26. Tarrés, Q.; Espinosa, E.; Domínguez-Robles, J.; Rodríguez, A.; Mutjé, P.; Delgado-Aguilar, M. The suitability of banana leaf residue as raw material for the production of high lignin content micro/nano fibers: From residue to value-added products. *Ind. Crop. Prod.* **2017**, *99*, 27–33. [CrossRef]

27. Daza Serna, L.V.; Solarte Toro, J.C.; Serna Loaiza, S.; Chacón Perez, Y.; Cardona Alzate, C.A. Agricultural Waste Management Through Energy Producing Biorefineries: The Colombian Case. *Waste Biomass Valorization* **2016**, *7*, 789–798. [CrossRef]

28. Velásquez-Arredondo, H.I.; Ruiz-Colorado, A.A.; De Oliveira, S., Jr. Ethanol production process from banana fruit and its lignocellulosic residues: Energy analysis. *Energy* **2010**, *35*, 3081–3087. [CrossRef]

29. Guerrero, A.B.; Ballesteros, I.; Ballesteros, M. The potential of agricultural banana waste for bioethanol production. *Fuel* **2018**, *213*, 176–185. [CrossRef]

30. Bello, R.H.; Linzmeyer, P.; Franco, C.M.B.; Souza, O.; Sellin, N.; Medeiros, S.H.W.; Marangoni, C. Pervaporation of ethanol produced from banana waste. *Waste Manag.* **2014**, *34*, 1501–1509. [CrossRef]

31. Duque, S.H.; Cardona, C.A.; Moncada, J. Techno-Economic and Environmental Analysis of Ethanol Production from 10 Agroindustrial Residues in Colombia. *Energy Fuels* **2015**, *29*, 775–783. [CrossRef]

32. Ingale, S.; Joshi, S.J.; Gupte, A. Production of bioethanol using agricultural waste: Banana pseudo stem. *Braz. J. Microbiol.* **2014**, *45*, 885–892. [CrossRef]

33. Espinosa, E.; Tarrés, Q.; Domínguez-Robles, J.; Delgado-Aguilar, M.; Mutjé, P.; Rodríguez, A. Recycled fibers for fluting production: The role of lignocellulosic micro/nanofibers of banana leaves. *J. Clean. Prod.* **2018**, *172*, 233–238. [CrossRef]

34. George, J.; Sabapathi, S.N. Cellulose nanocrystals: Synthesis, functional properties, and applications. *Nanotechnol. Sci. Appl.* **2015**, 45. [CrossRef]

35. Gómez H., C.; Serpa, A.; Velásquez-Cock, J.; Gañán, P.; Castro, C.; Vélez, L.; Zuluaga, R. Vegetable nanocellulose in food science: A review. *Food Hydrocoll.* **2016**, *57*, 178–186. [CrossRef]

36. Rambabu, N.; Panthapulakkal, S.; Sain, M.; Dalai, A.K. Production of nanocellulose fibers from pinecone biomass: Evaluation and optimization of chemical and mechanical treatment conditions on mechanical properties of nanocellulose films. *Ind. Crop. Prod.* **2016**, *83*, 746–754. [CrossRef]

37. Bettaieb, F.; Khiari, R.; Hassan, M.L.; Belgacem, M.N.; Bras, J.; Dufresne, A.; Mhenni, M.F. Preparation and characterization of new cellulose nanocrystals from marine biomass Posidonia oceanica. *Ind. Crop. Prod.* **2015**, *72*, 175–182. [CrossRef]

38. Lu, P.; Hsieh, Y.-L. Preparation and properties of cellulose nanocrystals: Rods, spheres, and network. *Carbohydr. Polym.* **2010**, *82*, 329–336. [CrossRef]

39. Ullah, M.W.; Ul-Islam, M.; Khan, S.; Kim, Y.; Park, J.K. Structural and physico-mechanical characterization of bio-cellulose produced by a cell-free system. *Carbohydr. Polym.* **2016**, *136*, 908–916. [CrossRef] [PubMed]

40. Chandra C.S., J.; George, N.; Narayanankutty, S.K. Isolation and characterization of cellulose nanofibrils from arecanut husk fibre. *Carbohydr. Polym.* **2016**, *142*, 158–166. [CrossRef]

41. Salas, C.; Nypelö, T.; Rodriguez-Abreu, C.; Carrillo, C.; Rojas, O.J. Nanocellulose properties and applications in colloids and interfaces. *Curr. Opin. Colloid Interface Sci.* **2014**, *19*, 383–396. [CrossRef]

42. Brinchi, L.; Cotana, F.; Fortunati, E.; Kenny, J.M. Production of nanocrystalline cellulose from lignocellulosic biomass: Technology and applications. *Carbohydr. Polym.* **2013**, *94*, 154–169. [CrossRef]

43. Tibolla, H.; Pelissari, F.M.; Menegalli, F.C. Cellulose nanofibers produced from banana peel by chemical and enzymatic treatment. *Food Sci. Technol.* **2014**, *59*, 1311–1318. [CrossRef]

44. Camacho, M.; Corrales Ureña, Y.R.; Lopretti, M.; Carballo, L.B.; Moreno, G.; Alfaro, B.; Vega-Baudrit, J.R. Synthesis and Characterization of Nanocrystalline Cellulose Derived from Pineapple Peel Residues. *J. Renew. Mater.* **2017**, *5*, 271–279. [CrossRef]

45. Oun, A.A.; Rhim, J.-W. Characterization of nanocelluloses isolated from Ushar (Calotropis procera) seed fiber: Effect of isolation method. *Mater. Lett.* **2016**, *168*, 146–150. [CrossRef]

46. Tibolla, H.; Pelissari, F.M.; Martins, J.T.; Vicente, A.A.; Menegalli, F.C. Cellulose nanofibers produced from banana peel by chemical and mechanical treatments: Characterization and cytotoxicity assessment. *Food Hydrocoll.* **2018**, *75*, 192–201. [CrossRef]

47. Getachew, A.; Woldesenbet, F. Production of biodegradable plastic by polyhydroxybutyrate (PHB) accumulating bacteria using low cost agricultural waste material. *BMC Res. Notes* **2016**, *9*, 509. [CrossRef] [PubMed]

48. Jiménez-Bonilla, P.; Salas-Arias, J.; Esquivel, M.; Vega-Baudrit, J.R. Optimization of Microwave-Assisted and Conventional Heating Comparative Synthesis of Poly(lactic acid) by Direct Melt Polycondensation from Agroindustrial Banana (Musa AAA Cavendish) and Pineapple (Ananas comosus) Fermented Wastes. *J. Polym. Env.* **2014**, *22*, 393–397. [CrossRef]

49. Navarro, M.; Michiardi, A.; Castaño, O.; Planell, J. Biomaterials in orthopaedics. *J. R. Soc. Interface* **2008**, *5*, 1137–1158. [CrossRef] [PubMed]

50. Singh, V.; Tiwari, M. Structure-Processing-Property Relationship of Poly(Glycolic Acid) for Drug Delivery Systems 1: Synthesis and Catalysis. *Int. J. Polym. Sci.* **2010**, *2010*, 1–23. [CrossRef]

51. Reddy, G.V.; Ravindra Babu, P.; Komaraiah, P.; Roy, K.R.R.M.; Kothari, I.L. Utilization of banana waste for the production of lignolytic and cellulolytic enzymes by solid substrate fermentation using two Pleurotus species (P. ostreatus and P. sajor-caju). *Process. Biochem.* **2003**, *38*, 1457–1462. [CrossRef]

52. Ortiz, L.; Dorta, E.; Gloria Lobo, M.; González-Mendoza, L.A.; Díaz, C.; González, M. Use of Banana (Musa acuminata Colla AAA) Peel Extract as an Antioxidant Source in Orange Juices. *Plant. Foods Hum. Nutr.* **2017**, *72*, 60–66. [CrossRef]

53. Kumar Rana, G.; Singh, Y.; Mishra, S.P.; Rahangdale, H.K. Potential Use of Banana and Its By-products: A Review. *Int. J. Curr. Microbiol. Appl. Sci.* **2018**, *7*, 1827–1832. [CrossRef]

54. Pathak, P.D.; Mandavgane, S.A.; Kulkarni, B.D. Utilization of banana peel for the removal of benzoic and salicylic acid from aqueous solutions and its potential reuse. *Desalin. Water Treat.* **2016**, *57*, 12717–12729. [CrossRef]

55. Kadirvelu, K.; Kavipriya, M.; Karthika, C.; Radhika, M.; Vennilamani, N.; Pattabhi, S. Utilization of various agricultural wastes for activated carbon preparation and application for the removal of dyes and metal ions from aqueous solutions. *Bioresour. Technol.* **2003**, *87*, 129–132. [CrossRef]

 *molecules*

Review

Electrosynthesis of Biobased Chemicals Using Carbohydrates as a Feedstock

Vincent Vedovato, Karolien Vanbroekhoven, Deepak Pant * and Joost Helsen *

Separation & Conversion Technology, Flemish Institute for Technological Research, Boeretang 200, 2400 Mol, Belgium; vincent.vedovato@vito.be (V.V.); karolien.vanbroekhoven@vito.be (K.V.)
* Correspondence: deepak.pant@vito.be (D.P.); joost.helsen@vito.be (J.H.);
  Tel.: +32-14-336-969 (D.P.); +32-14-336-940 (J.H.)

Academic Editors: Alejandro Rodríguez and Eduardo Espinosa
Received: 30 June 2020; Accepted: 11 August 2020; Published: 14 August 2020

Abstract: The current climate awareness coupled with increased focus on renewable energy and biobased chemicals have led to an increased demand for such biomass derived products. Electrosynthesis is a relatively new approach that allows a shift from conventional fossil-based chemistry towards a new model of a real sustainable chemistry that allows to use the excess renewable electricity to convert biobased feedstock into base and commodity chemicals. The electrosynthesis approach is expected to increase the production efficiency and minimize negative health for the workers and environmental impact all along the value chain. In this review, we discuss the various electrosynthesis approaches that have been applied on carbohydrate biomass specifically to produce valuable chemicals. The studies on the electro-oxidation of saccharides have mostly targeted the oxidation of the primary alcohol groups to form the corresponding uronic acids, with Au or TEMPO as the active electrocatalysts. The investigations on electroreduction of saccharides focused on the reduction of the aldehyde groups to the corresponding alcohols, using a variety of metal electrodes. Both oxidation and reduction pathways are elaborated here with most recent examples. Further recommendations have been made about the research needs, choice of electrocatalyst and electrolyte as well as upscaling the technology.

Keywords: electrosynthesis; biomass; carbohydrate; saccharides; electro-oxidation; electroreduction

1. Introduction

The increasing pressure on non-renewable fossil resources caused by the growing population has made increasingly urgent the need for the development of sustainable and alternative technologies to produce energy, food, and materials. In recent years, a significant number of environmentally friendly processes based on the utilization of renewable raw materials have been described, in the hope of replacing the petro-based industry [1]. This is especially true of Europe where bio-based industries are central to build a European circular economy [2].

Biomass is considered as an inexpensive feedstock and is the most abundant renewable resource on the planet. It is a mixture composed of organic and inorganic materials. One of the major components of the organic biomass materials is lignocellulose, itself being constituted primarily of cellulose (42–49%); hemicellulose (16–23%), and lignin (21–39%) [3]. Cellulose is thus considered to be the most important macromolecule on Earth. It is composed of long polymeric chains of at least 500 glucose molecules. This abundance explains the renewed interest towards saccharides and polysaccharides as an alternative to fossil resources in chemical, pharmaceutical, material, and energy industries. Even though carbohydrate chemistry has been a subject of study since the 19th century following Fisher's pioneering work [4], it has arguably received limited attention compared to other biomolecules such as proteins or amino-acids. While these compounds have found a plethora of applications in various industries, their production and derivatization are predominantly via chemical synthesis. Enzymatic processes have been studied as a greener alternative but remain yet underexplored [5,6]. Therefore, there is a great need for novel greener processes for the production of carbohydrate-based materials.

The field of electrochemistry has recently gained renewed traction as a clean and carbon-neutral way to promote chemical transformations. Electrochemistry can be employed in a large number of areas, from the synthesis of materials or chemicals to the development of sensors to detect and analyse specific compounds, as well as to generate power (cogeneration). In chemical synthesis, electrochemistry allows to avoid the use of chemicals to deliver the reductive or oxidative power and provides more safe processing conditions due to the decoupling of the chemical reaction in half reactions [7]. Although electrosynthesis satisfies most of the postulates of green chemistry, direct applications of such processes in industries remain rather scarce. However, recent advancements in the development of electrode materials and membrane technologies have improved the performances of electrochemical processes by reducing the energy consumption, improving the rates of reactions and selectivity and increasing the current density [8,9]. These novel technologies and the ability to use renewable electricity from wind or solar energy have made electrochemistry an intensively researched approach in recent years which has a potential to become less expensive and therefore a more economically and ecologically attractive alternative to chemical processes.

This review addresses the current knowledge on the electrochemical conversion of biobased chemicals, particularly saccharides, to commodity chemicals.

2. Electrochemical Oxidation of Biomass

The oxidation of carbohydrates has been a very active field of research for the last fifty years and is still intensively studied. This is mainly true for oxidation of carbohydrates to their uronic acid analogues. Chemical direct oxidation processes require the use of stochiometric amount of strong and toxic oxidizing agents, such as nitric acid and nitrogen dioxide. Due to the risks and cost of such protocols, catalytic processes were developed, employing atmospheric oxygen and noble metals. Although the catalytic systems were more selective and efficient, more sustainable systems were greatly sought. Therefore, many reports have demonstrated the capacity for electrochemical processes to transform saccharides—through direct or indirect oxidation—to their corresponding uronic acids, but also to other important derivatives (Scheme 1). Reducing and non-reducing saccharides have been studied. These two types of saccharides differ in the presence or the absence of a free aldehyde/hemiacetal functional group.

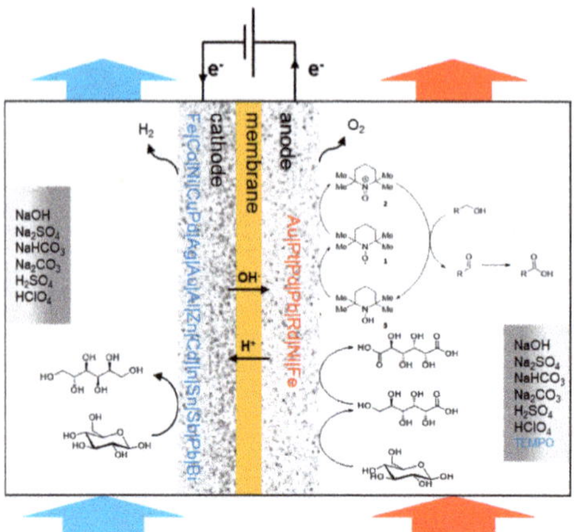

Scheme 1. Conceptualisation of oxidation and reduction of carbohydrates in an electrochemical cell. Components like electrocatalyst and membrane as well as processing parameters like choice of electrolytes indicate the variables for reaction and process efficiency.

2.1. Poly/Oligosaccharides

The organic stable nitroxyl radical 2,2,6,6-tetramethylpiperidin-1-oxyl (TEMPO, **1**) is used commonly in organic synthesis, mostly for its ability to oxidize primary alcohols with a high level of selectivity. Due to the high turnover frequency of TEMPO and its derivatives, as well as their high stability, such catalyst also found use in electrochemical processes, as homogeneous mediators for electrochemical oxidation of hydroxyl groups (Scheme 2).

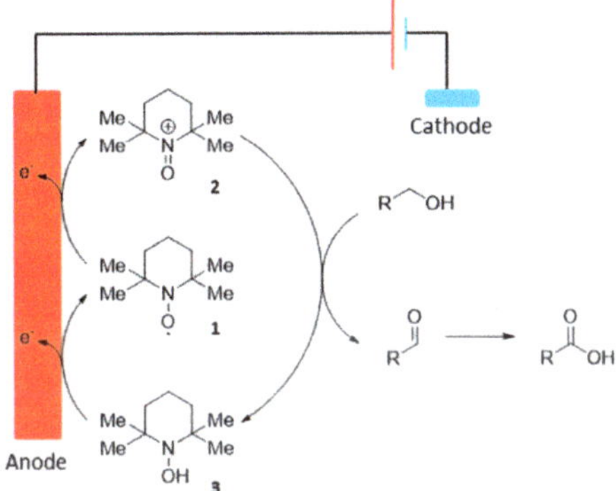

Scheme 2. TEMPO-mediated electrochemical oxidation of alcohol groups.

Polymeric materials with interesting physicochemical properties which, after their period of utilization, degrade to green components are highly sought. Natural polymers have been used as a starting platform to study the feasibility of such concepts by modifying their primary structure.

Being naturally present in most plants for energy storage, starch is produced on a multi-million tons scale per year and has found a great number of applications in the food industry (starch is naturally present in many basic cooking ingredients, but is also used as thickening agent) and mainly in the papermaking process as an adhesive.

The chemical oxidation of carbohydrates using nitrogen oxides is known since the mid-1900's, however, such transformation always requires the use of external, toxic co-oxidants, in order to regenerate the nitrosonium ion **2**, from the hydroxylamine **3**. In order to develop a more sustainable and user friendly process, Schafer and co. reported a selective TEMPO mediated electrochemical oxidation of potato starch **4** to its corresponding α-D-glucuronan **5** (Scheme 3) [10]. This electro-oxidation of potato starch was highly selective toward the primary hydroxyl group in the 6-position. The corresponding polyuronic acid was isolated in 63% yield as a crude product with a 93% conversion of primary alcohol groups.

Scheme 3. Anodic oxidation of potato starch.

The electrochemical transformation delivered a slightly lower yield of crude product compared to the chemical reaction (>95% selectivity and 98% crude product yield), however, it avoided the need to employ sodium hypochlorite, previously used in stoichiometric amount to regenerate the active TEMPO species.

The development of novel electrochemical processes also allows for new applications to be discovered. Dang et al. reported a one pot electrochemical oxidizing coupling of corn starch and natural gelatin [11]. This oxidized corn starch-gelatin material was then dehydrated and granulated with the spray-dying process. After full characterization, the authors reported that such material could be used for drug delivery applications.

Cellulose is the most abundant biopolymer on the planet and due to its high level of functionalization, it is a compound of great interest for a vast number of applications. Electrochemical transformations of cellulosic materials have been studied mainly toward the development of biobased fuel cells [12] in order to provide sustainable electricity. However, in such processes, the outcome of the electrochemical oxidation reaction is of minor importance and the selectivity of the reaction is in most cases not studied, therefore, such processes are beyond the scope of this review. Recently, Yang et al. documented a controlled electrochemical transformation of cellulose oligosaccharides **6** into glucose (**7**) [13]. A hydrothermal pre-treatment of cotton cellulose was first necessary to isolate the desired starting material oligosaccharides. The optimal reaction condition for the electrochemical transformation of the obtained oligosaccharides could then be determined. The process which showed the most reactivity and selectivity toward the formation of glucose was a direct electro-oxidation on a MnO_2/graphite/PTFE anode, in acid media (pH = 3.0). Under this optimized reaction conditions, glucose was formed in 72% yield with 100% selectivity (Scheme 4).

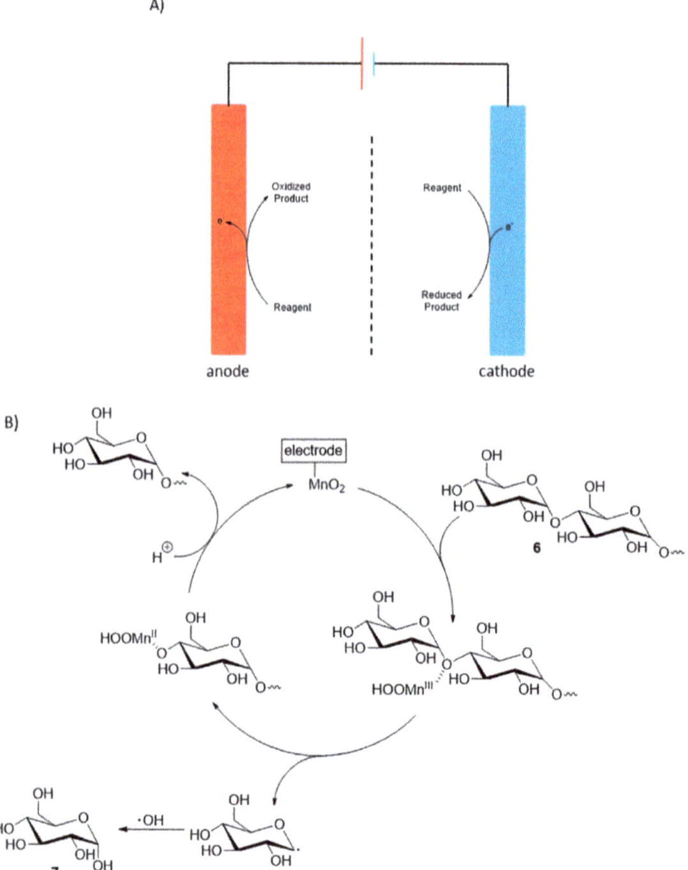

Scheme 4. (**A**) Concept of direct electrochemical transformations; (**B**) Proposed mechanism of MnO_2 catalysed electrooxidation of cellulose to glucose.

The electrochemical oxidation of cellulose has also been reported using a gold electrode, in alkaline media (1.3 M NaOH) [12,14]. Although the oxidized products were not clearly identified, it is worth noting that the authors proposed a detailed mechanism to explain the interaction between cellulose and the various gold species formed at the surface of the electrode. Indeed, each step of adsorption, oxidation, desorption of the cellulose could be identified and associated to specific gold species. The products generated from this reaction are presented as one water soluble material in which some hydroxyl groups were oxidized to carboxylic acids, and one water insoluble hybrid material made of cellulose and gold nanoparticles.

A similar study was then performed on the electrochemical oxidation of hemicelluloses materials at a gold electrode [14]. The authors analysed the responses of various hemicelluloses (xylan, arabinoxylan, glucomannan, xyloglucan, and glucuronoxylan) in alkaline solution (1.3 M NaOH solution). Although xylan, xyloglucan and glucuronoxylan were found to be electrochemically active, only xylan was studied in more details and was submitted to a long-term electrolysis. The analysis of the products form xylan electrolysis showed a water-soluble material in which OH-groups were oxidized to carboxylic groups and a xylan-based water insoluble material containing gold nanoparticles.

Raffinose (**8**), a trisaccharide mostly found in cotton seed, is constituted of galactose, glucose, and fructose units. The electrochemical oxidation of raffinose was reported using TEMPO as a mediator, to selectively oxidize the primary alcohol moieties, while leaving untouched all secondary alcohol groups present on the molecule. The electrochemically oxidized raffinose was then esterified to trimethyl D-raffinose trisuronate (**9**) and isolated in 63% yield (Scheme 5) [15].

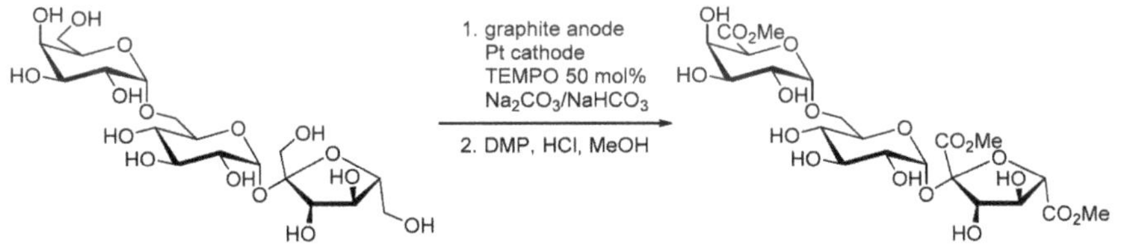

Scheme 5. TEMPO-mediated selective electro-oxidation of raffinose (**8**).

2.2. Disaccharides

Sucrose is a non-reducing disaccharide constituted of D-glucose and D-fructose units which are glycosidically bonded through their anomeric carbon atoms. Sucrose (or saccharose) is broadly present in plants and is the main disaccharide feedstock across the globe. It is an important carbohydrate reserve compound and one of the main energy sources required in the human's diet. It is produced from sugar cane and sugar beet (130×10^6 t/year) [16] and is present in honey, maple syrup, fruits, and vegetables. Sucrose itself is widely used in the food industry for its sweetening ability. Sucrose can be functionalized at various positions and therefore can be converted to various interesting additives. Monocarboxylic acids of sucrose are valuable intermediates in the synthesis of monoesters and monoamides of sucrose, which are used as for their tensioactive capacities. However, examples of chemical and electrochemical transformations of sucrose are scarce due to the poor selectivity of the reactions.

One of the most noticeable examples of electrosynthesis employing disaccharides as starting material was reported by Lamy and co-authors in 1993 [17]. The authors documented a thorough study of the electrochemical behaviour of sucrose in various reaction conditions, as well as the capacity of a range of electrodes to perform the desired oxidation reaction. Due to the known capacity of saccharose to hydrolyse under acidic condition an alkaline media was therefore used to conduct the experiments. A vast number of electrode materials were tested and evaluated by using voltammetry analysis. Ir, Fe, and Co showed no activity toward the oxidation of sucrose; Cu, Ag, Ni, Rh, and Pd displayed some activity but too weak to be explored further. Lastly, Au and Pt showed good reactivity for the oxidation of sucrose. Gold produced a much higher current density than Pt, but required a much higher working potential for the oxidation to occur. The authors then turned their attention to parameters such as temperature and initial sucrose concentration. They used a potential step program, designed to clean the electrode of species which could remain adsorbed to the electrode surface. With the optimal reaction conditions at hand, a long term electrolysis was performed on a large Au plate electrode with sucrose (10 mM) in NaOH solution (0.1 M) for 12.5 h, aiming for the specific oxidation of primary alcohol groups, without cleaving the C-O-C bond. After 12.5 h of electrolysis, 67% of sucrose was converted to a range of products (Table 1).

Table 1. Chromatographic analysis of long term electrolysis of sucrose on gold electrode.

Identified Product	Selectivity
Formic acid	2%
Glycolic acid	6%
Glyceric acid	4%
Hydroxymalonic acid	2%
Tartaric acid	1%
Gluconic acid	4%
Glucaric acid	7%
5-Ketogluconic acid	8%
6-Monocarboxylic acid of sucrose	**23%**
1′-Monocarboxylic acid of sucrose	**26%**
Unknown products	17%

In a follow up work, the same group was able to scale up their reaction by adapting it to a flow cell reactor [18]. The working electrode used in this novel system was a lead-modified platinum electrode. A similar potential step program as previously reported was employed in order to diminish to a maximum the electrode poisoning during the reaction, and therefore maintaining the high activity and selectivity of the electrocatalyst. After 8.5 h, the electrolysis was stopped and the products of the reaction analysed by different characterisation methods. The overall sucrose conversion was estimated to 60%, with a selectivity of 80% towards 6-monoacid of sucrose (**10**) and 10% toward the 1′-monoacid of sucrose (**11**). In order to confirm the structure of the reaction products, compounds **10** and **11** were hydrolysed and the resulting monosaccharide derivatives were analysed (Scheme 6).

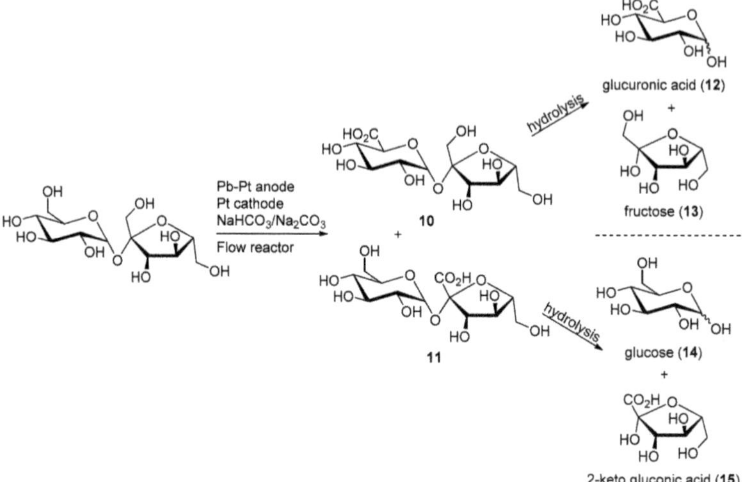

Scheme 6. Sucrose electro-oxidation and hydrolysis leading to the identification of the monomeric structure of the products.

Hence, the authors were able to assign with certainty the structure and the selectivity of the electrochemical oxidative process.

A mediated electro-oxidation of sucrose was also demonstrated by employing TEMPO as the homogenous mediator [10]. Within the structure of sucrose, the three primary alcohols groups are the most likely to be oxidised by TEMPO, due to its high selectivity toward primary alcohol oxidation. When the electrolysis was stopped after the consumption of 6970 C (4.8 F/mol), 83% sucrose was converted and the three possible sucrose monocarboxylic acid species (such as **10** and **11**) could be observed in various amounts (low to moderate yields). However, when the electrolysis was stopped after 28,450 C (19.7 F/mol) were consumed, only the tricarboxylic acid species could (**16**) be observed and isolated in 39% yield (Scheme 7).

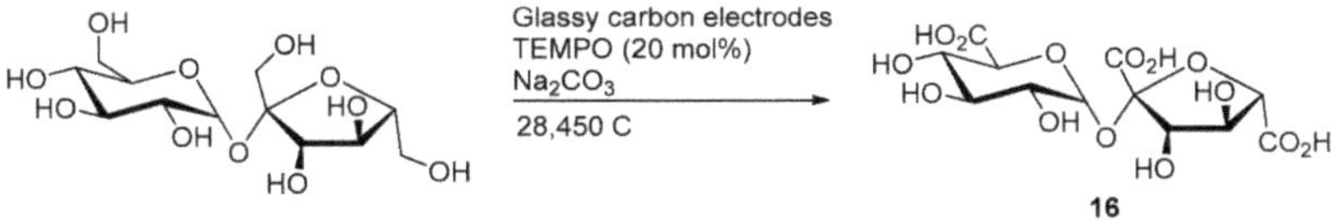

Scheme 7. Selective mediated electrochemical oxidation of sucrose to the corresponding tricarboxylic acid.

Other disaccharides were also submitted to electrochemical transformations. Trehalose is a non-reducing sugar constituted of two ᴅ-glucose units, linked by a $\alpha(1\rightarrow1)$ glycosidic bond. Both direct and indirect electrochemical oxidation of trehalose were studied. Schnatbaum et al. applied a similar reaction procedure than for the electrochemical oxidation of sucrose, and were able to demonstrate the ability of TEMPO to act as a selective mediator for the oxidation of α,α-trehalose (**17**) toward α,α-trehalose dicarboxylic acid (**18**) [10]. The desired product was isolated in 61% yield after consumption of 2010 C (8.8 F/mol). In a later study, Kokoh and co-workers were able to further improve the TEMPO-mediated electrochemical oxidation of trehalose by fine tuning the electrolysis potential They were also able to improve the analytical process by transforming **18** into its trimethylsilylated equivalent for a better detection by GC-MS [19]. Using this processes, the authors were able to report a full conversion of trehalose (**17**) to the dicarboxylic product (**18**).

Parpot et al. studied a direct electrochemical oxidation of trehalose using a gold electrode or a gold-nickel (70/30) alloy, in alkaline media [20]. It was postulated that combining gold and nickel in an alloy would produce a great selectivity towards the oxidation of primary alcohols (due to gold), and high current density because of the nickel atoms. Indeed, saccharide oxidation at nickel electrode is reported to occur at a high anodic potential, in NiOOH region, but generates low molecular weight carboxylic acid species. The authors first noticed that trehalose was not showing significant oxidation signals in carbonate buffer. They attributed this lack of reactivity to the absence of aldehyde/hemiacetal group, which are found in reducing sugars. Considering this aspect, the authors report changing the electrolyte from carbonate to NaOH (0.1 M) was significantly beneficial to the reactivity of trehalose with Au and Au/Ni electrodes. In order to maximise the efficiency of the electrodes, the authors used a pulse potential program for long-term electrolysis. The aim of this potential program was to perform the oxidation at potential 0.25 V vs. RHE for 60 s, followed by a cleaning pulse at 1.7 V vs. RHE for 1 s. This pulse was used to remove any species which might have remained adsorbed on the surface of the electrode after the oxidation reaction was done. The end products of the reactions were characterised by GC-MS and LC-MS. Independently of the electrode used, the main product was the trehalose monocarboxylic acid (**19**). Alongside this product, significant amount of trehalose dicarboxylic acid (**17**) and glucose were also detected, as well as other acids of low molecular masses (Table 2).

This study showed no significant difference between a pure gold electrode and a gold/nickel alloy electrode toward the selectivity or the efficiency of the reaction (Table 2) [21,22].

Table 2. Product distribution after electrolysis of trehalose in alkaline media at gold and gold/nickel (70/30) electrodes.

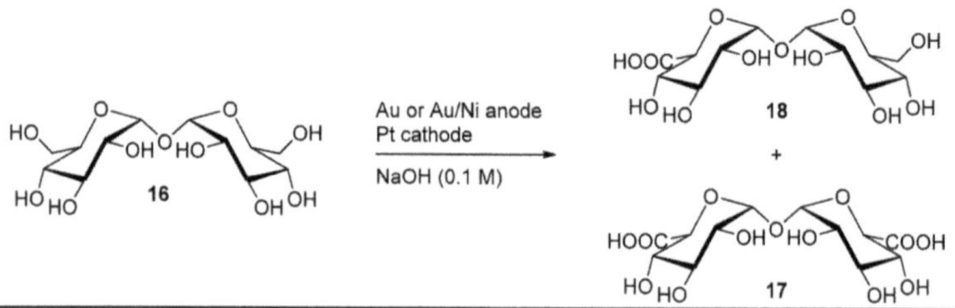

Compounds	Au	Au/Ni
	\multicolumn{2}{c}{% (Per Mole Oxidised Trehalose)}	
Conversion	65	61
18	70	66
17	9.5	5
D-glucose	6	10
Glucaric acid	1	0.5
Tartaric acid	0.5	1.2
Tartronic acid	0.5	1.5
Oxalic acid	1	0.2
Gluconic acid	2.5	1.5
2,3,4,5-Hydroxypentanoic acid	1.5	2
Glyceric acid	1.5	0.6
Glycolic acid	0.5	0.5

Reducing sugars are carbohydrates with an unprotected aldehyde or hemiacetal group. The reactivity order of the various groups present in reducing sugars toward an oxidative process are as follow: equatorial secondary hydroxy < axial secondary hydroxy < primary hydroxy < hemiacetal hydroxy. Therefore, the unprotected hemiacetal group present in disaccharides such as cellobiose, D-maltose, and D-lactose is the most likely to react first, which could bring further selectivity issues than with non-reducing sugars.

Cellobiose (**19**) is constituted by two glucose units, linked by a β(1→4) glycosidic bond. It can be obtained from materials such as cotton, jute, or paper after hydrolysis of the cellulose. D-maltose (**23**) is a stereoisomer of cellobiose, formed of two glucose units which are is this case linked by an α(1→4) bond. Due to their similar structure, cellobiose and maltose were often studied in parallel. Both disaccharides were submitted to similar electromediated oxidation processes by Liaigre et al. [23]. Both **19** and **23** were oxidised electrochemically using TEMPO as a homogeneous mediator, and in alkaline media (1.0 M carbonated buffer). However, the selectivity of these transformations toward the corresponding triacid products (**20** and **24** respectively) were very low when the reactions were performed at room temperature, and lead to compounds of low molecular masses such as tartaric acid and oxalic acid. Cooling the reaction mixture to 2 °C improved significantly the selectivity of the reaction and the desired tricarboxylic acid products **20** and **24** were reported in moderate to good yields (Scheme 8).

In a subsequent study Schämann et al. reported the scaling up of the electrochemical oxidation process of D-maltose (up to 67.5 mmol), and the further transformation of the tricarboxylic acid in the corresponding trimethyl ester in 50% yield [15,24]. In a comparative study between reductive and non-reductive sugars, Parpot et al. were able to demonstrate that the selectivity of TEMPO for the electro-oxidation of **23** was lower than the one observed for trehalose [19]. They attributed this to the presence of the free aldehyde group. During their investigation the authors were able to detect by GC-MS the low molecular weight compounds, but also mono-, di-, and tricarboxylic acid products, in significant amounts.

Scheme 8. Electromediated oxidation of D-cellobiose (**19**) and D-maltose (**23**).

D-Lactose (**25**) is also a reducing sugar, it is formed of a galactose and a D-glucose subunit, linked by a β(1→4) glycosidic bond. Lactose is naturally present in milk, in 2 to 8% by weight and is thus available in large quantities (about 300,000 t/year). Its TEMPO-electromediated oxidation was also studied by Shafer and co-workers [15]. D-Lactose was electro-oxidised in alkaline media (NaHCO₃/Na₂CO₃) at constant potential (E = 0.77 V vs. RHE). Under these reaction conditions, D-lactose was consumed quantitatively, however, the selectivity of the reaction was poor. Indeed, five different products could be identified by ESI-MS but were not isolated (Scheme 9). The expected mono, di, and tricarboxylic acid products **26-28** were present, as well as the diketo product **29**. The monosaccharide galactaric acid (**30**) was also identified, and arises from the cleavage of the β(1→4) glycosidic bond. This phenomenon was also observed during the TEMPO electromediated oxidation of D-cellobiose (**19**).

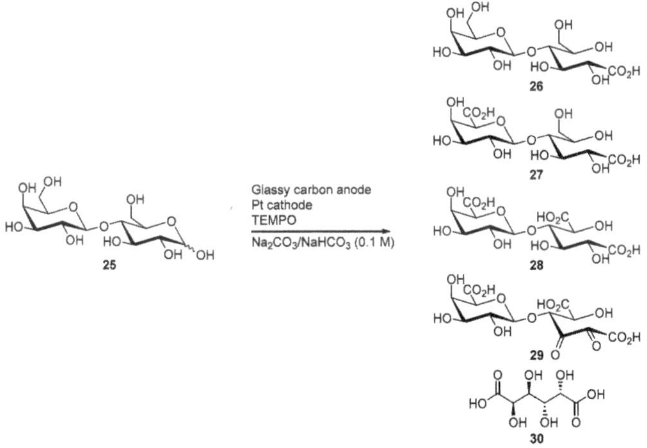

Scheme 9. TEMPO electro-mediated oxidation of D-lactose (**25**).

The electrochemical behaviour of disaccharides has also been studied in direct oxidation processes on rare metal electrodes. As previously mentioned, a direct electrochemical transformation gives more insights on the reaction itself, and in theory, allows a better control on the reaction by being able to tune the potential at which the transformation is occurring. Whilst a wide range of metal electrodes were tested for the electrochemical oxidation of disaccharides (Au, Pt, Ni, Cu, Co, Ru, Cd, Ir), most reports only documented voltammetric studies, evaluating the electrical response of disaccharides on the metal electrode. These types of studies were looking for application in sensors, and did not report the products of the reaction. Only few reports are giving information on the reaction products formed after a long term electrolysis.

Druliolle et al. studied the oxidation of D-Lactose (**25**) on platinum and on modified platinum electrodes in alkaline medium toward the formation of lactobionic acid (**26**) [25,26]. The authors first studied the electrocatalytic ability of a smooth platinum electrode to oxidise lactose in a sodium hydroxide solution (0.1 M), at room temperature using cyclic voltammetry (CV). However, from this preliminary tests, lactose did not demonstrate good electrochemical reactivity on platinum and no lactose was oxidised during a long term electrolysis. Then, they evaluated the effect of different underpotential deposited (upd) metal adatoms. Indeed, the ability of upd adatom to enhance the catalytic activity of noble metals for electro-transformation of organic compounds is well documented [27–30]. The authors studied the catalytic effect of Tl, Bi, and Pb adatoms on a platinum electrode in 10 mM of lactose and 0.1 M NaOH. The CV measurements showed some significant differences in terms of optimal oxidation potential and some drastic increase of current density.

Long term electrolysis of lactose was carried out with bismuth perchlorate salt in NaOH solution (0.1 M). A potential step program was used to maximise the rate of the oxidation reaction, while depositing the Bi metal adatoms onto the electrode, and minimising the poisoning of the electrode. After two hours of electrolysis, the recorded quantity of electricity passed was Q_{exp} = 20.7 C. After analysing the crude material of the reaction by HPLC, the authors reported that lactose was converted in 70% exclusively to lactobionic acid (**26**), and that no other products could be detected by their analytical method, therefore awarding a nearly 100% selectivity toward **26**, with a Faradaic yield of 75%. The lower Faradaic yield compared to the overall selectivity of the reaction can be explained by the slow decomposition of **26** under the reaction conditions.

Long term electrolysis of D-Lactose on lead adatoms modified platinum electrode in carbonated buffer was also studied [26]. A similar potential step program was used, with the adatom deposition plateau, the oxidative plateau, and the final step to remove the poisoning species from the surface of the electrode. After three hours of electrolysis, the recorded quantity of electricity passed was Q_{exp} = 15.1 C. After analysing the resulting mixture, the authors report that 78% of the starting lactose was converted to the desired lactobionic acid with nearly 100% selectivity. If the reaction was continued to reach four hours, 95% of lactose was converted, but the selectivity toward **26** decreased to 76%, which can be explained by the slow decomposition of lactobionic acid under the reaction conditions.

Druliolle et al. then turned their attention to gold electrode for the oxidation of D-Lactose (**25**) [31]. The ability of gold based electrocatalyst to oxidise selectively alcohol groups in alkaline medium was previously reported by Larew and co-workers [32]. Furthermore, gold is known to be less subject to poisoning than palladium, which explain the increase of attention which was directed toward this noble-metal for electrochemical transformations. To determine the optimal reaction conditions for the electro-oxidation of lactose on a gold electrode, the authors used voltammetric measurements. They report that a potential step program was used to maximise the oxidation of **25**, while minimising the poisoning of the gold electrode, under alkaline medium (carbonated buffer, 0.1 M). Then, the authors evaluated the impact of the starting concentration of D-Lactose, as well as the effect of small variation on the oxidative potential plateau. With the optimal reaction conditions ([lactose] = 10 mM) in hand, 95% of **25** was converted to lactobionic acid, with 98% selectivity after five hours of electrolysis.

By using two complementary IR analytical methods (SPAIRS and SNIFTIRS), the authors were able to identify the possible steps of the reaction, and therefore to propose a possible mechanism for the electro-oxidation of D-Lactose at a gold electrode (Scheme 10). During this investigation, the authors noticed that two different signals, depending on the oxidative potential applied. When the applied potential of the reaction is below 1.1 V vs. RHE, a characteristic signal of vibration mode was observed, which can be attributed to an asymmetric O-C-O bond stretching (Scheme 9, left hand side cycle). At higher potential, a signal characteristic of a symmetric O-C-O stretching (Scheme 9, right hand side cycle) bond could be identified. Gold being known for its oxygen adsorption properties, it was assumed that at low potential, a small quantity of hydroxyl species can be adsorbed on the gold electrode, and at higher potential, a much larger quantity of gold hydroxide species could be generated. Despite the differences observed by infra-red spectroscopy, only lactobionic acid was observed after the reaction, which indicate that at different potential, different mechanisms are taking places, yet delivering the same product. The authors postulated that once D-Lactose was adsorbed on the gold surface, either a hydroxide group could attack on the glycosidic carbon, therefore opening the ring (Scheme 9, left hand side cycle), or if the concentration of gold hydroxide species was sufficient, another Au-OH group could open the ring (Scheme 9, right hand side cycle). Both intermediate could then deliver the desired product **26**.

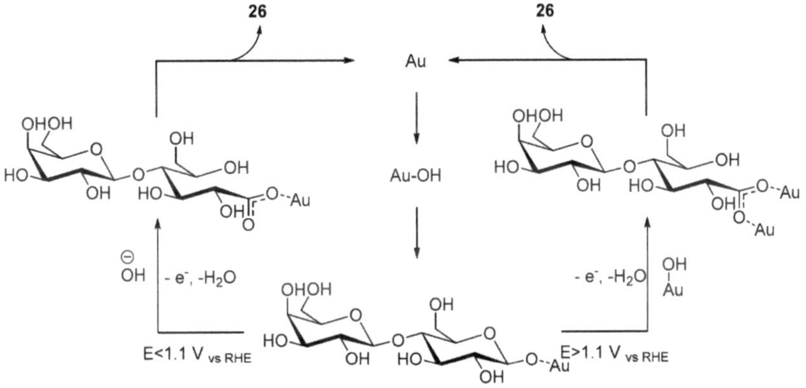

Scheme 10. Proposed mechanisms for the electro-oxidation of D-lactose at different potentials.

In a following study, Kokoh et al. documented the electrocatalytic oxidation of D-Lactose on nanoparticles of Au-colloids (5 nm), immobilised on a carbon felt electrode, in a flow reactor [33]. The optimal reaction conditions were first determined using CV, which demonstrated that the gold nanoparticles were able to generate a higher current density than when a plain gold electrode of three times the size was used. With the information gathered during the preliminary investigation, the authors determined that a two potential plateau program had to be employed to limit the poisoning of the electrode, as well as to maximise the oxidation reaction. The electro-oxidation was then carried out in carbonated buffer (0.1 M, pH = 10.2), and delivered lactobionic acid (**26**) in 91% yield.

Recently, Parpot et al. described the electrochemical oxidation of cellobiose (**19**) and D-maltose (**23**) on a gold electrode in alkaline media [20]. For each disaccharide, the author conducted preliminary investigations using CV. From these tests, it was possible to determine the optimal reaction conditions. As previously mentioned, gold electrodes are subject to poisoning. Therefore, in order to diminish this undesired phenomenon, a multi-potential plateau program was used for the long term electrolysis of both cellobiose and D-maltose.

With the optimal reaction conditions at hand, 95% of cellobiose (**19**) could be converted almost exclusively to its monocarboxylic acid equivalent, cellobionic acid (**27**), resulting from the oxidation of the aldehyde at carbon C1. The dicarboxylic acid product could also be detected in very small amount (3%) as well as trace amounts of the products resulting from C-C bond cleavage (Scheme 11A).

Then, D-maltose was submitted to its corresponding optimal reaction conditions for long term electrolysis. However, when D-maltose was submitted to electro-oxidation in NaOH solution (0.1 M), the reaction proved to be unselective, producing a variety of small carboxylic acids, such as tartaric acid, formic acid, etc., in non-negligible quantities. Furthermore only 50% of the initial D-maltose was converted after eight hours of electrolysis. A second electrolysis of D-maltose was then carried out in carbonated buffer (0.1 M, pH = 10), using the same potential plateau program. After twenty hours, 79% of the initial amount of D-maltose (**20**) was converted, with a selectivity up to 92% toward maltobionic acid (**28**), with only 3% of the dicarboxylic acid equivalent being detected, and no degradation products (Scheme 11B).

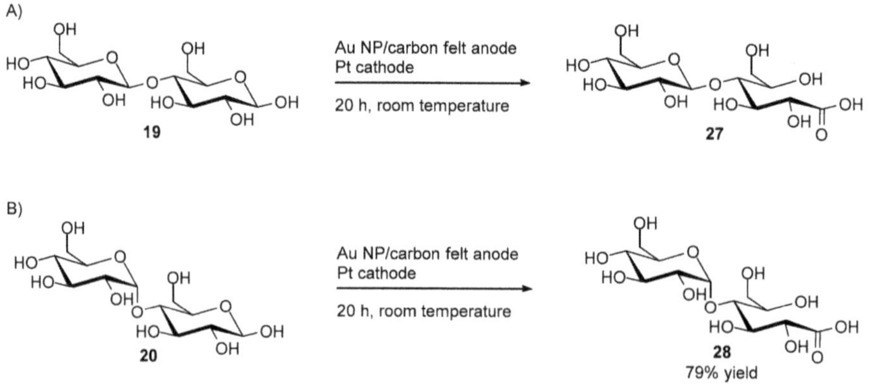

A)

Au NP/carbon felt anode
Pt cathode
20 h, room temperature

19

27

B)

Au NP/carbon felt anode
Pt cathode
20 h, room temperature

20

28
79% yield

Scheme 11. Direct electro-oxidation at on Au nanoparticles of (**A**) cellobiose; (**B**) D-maltose.

2.3. Monosaccharides

Significant interest has been drawn to the oxidation reaction of monosaccharides due to the variety of high-added-value chemicals which could potentially be obtained. Enzymatic and chemical processes, stoichiometric and catalytic, as well as homogeneous and heterogeneous catalytic systems have been reported for such transformations. However, these processes are beyond the scope of this work, and several reviews focusing on those aspects have been documented [34–36]. Concerning the electrochemical transformation of monosaccharides, the amount of reports is scarce. Most of the existing literature is focusing on the development of glucose biosensors to help for diabetic treatment, or as a reagent for fuel cells for clean energy production. However, in such studies, the structure of the products generated are generally not investigated. Amongst the electrosynthesis reports of the transformation of monosaccharides, D-glucose was the most widely investigated because it is one of the most abundant bio-based chemicals, and it can be converted to a variety of commodity chemicals [35,37]. In 1910, Lob described one of the first electrochemical oxidation of D-glucose, in sulfuric acid, with a lead anode and a platinum cathode [38]. The various products obtained were D-arabinonic acid and D-glucaric acid, as well as some fragmentation by products such as D-arabinose and formaldehyde. The slow degradation of D-glucose to lower aldoses during an electrolysis in non-aqueous media was investigated in more details by Hay et al. [39] although, the lack of detailed information such as yield, efficiency, and selectivity are limiting the comparative analysis.

Isbell and co-workers reported series of efficient indirect electrochemical oxidation of various monosaccharides into their corresponding aldonic acid salts [40–43]. Their process is based on the electro-oxidation of sodium or calcium bromide, generating free bromine at the anode, which reacts with the saccharide to give the corresponding aldonic acid. The calcium carbonate present in the reaction mixture quenches the acid, maintaining a constant pH (6.2) during the electrolysis (Table 3). The bromide is continuously regenerated, allowing the conversion of a large amount of sugars while using only a catalytic amount of bromide salt [40,41]. Most of the aldonic acid salts then crystallises and the acid form can then be recovered by reacidification and recrystallization [42]. The acid salt products could also be converted to the corresponding δ- or γ-lactones via an acid catalysed process.

Table 3. Bromide-mediated indirect electro-oxidation of monosaccharides.

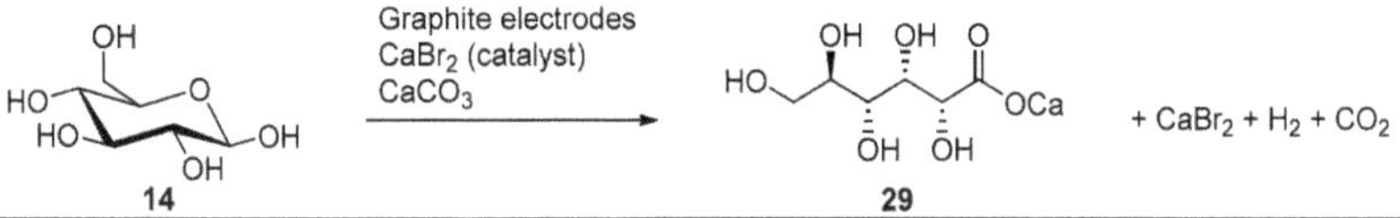

Entry	Monosaccharide	Product	Conversion	Isolated Yield
1	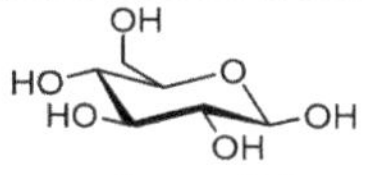 D-glucose (**14**)	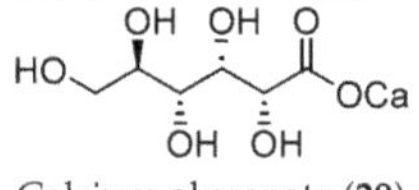 Calcium gluconate (**29**)	96%	77%
2	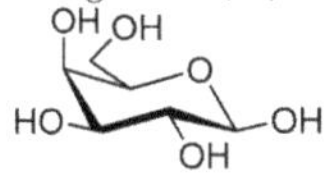 D-galactose (**30**)	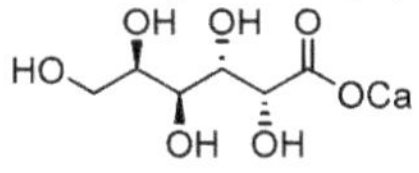 Calcium galactonate (**31**)	98%	54%
3	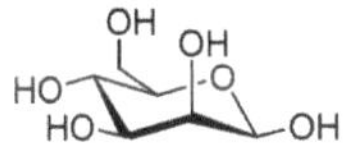 D-mannose (**32**)	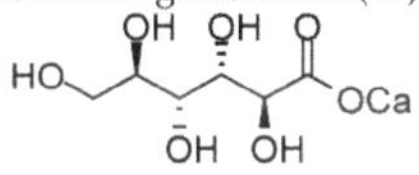 Calcium mannonate (**33**)	94%	69%
4	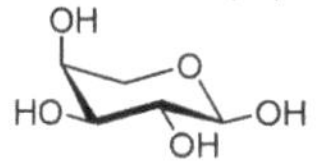 L-arabinose (**36**)	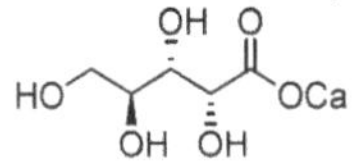 Calcium arabonate (**37**)	91%	40%
5	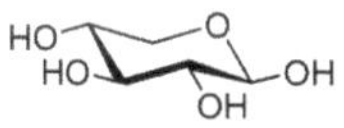 D-xylose (**34**)	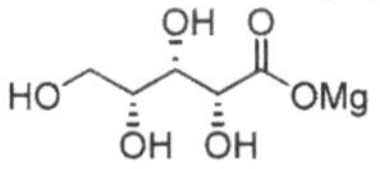 Calcium xylonate (**35**)	97%	80%

When xylose was submitted to the electro-oxidation, the current density and the conversion of the starting material were very promising. However, calcium xylonate appeared to be very soluble, and therefore could not be separated by crystallisation in an efficient manner. The authors then applied similar reaction conditions for the electro-oxidation of xylose, but in the presence of strontium or magnesium bromide and carbonate. This allowed them to isolate the corresponding xylonate salt in high yields 49% and 80% respectively (Table 3, entry 5) [43].

In a follow up study, the same authors, and others groups, proposed a semi-industrial procedure for the production of calcium D-gluconate [41,44]. In this process, the calcium D-gluconate was recovered after the addition of calcium hydroxide, which generate a less soluble basic salt species. Despite continuing the electrolysis longer than theoretically needed to oxidise glucose to gluconic acid, a reducing compound remained in solution. After repeating Isbell's procedure for an electrolysis, Cook et al. were able to isolate the calcium salt of 5-keto-gluconic acid (**36**) in approximately 5% yield [45]; therefore proving that gluconic acid was not the only product of the $CaBr_2$ mediated electro-oxidation of glucose (Scheme 12).

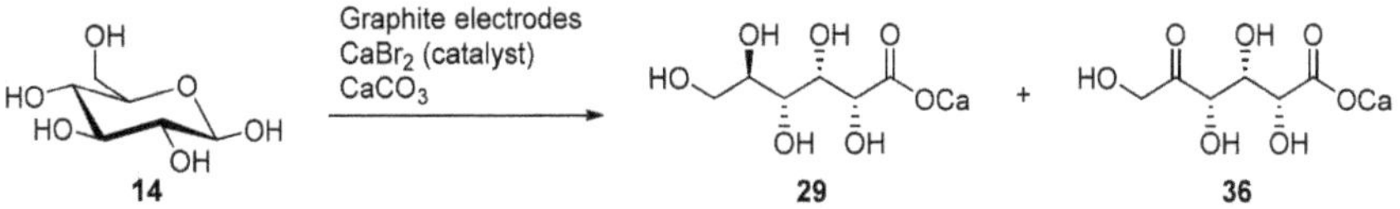

Scheme 12. Complete analysis of the by-products from the $CaBr_2$-mediated electro-oxidation of glucose.

Inspired by the examples of chemical oxidation of simple alcohols and carbohydrates by ruthenium complex catalysts [46]. Kokoh and co-workers decided to explore the possibility for a selective electro-oxidation of glucose using Ru-based complex catalysts [47]. Azopyridine (azpy) ruthenium complexes are known to maintain the ruthenium metal at a low redox potential state Ru(IV), which could be suitable for a selective electrochemical oxidation of carbohydrates [46,48–51]. The authors reported the synthesis of two isomers γ-RuCl$_2$(azpy)$_2$ and δ-RuCl$_2$(azpy)$_2$ and their immobilisation on carbon fibre materials prior to their electrochemical characterisation. Using CV measurements on crude RuCl$_2$(azpy)$_2$, the authors could determine that the redox couple observed in the positive scans was Ru(IV)/Ru(III). Voltammograms of the crude ruthenium complex recorded in the presence on D-glucose in alkaline medium (0.1 M NaHCO$_3$/Na$_2$CO$_3$) showed no reduction peak, thus suggesting a fast reaction between the Ru(IV) and glucose. After having proved that the ruthenium complex was able to react with glucose, the authors set out to perform long term electrolysis. Using their optimal reaction conditions, D-glucose was electro-oxidised during 45 h at fixed potential in presence of the crude RuCl$_2$(azpy)$_2$ catalyst. Analysis of the crude reaction mixture showed a 22% conversion of glucose toward gluconic acid (**29**, 11% yield, 50% selectivity) and 2-keto-gluconic acid (**37**, 7% yield, 33% selectivity). Trace amounts of a tricarboxylic acid could also be detected by LC-MS analysis, as well as oxalic and tartaric acid. The presence of these low molecular weights acids suggested that the tricarboxylic acid was hydrolysed which resulted in the C2-C3 bond cleavage [52]. The authors postulated that glucose is getting oxidised to gluconic acid (**29**) via the intermediate δ-gluconolactone, but that the gluconic acid can itself be further oxidised and therefore deliver 2-keto-gluconic acid (Scheme 13).

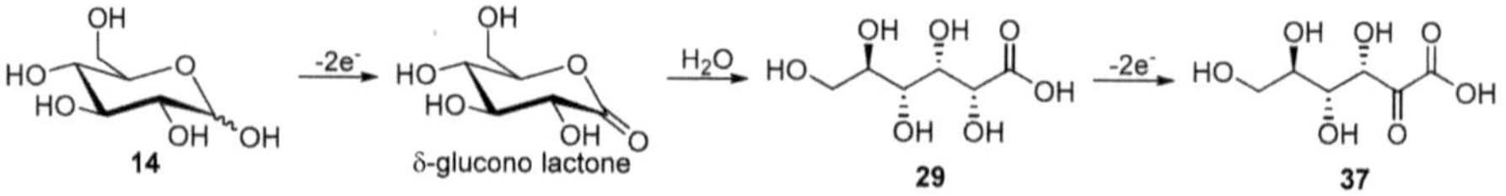

Scheme 13. Ruthenium-based, electro-mediated oxidation of glucose.

Long term electrolyses were carried out using pure γ-RuCl$_2$(azpy)$_2$ and δ-RuCl$_2$(azpy)$_2$, using the same reaction conditions as the one used for crude RuCl$_2$(azpy)$_2$ catalyst. When γ-RuCl$_2$(azpy)$_2$ was employed, the conversion of glucose increased to 50%, and delivered **29** and **37** in 30% and 18% yield, respectively. The mediated electro-oxidation of glucose by δ-RuCl$_2$(azpy)$_2$ gave only 10% conversion toward **29** mainly (70% selectivity), but gluconic, glucuronic, tartaric and oxalic acid could also be observed in small quantities. These results show that γ-RuCl$_2$(azpy)$_2$ catalyst is the most active isomer and can deliver 2-ketogluconic acid (**37**) in a selective manner.

As previously mentioned, 2,2,6,6-tetramethyl-1-piperidinyl free radical (TEMPO) and its derivatives are very selective for alcohol oxidation. Such organic homogeneous mediators provide a metal free option of alcohol oxidation and usually shows great reactivity in mild conditions. Therefore, after having thoroughly studied the chemical oxidation of glucose with TEMPO and several co-oxidants [53–55]. Ibert et al., decided to investigate the TEMPO mediated electro-oxidation of glucose in alkaline medium [56]. The authors first evaluated the ability of glassy carbon and platinum to oxidise TEMPO efficiently. Platinum showed much lower reactivity toward TEMPO than glassy carbon, which also showed great stability after 10 CV experiments. Glassy carbon was therefore used as the working electrode. The chemical oxidation of carbohydrates is known to be sensitive to pH changes [53], therefore, the authors turned their attention to the effect of the pH on the electrochemical oxidation of glucose by studying the current density during CV measurements. Various carbonated solutions were used with pH ranging from 7 to 12. The clear trend was that a basic pH is beneficial to the oxidation of glucose, demonstrating the important role played by the base in such transformation. With the optimal reaction conditions in hand, it was possible for the authors to attempt a preparative TEMPO mediated electrochemical oxidation of glucose, targeting glucaric acid (**38**) as the main product. The first attempts in undivided cells showed significant limitation due to a lack of control over the parameters of the reaction. This pushed the authors to move to a jacketed reactor with a control module to regulate the pH and the temperature of the reaction mixture. In these conditions, the full conversion of glucose was achieved with 20% faradaic excess (i.e., the reaction was stopped after the total current collected exceeded the theoretical value for full conversion by 20%), however, a large amount of by-products were detected (low molecular weight products from degradation reaction) and the desired glucaric acid was only formed in poor yield. One of the main by-products detected was a tricarboxylic acid compound which had never been observed in chemical oxidation of glucose. By using GC analysis, combined with NMR studies, Ibert et al. were able to unambiguously determine the exact structure of the tricarboxylic acid (**39**) [57]. This information, combined with various analytical information, made it possible for the authors to propose a mechanism explaining the formation of **39** (Scheme 14). They postulated that the triacid resulted from a benzylic rearrangement at the C-4, C-5, and C-6 centers. This rearrangement could occur via the formation of a diketone intermediate, due to the over-oxidation of the secondary alcohols of glucaric acid (**38**). In order to prove their hypothesis, the authors performed several electrochemical oxidations with compounds such as 2-keto-gluconate, which they postulated to be intermediates in the formation of **39**.

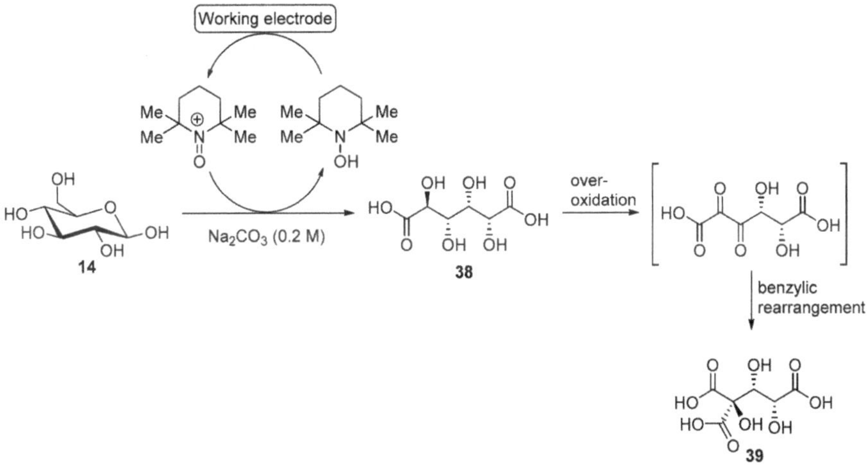

Scheme 14. TEMPO mediated electro-oxidation of D-glucose (**14**) leading to D-glucaric acid (**38**) and tricarboxylic acid (**39**).

The presence of **39** and degradation products (oxalic, tartaric, and malonic acid) in large amount, pushed the authors to reoptimize their reaction conditions. By lowering the temperature to 5 °C, and increasing the pH of the reaction to 12.2, most of the degradation and over-oxidation products formation could be inhibited (below 5% was generated). With these new optimal reaction conditions, an electrolysis of D-gluconic acid instead of D-glucose was performed, and the desired D-glucaric acid was delivered in 85% yield.

Due to its large availability and the added value of the potential reaction products, glucose has been the most studied monosaccharide for direct electrochemical transformation. Although the anodic oxidation of glucose has been studied with a range of metal (Ag, Hg, Ru, Rh, Cu, Ni, Ir, Co, Mg), most of the literature arguably focuses on the use of Au and Pt working electrodes. Indeed, both metals reactivity and selectivity for glucose electro-oxidation have first been thoroughly studied and compared via voltammetry experiments [58–62]. These studies gave insights on various important parameters, such as the adsorption/desorption ability of the metal electrodes [60–62], as well as the potential mechanisms for the direct-electro-oxidation of glucose at the surface of the electrodes [58,59]. It was found that in neutral and alkaline medium, gold electrodes displayed a greater electro-oxidation rate than any other metals, however, gold was also very prone to poisoning. To gain a deeper understanding of the electro-oxidation process and improve the performance of the electrodes, the metal surface was covered by various oxide species, or via underpotential-deposited adatoms (Pb, Tl, Bi) [27,30].

By partially covering the surface of the electrodes, Pt and Au were less sensitive to poisoning, therefore a higher current response could be observed when put in presence of glucose. However, these preliminary studies only used voltammetric measurements and did not perform long term electrolysis to test the stability and the selectivity of the electrodes for glucose oxidation. By basing their process on the results gathered by these preliminary studies, Belgsir et al. demonstrated that D-glucose could efficiently be electro-oxidised by a gold anode in alkaline medium [63]. At first, the authors employed a bare gold electrode using a three-step pulsed potential program to avoid poisoning by glucose or the reaction products. After 25 h of electrolysis, gluconic acid (**29**) could be observed in 77% yield, and glucaric acid (**38**) in trace amounts. Glucaric acid being a chemical with high added value, the authors decided to modify the surface of the gold electrode with Pb-adatoms, hoping to increase the reactivity of the electrode and thus the selectivity toward glucaric acid. The authors tested two different oxidation potential plateaus, and showed that the selectivity can clearly be changed by the addition of Pb adatoms and by tuning the potential used for the reaction (Table 4).

Table 4. Chemical and Faradaic yields for the electro-oxidation of glucose on gold and -modified gold electrodes.

Product	Electrodes (Potentials Vs. RHE)		
	Au (0.6 V)	Au-Pb (0.6 V)	Au-Pb (0.9 V)
Gluconic acid	77%	40%	26%
Glucaric acid	<1%	25%	38%
Oxalic acid	-	2%	3%
Formic acid	-	3%	4%
Tartaric acid	-	-	16%
Glyoxylic acid	<1%	-	-
Glycolic acid	-	2%	3%
Glyceric acid	-	<1%	1%
Faradaic yield	89%	80%	78%

Although the faradaic yield decreased and more degradation products could be observed, these long term electrolysis showed that it was possible to modify the selectivity and reactivity of a gold electrode for glucose electro-oxidation by using a suitable potential program and by modifying the electrode surface with foreign metal adatoms.

Kokoh and co-workers then set out to fully investigate the influence of upd adatoms (Bi, Tl, and Pb) on both Au and Pt electrodes, in acidic, neutral, and basic mediums for the electrochemical oxidation of D-glucose [64,65]. The voltammograms recorded in acidic medium (HClO$_4$, 0.1 M) showed that regardless of the electrode metal and of the adatom used, the current densities of glucose oxidation are very low, suggesting that the reactivity of glucose in acidic conditions is limited. Indeed, when the reaction mixtures were analysed after 25 h electrolysis, only a very low amount of glucose was converted to a variety of oxidation and degradation products. When voltammograms were recorded on Pt and Au anodes in neutral medium (KH$_2$PO$_4$/NaOH, pH = 7.3), a much greater current density was observed than under acidic conditions. However, only a limited amount of glucose was converted (up to 15%) to gluconic acid (up to 12% yield) and traces of glucaric acid with Au anode or traces of glucuronic acid with Pt electrodes. The authors then moved to investigate the behaviour of Au and Pt in presence of glucose in basic medium (NaOH 0.1 M) [64]. It is known that Au is the most efficient electrocatalyst for glucose oxidation in basic conditions, because it generates higher current densities and is less sensitive to poisoning than Pt [32,66,67]. Kokoh et al. studied the influence of the concentration of glucose (from 1 mM to 50 mM) on a pure gold anode to gain insights on the rate of the reaction and on its mechanism using voltammetry. At low initial concentration of glucose, the reaction is of order one, but at high concentration, a saturation phenomenon is observed and the overall order drops to zero. The effect of upd Tl, Pb, and Bi adatoms was also studied on a gold electrode in alkaline media. Although the voltammograms remained unchanged with and without adatoms, the current density increased during the electrolysis of glucose in presence of upd adatoms. When comparing the results of long electrolysis of glucose (**14**) performed on pure Au, Au-Bi, Au-Tl, and Au-Pb, gluconic acid (**29**) appeared to be the main product for every system. However, the Au-Pb adatom electrocatalyst also produced a significant amount of glucaric acid (**38**). This system was then studied further, by modifying the potential at which the oxidation reaction was occurring (0.5, 0.6, and 0.9 V vs. RHE). The trends observed were that the highest concentration of gluconic acid was obtained at lower potential (75% of the oxidised glucose), and that at higher potential, glucaric acid could was becoming the major product of the reaction (37% of the oxidised glucose) after 24 h of electrolysis. However, at high potential more degradation products (tartaric, oxalic, formic, and glyoxylic acid) could be observed in small quantities. The authors then looked at the glucose electro-oxidation on a Pt anode in basic conditions. After 21 h of electrolysis, 70% of glucose was converted to gluconic acid (**29**) as the main product (50% yield), and other degradation products. They then moved to electro-oxidation of glucose on a Pt electrode with Pb adatoms, using the same reaction conditions as the one used for the pure Pt electrode. A significant increase in current density was observed during the 24 h electrolysis. Within 4 h, the conversion of glucose (**14**) was nearly complete, and the concentration of gluconic acid (**29**) reached its maximum (70% yield). After the first 4 h, the gluconic acid concentration dropped to almost zero and the concentration of by-products (oxalic acid: 60%, 2-ketogluconic acid (**37**): 30%, other degradation products: traces) increased. The difference observed in the distribution of the by-products of the reaction when Pb adatoms were added to a Pt anode, lead the authors to propose possible mechanisms. For the electro-oxidation of glucose on pure Au or Pt, Kokoh and co-workers suggest that the mechanism follow similar steps (Scheme 15A). First is the adsorption/dehydrogenation of glucose at the electrode surface. Then, the free sites of the electrodes oxidise to metal-hydroxide species. The interaction of the two chemisorbed intermediates can then react and release gluconic acid.

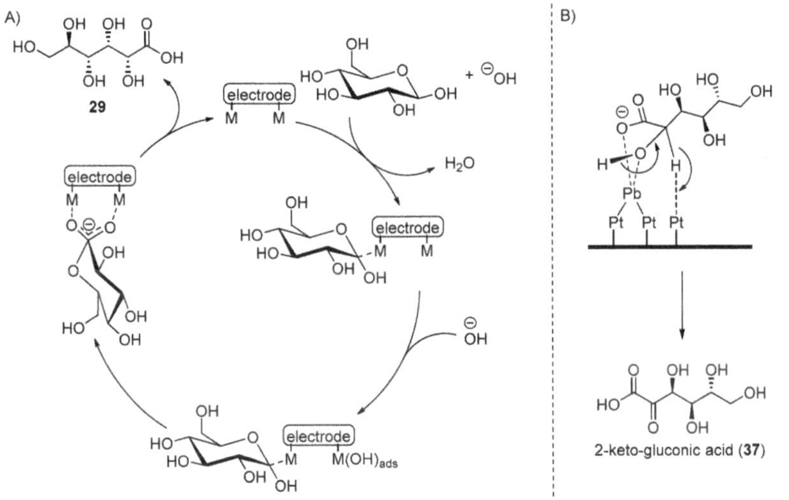

Scheme 15. (**A**) Proposed mechanism for Au or Pt anodic oxidation of D-glucose; (**B**) Proposed mechanism for the formation of 2-ketogluconic acid for gluconic acid.

When Pb adatoms were adsorbed on the electrode surface, D-gluconic acid (**29**) was generated rapidly and then consumed, leading to the formation of by-products, and more particularly 2-keto-gluconic acid. Taking this phenomenon into account, the authors generated the following hypothesis to explain the formation of this by-product mostly in the presence of Pb adatoms (Scheme 15B). It is postulated that the formation of gluconic acid follows the same pathway as it does with pure metal electrodes. However, once the gluconic acid is desorbed from the surface, it is possible that a new surface/chemical interaction occurs. Indeed, due to their free p-orbitals, Pb adatoms are able to coordinate with one oxygen from the carboxylic group and the oxygen from the C2 carbon. Then a free site the Pt electrode surface can dehydrogenate the C2 carbon, generating the 2-ketogluconic acid. Kokoh et al. then studied the electrochemical oxidation of D-gluconic acid at Pt and Pt-Pb adatoms anodes under basic conditions (0.1 M NaOH) [68]. After long term electrolysis with either pure Pt or Pt-Pb adatoms electrode, the distribution of products was significantly impacted by the presence of the Pb on the electrode (Scheme 16). The oxidation of gluconic acid oxidation on a pure Pt electrode appeared to be very slow and yielded mainly glucuronic acid (**40**). When Pb adatoms were adsorbed on the surface of the anode, much higher current density could be reached, and the main product of the reaction was 2-ketogluconic acid (**37**), while glucuronic acid was not observed. These observations were confirming the hypothesis that Pb was orienting the oxidation of gluconic acid on the C2 carbon by coordinating to the carboxylic acid group and the C2 hydroxyl group as shown in Scheme 15B.

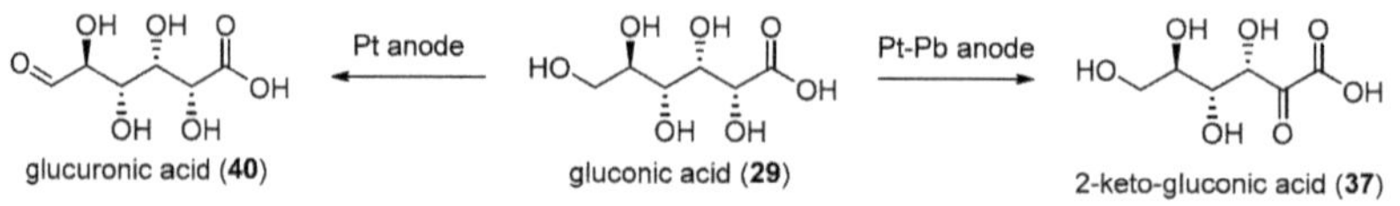

Scheme 16. Product dependence on the electrode surface.

Similar results were obtained when Tl adatoms were adsorbed on a Pt anode for glucose electro-oxidation [69]. Glucuronic acid (**40**) was detected in very small amount, gluconic acid was the major product of the reaction, and the concentration of 2-keto-gluconic acid (**37**) increased with the potential of the reaction (1% yield at E = −0.15 V vs. RHE; 14% yield at E = −0.02.V vs. RHE).

To deepen the understanding of the reactivity of D-glucose on a Pt electrode for anodic oxidation, Largeaud et al. set out to study the difference in behaviour of the two main dissolved anomeric forms of glucose (α and β) [70]. Indeed, the free linear aldehyde form of D-glucose is known to be the minor form in solution, and the α and β-glucopyranose forms are in an equilibrium of approximately 33% and 67% respectively. The authors evaluated the stability of α D-glucose and β D-glucose at low temperature in acidic, neutral, and basic conditions by polarimetric measurements. In acidic and neutral medium, the starting form of α or β-glucose used to make the solution remained the major anomer in solution after 5 h in solution. However, in alkaline medium, β D-glucose rapidly became the main anomer in solution, independently of the crystalline form used to make the solution. The authors then performed voltammetric measurements, in acidic and basic reaction conditions at low temperature. In basic medium, the β D-glucose being the major form in solution, it is difficult to draw conclusions on the reactivity of the α anomeric form. However, the CV measurements in acidic condition of the α and of the β D-glucose forms showed some clear difference of reactivity. Indeed, the β D-glucose showed much higher current density than the α D-glucose, suggesting a much higher reactivity of the β-anomer on a Pt anode. The hypothesis proposed by the authors is that the β conformation allows the most planar approach of the glucose molecule to the electrode surface, leading to a better interaction and a more rapid reaction. By analysing the solution during and after long term electrolysis experiments in basic medium, gluconic acid seems to be the final product of the reaction, but δ-gluconolactone could be detected and was assumed to be an intermediate during the transformation.

By using in situ reflectance infrared spectroscopic techniques (SPAIRS and SNIFTIRS), Beden et al. were able to give more details to the electro-oxidation of the β D-glucose anomer on a Pt anode in alkaline media [71]. With these investigations, the authors were able to propose a detailed mechanism for the electro-oxidation of β D-glucose in alkaline medium on a Pt anode. The first step is the chemisorption of glucose onto the surface at a free platinum site. At a potential lower than 0.6 V vs. RHE, the adsorbed dehydrated species can then be further oxidized as a weakly adsorbed gluconate via two different possible mechanisms, which depend on the potential applied to the reaction. The gluconate can easily desorb from the surface as a gluconate species (Scheme 17A). At potential higher than 0.6 V vs. RHE, the adsorbed dehydrated glucose species generates by oxidation a δ-gluconolactone species.

This species can desorb and by hydrolysed to deliver gluconic acid or can remain weakly bonded to the surface while being oxidized, leading to a weakly bonded gluconate (Scheme 17B).

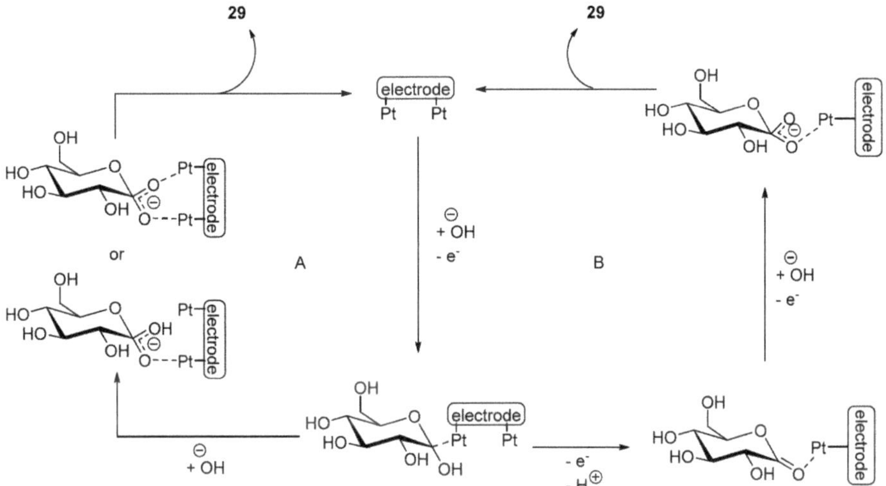

Scheme 17. Proposed mechanisms for the Pt electro-oxidation of glucose to gluconic acid (**29**) at low potential (**A**) and at high potential (**B**).

The ability to change the selectivity of the electro-oxidation of glucose by tuning the reaction potential was also exemplified on Au electrode by Moggia et al. [72]. Indeed, the authors studied the reactivity of D-glucose in alkaline media on a pure gold electrode, targeting D-glucaric acid as the main product of the reaction, using voltammetric measurements. In order to assign the peaks observed during CV experiments with D-glucose, they studied the response of the potential intermediates of the reaction, glucuronic acid (**40**), gluconic acid (**29**), and glucaric acid (**38**).

These measurements allowed them to clearly identify the potential at which glucose was oxidised to gluconic acid (0.55 V vs. RHE), and the potential for the oxidation of gluconic to glucaric acid (1.3 V vs. RHE). Then, by applying the appropriate potential for gluconic acid, the authors report 87% selectivity toward **29** and only trace amounts of **38** detected. When the potential was then increased to favour glucaric acid formation, the selectivity reached 14% and the selectivity toward gluconic acid decreased to 66%, thus demonstrating the relationship between the potential of the reaction and the selectivity of the outcome.

Due to the scarcity and the high price of noble metals such as Pt and Au, pure metal electrodes are not easily scalable because high costs that are involved. In order to lower the amount of metal loading on the anode, whilst preserving the selectivity and the reactivity of the metal, Holade et al. reported a selective electro-oxidation of glucose using Au nanoparticles on carbon as the anode material [73,74]. The aim of their work was to develop a highly efficient fuel cell which harvests electrical power from the oxidation of glucose, while producing added value chemicals. At first, the authors studied the reactivity of glucose toward the Au nanoparticles via CV paired with infrared analysis. These analyses revealed that glucose was successfully adsorbed on the Au nanoparticles, and could be oxidised rapidly without generating any C-C bond cleavage, and a fast hydrolysis of gluconolactone, which proved to be an inevitable intermediate. Long term electrolysis were then performed for 7.5 h under basic conditions (0.1 M NaOH). Gluconate was detected as the major product of the reaction (65% yield) with a nearly 100% faradaic efficiency. The reaction was followed by IR spectroscopy to verify that at any stage of the reaction, no C-C bond cleavage occurred.

Bimetallic nano-objects were also investigated as electro-catalyst for glucose electro-oxidation in NaOH solution (0.1 M). Rafaideen et al. recently reported the study of carbon-supported PdAu nanomaterials for selective oxidation of monosaccharides [75]. At first, the authors evaluated separately the activity of Pd and Au on carbon. They noticed that Au was significantly more active toward glucose oxidation. The authors then set out to test the electrocatalytic activity of the bimetallic Pd_xAu_{10-x}/C for monosaccharide oxidation in alkaline medium. By using voltammetric measurements, it was possible to prove that Pd rich catalysts and Au rich catalysts behaved similarly to their respective pure metal equivalents. However, the changes detected were believed to be related to the surface composition rather than to the bulk composition. It was observed that at low potential (<0.9 V vs. RHE), the activity of the catalyst increases with the increase of Au content up to 70% and then decreases, with Pd_3Au_7/C showing the highest catalytic activity. With the optimised reaction conditions, D-glucose (14) was electrolysed at a Pd_3Au_7/C anode in basic medium. After 6 h of electrolysis, 67% of glucose was consumed, and gluconate (29) was detected in 58% yield (selectivity towards gluconate of 87%), and very small amount of over-oxidation products could be detected. The authors then applied the same process to the electro-oxidation of D-xylose (34), which consumed up to 52% of xylose after 6 h of electrolysis, to deliver xylonate (35) in 48% yield and 92% selectivity.

Although most of the research arguably focused on the electrochemical behaviour of D-glucose, other monosaccharides could also be successfully electro-oxidised on Au or Pt electrodes. Xylose, as previously mentioned, has been electrochemically oxidised on Au and Pt electrode by Governo et al. [76]. The authors studied the catalytic activity of both metals for anodic oxidation of D-xylose (34) in alkaline medium using voltammetric measurements. They evaluated the effect of the concentration of xylose and of the temperature. By studying the effect of the pH, it was possible to show that on both Au and Pt, a high concentration of NaOH could contribute to the inhibition of the reaction. To avoid the electrode poisoning, a program of potentials was designed for each electrode to perform electrolysis. Au showed significantly better reactivity than Pt for the electro-oxidation of xylose (98% and 26% conversion respectively) and selectivity toward xylonic acid (35) (Table 5).

Table 5. Direct electro-oxidation of xylose at Au and Pt anodes.

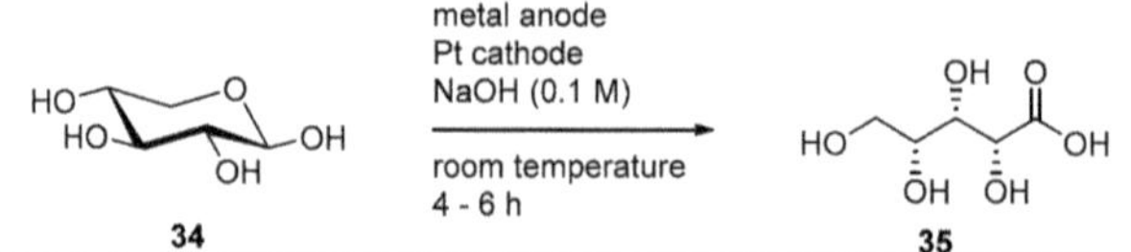

Product	Chemical Yield (%)	
	Au [a]	Pt [b]
Xylonic acid (35)	61	5
Glioxylic acid	-	4
Tartronic acid	4	1
Glyceric acid	6	3
Oxalic acid	7	-
Glicolyc acid	6	3
Formic acid	3	2
Tartaric acid	5	-

Chronoamperometry carried out for a: 6 h; b: 4 h.

The oxidation of D-galactose (30), a C-4 epimer of D-glucose, has been reported under homogeneous and heterogeneous catalysis [77,78]. However, its electrochemical oxidation has been scarcely investigated. Parpot et al. documented the electrochemical transformation of D-galactose (30), and compared the selectivity and reactivity of Pt, Pt-Pb, and Au electrodes in alkaline medium [79]. For the long term electrolysis, the appropriate potential step programs were used. These potential programs were optimised with the information gathered with preliminary voltammetric measurements with the corresponding electrode. The electro-oxidation of D-galactose was carried out for 25 h at room temperature in 0.1 M NaOH solution. For the three electrodes, D-galactonic acid (31) was the main product of the reaction, however, Pt-Pb and Au electrode were significantly more selective toward **31** than the smooth Pt electrode, and Au was also more reactive, converting a greater amount of **30** than any other electrode (Table 6).

Table 6. Direct electro-oxidation of D-galactose.

Product	Chemical Yield (%)		
	Pt	Pt-Pb	Au
Galactonic acid (**31**)	34	67	86
Galactaric acid	1	1	3
Glycolic acid	10	1	1
Oxalic acid	2	Traces	Traces
Glyoxylic acid	2	Traces	-
Glyceric acid	4	Traces	1
Formic acid	10	Traces	Traces
Tartaric acid	1	Traces	1
Tartronic acid	3	Traces	Traces

In this study, the authors propose a mechanism which follows the same steps as the ones described for the electrochemical oxidation of D-glucose at Pt, Pt-Pb and Au electrode (see Scheme 15A,B). The amount of galactaric acid generated during the electro-oxidation of galactose could be significantly increased by readjusting the oxidation plateau potential used [22]. Parpot et al. showed by employing voltammetry experiments that the enediol tautomer plays an important role in the oxidation of the primary alcohol of D-galatose on Au electrodes. Indeed, without this enediol form, the electro-oxidation of the primary alcohol is significantly disfavoured compared to the electro-oxidation of the aldehyde/hemiacetal group. It was also possible to show that the adsorption of the enediol species occurs at higher potential than the one used for aldehyde adsorption. Furthermore this study demonstrated that the relative position of the hydroxyl groups has an influence on the current density. Indeed, *trans*-diols are known to be more reactive than their *cis*-isomer on at Au electrode [80]. which seems to also apply for saccharides electro-oxidation. Therefore, by modifying the potential program to favour the enediol adsorption/oxidation, the authors were able to generate galactaric acid in 20% yield after 7 h of electrolysis under basic conditions (0.1 M NaOH). However, more degradation products, resulting from C-C bond cleavage, could also be observed.

The direct electrochemical oxidation of D-mannose (**32**) was also studied on Pt and Au electrodes [81]. Parpot et al. studied the behaviour of D-mannose to acidic, neutral, and basic conditions on each electrode. They found that, like D-glucose, D-mannose showed very limited reactivity toward Au or Pt in acidic media, and that the current density increased with the increase of the pH. Therefore, for the rest of the study, D-mannose was only submitted to basic conditions (0.1 M NaOH). The authors then looked at the influence of temperature and initial concentration of D-mannose. The smooth Pt electrode displayed low current density compared to Au, therefore upd Pb adatoms were also studied and proved to be very beneficial to the current density displayed. With the optimal reaction conditions in hand, D-mannose was then submitted to long term electrolysis at Au and Pt-Pb adatoms electrodes. At a Au electrode, 85% of the initial D-mannose was converted in 6 h of electrolysis and showed 49% selectivity toward D-mannonic acid (**33**, 42% yield). Beside **33**, a few low molecular weight acids arising from C-C bond cleavage could be detected in small amount (Table 7). For the direct electrolysis of D-mannose at the Pt-Pb adatoms electrode, 80% conversion of D-mannnose (**32**) was reached in 11 h. The main product was also D-mannonic acid (**33**, 50% yield), showing a slightly better selectivity than the pure Au electrode (Table 7).

Table 7. Direct electro-oxidation of D-mannose (**32**).

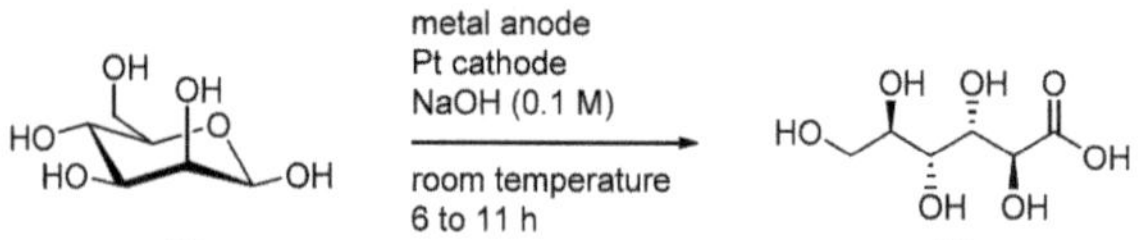

	metal anode Pt cathode NaOH (0.1 M) ⟶ room temperature 6 to 11 h	
32		**33**

Product	Chemical Yield (%)	
	Au	Pt-Pb
Mannonic acid (**33**)	42	50
Mannaric acid	2	2
Mannuronic acid	2	-
2,3,4,5-Hydroxypentanoic acid	2	4
Formic acid	2	1
Glycolic acid	2	4
Tartronic acid	4	2
Glyceric acid	1	1
Oxalic acid	6	1
Glyoxylic acid	4	3
Tartaric acid	4	4

Although Au- and Pt-based electrodes have been the most widely studied, other metal-based electrodes have been reported for the selective, direct electrochemical oxidation of monosaccharides. For instance, the electrochemical oxidation of monosaccharides at various copper-oxide-modified electrodes has been investigated, mainly for detection purposes [21,82,83]. However, some studies do report the products resulting from prolonged electrolysis [84–86]. The long term electrolysis of glucose on a Cu based electrode (plain Cu plate, Cu-oxides or Cu alloys) mostly resulted in the formation of one product: formate (Scheme 18). Indeed, each mol of initial glucose could be transformed via C-C bond cleavage to 6 moles of formic acid in basic media (0.1 M NaOH). Sorbitol and xylose appeared to follow the same process, generating 6 moles and 5 moles, respectively.

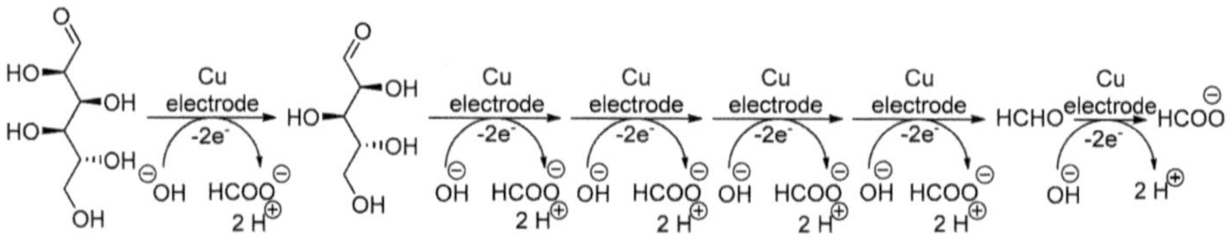

Scheme 18. Proposed mechanism for Cu based electro-oxidation of glucose.

Certain kind of Rh porphyrins, adsorbed on a carbon electrode have also been reported for their high activity and selectivity toward glucose electro-oxidation [87]. In their study, Yamazaki et al. have employed four different porphyrin ligands to generate Rh complexes. They evaluated the performance of each catalyst by CV in alkaline medium (0.1 M NaOH) and selected the most active form of the Rh porphyrin complex, [Rh^{II}(DPDE)]$^+$. With the appropriate catalyst, glucose was electro-oxidised for 3 h in 0,1 M NaOH. Gluconate was the desired product in this reaction, and its concentration was determined by enzymatic method using gluconate kinase and 6-phosphogluconic dehydrogenase. Only a very small amount of gluconate could be detected (<1% yield), however a faradaic yield up to 73% was achieved due to the very low current running through the electrochemical cell.

To control the electro-oxidation of glucose to gluconic acid (**29**) and glucaric acid (**38**), Bin et al. employed an electrocatalytic reactor with a tubular porous Ti anode loaded by sol-gel method with nano-sized MnO_2 [88]. The flow rate used during the electrolysis was adjusted to ensure that the maximum amount of d-glucose could be oxidised, while maintaining a high selectivity and avoiding over-oxidation. The effect of the loading of MnO_2 on the Ti electrode was also investigated, as well as the pH. Surprisingly, the conversion of glucose was not strongly dependent on the pH. Indeed, from acidic (pH = 2) to basic (pH = 10), the conversion was always above 90%. However, the selectivity toward **29** and **38** appeared to be improved when the pH was neutral. The temperature also had an impact on the selectivity but also on the overall conversion of glucose, which significantly decreased when the reaction was performed at a temperature above 30 °C. The authors explain this observation by the fact that the adsorption process is unfavoured with high temperatures. Performing the electrolysis at the most favourable conditions, 93% of the initial D-glucose was converted mainly to both gluconic acid (42% yield) and glucaric acid (44% yield).

Nickel-based electrodes have also been investigated for the electro-oxidation of mono-saccharides. Indeed, Parpot et al. first documented the direct electrochemical oxidation of D-galactose (**30**) on a nickel electrode in alkaline medium (0.1 M NaOH) [79]. Two types of long term electrolysis were carried out: one with a constant potential and one with a step potential program. In both cases, the conversion of D-galactose was low (17% and 43% respectively). Although D-galactonic acid (**31**) was detected as the major product of the reaction, only small amounts were generated (9% at constant potential and 7% with the plateaus potential program). These electrolysis also produced various low molecular weight acids (formic, glycolic, oxalic acid). The poor yield of **31** can be explained by the competition reaction between the desired oxidation, the degradation responsible for low molecular weight compounds formation, and water oxidation. The oxidation of galactose on a Ni anode occurs near the oxygen evolution potential region. This suggests that the surface is covered with NiOH and NiO species which can react with galactose (or galactonic acid) and generate the C-C bond cleavage products (Scheme 19).

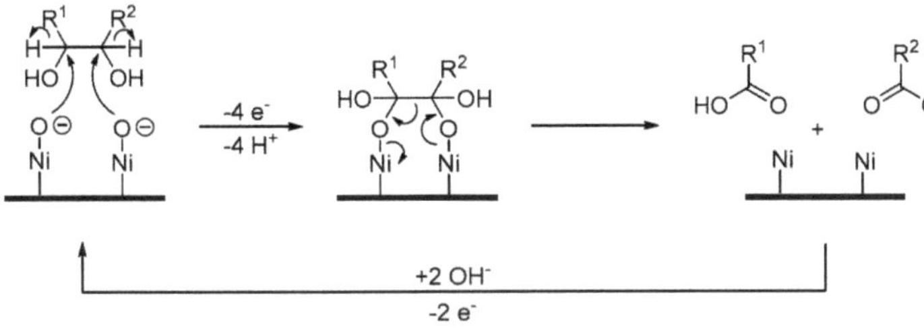

Scheme 19. Proposed mechanism for degradation of galactose on a Ni anode.

Similar results were observed for the electrochemical oxidation of D-mannose (**32**) on a Ni electrode under basic conditions (0.1 M NaOH) at constant potential [81]. The conversion of D-mannose was very low (35%), and a significant amount of degradation products were presents, some in larger quantities than the desired D-mannonic acid (**33**, 3% yield).

More recently, Liu et al. reported that nanostructured NiFe oxides ($NiFeO_x$) and NiFe nitrides ($NiFeN_x$) catalysts exhibit a great activity and selectivity toward the electrochemical oxidation of D-glucose (**14**) [89]. The authors used nickel foam as a support as a Ni source for the catalyst and to ensure a 3D structure to the electrode. Then, the resulting oxides and nitride electrodes were tested in 1.0 M KOH solution (pH = 13.9), for D-glucose oxidation. The preliminary tests showed that the electro-oxidation of glucose happened at a lower potential than the oxygen evolution reaction, which was presumed to be the main undesired side reaction. $NiFeO_x$ appeared to be more active than $NiFeN_x$ toward glucose oxidation. By analysing the electrode materials after electrolysis of glucose, NiOOH and FeOOH species were detected. The authors postulated that these species are the catalytic active sites of the electrodes responsible for the glucose electro-oxidation reaction. Then, they turned their attention to evaluating various parameters such as the reaction potential and the initial glucose concentration.

The authors were able to carry out long term electrolysis with both catalysts. Both NiFeO$_x$ and NiFeN$_x$ converted glucose very efficiently (91% and 93% respectively), and with great selectivity toward glucaric acid (**38**, 71% and 64% yield respectively) after 18 h. The monitoring of the reaction allowed the authors to see the initial formation of gluconic acid (**29**) and then its conversion to glucaric acid (**38**). The catalysts also showed great stability, when five chronoamperometric measurements were carried out the conversion of glucose slightly decreased but the faradaic efficiency remained almost unchanged. By using IR and NMR spectrometry, the authors were able to identify various intermediates and products of the reaction, and proposed the following reaction pathway for the NiFe anodic oxidation of D-glucose (Scheme 20).

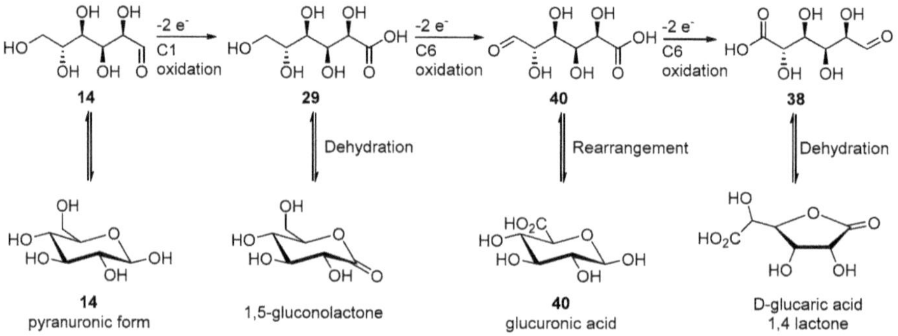

Scheme 20. Possible reaction pathway for NiFe anodic oxidation of D-glucose.

2.4. Summary

The electrochemical oxidation of saccharides has attracted a lot of attention in the past century. Although it is difficult to compare different studies between themselves because they all use different reaction parameters, it seems necessary to clarify which electro-catalysts seems promising and for which time of transformation. The time of reaction being different in each example is not taken into consideration, as well as the pH of the reaction. However, it is worth noting that the electrochemical oxidations of saccharides are almost exclusively performed at high pH. Indeed, at low pH, the saccharide are likely to degrade through cleavage of the C-O-C bond or the opening of the ring. The few electro-oxidations documented to happen in acidic medium mainly lead to the corresponding hydration products (for poly- and di-saccharides) as well as degradation products.

For the indirect electro-chemical oxidation of saccharides, the most explored and promising mediator is TEMPO. It has been successfully used on poly-, di-, and monosaccharides. TEMPO selectively oxidises primary alcohols and is able to deliver triacids when used on disaccharides such as D-raffinose with 63% yield, but also with cellobiose and D-maltose, although less selectively. Dicarboxylic acid could also be obtained in 61% when TEMPO was used to oxidise trehalose. The TEMPO mediated electrochemical oxidation of monosaccharides such as D-glucose yielded the corresponding mono- and disaccharide products and the selectivity could be tuned to favor one or the other.

A vast number of metal electrodes has been investigated for the direct electro-oxidation of saccharides, and it is therefore complex to identify which is the most promising, the most reactive, or the most selective. However, it is very clear that Au is the most investigated metal. It was tested with poly-, di-, and monosaccharides, with and without adatoms. Pure Au plates were used as well as nanoparticles. In all reports, the major product isolated with Au as the electrocatalyst is the mono carboxylic acid equivalent of the starting saccharide. Pt was also extensively studied and demonstrated similar selectivity as gold, but usually lower reactivity. When a NiFe electrocatalyst was used on D-glucose, the major product isolated was D-glucaric acid (71% yield), therefore showing a different selectivity than the other catalysts. Finally, when Cu was used as electrocatalyst with monosaccharides, formate appeared to be the only product formed quantitatively (Table 8).

Table 8. Summary of the direct electro-oxidation of saccharides.

Starting Material	Catalyst	pH	Main Reaction Product	Yield	Selectivity
Cellulose	Au	Alkaline	Polyoxidised cellulose		
	MnO_2	Acidic	Glucose	72%	100%
Hemicellulose	Au	Alkaline	Polyoxidised xylan		
Sucrose	Au	Alkaline	1'- and 6-mono acid	17% and 15%	26% and 23%
	Pt-Pb	Alkaline	6-mono acid	48%	80%
Trehalose	Au	Alkaline	Mono acid (**18**)	46%	70%
Lactose	Au	Alkaline	Lactobionic acid	93%	98%
	Pt-Pb	Alkaline	Lactobionic acid	78%	100%
Cellobiose	Au NP	Alkaline	Cellobionic acid	>90%	>95%
Maltose	Au	Alkaline	Maltobionic acid	73%	92%
Glucose	Au	Alkaline	Gluconic acid	77%	90%
	Au NP	Alkaline	Gluconic acid	65%	100%
	PdAu NP	Alkaline	Gluconic acid	58%	87%
	Pt-Pb	Alkaline	Gluconic acid	70%	75%
	Cu	Alkaline	Formate	99%	100%
	MnO_2	Neutral	Glucaric acid	42%	47%
	NiFe	Alkaline	Glucaric acid	71%	78%
Xylose	Au	Alkaline	Xylonic acid	61%	62%
	PdAu NP	Alkaline	Xylonic acid	48%	92%
Galactose	Au	Alkaline	Galactonic acid	86%	92%
	Pt-Pb	Alkaline	Galactonic acid	67%	>95%
	Ni	Alkaline	Galactonic acid	9%	53%
Mannose	Au	Alkaline	Mannonic acid	42%	49%
	Pt-Pb	Alkaline	Mannonic acid	50%	63%
	Ni	Alkaline	Mannonic acid	3%	9%

3. Electrochemical Reduction of Biomass

3.1. From Polysaccharides to Monosaccharides

The electrochemical reduction of saccharides has been studied extensively since the early 1900s., but arguably, most of the effort in this domain has been focusing towards the electroreduction of D-glucose (**14**) toward the corresponding alditol: sorbitol (**41**). Alditols have found a wide range of applications across various industries [90]. nevertheless, their current way of production is via catalytic hydrogenation processes which require high temperature and high pressures [91]. Therefore, the development of a greener, more user friendly, and sustainable way to produce such polyols is greatly sought after.

Amongst the first examples of electrochemical reduction of glucose, it was found that under acidic reaction conditions, a lead cathode does not generate the desired alditol but D-arabinose and formaldehyde [38]. Other studies focused on monosaccharides electroreduction under basic or neutral reaction conditions. Creighton et al. reported the successful electrochemical transformation of D-glucose (**14**), D-mannose (**32**), D-fructose (**13**), and D-xylose (**34**) to the corresponding alditols in neutral medium (Na_2SO_4) [92]. Preliminary tests were performed on the electroreduction of D-mannose (**32**) on a Hg cathode. The authors evaluated various parameters such as the temperature, the initial concentration and the pH. Mannitol (**42**) could be isolated in 62% yield. Despite these promising results, the use of an amalgamated Pb cathode instead of Hg was explored, due to the numerous disadvantages Hg brings. D-glucose was then used for the preliminary tests. Due to isomerisation equilibrium between glucose, fructose and mannose, the authors anticipated that it could be possible to generate not only sorbitol from glucose, but also mannitol (Scheme 21). After screening the same parameters as previously used, it was clear that the hydroxyl ion concentration in the catholyte influenced greatly the selectivity of the reaction. Indeed, at low concentration of NaOH (2.5 mM), no mannitol was generated and sorbitol was isolated in 59% yield. However, with a high concentration of NaOH (0.75 M), **42** was isolated in 13% yield, and **41** was isolated in 47% yield. The authors also reported the successful conversion of xylose (**34**) to xylitol, galactose (**30**) to dulcitol, and lactose (**25**) to a mixture of dulcitol and sorbitol, however, no information was given on the selectivity and the efficiency of these electrochemical reductions. From the range of parameters tested and by varying the reduced saccharides, the authors postulated that Pb and Hg were good cathode materials for poorly reducible compounds due to their high overpotential for hydrogen evolution.

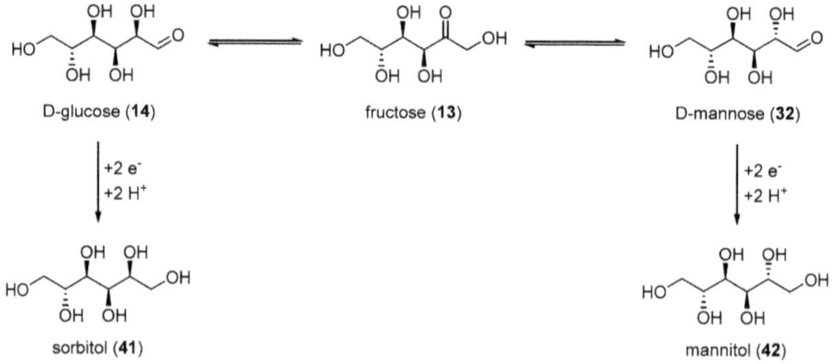

Scheme 21. Possible mechanism for electroreduction of D-glucose (**14**) to sorbitol (**41**) and mannitol (**42**).

The electrochemical reduction of saccharides was used industrially to generate the corresponding alditols. However, a range of undesired by-products were also present, depending on the reaction conditions. Wolfrom et al. documented a very thorough study to characterise the various products present after the electro-reduction of glucose on a Hg cathode [93–97]. From the commercial sorbitol (**41**) in weakly alkaline medium (pH = 7–10) and at a temperature below 30 °C, D-mannitol (1% yield), and 2-deoxysorbitol (**43**, 5% yield) could be isolated. When glucose was electro-reduced to sorbitol in highly basic conditions, (pH = 10–13), a more complex mixture was present at the end of the reaction. As expected, sorbitol (**41**) and mannitol (**42**) were isolated, as well as 2-deoxysorbitol (**43**) and 1-deoxymannitol (**44**). D,L-glucitol [96] and a dodecitol [97] were also detected. The formation of dodecitol was assumed to arise from the electroreduction of the aldehyde group of two glucose molecules.

The use of alkaline electrolytes is beneficial for the reactivity of the electrodes for the reduction of monosaccharides but is detrimental to the selectivity of the reaction because it favours the isomerisation reaction of glucose to mannose and fructose. By using sulphate solutions as electrolyte, this problem was solved. Despite their nearly neutral pH, the hydrogen evolution reaction at the electrode surface generates a layer of higher alkalinity, which is favourable for the electroreduction of glucose to sorbitol. Rice and co-workers reported that, in such condition and on a Pb cathode, sorbitol was the major product detected after only 4 h of electrolysis, and the concentration of the undesired fructose and maltose started increasing only after longer electrolysis [98]. The authors reported in their study the influence of various parameters such as the pH, the initial concentration of glucose, and the flow rate on the current density at 20 °C. The addition of Zn^{2+} in the catholyte had been used industrially to improve the rate of the reaction. The authors investigated how the Zn^{2+} ions impacted the reaction. They observed that the Zn was deposited on the electrode surface, which increased the overall surface area of the electrode, thus the rate of glucose electroreduction will increase. The authors also hypothesized that Zn could have a dual effect by acting as a co-catalyst for the electroreduction reaction.

The electrochemical reduction of glucose on Zn on carbon nanotubes (CNTs) grown on a graphite electrode was documented by Fei et al. [99]. At first, the authors studied the variation of reactivity between a flat Zn, a Zn/graphite, and a Zn/CNTs electrodes in weakly alkaline medium (0.1 M Na_2SO_4, pH = 11) using CV measurements. The flat Zn and the Zn/graphite electrodes displayed similar behaviours, but the Zn/CNTs electrode showed an increase of the reduction current in the presence of glucose, due to the high surface area provided by the CNTs. The 3D structure of the CNTs provided a much larger zone of higher alkalinity in the solution than on a flat electrode, which is known to be beneficial for the conversion of the cyclic hemiacetal form of glucose to its electroreducible aldehyde form, thus explaining the higher efficiency of the Zn/CNTs electrode to reduce glucose. However, the electrode stability was very low (50% loss of efficiency in 4 h), probably due to the dissolution of Zn on CNTs which happens during the electrolysis of glucose. In order to improve the electrode stability, Zn-Ni and Zn-Fe alloys were deposited on CNTs. Both electrodes display similar reactivity as the Zn/CNTs electrode, but with slightly higher hydrogen evolution current. The Zn-Fe/CNTs and Zn-Ni/CNTs electrodes showed better stability than the Zn/CNTs electrode. Indeed, after 8 h of electrolysis, the current decreased but significantly less than for the Zn/CNTs electrode, proving that the activity of the electrode was maintained. The authors noticed that the current efficiency of the Zn-Fe/CNTs electrode was higher for glucose electroreduction than all the other electrodes, thus they investigated further the reaction with the Zn-Fe/CNTs electrode only. They studied the effect of the pH on the electrode response toward glucose, as well as the temperature. Although an increase of pH (from 7 to 13) shows a drastic increase in current, the authors acknowledge that the isomerisation of glucose to fructose and mannose will have a negative impact on the current efficiency. They report the highest current efficiency at 40 °C. Their study was thus able to prove the benefices of the utilisation of Zn in combination with CNTs for the electro-transformation of glucose to sorbitol.

The most commonly used metals for the electrochemical reduction of glucose are the metals which suppress the hydrogen evolution reaction (HER), where the chemisorbed hydrogen on the surface of the electrode is produced by the electro-reduction of water and then produce hydrogen gas, which is in direct competition with the hydrogenation of the adsorbed glucose to yield sorbitol or deoxygenate to deoxysorbitol. Kwon et al. documented a very thorough study where a vast number of pure metal electrode catalysts were screened in order to comprehend the activity and the selectivity of catalysts for the electro-reduction of glucose [100]. All the tests were performed in neutral electrolyte, and the metals were separated in three groups, metals that generate sorbitol (**41**) and hydrogen gas (Fe, Co, Ni, Cu, Pd, Ag, Au, and Al); metals forming H_2, sorbitol (**41**) and 2-deoxysorbitol (**43**) (Zn, Cd, In, Sn, Sb, Pb, and Bi); and metals forming only hydrogen gas (Ti, V, Cr, Mn, Zr, Nb, Mo, Hf, Ta, W, Re, Ru, Rh, Ir, and Pt). The metals were tested via voltammetric measurements, while the products in solution were analysed via an online HPLC analysis (Table 9).

Table 9. Screening of pure metal electrodes for the electro-reduction of glucose to sorbitol.

Entry	Metal	Onset Potential (vs. RHE)	[Sorbitol] (mM·cm⁻²)	[2-Deoxysorbitol] (mM·cm⁻²)
1	Fe	−0.75 V	0.03	-
2	Co	−1.0 V	0.06	-
3	Ni	−0.41 V	0.075	-
4	Cu	−1.3 V	0.15	-
5	Pd	-0.87 V	0.09	-
6	Ag	−1.05 V and −1.41 V	0.06	-
7	Au	−1.05 V	0.15	-
8	Al	−1.05 V	0.01	-
9	Zn	−1.4 V	0.06	0.07
10	Cd	−1.7 V	0.35	0.22
11	In	−2 V	0.8	0.8
12	Sn	−1.5 V	traces	0.05
13	Sb	−1.0 V	0.12	0.03
14	Pb	−2.0 V	0.7	0.6
15	Bi	−1.5 V	0.4	0.05

First, the reactivity and selectivity of Fe, Co, Ni, Cu, Pd, Ag, Au, and Al electrodes were investigated in 0.1 M Na_2SO_4 (entries 1 to 8). All these metals did convert glucose to sorbitol exclusively, but also performed the hydrogen evolution reaction. Most of these metals showed a lower reduction current when put in presence of glucose. This observation indicates that the adsorption of glucose on the surface of the electrode blocks the hydrogen evolution reaction by competing for the same active sites on the electrodes. Au and Cu electrodes were the two exceptions and displayed a higher reductive current in presence of glucose. These two metal electrodes also appeared to be the most efficient at converting glucose to sorbitol. For the formation of sorbitol, Ni is the metal which has the lowest onset potential compared to all the other metals (entry 3), and Cu gave the highest conversion to **41** of this series of metals (entry 4).

The second series of metals investigated (entries 9 to 15) were the metals able to generate sorbitol (**41**), as well as 2-deoxysorbitol (**43**). Higher cathodic currents could be applied to these metals, and therefore a higher conversion of glucose could be reached. The cathodic current generated by Zn was higher than the one generated by cadmium. Cd proved to be much more effective at converting glucose to **41** and **43** (entries 9 and 10, respectively). This observation suggests that Cd is more effective than Zn to hydrogenate glucose by suppressing the hydrogen evolution reaction and has a higher selectivity toward sorbitol. Sb showed the lowest onset potential (entry 13), Pb the best yield (entry 14), and Sn gave the best selectivity toward sorbitol (entry 12).

The final series comprise no active metals for the electrochemical reduction of glucose and are therefore not discussed here.

Acidic reaction condition (0.5 M H_2SO_4) proved to be inadequate for glucose hydrogenation despite the use of metals such as Co, Pb, or Cd, which were active in neutral medium. Therefore, the presence of OH- species is necessary for the electrochemical hydrogenation of glucose. However, having an alkaline electrolyte is detrimental to the selectivity of the reaction by promoting the isomerisation of glucose to fructose and mannose, and therefore to various side products, proving that using an unbuffered neutral electrolyte is the best reaction conditions for efficient and selective electrocatalytic glucose reduction.

When high cathodic currents were used, fructose (**13**) could also be detected in the solutions, which emphasizes the important of local pH at the electrode surface. Indeed, the rate of formation of OH^- species is controlled by the current intensity. With the information gathered during the investigation on the metal electrode screening, the authors were able to propose a mechanism explaining the formation of the various products and by-products (Scheme 22). The authors based their assumption on previous work which described that the cleavage of C-OH bond was favoured during the deoxygenation reaction when the hydroxyl group was vicinal to a C=O bond [101]. From this fact, glucose (**14**) itself could be a direct precursor to 2-deoxysorbitol (**43**), and would not require the formation of fructose (**13**) as an intermediate as it was previously postulated. Furthermore, the absence of mannitol (**42**) supports the scenario supporting the direct deoxygenation of glucose (**14**) to generate 2-deoxysorbitol (**43**). Indeed, the prochiral C2 position in fructose (the ketone) can be hydrogenated to two isomeric hydroxyl groups in almost equivalent ratios 40% sorbitol, 60% mannitol) [102].

Although most of the work on saccharide electro-reduction has been directed to the reduction of D-glucose, other monosaccharides have also been studied. D-xylose was successfully converted into its corresponding alditol, xylitol [103]. Piszczek et al. first optimised the reaction conditions by variating the pH and the temperature of the reaction. They reported that by electrolysing xylose (**34**) on an amalgamated Pb cathode in a alkaline solution (Na_2SO_4, pH = 11) and at 45 °C, the desired xylitol (**44**) could be isolated in 75% yield. Small amounts of 2-deoxyxylitol could also be detected (Scheme 23).

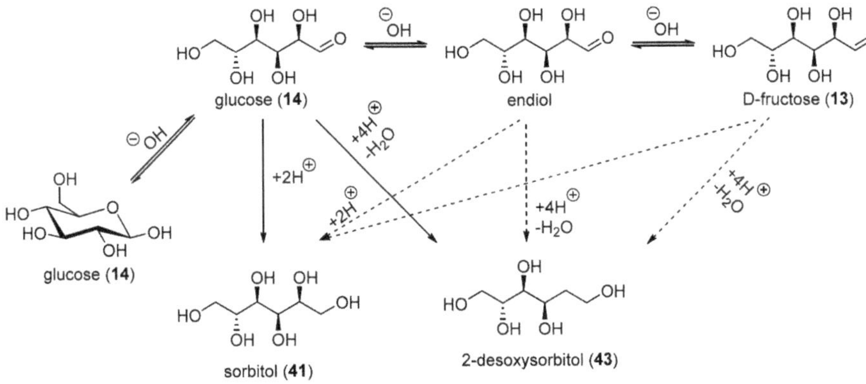

Scheme 22. Proposed mechanism for the direct electro-reduction of glucose.

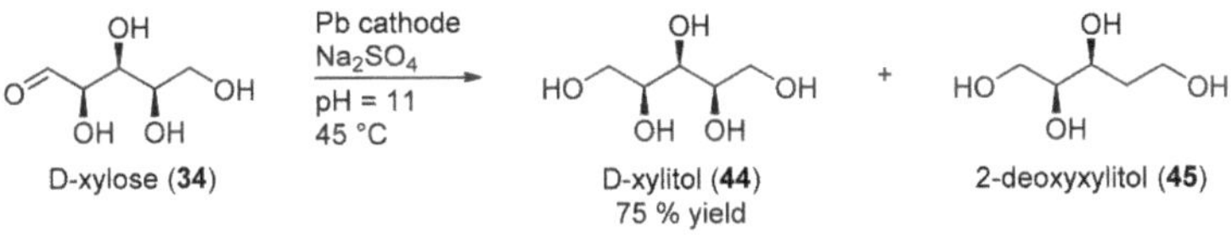

Scheme 23. Electrochemical reduction of D-xylose (**34**).

Hamann et al. documented the formation of ester derivatives of sucrose and various protected monosaccharides [104,105]. The reaction was a two-step process, first the saccharides were electrochemically reduced to their corresponding anion, and then were reacted with the alkyl halide to deliver the desired saccharide ester derivative (Table 10).

Table 10. Electrochemical alkylation of sucrose.

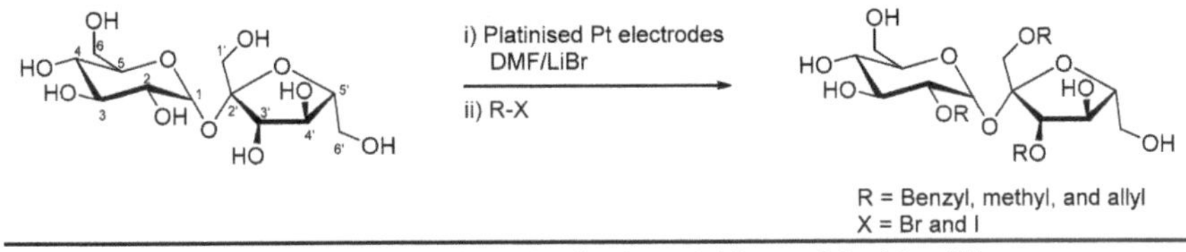

R = Benzyl, methyl, and allyl
X = Br and I

Entry	Alkyl Halide (R-X)	Product Distribution at Different C Position (%)			Faradaic Yield
		2	1′	3′	
1	Benzyl bromide	48	39	13	48%
2	Methyl iodide	54	27	19	63%
3	Allyl bromide	39	61	-	28%

The reaction took place in DMF in presence of LiBr. The Li ion was necessary to stabilise the anionic saccharide species generated during the electrolysis. The authors screened other metal counter cations, (Mg and Zn), however, the deprotonation did not take place. When sucrose was submitted to such reaction conditions, a poor selectivity was observed. Indeed, three hydroxyl groups reacted and therefore three different products could be detected. Nevertheless, the substitution at the C2 position appeared to be the most favoured one.

3.2. Summary

The electrochemical reduction of saccharides has arguable been less documented than the electro-oxidation process. However, the various reports demonstrate the feasibility and scalability of such processes. Furthermore, a range of product selectivity can be obtained by modifying the reaction parameters such as the electrode material, the supporting electrolyte, and the pH (Table 11).

Table 11. Summary of the saccharide electro-reduction processes.

Starting Material	Catalyst	pH	Main Reaction Products	Yield	Selectivity
Mannose	Hg	Neutral	Mannitol	62%	
Glucose	Pb	Alkaline (pH = 11)	Sorbitol	59%	>95%
	Pb	Alkaline (pH = 14)	Sorbitol and mannitol	47% and 13%	
	Fe	Alkaline	Sorbitol		100%
	Co	Alkaline	Sorbitol		100%
	Ni	Alkaline	Sorbitol		100%
	Cu	Alkaline	Sorbitol		100%
	Pd	Alkaline	Sorbitol		100%
	Ag	Alkaline	Sorbitol		100%
	Au	Alkaline	Sorbitol		100%
	Al	Alkaline	Sorbitol		100%
	Zn	Alkaline	Sorbitol and 2-deoxysorbitol		35% and 65%
	Cd	Alkaline	Sorbitol and 2-deoxysorbitol		61% and 39%
	In	Alkaline	Sorbitol and 2-deoxysorbitol		35% and 65%
	Sn	Alkaline	Sorbitol and 2-deoxysorbitol		10% and 90%
	Sb	Alkaline	Sorbitol and 2-deoxysorbitol		80% and 20%
	Bi	Alkaline	Sorbitol and 2-deoxysorbitol		75% and 25%
Xylose	Pb	Alkaline	Xylitol	75%	
Galactose	Pb	Alkaline	Dulcitol		
Lactose	Pb	Alkaline	Dulcitol and sorbitol		
Sucrose	Pt	Alkaline	2-, 1′-, and 3′-sucrose ester		5:3:2

4. Combined Electrolysis

During electrolysis, oxidation and reduction reactions are happening simultaneously on both the anode and cathode respectively. When oxidation and reduction reactions are coordinated in such way that they both yield targeted organic compounds, this is often referred to as 'paired electrolysis'. This can be performed in an undivided reactor by using the same starting material which can be reduced and oxidised and deliver two different products, or in a divided reactor, with one or two different starting materials, thus delivering one oxidised and one reduced product (Scheme 24). This type of electrochemical reactions are intensely pursued due to the high energy efficiency relative to the amount of chemicals generated.

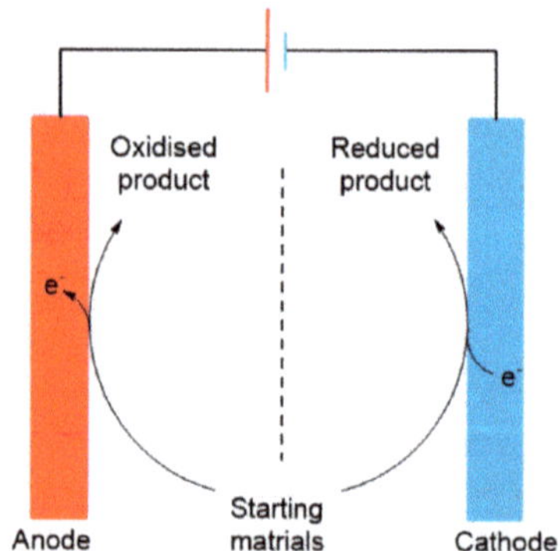

Scheme 24. Illustration of a paired electrolysis system.

Park et al. documented a paired electrolysis of D-glucose to gluconate (**29**) at the anode and simultaneously to sorbitol (**41**) at the cathode in an undivided flow reactor [106,107]. Graphite chip was used as the anode material for an indirect electro-oxidation and either a Zn(Hg) shot or Raney Ni as the cathode for a direct electro-reduction reaction. The anodic oxidation of glucose to gluconate was mediated by NaBr as previously described by Isbell and co-workers [40–42]. To push the reaction toward the consumption of the intermediate δ-gluconolactone, the paired electrosynthesis was carried out at 58 °C with a residence time of 10 min in the reactor and at pH 7. The use of a Raney-Ni electrode allowed to reduce the overall cell potential compared to the Zn(Hg) electrode. Furthermore, when Raney-Ni and graphite were associated as the cathode and the anode material respectively, the current efficiency for both transformations could reach 100%, but the glucose conversion remained low (<30%). The authors also observed that Raney-Ni get poisoned during the electro-reduction of glucose due to the hydrogen evolution reaction. An in situ catalyst regeneration was then possible by washing the electrode with a 17 wt% NaOH solution at 60 °C for 90 min [107]. This procedure allowed to restore the catalytic activity of the cathode and to restore the current efficiency for sorbitol from 35–45% to 70–100%.

In a similar process, Bardot and co-workers developed a paired electrolysis of fructose (**13**) to gluconic acid (**29**) at the anode and to sorbitol (**41**) and mannitol (**42**) at the cathode in a membrane separated reactor [102]. In this case, the anode was a graphite electrode and the cathode was a graphite electrode coated with Pt and Rh metals. This study only focused on the electrochemical reduction of fructose to sorbitol and mannitol, and the authors report that fructose was converted up to 60% and was transformed exclusively to sorbitol (40% selectivity, 24% yield) and mannitol (60% selectivity, 36% yield).

Finally, D-xylose (**34**) was successfully anodically oxidised to xylonic acid (**35**) and reduced to xylitol (**44**) at the cathode in a membrane separated reactor [108]. First, the authors investigated the cathodic reduction of xylose on an amalgamated Zn electrode. However, for the voltammetric measurements performed, they could deduct that the reduction of xylose was heterogeneously catalysed by adsorbed H-adatoms on the surface of the electrode. To promote this catalytic process, a variety of metal coating were screened ($WFe_{3-x}Pt_x$ and $MoFe_{3-x}Pt_x$, and other coating with Co or Ni instead of Fe or with Ru instead of Pt). With such coating, the H-adatoms catalytic reduction of xylose was significantly faster, the cathodic potential was reduced compared to the Zn(Hg) electrode and the Faradaic yield for xylitol formation was exceeding 90%. For the anode reaction, RuO_2/TiO_2 and Pb/Sn were used as intermetallic coating on a Ti electrode. These modified electrodes were used in combination with bromine as mediator for the electro-oxidation of xylose. This process displayed high anodic current efficiency (>97%) and the Pb/Sn coating shifted the potential for oxygen evolution to higher potential, improving the selectivity toward the anodic oxidation of xylose. Bromide species which are essential to the anodic formation of **35** and detrimental to the cathodic reaction reduced the cathodic Faradaic yield below 60%. Therefore, in order to maintain the highest efficiency possible of both side of the reaction, the authors emphasised on the necessity to separate this process with a membrane.

5. Conclusions

Research on electrosynthesis in general has come a long way, with some applications more advanced than others and a huge acceleration due to the emerging electrification of the chemical industry [109]. Electrosynthesis of biomass-based feedstocks is not very straightforward and depends on several process parameters such as pH, temperature, concentration, and potential as well as on the choice and availability of components like stable electrocatalysts and membranes. In this review, we have summarized the recent state of the art for oxidation and reduction of mono-, oligo- and polysaccharides. The main electrocatalyst used in these conversions were summarized and a basic screening reveals that most often used ones are based on pure metals like gold and platinum. This trend is changing in recent years when more metals, carbons, alloys, oxides and bimetallic catalysts are being explored. The need of the hour is moving from brute screening to smart designing and screening of the electrocatalysts coupling density functional theory (DFT)-based modelling with high throughput screening. Currently this is still based on a trial and error approach which is costly and time consuming due to a myriad of possible catalysts and shapes thereof. Besides, we noticed that most reported studies focus on one or a few 'promising' metals. Although electrolysis is often hyped as providing a 'highly selective' reaction pathway, the review shows that this statement is flawed in many cases. Even when starting from pure substances (i.e., disregarding the complexities of real-life biobased feedstock), formation of side-products is oftentimes difficult to avoid, resulting in poor selectivities and ultimately a process that is challenging to become economically viable.

Standardisation in terms of reporting the results is an important issue to be tackled and should address which are the most important performance numbers for evaluation of the electrosynthesis process. More standardization is needed in terms of reporting on yields, efficiencies, selectivity and other process parameters, the absence of which makes the comparison between different studies difficult. Even among two similar studies, the reported results are not always consistent. The research on this topic is rather fragmented research with many fundamental knowledge gaps (such as missing reaction mechanisms). Even though electrosynthesis has shown promise, it still needs to conquer a lot of ground in terms of performance and scalability.

Author Contributions: Conceptualization, K.V. and D.P.; resources, V.V., D.P., J.H.; writing—original draft preparation, V.V.; writing—review and editing, K.V., D.P. and J.H.; visualization, V.V., D.P.; supervision, D.P., J.H.; project administration, J.H.; funding acquisition, K.V., D.P. and J.H. All authors have read and agreed to the published version of the manuscript.

Funding: This research was funded by the project PERFORM funded by the European Union's Horizon 2020 Research and Innovation Programme, under Grant Agreement No. 820723.

Conflicts of Interest: The authors declare no conflict of interest.

References

1. De Luna, P.; Hahn, C.; Higgins, D.; Jaffer, S.A.; Jaramillo, T.F.; Sargent, E.H. What would it take for renewably powered electrosynthesis to displace petrochemical processes? *Science* **2019**, *364*, eaav3506. [CrossRef] [PubMed]
2. Pant, D.; Misra, S.; Nizami, A.S.; Rehan, M.; van Leeuwen, R.; Tabacchioni, S.; Goel, R.; Sarma, P.; Bakker, R.; Sharma, N.; et al. Towards the development of a biobased economy in Europe and India. *Crit. Rev. Biotechnol.* **2019**, *39*, 779–799. [CrossRef] [PubMed]
3. Li, J.; Qiao, Y.; Zong, P.; Wang, C.; Tian, Y.; Qin, S. Thermogravimetric Analysis and Isoconversional Kinetic study of biomass pyrolysis derived from land, coastal zone, and marine. *Energy Fuels* **2019**, *33*, 3299–3310. [CrossRef]
4. Lichtenthaler, F.W. Emil Fischers Beweis Der konfiguration von zuckern: Eine würdigung nach hundert jahren. *Angew. Chem.* **1992**, *104*, 1577–1593. [CrossRef]
5. Chiranjeevi, P.; Bulut, M.; Breugelmans, T.; Patil, S.A.; Pant, D. Current trends in enzymatic electrosynthesis for CO_2 reduction. *Curr. Opin. Green Sustain. Chem.* **2019**, *16*, 65–70. [CrossRef]
6. Wu, R.; Ma, C.; Zhu, Z. Enzymatic electrosynthesis as an emerging electrochemical synthesis platform. *Curr. Opin. Electrochem.* **2020**, *19*, 1–7. [CrossRef]
7. Yuan, Y.; Lei, A. Is electrosynthesis always green and advantageous compared to traditional methods? *Nat. Commun.* **2020**, *11*, 2018–2020. [CrossRef]
8. Siu, J.C.; Fu, N.; Lin, S. Catalyzing electrosynthesis: A homogeneous electrocatalytic approach to reaction discovery. *Acc. Chem. Res.* **2020**, *53*, 547–560. [CrossRef]
9. Neha, N.; Kouamé, B.S.R.; Rafaïdeen, T.; Baranton, S.; Coutanceau, C. Remarkably efficient carbon-supported nanostructured platinum-bismuth catalysts for the selective electrooxidation of glucose and methyl-glucoside. *Electrocatalysis* **2020**, 1–14. [CrossRef]
10. Schnatbaum, K.; Schäfer, H.J. Electroorganic Synthesis 66: Selective anodic oxidation of carbohydrates mediated by TEMPO. *Synthesis (Stuttg)* **1999**, *1999*, 864–872. [CrossRef]
11. Dang, X.; Chen, H.; Dai, R.; Wang, Y.; Shan, Z. Electrochemical-assisted synthesis, spray granulation and characterization of oxidized corn starch-gelatin. *Ind. Eng. Chem. Res.* **2018**, *57*, 7099–7104. [CrossRef]
12. Sugano, Y.; Vestergaard, M.; Yoshikawa, H.; Saito, M.; Tamiya, E. Direct electrochemical oxidation of cellulose: A cellulose-based fuel cell system. *Electroanalysis* **2010**, *22*, 1688–1694. [CrossRef]

13. Yang, F.; Zhang, Q.; Fan, H.X.; Li, Y.; Li, G. Electrochemical control of the conversion of cellulose oligosaccharides into glucose. *J. Ind. Eng. Chem.* **2014**, *20*, 3487–3492. [CrossRef]

14. Sugano, Y.; Saloranta, T.; Bobacka, J.; Ivaska, A. Electro-catalytic oxidation of hemicelluloses at the au electrode. *Phys. Chem. Chem. Phys.* **2015**, *17*, 11609–11614. [CrossRef] [PubMed]

15. Schämann, M.; Schäfer, H.J. Reaction of enamines and mediated anodic oxidation of carbohydrates with the 2,2,6,6-Tetramethylpiperidine-1-Oxoammonium ion (TEMPO+). *Electrochim. Acta* **2005**, *50*, 4956–4972. [CrossRef]

16. Lichtenthaler, F.W.; Peters, S. Carbohydrates as green raw materials for the chemical industry. *C. R. Chim.* **2004**, *7*, 65–90. [CrossRef]

17. Parpot, P.; Kokoh, K.B.; Beden, B.; Lamy, C. Electrocatalytic oxidation of saccharose in alkaline medium. *Electrochim. Acta* **1993**, *38*, 1679–1683. [CrossRef]

18. Parpot, P.; Kokoh, K.B.; Belgsir, E.M.; Léger, J.-M.; Beden, B.; Lamy, C. Electrocatalytic oxidation of sucrose: Analysis of the reaction products. *J. Appl. Electrochem.* **1997**, *27*, 25–33. [CrossRef]

19. Parpot, P.; Servat, K.; Bettencourt, A.P.; Huser, H.; Kokoh, K.B. TEMPO Mediated oxidation of carbohydrates using electrochemical methods. *Cellulose* **2010**, *17*, 815–824. [CrossRef]

20. Parpot, P.; Muiuane, V.P.; Defontaine, V.; Bettencourt, A.P. Electrocatalytic oxidation of readily available disaccharides in alkaline medium at gold electrode. *Electrochim. Acta* **2010**, *55*, 3157–3163. [CrossRef]

21. Torto, N.; Ruzgas, T.; Gorton, L. Electrochemical oxidation of mono- and disaccharides at fresh as well as oxidized copper electrodes in alkaline media. *J. Electroanal. Chem.* **1999**, *464*, 252–258. [CrossRef]

22. Parpot, P.; Nunes, N.; Bettencourt, A.P. Electrocatalytic oxidation of monosaccharides on gold electrode in alkaline medium: Structure-reactivity relationship. *J. Electroanal. Chem.* **2006**, *596*, 65–73. [CrossRef]

23. Liaigre, D.; Breton, T.; Belgsir, E.M. Kinetic and selectivity control of TEMPO electro-mediated oxidation of alcohols. *Electrochem. commun.* **2005**, *7*, 312–316. [CrossRef]

24. Schäfer, H.J.; Harenbrock, M.; Klocke, E.; Plate, M.; Weiper-Idelmann, A. Electrolysis for the benign conversion of renewable feedstocks. *Pure Appl. Chem.* **2007**, *79*, 2047–2057. [CrossRef]

25. Druliolle, H.; Kokoh, K.B.; Beden, B. Electro-oxidation of lactose on platinum and on modified platinum electrodes in alkaline medium. *Electrochim. Acta* **1994**, *39*, 2577–2584. [CrossRef]

26. Druliolle, H.; Kokoh, K.B.; Beden, B. Selective oxidation of lactose to lactobionic acid on lead-adatoms modified platinum electrodes in Na_2CO_3 + $NaHCO_3$ buffered medium. *J. Electroanal. Chem.* **1995**, *385*, 77–83. [CrossRef]

27. Shao, M.-J.; Xing, X.-K.; Liu, C.-C. Cyclic voltammetric study of glucose oxidation on an oxide-covered platinum electrode in the presence of an underpotential-deposited thallium layer. *Bioelectrochemistry Bioenerg.* **1987**, *17*, 59–70. [CrossRef]

28. Papoutsis, A.; Kokkinidis, G. Electrocatalytic reduction of nitro compounds on gold upd modified electrodes. *J. Electroanal. Chem.* **1994**, *371*, 231–239. [CrossRef]

29. Kokkinidis, G.; Papanastasiou, G. Some specific features of the catalytic role of upd foreign metal adatoms and water on the anodic oxidation of absolute methanol on Pt. *J. Electroanal. Chem.* **1987**, *221*, 172–186. [CrossRef]

30. Kokkindis, G.; Leger, J.M.; Lamy, C. Structural effects in electrocatalysis. oxidation of d-glucose on Pt (100), (110) and (111) single crystal electrodes and the effect of upd adlayers of Pb, Tl and Bi. *J. Electroanal. Chem.* **1988**, *242*, 221–242. [CrossRef]

31. Druliolle, H.; Kokoh, K.B.; Hahn, F.; Lamy, C.; Beden, B. On some mechanistic aspects of the electrochemical oxidation of lactose at platinum and gold electrodes in alkaline medium. *J. Electroanal. Chem.* **1997**, *426*, 103–115. [CrossRef]

32. Larew, L.A.; Johnson, D.C. Concentration Dependence of the mechanism of glucose oxidation at gold electrodes in alkaline media. *J. Electroanal. Chem.* **1989**, *262*, 167–182. [CrossRef]

33. Kokoh, K.B.; Alonso-Vante, N. Electrocatalytic oxidation of lactose on gold nanoparticle modified carbon in carbonate buffer. *J. Appl. Electrochem.* **2006**, *36*, 147–151. [CrossRef]

34. Lee, D.; Taylor, M.S. Catalyst-controlled regioselective reactions of carbohydrate derivatives. *Synthesis (Stuttg)* **2012**, *44*, 3421–3431. [CrossRef]

35. Mehtiö, T.; Toivari, M.; Wiebe, M.G.; Harlin, A.; Penttilä, M.; Koivula, A. Production and applications of carbohydrate-derived sugar acids as generic biobased chemicals. *Crit. Rev. Biotechnol.* **2016**, *36*, 904–916. [CrossRef]

36. Esteban, J.; Yustos, P.; Ladero, M. Catalytic Processes from biomass-derived hexoses and pentoses: A recent literature overview. *Catalysts* **2018**, *8*, 637. [CrossRef]
37. Mangiameli, M.F.; González, J.C.; Bellú, S.; Bertoni, F.; Sala, L.F. Redox and complexation chemistry of the crvi/cr v-d-glucaric acid system. *Dalt. Trans.* **2014**, *43*, 9242–9254. [CrossRef]
38. Lob, W. Zur Kenntnis Der Elektrolyse Des Traubenzuckers, Glyzerins Und Glykols. *Zeitschrift Elektrotechnik Elektrochemie* **1910**, *16*, 1–9. [CrossRef]
39. Hay, G.W.; Smith, F. Electrolysis of low molecular weight carbohydrates in non-aqueous media. i. the products of electrolysis of monosaccharides. *Can. J. Chem.* **1969**, *47*, 417–421. [CrossRef]
40. Isbell, H.S.; Frush, H.L. The Oxidation of Sugars. I. The electrolytic oxidation of aldose sugars in the presence of a bromide and calcium carbonate. *Bur. Stand. J. Res.* **1931**, *6*, 1145. [CrossRef]
41. Isbell, H.S.; Frush, H.L.; Bates, F.J. Manufacture of calcium gluconate by electrolytic oxidation of dextrose. *Ind. Eng. Chem.* **1932**, *24*, 375–378. [CrossRef]
42. Isbell, H.S.; Frush, H.L. Preparation and properties of aldonic acids and their lactones and basic calcium salts. *Bur. Stand. J. Res.* **1933**, *11*, 649. [CrossRef]
43. Isbell, H.S.; Frush, H.L. Electrolytic oxidation of xylose in the presence of alkaline earth bromides and carbonates. *J. Res. Natl. Bur. Stand.* **1935**, *14*, 359. [CrossRef]
44. Fink, C.G.; Summers, D.B. Electrolytic preparation of calcium gluconate and other salts of aldonic acids. *Trans. Electrochem. Soc.* **1938**, *74*, 625. [CrossRef]
45. Cook, E.W.; Major, R.T. Preparation of 5-Keto-gluconic acid by bromine oxidation. *J. Am. Chem. Soc.* **1935**, *57*, 773. [CrossRef]
46. Jorna, A.M.J.; Boelrijk, A.E.M.; Hoorn, H.J.; Reedijk, J. Heterogenization of a ruthenium catalyst on silica and its application in alcohol oxidation and stilbene epoxidation. *React. Funct. Polym.* **1996**, *29*, 101–114. [CrossRef]
47. Bamba, K.; Léger, J.M.; Garnier, E.; Bachmann, C.; Servat, K.; Kokoh, K.B. Selective electro-oxidation of d-glucose by rucl2(azpy) 2 complexes as electrochemical mediators. *Electrochim. Acta* **2005**, *50*, 3341–3346. [CrossRef]
48. Bailar, J.C. "Heterogenizing" homogeneous catalysts. *Catal. Rev.* **1974**, *10*, 17–36. [CrossRef]
49. Hardee, J.R.; Tunney, S.E.; Frye, J.; Stille, J.K. Synthesis and incorporation of 4-triethoxysilyl-1,2-bis(diphenylphosphino)benzene into silica gel and solgel polymers: Preparation and utilization of recyclable catalysts for alkene hydroformylation and methanol homologation. *J. Polym. Sci. Part A Polym. Chem.* **1990**, *28*, 3669–3684. [CrossRef]
50. Kaim, W. Complexes with 2,2′-Azobispyridine and related "s-frame" bridging ligands containing the azo function. *Coord. Chem. Rev.* **2001**, *219–221*, 463–488. [CrossRef]
51. Geneste, F.; Moinet, C.; Jezequel, G. First covalent attachment of a polypyridyl ruthenium complex on a graphite felt electrode. *New J. Chem.* **2002**, *26*, 1539–1541. [CrossRef]
52. Boelrijk, A.E.M.; Jorna, A.M.J.; Reedijk, J. Oxidation of octyl α-d-glucopyranoside to octyl α-d-glucuronic acid, catalyzed by several ruthenium complexes, containing a 2-(phenyl)azopyridine or a 2-(nitrophenyl)azopyridine ligand. *J. Mol. Catal. A Chem.* **1995**, *103*, 73–85. [CrossRef]
53. Thaburet, J.F.; Merbouh, N.; Ibert, M.; Marsais, F.; Queguiner, G. TEMPO-mediated oxidation of maltodextrins and d-glucose: Effect of ph on the selectivity and sequestering ability of the resulting polycarboxylates. *Carbohydr. Res.* **2001**, *330*, 21–29. [CrossRef]
54. Merbouh, N.; Francois Thaburet, J.; Ibert, M.; Marsais, F.; Bobbitt, J.M. Facile nitroxide-mediated oxidations of d-glucose to d-glucaric acid. Bobbitt, J.M. facile nitroxide-mediated oxidations of d-glucose to d-glucaric acid. *Carbohydr. Res.* **2001**, *336*, 75–78. [CrossRef]
55. Ibert, M.; Marsais, F.; Merbouh, N.; Brückner, C. Determination of the side-products formed during the nitroxide-mediated bleach oxidation of glucose to glucaric acid. *Carbohydr. Res.* **2002**, *337*, 1059–1063. [CrossRef]
56. Marsais, F.; Feasson, C.; Fiol-Petit, C.; Merbouh, N.; Fuertès, P.; Ibert, M. improved preparative electrochemical oxidation of d-glucose to d-glucaric acid. *Electrochim. Acta* **2010**, *55*, 3589–3594. [CrossRef]
57. Ibert, M.; Fuerts, P.; Merbouh, N.; Feasson, C.; Marsais, F. Evidence of benzilic rearrangement during the electrochemical oxidation of d-glucose to d-glucaric acid. *Carbohydr. Res.* **2011**, *346*, 512–518. [CrossRef]
58. Vassilyev, Y.B.; Khazova, O.A.; Nikolaeva, N.N. Kinetics and mechanism of glucose electrooxidation on different electrode-catalysts: Part I. Adsorption and oxidation on platinum. *J. Electroanal. Chem.* **1985**, *196*, 105–125. [CrossRef]

59. Vassilyev, Y.B.; Khazova, O.A.; Nikolaeva, N.N. Kinetics and mechanism of glucose electrooxidation on different electrode-catalysts part ii. effect of the nature of the electrode and the electrooxidation mechanism. *J. Electroanal. Chem.* **1985**, *196*, 127–144. [CrossRef]

60. Neuburger, G.G.; Johnson, D.C. Pulsed amperometric detection of carbohydrates at gold electrodes with a two-step potential waveform. *Anal. Chem.* **1987**, *59*, 150–154. [CrossRef]

61. Neuburger, G.G.; Johnson, D.C. Comparison of the pulsed amperometric detection of carbohydrates at gold and platinum electrodes for flow injection and liquid chromatographic systems. *Anal. Chem.* **1987**, *59*, 203–204. [CrossRef] [PubMed]

62. Neuburger, G.G.; Johnson, D.C. Pulsed coulometric detection with automatic rejection of background signal in surface-oxide-catalyzed anodic detections at gold electrodes in flow-through cells. *Anal. Chem.* **1988**, *60*, 2288–2293. [CrossRef]

63. Belgsir, E.M.; Bouhier, E.; Essis Yei, H.; Kokoh, K.B.; Beden, B.; Huser, H.; Leger, J.-M.; Lamy, C. Electrosynthesis in aqueous medium: A kinetic study of the electrocatalytic oxidation of oxygenated organic molecules. *Electrochim. Acta* **1991**, *36*, 1157–1164. [CrossRef]

64. Kokoh, K.B.; Léger, J.M.; Beden, B.; Lamy, C. "On line" chromatographic analysis of the products resulting from the electrocatalytic oxidation of d-glucose on Pt, Au and adatoms modified Pt electrodes-Part I. Acid and neutral media. *Electrochim. Acta* **1992**, *37*, 1333–1342. [CrossRef]

65. Kokoh, K.B.; Léger, J.-M.; Beden, B.; Huser, H.; Lamy, C. "On line" chromatographic analysis of the products resulting from the electrocatalytic oxidation of d-glucose on pure and adatoms modified Pt and au electrodes—part II. alkaline medium. *Electrochim. Acta* **1992**, *37*, 1909–1918. [CrossRef]

66. .De Mele, M.L.F.; Videla, H.A.; Arvia, A.J. The influence of glucose and electrolyte composition on the voltammetric response and open-circuit potential decay of bright polycrystalline gold electrodes. *Bioelectrochem. Bioenerg.* **1986**, *16*, 213–233. [CrossRef]

67. Adzic, R.R.; Hsiao, M.W.; Yeager, E.B. Electrochemical oxidation of glucose on single crystal gold surfaces. *J. Electroanal. Chem.* **1989**, *260*, 475–485. [CrossRef]

68. Kokoh, K.B.; Parpot, P.; Belgsir, E.M.; Léger, J.-M.; Beden, B.; Lamy, C. Selective oxidation of d-gluconic acid on platinum and lead adatoms modified platinum electrodes in alkaline medium. *Electrochim. Acta* **1993**, *38*, 1359–1365. [CrossRef]

69. Bamba, K.; Kokoh, K.B.; Servat, K.; Léger, J.M. Electrocatalytic oxidation of monosaccharides on platinum electrodes modified by thallium adatoms in carbonate buffered medium. *J. Appl. Electrochem.* **2006**, *36*, 233–238. [CrossRef]

70. Largeaud, F.; Kokoh, K.B.; Beden, B.; Lamy, C. On the electrochemical reactivity of anomers: Electrocatalytic oxidation of α- and β-d-glucose on platinum electrodes in acid and basic media. *J. Electroanal. Chem.* **1995**, *397*, 261–269. [CrossRef]

71. Beden, B.; Largeaud, F.; Kokoh, K.B.; Lamy, C. Fourier transform infrared reflectance spectroscopic investigation of the electrocatalytic oxidation of d-glucose: Identification of reactive intermediates and reaction products. *Electrochim. Acta* **1996**, *41*, 701–709. [CrossRef]

72. Moggia, G.; Kenis, T.; Daems, N.; Breugelmans, T. Electrochemical oxidation of d-glucose in alkaline medium: Impact of oxidation potential and chemical side reactions on the selectivity to d-gluconic and d-glucaric acid. *ChemElectroChem* **2020**, *7*, 86–95. [CrossRef]

73. Holade, Y.; Servat, K.; Napporn, T.W.; Morais, C.; Berjeaud, J.M.; Kokoh, K.B. Highly selective oxidation of carbohydrates in an efficient electrochemical energy converter: Cogenerating organic electrosynthesis. *ChemSusChem* **2016**, *9*, 252–263. [CrossRef] [PubMed]

74. Holade, Y.; Engel, A.B.; Servat, K.; Napporn, T.W.; Morais, C.; Tingry, S.; Cornu, D.; Boniface Kokoh, K. Electrocatalytic and electroanalytic investigation of carbohydrates oxidation on gold-based nanocatalysts in alkaline and neutral phs. *J. Electrochem. Soc.* **2018**, *165*, H425–H436. [CrossRef]

75. Rafaïdeen, T.; Baranton, S.; Coutanceau, C. Highly efficient and selective electrooxidation of glucose and xylose in alkaline medium at carbon supported alloyed pdau nanocatalysts. *Appl. Catal. B Environ.* **2019**, *243*, 641–656. [CrossRef]

76. Governo, A.T.; Proença, L.; Parpot, P.; Lopes, M.I.S.; Fonseca, I.T.E. Electro-oxidation of d-xylose on platinum and gold electrodes in alkaline medium. *Electrochim. Acta* **2004**, *49*, 1535–1545. [CrossRef]

77. Singh, A.K.; Chopra, D.; Rahmani, S.; Singh, B. Kinetics and mechanism of Pd(II) catalysed oxidation of d-arabinose, d-xylose and d-galactose by n-bromosuccinimide in acidic solution. *Carbohydr. Res.* **1998**, *314*, 157–160. [CrossRef]
78. Williams, S.R.; Maynard, H.D.; Chmelka, B.F. Synthesis of 13C-enriched pyrrole from 2-13C d-galactose. *J. Label. Compd. Radiopharm.* **1999**, *42*, 927–936. [CrossRef]
79. Parpot, P.; Pires, S.G.; Bettencourt, A.P. Electrocatalytic oxidation of d-galactose in alkaline medium. *J. Electroanal. Chem.* **2004**, *566*, 401–408. [CrossRef]
80. Luczak, T.; Holze, R.; Beltowska-Brzezinska, M. The oxidation of cyclic diols on a gold elecrode structure-reactivity relationships. *Electroanalysis* **1994**, *6*, 773–778. [CrossRef]
81. Parpot, P.; Santos, P.R.B.; Bettencourt, A.P. Electro-oxidation of d-mannose on platinum, gold and nickel electrodes in aqueous medium. *J. Electroanal. Chem.* **2007**, *610*, 154–162. [CrossRef]
82. Luo, P.; Prabhu, S.V.; Baldwin, R.P. Constant potential amperometric detection at a copper-based electrode: Electrode formation and operation. *Anal. Chem.* **1990**, *62*, 752–755. [CrossRef]
83. Torto, N. Recent progress in electrochemical oxidation of saccharides at gold and copper electrodes in alkaline solutions. *Bioelectrochemistry* **2009**, *76*, 195–200. [CrossRef]
84. Mho, S.I.; Johnson, D.C. Electrocatalytic response of carbohydrates at copper-alloy electrodes. *J. Electroanal. Chem.* **2001**, *500*, 524–532. [CrossRef]
85. Luo, M.Z.; Baldwin, R.P. Characterization of carbohydrate oxidation at copper electrodes. *J. Electroanal. Chem.* **1995**, *387*, 87–94. [CrossRef]
86. Kano, K.; Torimura, M.; Esaka, Y.; Goto, M.; Ueda, T. Electrocatalytic oxidation of carbohydrates at copper(II) -modified electrodes and its application to flow-through detection. *J. Electroanal. Chem.* **1994**, *372*, 137–143. [CrossRef]
87. Yamazaki, S.I.; Fujiwara, N.; Takeda, S.; Yasuda, K. Electrochemical oxidation of sugars at moderate potentials catalyzed by Rh porphyrins. *Chem. Commun.* **2010**, *46*, 3607–3609. [CrossRef]
88. Bin, D.; Wang, H.; Li, J.; Wang, H.; Yin, Z.; Kang, J.; He, B.; Li, Z. Controllable oxidation of glucose to gluconic acid and glucaric acid using an electrocatalytic reactor. *Electrochim. Acta* **2014**, *130*, 170–178. [CrossRef]
89. Liu, W.J.; Xu, Z.; Zhao, D.; Pan, X.Q.; Li, H.C.; Hu, X.; Fan, Z.Y.; Wang, W.K.; Zhao, G.H.; Jin, S.; et al. Efficient electrochemical production of glucaric acid and H2 via glucose electrolysis. *Nat. Commun.* **2020**, *11*, 1–11. [CrossRef]
90. van Velthuijsen, J.A. Food additives derived from lactose: Lactitol and lactitol palmitate. *J. Agric. Food Chem.* **1979**, *27*, 680–686. [CrossRef]
91. Gallezot, P.; Cerino, P.J.; Blanc, B.; Flèche, G.; Fuertes, P. Glucose hydrogenation on promoted raney-nickel catalysts. *J. Catal.* **1994**, *146*, 93–102. [CrossRef]
92. Creighton, H.J. The electrochemical reduction of sugars. *Trans. Electrochem. Soc.* **1939**, *75*, 289. [CrossRef]
93. Wolfrom, M.L.; Konigsberg, M.; Moody, F.B.; Goepp, R.M. Sugar interconversion under reducing conditions. I. *J. Am. Chem. Soc.* **1946**, *68*, 122–126. [CrossRef]
94. Wolfrom, M.L.; Moody, F.B.; Konigsberg, M.; Goepp, R.M. Sugar interconversion under reducing conditions. II. *J. Am. Chem. Soc.* **1946**, *68*, 578–580. [CrossRef]
95. Wolfrom, M.L.; Lew, B.W.; Goepp, R.M. Sugar interconversion under reducing conditions. III. *J. Am. Chem. Soc.* **1946**, *68*, 1443–1448. [CrossRef]
96. Wolfrom, M.L.; Lew, B.W.; Hales, R.A.; Goepp, R.M. Sugar interconversion under reducing conditions. IV. D,L-glucitol. *J. Am. Chem. Soc.* **1946**, *68*, 2342–2343. [CrossRef]
97. Wolfrom, M.L.; Binkley, W.W.; Spencer, C.C.; Lew, B.W. A dodecitol from the alkaline electroreduction of d-glucose. *J. Am. Chem. Soc.* **1951**, *73*, 3357–3358. [CrossRef]
98. Bin Kassim, A.; Rice, C.L.; Kuhn, A.T. Formation of sorbitol by cathodic reduction of glucose. *J. Appl. Electrochem.* **1981**, *11*, 261–267. [CrossRef]
99. Fei, S.; Chen, J.; Yao, S.; Deng, G.; Nie, L.; Kuang, Y. Electroreduction of α-glucose on CNT/graphite electrode modified by Zn and Zn-Fe alloy. *J. Solid State Electrochem.* **2005**, *9*, 498–503. [CrossRef]
100. Kwon, Y.; Koper, M.T.M. Electrocatalytic hydrogenation and deoxygenation of glucose on solid metal electrodes. *ChemSusChem* **2013**, *6*, 455–462. [CrossRef]
101. Kwon, Y.; Birdja, Y.; Spanos, I.; Rodriguez, P.; Koper, M.T.M. Highly selective electro-oxidation of glycerol to dihydroxyacetone on platinum in the presence of bismuth. *ACS Catal.* **2012**, *2*, 759–764. [CrossRef]

102. Owobi-Andely, Y.; Fiaty, K.; Laurent, P.; Bardot, C. Use of electrocatalytic membrane reactor for synthesis of sorbitol. *Catal. Today* **2000**, *56*, 173–178. [CrossRef]
103. Piszczek, L.; Gieldon, K.; Wojtkowiak, S. Galvanostatic electrochemical reduction of pentoses. *J. Appl. Electrochem.* **1992**, *22*, 1055–1059. [CrossRef]
104. Hamann, C.H.; Fischer, S.; Polligkeit, H.; Wolf, P. The alkylation of mono-and disaccharides via an initialising electrochemical step. *J. Carbohydr. Chem.* **1993**, *12*, 173–190. [CrossRef]
105. Hamann, C.H.; Koch, R.; Pleus, S. New results on the benzylation of saturated mono- and disaccharides-a semiempirical study. *J. Carbohydr. Chem.* **2002**, *21*, 53–63. [CrossRef]
106. Park, K.; Pintauro, P.N.; Baizer, M.M.; Nobe, K. Flow reactor studies of the paired electro-oxidation and electroreduction of glucose. *J. Electrochem. Soc.* **1985**, *132*, 1850–1855. [CrossRef]
107. Park, K.; Pintauro, P.N.; Baizer, M.M.; Nobe, K. Current efficiencies and regeneration of poisoned raney nickel in the electrohydrogenation of glucose to sorbitol. *J. Appl. Electrochem.* **1986**, *16*, 941–946. [CrossRef]
108. Jokic, A.; Ristic, N.; Jaksic, M.M.; Spasojevic, M.; Krstajic, N. Simultaneous electrolytic production of xylitol and xylonic acid from xylose. *J. Appl. Electrochem.* **1991**, *21*, 321–326. [CrossRef]
109. Kawamata, Y.; Baran, P.S. Electrosynthesis: Sustainability is not enough. *Joule* **2020**, 1–4. [CrossRef]

Article

Biocomposites of Bio-Polyethylene Reinforced with a Hydrothermal-Alkaline Sugarcane Bagasse Pulp and Coupled with a Bio-Based Compatibilizer

Nanci Vanesa Ehman [1,*], Diana Ita-Nagy [2], Fernando Esteban Felissia [1], María Evangelina Vallejos [1], Isabel Quispe [2], María Cristina Area [1] and Gary Chinga-Carrasco [3,*]

1   IMAM, UNaM, CONICET, FCEQYN, Programa de Celulosa y Papel (PROCYP), Misiones, Félix de Azara 1552, Posadas, Argentina; ffelissia@gmail.com (F.E.F.); mariaxvallejos@gmail.com (M.E.V.); cristinaarea@gmail.com (M.C.A.)
2   Peruvian LCA and Industrial Ecology Network (PELCAN), Department of Engineering, Pontificia Universidad Católica del Perú (PUCP), 1801 Avenida Universitaria, San Miguel, Lima 15088, Peru; dita@pucp.edu.pe (D.I.-N.); iquispe@pucp.edu.pe (I.Q.)
3   RISE PFI, NO-7491 Trondheim, Norway
*   Correspondence: nanciehman@gmail.com (N.V.E.); gary.chinga.carrasco@rise-pfi.no (G.C.-C.); Tel.: +47-908-36-045 (G.C.-C.)

Academic Editors: Alejandro Rodríguez, Eduardo Espinosa and Sylvain Caillol
Received: 2 April 2020; Accepted: 30 April 2020; Published: 5 May 2020

Abstract: Bio-polyethylene (BioPE, derived from sugarcane), sugarcane bagasse pulp, and two compatibilizers (fossil and bio-based), were used to manufacture biocomposite filaments for 3D printing. Biocomposite filaments were manufactured and characterized in detail, including measurement of water absorption, mechanical properties, thermal stability and decomposition temperature (thermo-gravimetric analysis (TGA)). Differential scanning calorimetry (DSC) was performed to measure the glass transition temperature (Tg). Scanning electron microscopy (SEM) was applied to assess the fracture area of the filaments after mechanical testing. Increases of up to 10% in water absorption were measured for the samples with 40 wt% fibers and the fossil compatibilizer. The mechanical properties were improved by increasing the fraction of bagasse fibers from 0% to 20% and 40%. The suitability of the biocomposite filaments was tested for 3D printing, and some shapes were printed as demonstrators. Importantly, in a cradle-to-gate life cycle analysis of the biocomposites, we demonstrated that replacing fossil compatibilizer with a bio-based compatibilizer contributes to a reduction in CO_2-eq emissions, and an increase in CO_2 capture, achieving a CO_2-eq storage of 2.12 kg CO_2 eq/kg for the biocomposite containing 40% bagasse fibers and 6% bio-based compatibilizer.

Keywords: bio-based filament; 3D printing; sugarcane bagasse pulp

1. Introduction

Three-dimensional (3D) printing allows the manufacturing of custom pieces that usually demand higher costs and production time when manufactured by conventional methods. In addition, it offers unparalleled flexibility in achieving controlled composition, geometric shape, functions, and complexity [1]. Numerous studies about ink formulations for 3D printing have shown that bioplastics are promising materials for 3D printing applications [2–6].

Bioplastics are obtained from first-generation (1G) resources (starches or sugars, such as corn, sugarcane, wheat, and soy) or second-generation (2G) resources (cellulose from crops or industrial processes). One of the bioplastics that are expected to grow greatly in the next years is bio-polyethylene (BioPE) 1G for flexible and rigid packaging applications [7].

The main advantage of BioPE with respect to traditional polyethylene (PE) (fossil-based) is the utilization of renewable raw materials to reduce greenhouse gas emissions [8]. However, BioPE has relatively low mechanical properties (modulus and strength) compared to other commodity materials. Hence, the inclusion of dispersed phases (e.g., cellulosic fibers) with high stiffness and strength is commonly applied to enhance its mechanical properties [9]. Studies evaluating the effect of natural fibers include raw materials such as wood sawdust, bleached and unbleached pine pulps [10], eucalyptus pulp [11], and nonwoods such as bamboo fibers [12], flax [13] and kenaf [14]. The previous studies showed that the addition of fibers or pulps increased the tensile strength and Young's modulus and caused some variations in thermal properties.

Sugarcane bagasse is a lignocellulosic agro-industrial waste generated by the sugar and alcohol industries and is usually burned in the sugar mill. The valorization of sugarcane bagasse implies its integral use in a lignocellulosic biorefinery scheme [15]. Figure 1 shows different routes to obtain products with an integrated approach concept. Hydrothermal treatment allowed the extraction of the hemicelluloses fraction to produce polyhydroxialkanoates (PHAs) [16], which are a group of bioplastics. The solid fraction after treatments contains sugarcane bagasse fibers. Fibers can take two alternative routes: they can be subjected to an enzymatic treatment followed by fermentation, chemical conversion, and finally, polymerization to obtain second generation BioPE [17], or they can be used as reinforcement for biocomposites and 3D printing [18].

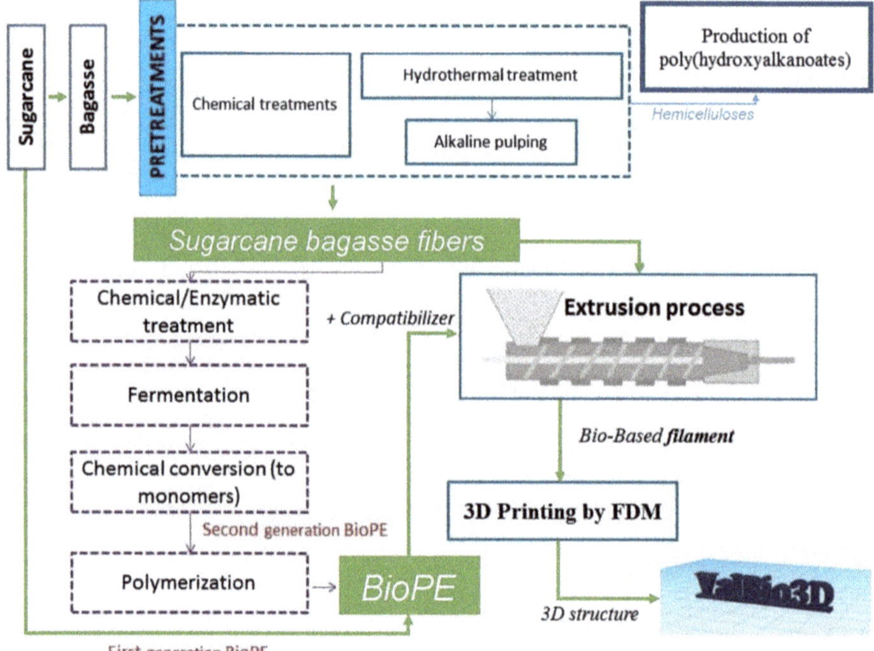

Figure 1. Scheme for the integral use of sugarcane, including the production of Bio-polyethylene (BioPE) and bagasse fibers.

A good interfacial bonding is required to achieve optimum reinforcement. Hence, the bonding between fibers and matrix plays a vital role in determining the mechanical properties of the biocomposites. For plant-based fiber composites, there is limited interaction between the hydrophilic fiber and the usually hydrophobic matrices, limiting their mechanical performance [19]. This effect could be reduced by introducing a compatibilizer. The most common compatibilizer used for the combination of BioPE with fibers is the maleic anhydride (MA). MA is grafted on polyethylene (PE) to form maleate polyethylene (MAPE), which is commercialized as a compatibilizer. MAPE covalently couples PE and fibers [20]. Studies that included 5 wt% of MAPE to coupling PE with bamboo fibers showed increases in flexural and tensile strength [21], respect to the composite without compatibilizer. Yuan et al. [22] used only 3 wt% of MAPE as compatibilizer between PE and fibers of maple wood and found that the use of the coupling agent increased the tensile strength. Tarrés et al. [23] found an optimal load of 6 wt% of MAPE to maximize the tensile properties of BioPE reinforced with thermo-mechanical pulp (TMP) fibers. Additionally, Mendez et al. [24] used the same load of MA-grafted polypropylene to evaluate the mechanical properties of polypropylene reinforced with groundwood pulp from pine. However, MAPE is conventionally derived from fossil resources; therefore, it does not contribute to obtaining a 100% bio-based product.

Recent studies have evaluated filaments of BioPE and wood fibers for 3D printing applications. Filgueira et al. [18] obtained filaments of BioPE with TMP fibers modified enzymatically. This approach reduced the water absorption of the filaments and improved the 3D printability of structures. Similarly, Tarrés et al. [23] obtained 3D structures by 3D printing (Fused Deposition Modeling, FDM) of BioPE biocomposites, evaluating the impact of different TMP fiber loads on the mechanical properties.

This work aims to assess the potential advantages of biocomposites filaments containing BioPE, sugarcane bagasse pulp and a bio-based compatibilizer (Figure 1), compared to neat BioPE or in combination with a fossil-based compatibilizer. For filaments composed of BioPE, two percentages of pulp fibers and two compatibilizers (fossil and bio-based) were manufactured. Mechanical properties and water absorption of the filaments were determined. Scanning electron microscopy (SEM) images from the fracture area of the tensile tested filaments were acquired. The thermal stability and decomposition temperature of the BioPE and biocomposites were determined with thermo-gravimetric analysis (TGA) and differential scanning calorimetry (DSC) and the glass transition temperature (Tg) was measured. The capability of the filaments to be extruded and deposited on a substrate to form a 3D shape was evaluated and demonstrated. Importantly, we performed a cradle-to-gate study to assess the impacts of replacing the fossil-based compatibilizer with a bio-based alternative.

2. Results

The chemical composition of the raw material and pulp are shown in Table 1.

Table 1. The chemical composition of raw material and hydrothermal treatment (HT)/Soda pulp (% oven-dried, od).

Chemical Composition (% od)	Sugarcane Bagasse	HT/Soda Pulp
Glucans	40.4 ± 0.04	90.8 ± 0.18
Hemicelluloses	35.0	5.1
Xylans	26.7 ± 0.32	5.1 ± 0.07
Arabinans	5.3 ± 0.28	-
Acetyl Groups	3.0 ± 0.02	-
Lignin	20.6 ± 0.12	2.1 ± 0.08
Extractives	3.2 ± 0.01	-

od: on oven dry base.

Bagasse, remaining from BioPE production, can be used to obtain sugarcane bagasse pulp, which can be further used to reinforce BioPE and thus close the loop in a biorefinery. In this study we applied a hydrothermal treatment (HT)/Soda treatment on the raw material to allow the release of the lignocellulosic fibers (HT/Soda fibers). Examples of milled HT/Soda fibers are provided in Figure 2. The quantification of the fiber morphology with a Fiber Tester device revealed an average fiber length of 367 μm, fiber width of 24.3 μm and a fraction of fines of 50%.

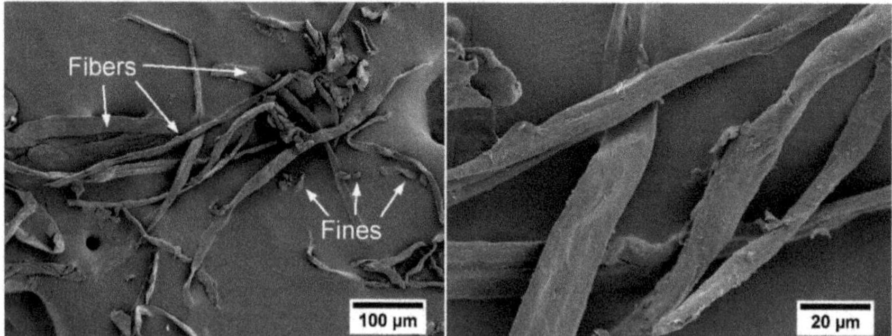

Figure 2. SEM images of HT/Soda fibers. Left) Fibers and fines are exemplified (magnification 200×). Right) Fibers observed at 1000× magnification.

Bagasse, remaining from BioPE production, can be used to obtain sugarcane bagasse pulp, which can be further used to reinforce BioPE and thus close the loop in a biorefinery. In this study we applied an HT/Soda treatment on the raw material to allow the release of the lignocellulosic fibers (HT/Soda fibers). Examples of milled HT/Soda fibers are provided in Figure 2. The quantification of the fiber morphology with a Fiber Tester device revealed an average fiber length of 367 µm, fiber width of 24.3 µm and a fraction of fines of 50%.

The glucans content increased from 40.4% oven dry (od) material to 90.8% od pulp after the HT/Soda treatments. The hemicelluloses are mainly eliminated during hydrothermal stage from 35.0% od material to 5.1% od pulp [25], whilst during alkaline treatment lignin is mainly dissolved, which is then eliminated with pulp liquor (from 20.6% to 2.1% od). HT/Soda pulp total yield and kappa number were 32.4% and 16.8%, respectively.

The HT/Soda fibers were compounded with BioPE and the corresponding compatibilizers to form filaments for 3D printing (Figure S1). Various properties of the filaments were assessed as we will explore in the next sections.

2.1. Water Absorption Behavior

Water absorption was determined by immersing the filaments in water for 7 days (Figure 3). The importance of this analysis is that the exposure of cellulose-reinforced biocomposites to moisture can cause the cellulosic elements to swell, a phenomenon that can lead to the weakening of the structure [26]. Several authors have studied the effect of hemicellulose content on water absorption and found that reducing the amount of hemicelluloses reduces the swelling capacity of the fibers [27–30]. The chemical composition for HT/Soda pulp used as reinforcement showed low hemicellulose content. Therefore, the increase in water absorption was expected to be relatively low.

ANOVA analysis indicated significant differences in water absorption values in respect to the fiber load ($p < 0.05$). The increase in fiber content in the filaments caused an increase in water absorption due to the hydrophilic nature of the fiber. Compared to the BioPE filament, the samples containing 20% and 40% fiber showed an increase in water absorption of approx. 6% and 10%, irrespective of the used compatibilizer. It is worth mentioning that the fibers applied in this study were not modified to increase their hydrophobicity. Thus, the increase in fiber content produces an increase in water absorption due to the hydrophilic and hygroscopic nature of the fibers used as reinforcement.

Previously, we have demonstrated that using laccase enzymes to graft gallate compounds on the surface of lignocellulosic fibers could reduce the water absorption considerably [18].

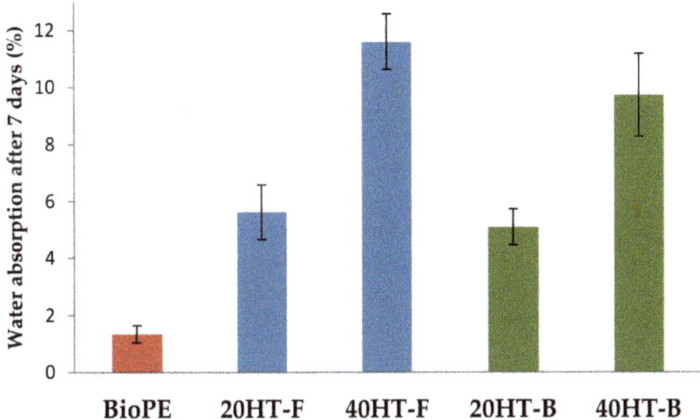

Figure 3. Water absorption of filaments with different percentages of fibers, using fossil- and bio-based compatibilizers.

An interesting feature observed was the presence of bubbles on the surface of all the fiber-containing filaments after 24 h of immersion in water. The presence of bubbles can be due to the roughness of the filament surface that generates empty spaces containing air, which is displaced by water. The effect of the filament roughness on the water absorption was previously demonstrated by Filgueira et al. [18], who found that a high filament roughness leads to a high specific surface area that increases the contact between the filament and water.

2.2. Mechanical Properties

The addition of fibers as reinforcement in polymer matrices yields an increase in mechanical properties, such as strength and stiffness. The mechanical properties of the filaments are presented in Table 2.

Table 2. Tensile strength (σ), Young's modulus (E), and elongation at maximum strength (εmáx) of filaments.

Code	σ (MPa)	E (MPa)	εmáx (%)
BioPE	20.4 ± 3.1	800 ± 40	23.9 ± 2.1
20HT-F	24.8 ± 1.8	970 ± 100	18.2 ± 2.1
40HT-F	33.0 ± 2.4	1480 ± 200	21.4 ± 2.9
20HT-B	23.8 ± 3.9	1020 ± 60	19.1 ± 1.8
40HT-B	32.3 ± 2.0	1280 ± 130	19.9 ± 1.8

Tensile properties provide useful information about the microstructure and the interface between the different materials [23]. The filaments with 40 wt% fibers presented the highest tensile strength, which was significantly different ($p < 0.01$) compared to the neat BioPE and the sample with 20% fiber fraction. For the two compatibilizers, increments of roughly 20% and 60% were found for fiber loads of 20 wt% and 40 wt%, respectively. Higher values of tensile strength were found by other authors using TMP fibers, reaching increments of up to 74%, adding 20% of fibers [23]. It was presumed that the quantified difference in tensile strength between our study and Tarrés et al. [19] was due to the length of the TMP and bagasse fibers. In this study the bagasse fibers were milled in order to ease the blending of the fibers and BioPE, with the equipment used in this study (Figure 2). However, Tarrés et al. [19] quantified the length of the TMP fibers in the biocomposite, and this was in the same range as the sugarcane bagasse fibers used in this study (~200–400 µm). Hence, the higher value in tensile strength obtained by Tarrés et al. [19] may be due to two factors: i) the morphology of the TMP fibers which are more fibrillated (larger fraction of split fibers) compared to the more intact chemical pulp fibers [31] and ii) a more effective compounding which probably leads to a more homogeneous fiber spatial distribution. It is expected that the more fibrillated TMP fibers have a large surface area which may facilitate the anchoring of the fibers in the BioPE matrix. Additionally, keep in mind that the mechanical testing in the present study was performed on filaments, while the quantification performed by Tarrés et al. [19] was based on injection molding samples. This is also expected to affect the mechanical performance of the biocomposite materials.

Similar to tensile strength, Young's modulus increased when increasing fiber loadings. Fibers are located between the polymer chains of the BioPE, reducing their mobility [24], and thus increasing the stiffness. The increment in stiffness with the addition of 20 wt% and 40 wt% fibers and both compatibilizers was about 25% and 75% compared to neat BioPE, respectively. The strain at break for all the produced biocomposites decreased with respect to the BioPE filament. An ANOVA analysis was performed to evaluate the effect of the compatibilizer on the mechanical properties. Statistical differences for the measured mechanical properties were not significant.

2.3. SEM Assessment

The SEM images of the filaments after the mechanical testing exhibit the fracture surface details. Figure 4a shows the enlarged image of the 40HT-F sample, where a large number of fibers are homogeneously distributed throughout the filament. A lower amount of fibers can be observed in the fracture area of the filament 20HT-F (Figure 4b). Besides, Figure 4b shows the high roughness of the filament surface. A similar effect with respect to fiber content was observed for the samples when a bio-based compatibilizer was used (40HT-B in Figure 4c and 20HT-B in Figure 4d). Most SEM images of the filaments showed that fibers were oriented longitudinally (0° in respect to tensile strength test direction). However, some samples presented several fibers distributed at different angles. Fiber orientation is important because it influences the final mechanical strength of filaments [32]. In addition to fiber orientation, the length of fibers is also affected by the manufacturing processes. Joffre 2014 found that fibers are drastically shortened during the manufacturing process and lose their interesting aspect ratios [33].

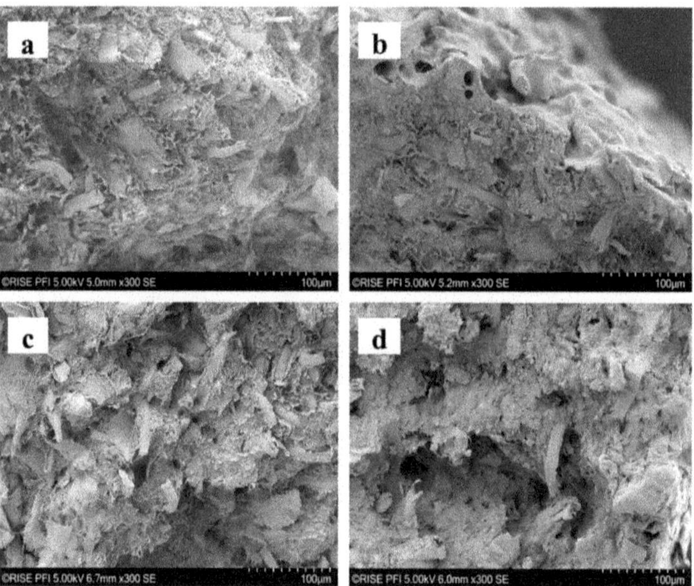

Figure 4. SEM microphotograph of the tensile breaking area: (**a**) 40HT-F filament, (**b**) 20HT-F filament, (**c**) 40HT-B filament, and (**d**) 20HT-B.

2.4. TGA and DSC

The TGA and DSC results are shown in Table 3. The BioPE filament degraded in a single step that started at 464.4 °C as shown in Figure 5a; this process takes place rapidly and the quantity of residue is very small (0.26%). However, the BioPE composites showed two-step decomposition, where the first starts at 332–356 °C (onset temperature range), corresponding to the cellulose decomposition (around 330 °C) [34]. The second step begins at a similar onset temperature (468–469 °C), involving a fast and significant degradation attributed to the thermal cracking of the hydrocarbon chains of BioPE, which ends approximately around 510 °C [35].

Table 3. Thermo-gravimetric analysis (TGA) and differential scanning calorimetry (DSC) analysis.

TGA Analysis			
Code	**T-Onset 1 (°C)**	**T-Onset 2 (°C)**	**Residue (%)**
BioPE	-	464	0.26
20HT-F	356	468	4.80
40HT-F	335	468	5.85
20HT-B	332	468	1.77
40HT-B	333	469	2.59

DSC Analysis				
Code	**Tm (°C)**	**ΔHm (J/g)**	**Tpeak 1 (°C)**	**Tpeak 2 (°C)**
BioPE	146	135	-	487
20HT-F	137	120	351	486
40HT-F	137	103	354	485
20HT-B	137	122	356	487
40HT-B	139	115	356	489

The DSC thermograph is given in Figure 5b. The BioPE filament presented a first peak (146 °C) which corresponds to the melting point of the crystalline domains of the BioPE. The Tm values of all the composite samples were the same (about 130 °C) regardless of the fiber and MAPE contents. This indicated that the size of the crystalline domains, which was directly related to the Tm, was retained in the matrix. However, the melting enthalpy decreased with the increase of the fiber fraction, as expected.

The second peak can be attributed to the degradation of the hydrocarbon chains of BioPE (488 °C). The BioPE biocomposites showed three peaks. The first and third peak (137–139 °C and 485–489 °C, respectively) are associated with the melting point and degradation of BioPE [35], whilst the second peak (351–357 °C) corresponds to the thermal degradation of the fraction of bagasse fiber reinforcement, mainly due to cellulose degradation.

The addition of fibers to the matrix decreased the onset decomposition temperature of the filaments in comparison to those of pure BioPE. The weight loss at the peak around 330 °C was higher for the biocomposite filaments with high fiber contents (40%). It was lower when the bio-based compatibilizer substituted the fossil compatibilizer. This is most probably due to the different composition of the bio-based compatibilizer (97% BioPE and 2% MA), compared to the fossil-based compatibilizer (<93% fossil PE and 7% MA).

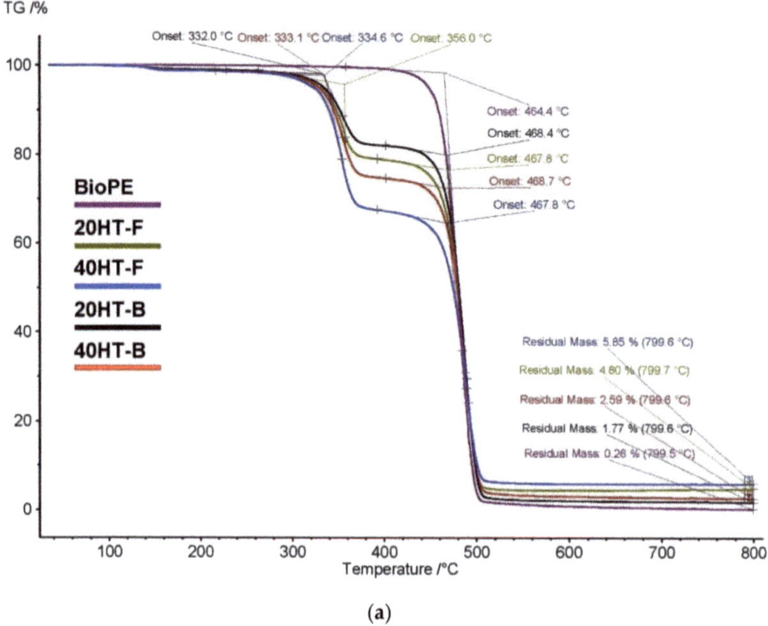

(a)

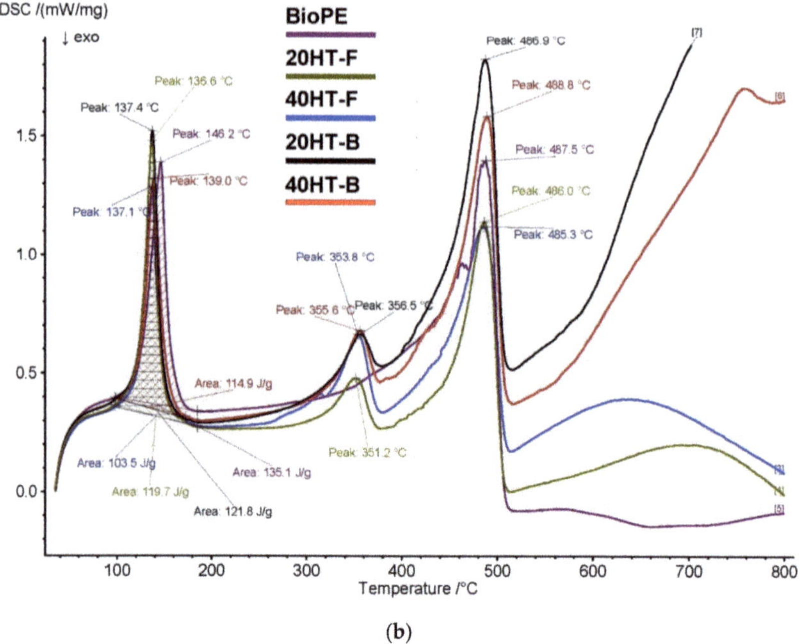

(b)

Figure 5. Thermo-gravimetric analysis (**a**) and differential scanning calorimetric (**b**) of the filaments.

2.5. Printability of the Filaments

It was previously demonstrated that the addition of TMP fibers to BioPE improved printability, yielding more homogeneous structures [6]. The capability of the filaments composed of BioPE and bagasse fibers for 3D printing by FDM was demonstrated in this study (Figure 6).

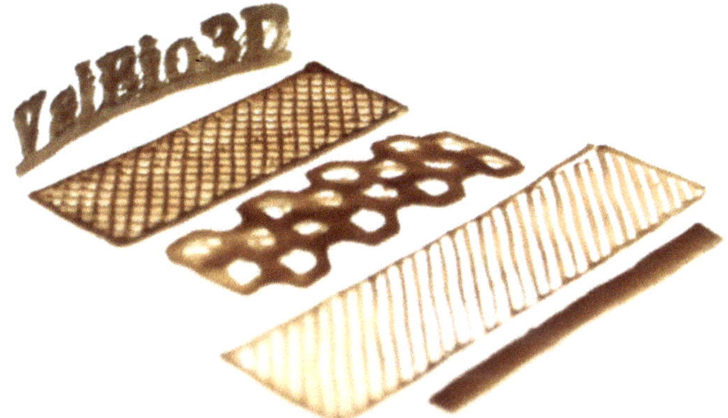

Figure 6. 3D shapes printed with filaments containing BioPE and bagasse fibers (ValBio-3D project).

Various shapes were modeled and printed, which exemplified the potential of the biocomposites for 3D printing by FDM technology. No remarkable difference was observed with respect to the printability of the different filaments, considering fiber content and type of compatibilizer (Supplementary Materials, Figure S2). It is worth emphasizing that in this study we have developed a recipe composed of 40% bagasse fibers, 54% BioPE and 6% bio-based compatibilizer. The bio-based compatibilizer was composed of 97% BioPE and only 2% MA. This implies that the biocomposite material that is printable contained less than 0.18% MA which was not bio-based. MA can also be derived from bio-based components and future initiatives could focus on developing MA from HMF, furans and furfural, that can be obtained from carbohydrates [36].

2.6. Environmental Aspects of Bagasse Fiber-Reinforced Biocomposites

The application of sugarcane bagasse pulp as filler in BioPE matrices for biocomposite production is mainly driven by the current need to use more environmentally friendly materials, taking into account the cost of raw materials and production processes. Thus, we performed a life cycle assessment (LCA) of bagasse fiber-reinforced BioPE pellets, comparing the environmental performance with pure sugarcane- and petroleum-based PE pellets, using as a FU 1 kg of (bio)plastic pellets. The utilization of HT bagasse fibers in a BioPE matrix decreased the effect on global warming, fossil resource scarcity, ozone formation, terrestrial acidification, and freshwater eutrophication, with respect to the use of 100% BioPE [37].

In this study, we complemented the LCA considering the comparison between the bio-based and fossil compatibilizers. The results showed reductions of 3% emissions of GHG when the biocomposite uses a bio-based compatibilizer (bMAPE) rather than the fossil one (fMAPE). Additionally, when increasing the amount of fibers from 20% to 40%, reductions of GHG emissions are more notable, reaching 18%, without considering the amount of carbon storage in the final polymer. Additionally, 20HT-F, 40HT-F, 20HT-B and 40HT-B resulted in an average CO_2-eq storage of 1.71 kg CO_2-eq/kg, 1.89 kg CO_2-eq/kg, 1.94 kg CO_2-eq/kg, and 2.12 kg CO_2-eq/kg respectively (see Table 4 and Figure 7). However, it is important to mention that this comparison is based on a replacement ratio of 1:1 between these biocomposites, which may not be the case during industrial production. Figure 8 showed that, when comparing the different stages of biocomposite production, around 40% of all impacts are related to the cultivation and harvesting stage, followed by the production of bioethylene from fermented bioethanol. Increasing the amount of fibers on the composition of the material, reductions are observed even after considering the impacts of processing raw bagasse to separate and obtain cellulose fibers. Furthermore, using a bio-based compatibilizer reduces the impacts of the production of the composite, but to a lesser extent.

Table 4. Environmental burdens per impact category for evaluated materials *.

Impact Category	Unit	Fossil PE	BioPE	20HT-F	40HT-F	20HT-B	40HT-B
Global warming without CO_2 capture	kg CO_2 eq	2.01	1.38	1.24	1.06	1.20	1.02
Global warming with CO_2 capture	kg CO_2 eq	2.01	−1.76	−1.71	−1.89	−1.94	−2.12
Ozone formation, Terrestrial ecosystems	kg NO_x eq	4.52×10^{-3}	8.40×10^{-3}	7.36×10^{-3}	6.56×10^{-3}	7.59×10^{-3}	6.78×10^{-3}
Terrestrial acidification	kg SO_2 eq	5.27×10^{-3}	2.17×10^{-2}	1.88×10^{-2}	1.69×10^{-2}	1.98×10^{-2}	1.78×10^{-2}
Freshwater eutrophication	kg P eq	2.79×10^{-5}	4.80×10^{-4}	4.05×10^{-4}	3.55×10^{-4}	4.30×10^{-4}	3.80×10^{-4}
Fossil resource scarcity	kg oil eq	1.57	0.41	0.43	0.38	0.36	0.31

* reported per functional unit: 1 kg of pellets.

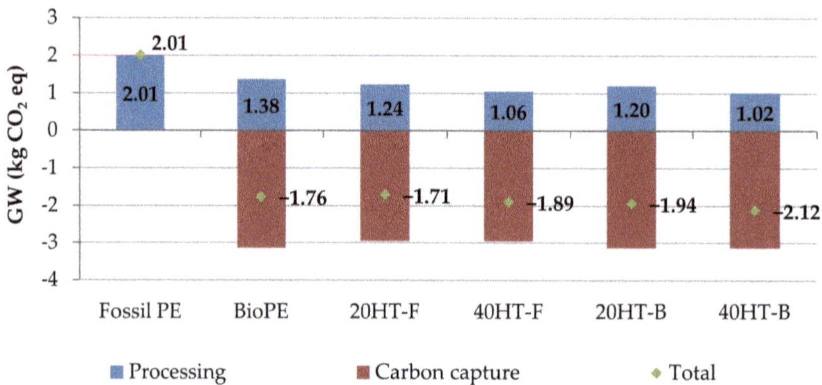

Figure 7. Greenhouse gas (GHG) emissions of 1 kg of fossil-based polyethylene (PE), BioPE and the corresponding biocomposites, from a cradle-to-gate perspective.

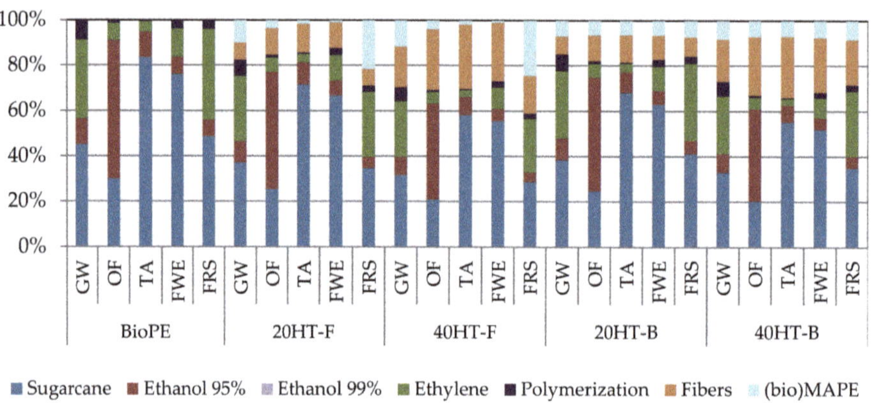

Figure 8. Relative contribution per life cycle stages for the production of fossil PE, BioPE and biocomposites. Results do not include carbon sequestration. Reported per functional unit: 1 kg of pellets.

When analyzing additional environmental impacts, it can be observed that, even though potential reductions of GHG emissions are obtained when utilizing bio-based MAPE instead of fossil MAPE, the environmental impact categories of ozone formation (OF), terrestrial acidification (TA) and freshwater eutrophication (FWE) show an increase of around 3% to 6%. In contrast, when comparing a biocomposite with higher fiber content, reduction on all impacts is observed.

Finally, we have demonstrated that the biocomposites developed in this study are suitable for manufacturing filaments for 3D printing. Similar recipes can be used for injection molding applications [23], which is a technology for high-volume manufacturing of e.g., automotive, furniture and packaging products.

3. Materials and Methods

3.1. Materials

Sugarcane bagasse was provided by a local mill and used in this study. A hydrothermal treatment (HT) with a liquid/bagasse ratio of 7/1, 180 °C, and 30 min at maximum temperature was performed, followed by a soda treatment using a liquid/bagasse ratio of 10/1; 170 °C, 60 min at maximum temperature, and 18% sodium hydroxide (NaOH) on oven-dry (od) bagasse (HT/Soda pulp). The determination of structural carbohydrates and lignin in biomass was carried out according to the NREL/TP–510–42618 procedure [38]. The carbohydrates were analyzed by high-performance liquid chromatography using a SHODEX SP810 (Showa Denko America, Inc., New York, NY, USA) column connected in series to a Bio-Rad (Hercules, CA, USA) deionizing pre-column. The chromatographic conditions were: water as eluent, a flow rate of 0.6 mL/min, 85 °C, and a refractive index detector. The sample was placed in a vial and frozen until the moment of analysis.

The polymer matrix was a sugarcane bio-based polyethylene (BioPE) provided by Braskem (Sao Paulo, Brazil). The BioPE had a relatively low melt flow index (MFI, 4.5 g/10 min) and a molecular weight of 92.9 g/mol.

Two compatibilizers were used to improve the compatibility between fibers and the BioPE. The fossil-based compatibilizer, maleic anhydride grafted polyethylene (denominated fMAPE), was provided in powder form by Clariant (product Licocene PE MA 4351, Clariant Plastics & Coatings (Nordic) AB, Malmö, Sweden). According to the supplier, the product is a metallocene-catalyzed PE with a grafted MA content of approx. 7 to 9 wt%. The bio-based compatibilizer was provided by YPAREX BV (Geleen, Netherlands). According to the supplier (YPAREX), the compatibilizer is derived from sugarcane BioPE with a bio-based content >97% and with a MA content <2% and was denominated bMAPE in this study.

3.2. Filaments Elaboration

The HT/Soda pulp was ground to a size that passed a 30 mesh sieve, and was dried for 1 h at 105 °C, whereas the BioPE and the bio-based compatibilizer were milled until passing a 10 mesh sieve.

The milled pulp fibers were assessed with a FiberTester device (L&W FiberTester Plus, Code 912. Software: Version 4.0–3.0, ABB AB/Lorentzen & Wettre, SE-164 93 Kista, Sweden). The fiber width, length and fraction of fines (objects smaller than 200 μm) were quantified. The results are based on 4037 objects that were quantified.

The pulps, BioPE, and compatibilizers were mixed as reported in Table 5. The series included two HT/Soda pulp contents (20% *w/w* and 40% *w/w*) and 6% *w/w* of each compatibilizer (fMAPE and bMAPE) loads. The fraction of compatibilizer (6 wt%) was selected following a previous optimization [23].

Table 5. Biocomposites composition.

Code	BioPE (wt%)	Fibers (wt%)	Compatibilizer (wt%)	Compatibilizer Type
BioPE	100	0	0	-
20HT-F	74	20	6	fMAPE
40HT-F	54	40	6	fMAPE
20HT-B	74	20	6	bMAPE
40HT-B	54	40	6	bMAPE

Filaments for 3D printing were manufactured as described by Filgueira et al. [18]. The blends were extruded twice in a Noztek Xcalibur filament extrusion system (Shoreham, UK). After the first extrusion the filaments were pelletized, and the pellets were used to extrude the final filament. The filament extruder had a single screw, and the filaments were extruded at a speed of 15 mm/s using three temperatures in the three sections of the extruder; 165 °C, 170 °C, and 175 °C. The speed of the extruder was determined to obtain an average filament diameter of 1.75 mm.

3.3. Mechanical Characterization of Filaments

Ten 60 mm length test specimens were used. The tensile mechanical properties of the filaments were measured with a Zwick/Roell (Ulm, Germany) universal tensile machine following ASTM D 5937–1996 (West Conshohocken, PE, USA). The crosshead speed was set at 100 mm/min with a 2.5 kN load cell. All tests were conducted at ambient temperature, and the means of 10 replicas were reported for each sample.

3.4. Water Absorption Experiments

Three filaments (60 mm) of each sample were dried for 4 h at 50 °C, and the initial weight was determined. Samples were immersed in containers with distilled water at 25 °C. After 7 days, the samples were taken out from the chambers and weighed using an analytical balance (precision of 0.1 mg). Water absorption was calculated according to the following equation:

$$\text{Water absorption } (\%) = \frac{(\text{Wh} - \text{Wo})}{\text{Wo}} \times 100 \tag{1}$$

where Wh is the weight of filaments after 7 days of immersion, and Wo is the initial weight of filaments.

3.5. SEM Observations

The morphology of milled fibers was assessed with scanning electron microscopy (SEM). Before SEM observation, the sample was sputter-coated with a thin layer of gold to avoid electrical charging. The applied working distance and acceleration voltage during image acquisition were 5–7 mm and 5 kV, respectively. The same settings were used to assess the fracture surfaces after tensile testing.

3.6. Thermo-Gravimetric Analysis (TGA) and Differential Scanning Calorimetry (DSC)

The thermal stability and decomposition temperature of the BioPE and biocomposites were determined with TGA, using Nitrogen gas, inert atmosphere, in a Netzsch Jupiter F3 equipment (Selb, Bavaria, Germany). The applied temperature range was from 30 °C to 800 °C with a heating rate of 10 °C/min. DSC experiments were performed to measure the glass transition temperature (Tg) using the same conditions and equipment.

3.7. 3D Printing

A REGEMAT 3D V1 printer (Granada, Spain) equipped with a 0.6 mm nozzle was used. The temperature of the nozzle was adjusted to 180 °C. The REGEMAT 3D Designer software (Version 1.0,

Granada, Spain) and TinkerCad (San Francisco, CA, USA.) were used for designing the 3D models. Various shapes were printed to demonstrate the potential of the filaments in 3D printing operations.

3.8. Statistical Analysis

Statistical analyses were performed using the Statgraphics Centurion XV software (Statgraphics Technologies, Inc., The Plains, VA, USA). ANOVA tests were applied at a significance level $p < 0.05$.

3.9. Impact Assessment

An impact assessment was developed considering the production of 1 kg of biocomposite pellets as the functional unit (FU). The environmental impacts of the studied materials were analyzed using the IPCC methodology 2013 for the GHG emissions considering a 100-year period [39]. The used methodology is currently the most robust and recommended model for the estimations of global warming potential [40,41]. Additionally, considering the potential impacts related to cultivation and harvesting of biomass, impact categories related to the use of agrochemicals were also considered. The additional categories include ozone formation (OF), terrestrial acidification (TA), freshwater eutrophication (FWE) and fossil resource scarcity (FRS) at the midpoint level. These categories were evaluated using the ReCiPe methodology (Table 6). For further details on the impact assessment see Supplementary Materials and the corresponding methodology described elsewhere [42–49].

Table 6. Environmental impact categories considered for the environmental assessment.

Methodology	Impact Category	Unit
IPCC 2013	Global warming 100 year (GW)	kg CO2 eq
	Ozone formation, terrestrial ecosystems (OF)	kg NOx eq
ReCiPe 2016	Terrestrial acidification (TA)	kg SO2 eq
	Freshwater eutrophication (FWE)	kg P eq
	Fossil resource scarcity (FRS)	kg oil eq

The evaluated materials in this section include a biocomposite containing 74% BioPE, 20% bagasse fibers (hydrothermal treatment) and 6% of fossil MAPE as a compatibilizer (20HT-F), a biocomposite containing 54% BioPE, 40% bagasse fibers and 6% of fossil MAPE (40HT-F), a biocomposite containing 74% BioPE, 20% bagasse fibers and 6% of bio-based MAPE (20HT-B) and a biocomposite containing 54% BioPE, 40% bagasse fibers and 6% of bio-based MAPE (40HT-B). Additionally, fossil PE and bioPE from sugarcane were also included as a baseline for comparison.

4. Conclusions

Filaments for 3D printing were manufactured using 100% bio-based PE, hydrothermal-soda sugarcane bagasse pulp, and bio- and fossil-based compatibilizers, demonstrating that the bagasse remaining from BioPE production can be used to obtain sugarcane bagasse pulp with adequate characteristics to reinforce BioPE, closing the loop in a biorefinery.

Increasing the fiber content caused an increase in the water absorption of the filament. The samples with 40 wt% fiber showed the highest water absorption compared to the neat BioPE filament (water absorption ~10%).

The filaments with 40 wt% fibers presented the highest tensile strengths. Increments in tensile strength of about 60% and 20% were found with fiber loads of 40 wt% and 20 wt%, respectively, using both compatibilizers. Similar behavior was found for stiffness values. The Young's modulus reached its highest value in the filaments with 40 wt% fibers. Compared with 100% BioPE filaments, the increment in stiffness was about 75% and 25% in filaments with 40 wt% and 20 wt% fibers and both compatibilizers, respectively. Lower elongations were obtained in all cases. For all the evaluated mechanical properties, no statistically significant differences were found between filaments with fossil or bio-based compatibilizers. According to TGA analysis, the effect of compatibilizers and fiber loads on the temperature of thermal degradation is similar. TGA showed that the weight loss at the peak around 330 °C was higher for the biocomposite filaments with high fiber contents (40%).

Importantly, we demonstrated that replacing fossil compatibilizer with a bio-based compatibilizer contributes to an increase in CO_2-capture, achieving a CO_2-eq storage of 2.12 kg CO_2-eq/kg for the biocomposite containing 40% bagasse fibers and 6% bio-based compatibilizer.

Supplementary Materials: The following are available online. Figure S1: Bagasse fibres, fracture areas of filaments (bagasse fibres and BioPE) and 3D printed tracks; Figure S2: Optical images of part of 3D printed layers. The images were acquired with an Epson Perfection scanner in transmission mode, 4800 dots per inch. From left to right) 20HT-F, 40HT-F, 20HT-B and 40HT-B; Figure S3: Graphical representation of the system boundaries for the production of sugarcane bioPE and fiber reinforced biocomposites. * Hemicelluloses are a by-product of the bagasse hydrothermal treatment.

Author Contributions: Conceptualization: N.V.E. and G.C.-C. Methodology: N.V.E., G.C.-C., D.I.-N. and I.Q. Investigation: N.V.E., D.I.-N., I.Q. and G.C.-C. Validation: N.V.E., M.C.A. and G.C.-C. Formal analysis: N.V.E., D.I.-N., I.Q. and G.C.-C. Visualization: N.V.E., D.I.-N., I.Q., F.E.F., M.E.V., M.C.A. and G.C.-C. Writing—Original draft preparation: N.V.E., D.I.-N., I.Q., M.C.A. and G.C.-C. Writing—Review and editing: N.V.E., D.I.-N., I.Q., F.E.F., M.E.V., M.C.A. and G.C.-C. Resources: M.C.A. and G.C.-C. Project administration: M.C.A. and G.C.-C. All authors have read and agreed to the published version of the manuscript."

Funding: The authors acknowledge the financial support from the ValBio-3D project Grant ELAC2015/T03–0715 (Ministry of Science, Technology and Innovation Production of Argentina, and Research Council of Norway, Grant no. 271054), CONICET, UNaM, and RISE-PFI.

Acknowledgments: Thanks to Kenneth Aasarød and Johnny Kvakland Melbø (RISE PFI) for performing the TGA and DSC analyses and laboratory assistance.

Conflicts of Interest: The authors declare no conflict of interest.

References

1. Compton, B.; Lewis, J. 3D-printing of lightweight cellular composites. *Adv. Mater.* **2014**, *26*, 5930–5935. [CrossRef]
2. Dai, L.; Cheng, T.; Duan, C.; Zhao, W.; Zhang, W.; Zou, X.; Aspler, J.; Ni, Y. 3D printing using plant-derived cellulose and its derivatives: A review. *Carbohydr. Polym.* **2019**, *203*, 71–86. [CrossRef]
3. Ladd, C.; So, J.; Muth, J.; Dickey, M. Microstructures:3D printing of free standing liquid metal microstructures. *Adv. Mater.* **2013**, *25*, 5081–5085. [CrossRef] [PubMed]
4. Jakus, A.E.; Taylor, S.L.; Geisendorfer, N.R.; Dunand, D.C. Metallic architectures from 3D-printed powder-based liquid inks. *Adv. Funct. Mater.* **2015**, 1–11. [CrossRef]
5. Curodeau, A.; Sachs, E.; Caldarise, S. Design and fabrication of cast orthopedic implants with freeform surface textures from 3-D printed ceramic shell. *J. Biomed. Mater. Res.* **2000**, *53*, 525–535. [CrossRef]
6. Filgueira, D.; Holmen, S.; Melbø, J.K.; Moldes, D.; Echtermeyer, A.T.; Chinga-Carrasco, G. Enzymatic-assisted modification of TMP fibres for improving the interfacial adhesion with PLA for 3D printing. *ACS Sustain. Chem. Eng.* **2017**, *5*, 9338–9346. [CrossRef]
7. Chinthapalli, R.; Skoczinski, P.; Carus, M.; Baltus, W.; De Guzman, D.; Käb, H.; Raschka, A.; Ravenstijn, J. *Biobased Building Blocks and Polymers—Global Capacities, Production and Trends, 2018–2023*; Nova Institute for Ecology and Innovation: Hürt, Germany, 2020.
8. Mendieta, C.M.; Vallejos, M.E.; Felissia, F.E.; Chinga-Carrasco, G.; Area, M.C. Review: Bio-polyethylene from wood wastes. *J. Polym. Environ.* **2020**, *28*. [CrossRef]
9. Kakroodi, A.; Kazemi, Y.; Cloutier, A.; Rodrigue, D. Mechanical performance of polyethylene (PE)-based biocomposites. In *Biocomposites: Design and Mechanical Performance*; Elsevier Ltd.: Cambridge, UK, 2015; pp. 237–256.
10. Abdelmouleh, M.; Boufi, S.; Belgacem, M.; Dufresne, A. Short natural-fibre reinforced polyethylene and natural rubber composites: Effect of silane coupling agents and fibres loading. *Compos. Sci. Technol.* **2007**, *67*, 1627–1639. [CrossRef]
11. Mengeloğlu, F.; Karakuş, K. Some properties of Eucalyptus wood flour filled recycled high density polyethylene polymer-composites. *Turk. J. Agric.* **2008**, *32*, 537–546.
12. Suhaily, S.S.; Khalil, H.P.S.A.; Nadirah, W.O.W.; Jawaid, M. Bamboo based biocomposites material, design and applications. In *Materials Science-Advanced Topics*; Mastai, Y., Ed.; Intech: Rijeka, Croatia, 2013; pp. 489–517.

13. Kakroodi, A.R.; Bainier, J.; Rodrigue, D. Mechanical and morphological properties of flax fiber reinforced high density Polyethylene/recycled rubber composites. *Int. Polym. Proc.* **2012**, *27*, 196–204. [CrossRef]

14. Tajeddin, B.; Abdulah, L.C. Thermal properties of high density polyethylene-kenaf cellulose composites. *Polym. Polym. Compos.* **2010**, *18*, 257–261. [CrossRef]

15. Clauser, N.; Felissia, F.E.; Vallejos, M.E. Small-sized biorefineries as strategy to add value to sugarcane bagasse. *Chem. Eng. Res. Des.* **2015**, *107*, 137–146. [CrossRef]

16. Ramsay, J.A.; Hassan, M.A.; Ramsay, B.A. Hemicellulose as a potential substrate for production of poly(β-hydroxyalkanoates). *J. Microbiol.* **1995**, *41*, 262–266. [CrossRef]

17. Brodin, M.; Vallejos, M.; Tanase, M.; Area, M.C.; Chinga-carrasco, G. Lignocellulosics as sustainable resources for production of bioplastics A review. *J. Clean. Prod.* **2017**, *162*, 646–664. [CrossRef]

18. Filgueira, D.; Holmen, S.; Melbo, J.; Moldes, D.; Echtermeyer, A.; Chinga-Carrasco, G. 3D printable filaments made of biobased polyethylene biocomposites. *Polymers* **2018**, *10*, 1–15. [CrossRef]

19. Pickering, K.L.; Efendy, M.G.A.; Le, T.M. A review of recent developments in natural fibre composites and their mechanical performance. *Compos. Part. A* **2016**, *83*, 98–112. [CrossRef]

20. Lu, J.Z.; Negulescu, I.I.; Wu, Q. Maleated wood-fiber/high-density-polyethylene composites: Coupling mechanisms and interfacial characterization. *Compos. Interfaces* **2005**, *12*, 125–140. [CrossRef]

21. Aggarwal, P.K.; Chauhan, S.; Raghu, N.; Karmarkar, S.; Shashidhar, G.M. Mechanical properties of bio-fibers-reinforced high-density polyethylene composites: Effect of coupling agents and bio-fillers. *J. Reinf. Plast. Compos.* **2013**, *32*, 1722–1732. [CrossRef]

22. Yuan, Q.; Wu, D.; Gotama, J.; Bateman, S. Wood fiber reinforced polyethylene and polypropylene composites with high modulus and impact strength. *J. Thermoplast. Compos. Mater.* **2008**, *21*, 195–208. [CrossRef]

23. Tarrés, Q.; Melbø, J.; Delgado-Aguilar, M.; Espinach, F.X.; Mutjé, P.; Chinga-Carrasco, G. Bio-polyethylene reinforced with thermomechanical pulp fibers: Mechanical and micromechanical characterization and its application in 3D-printing by fused deposition modelling. *Compos. Part B Eng.* **2018**, *153*, 70–77. [CrossRef]

24. Méndez, J.; Vilaseca, F.; Pélach, M.Á.; López, J.; Barberá, L.; Turon, X.; Gironés, J.; Mutjé, P. Evaluation of the reinforcing effect of ground wood pulp in the preparation of polypropylene-based composites coupled with Maleic Anhydride grafted Polypropylene. *J. Appl. Polym. Sci.* **2007**, *105*, 3588–3596. [CrossRef]

25. Ehman, N.; Rodriguez Rivero, G.; Area, M.C.; Felissia, F. Dissolving pulps by oxidation of the Cellulosic fraction of lignocellulosic waste. *Cellul. Chem. Technol.* **2017**, *51*.

26. Kargarzadeh, H.; Ahmad, I.; Thomas, S.; Dufresne, A. *Handbook of Nanocellulose and Cellulose Nanocomposites*; Wiley-VCH: Weinheim, Germany, 2012.

27. Shahzad, A. Effects of water absorption on mechanical properties of hemp fiber composites. *Polym. Compos.* **2011**, *33*, 120–128. [CrossRef]

28. Cheng, H.; Zhan, H.; Fu, S.; Lucia, L. Alkali extraction of hemicellulose from depithed corn stover and effects on Soda-AQ pulping. *Bioresources* **2010**, *11*, 196–206.

29. Tajvidi, M.; Ebrahimi, G. Water uptake and mechanical characteristics of natural filler-polypropylene composites. *J. Appl. Polym. Sci.* **2003**, *88*, 941–946. [CrossRef]

30. Zhang, Y.; Hosseinaei, O.; Wang, S. Influence of hemicellulose extraction on water uptake behaviour of wood strands. *Wood Fiber Sci.* **2011**, *43*, 244–250.

31. Chinga-Carrasco, G.; Johnsen, P.O.; Øyaas, K. Structural quantification of wood fibre surfaces—Morphological effects of pulping and enzymatic treatment. *Micron* **2010**, *41*, 648–659. [CrossRef]

32. Cordin, M.; Bechtold, T.; Pham, T. Effect of fibre orientation on the mechanical properties of polypropylene–lyocell composites. *Cellulose* **2018**, *25*, 7197–7210. [CrossRef]

33. Joffre, T.; Miettinen, A.; Berthold, F.; Gamstedt, E.K. X-ray micro-computed tomography investigation of fibre length degradation during the processing steps of short-fibre composites. *Compos. Sci. Technol.* **2014**, *105*, 127–133. [CrossRef]

34. Lomakin, S.M.; Rogovina, S.Z.; Grachev, A.V.; Prut, E.V.; Alexanyan, C. V Thermal degradation of biodegradable blends of polyethylene with cellulose and ethylcellulose. *Thermochim. Acta* **2011**, *521*, 66–73. [CrossRef]

35. Banat, R.; Fares, M.M. Thermo-gravimetric stability of high density polyethylene composite filled with olive shell flour. *Am. J. Polym. Sci.* **2015**, *5*, 65–74.

36. Wojcieszak, R.; Santarelli, F.; Paul, S.; Dumeignil, F.; Cavani, F.; Gonçalves, R. V Recent developments in maleic acid synthesis from bio-based chemicals. *Sustain. Chem. Process.* **2015**, *3*, 1–11. [CrossRef]

37. Ita-Nagy, D.; Vázquez-Rowe, I.; Kahhat, R.; Quispe, I.; Chinga-Carrasco, G.; Clauser, N.M.; Area, M.C. Life cycle assessment of bagasse fiber reinforced biocomposites. *Sci. Total Environ.* **2020**, *720*, 137586. [CrossRef] [PubMed]

38. Sluiter, A.; Hames, B.; Ruiz, R.; Scarlata, C. *Determination of Structural Carbohydrates and Lignin in Biomass Determination of Structural Carbohydrates and Lignin in Biomass*; National Renewable Energy Laboratory USA: Golden, CO, USA, 2012; pp. 1–15.

39. Stocker, T.F.; Qin, D.; Plattner, G.K.; Tignor, M.M.B.; Allen, S.K.; Boschung, J.; Nauels, A.; Xia, Y.; Bex, V.; Midgley, P.M. *Climate Change 2013 the Physical Science Basis: Working Group I Contribution to the Fifth Assessment Report of the Intergovernmental Panel on Climate Change*, 1st ed.; Cambridge University Press: Cambridge, UK; New York, NY, USA, 2013; pp. 1–1535.

40. Hauschild, M.Z.; Goedkoop, M.; Guinée, J.; Heijungs, R.; Huijbregts, M.; Jolliet, O.; Margni, M.; De Schryver, A.; Humbert, S.; Laurent, A.; et al. Identifying best existing practice for characterization modeling in life cycle impact assessment. *Int. J. Life Cycle Assess.* **2013**, *18*, 683–697. [CrossRef]

41. Jolliet, O.; Antón, A.; Boulay, A.M.; Cherubini, F.; Fantke, P.; Levasseur, A.; McKone, T.E.; Michelsen, O.; Milà i Canals, L.; Motoshita, M.; et al. Global guidance on environmental life cycle impact assessment indicators: Impacts of climate change, fine particulate matter formation, water consumption and land use. *Int. J. Life Cycle Assess.* **2018**, *23*, 2189–2207. [CrossRef]

42. Guinée, J.B.; Heijungs, R. Life Cycle Assessment. In *Kirk-Othmer Encyclopedia of Chemical Technology*; John Wiley & Sons, Inc.: Hoboken, NJ, USA, 2005.

43. Curran, M.A. Strengths and Limitations of Life Cycle Assessment. In *Background and Future Prospects in Life Cycle Assessment. LCA Compendium-The Complete World of Life Cycle Assessment*; Klöpffer, W., Ed.; Springer: Dordrecht, The Netherlands, 2014; pp. 189–206.

44. International Standardization Organization ISO Technical Committee ISO/TC 207. *ISO 14040:2006 Environmental Management—Life Cycle Assessment–Principles and Framework*, 2nd ed.; International Organization for Standardization: Geneva, Switzerland, 2006.

45. International Standardization Organization ISO Technical Committee ISO/TC 207. *14044. Environmental Management—Life Cycle Assessment–Requirements and Guidelines*, 1st ed.; International Organization for Standardization: Geneva, Switzerland, 2006.

46. Guinée, J.; Gorrée, M.; Heijungs, R.; Huppers, G.; Kleijn, R.; De Koning, A.; Van Oers, L.; Weegener Sleeswijk, A.; Suh, S.; De Haes, H.A.U.; et al. *Handbook on Life Cycle Assessment: Operational Guide to the ISO Standards*; Kluwer Academic Publishers: Dordrecht, The Netherlands, 2002; Volume 7.

47. European Commission—Joint Research Centre (EC-JRC). *General Guide for Life Cycle Assessment—Detailed Guidance*, 1st ed.; Publications Office of the European Union: Luxembourg, 2010.

48. Ecoinvent Ecoinvent v3.4 Database. Ecoinvent Centre. Available online: http://www.ecoinvent.org/ (accessed on 20 May 2019).

49. Product Ecology Consultants. *PRé-Product Ecology Consultants SimaPro*; Product Ecology Consultants: Amersfoort, The Netherlands, 2018.

Sample Availability: Not available.

 *molecules*

Article

Feasibility of Barley Straw Fibers as Reinforcement in Fully Biobased Polyethylene Composites: Macro and Micro Mechanics of the Flexural Strength

Ferran Serra-Parareda [1,*], Fernando Julián [1], Eduardo Espinosa [2,*], Alejandro Rodríguez [2], Francesc X. Espinach [1] and Fabiola Vilaseca [3,4]

1   LEPAMAP+PRODIS research group, University of Girona, Maria Aurèlia Capmany, 61, 17003 Girona, Spain; fernando.julian@udg.edu (F.J.); francisco.espinach@udg.edu (F.X.E.)
2   Chemical Engineering Department, Bioagres Group, Universidad de Córdoba, 14014 Córdoba, Spain; a.rodriguez@uco.es
3   Advanced Biomaterials and Nanotechnology, Dept of Chemical Engineering, University of Girona, 17003 Girona, Spain; fabiola.vilaseca@udg.edu
4   Department of Industrial and Materials Science, Chalmers University of Technology, SE 412 96 Gothenburg, Sweden
*   Correspondence: ferran.serrap@udg.edu (F.S.-P.); eduardo.espinosa@uco.es (E.E.); Tel.: +34-616239050 (F.S.-P); +34-957218586 (E.E.)

Academic Editor: Fabrizio Sarasini
Received: 14 April 2020; Accepted: 9 May 2020; Published: 10 May 2020

Abstract: Awareness on deforestation, forest degradation, and its impact on biodiversity and global warming, is giving rise to the use of alternative fiber sources in replacement of wood feedstock for some applications such as composite materials and energy production. In this category, barley straw is an important agricultural crop, due to its abundance and availability. In the current investigation, the residue was submitted to thermomechanical process for fiber extraction and individualization. The high content of holocellulose combined with their relatively high aspect ratio inspires the potential use of these fibers as reinforcement in plastic composites. Therefore, fully biobased composites were fabricated using barley fibers and a biobased polyethylene (BioPE) as polymer matrix. BioPE is completely biobased and 100% recyclable. As for material performance, the flexural properties of the materials were studied. A good dispersion of the reinforcement inside the plastic was achieved contributing to the elevate increments in the flexural strength. At a 45 wt.% of reinforcement, an increment in the flexural strength of about 147% was attained. The mean contribution of the fibers to the flexural strength was assessed by means of a fiber flexural strength factor, reaching a value of 91.4. The micromechanical analysis allowed the prediction of the intrinsic flexural strength of the fibers, arriving up to around 700 MPa, and coupling factors between 0.18 and 0.19, which are in line with other natural fiber composites. Overall, the investigation brightness on the potential use of barley straw residues as reinforcement in fully biobased polymer composites.

Keywords: barley straw; composite; flexural strength; biobased polyethylene

1. Introduction

The agri-food industry is becoming increasingly important in the world. In 1950, the world population was estimated to be around 2.6 billion people according to United Nations. Seventy years later, this number is still rising (7.7 billion) and is expected to reach 10 billion by 2050 [1]. This enormous increase in population brings with it major challenges to be faced, two of which are: to provide food, and to reduce as much as possible the depletion of natural resources. In addition, socially, in recent years there has been a change in the way of life, with an increase in the population in the cities, to the detriment of rural areas, leading to depopulation that undoubtedly affects the natural environment.

The agri-food activity becomes one of the pillars on which to sustain an economic model and sustainable development, environmentally, economically, and socially. If society really wants to approach a sustainable development, it is necessary to leave the linear economic model and evolve to a circular one, where each of the inputs is valued, so that the amount of waste tends to zero. In the agricultural activity the great amount of resources that are used, human and material, do it not only in the growth of the grain or fruit, but also in the growth of the plant. This therefore generates a considerable amount of waste, also called lignocellulosic biomass, the recovery of which would bring great benefits to the agricultural economic cycle, which is sometimes in need of subsidies. In fact, if a product with added value is obtained from a waste, an economic return can be obtained from it.

World cereal production in 2018 was 2,968 MM tons, with a cultivated area of 728 MM hectares. Barley contributed 4.75% of total production, with 141 million tons, representing production in the countries of the European Union a 40% (56 million tons), according to the FAOSTAT (Food and Agriculture organization of the United Nations). It can be deduced, considering a straw/grain ratio around 1 [2,3], the enormous amount of waste that this activity generates every year.

Using a byproduct from any agri-food or industrial process to obtain products with added value is one of the goals of the circular economy and it is also in line with the principles of green chemistry [4]. In some cases, cereal straws are left in the fields to be incinerated or decomposed as fertilizer for the next harvest [5]. These practices provide undoubted benefits but also produce CO_2 emissions and can be impractical for long straws and useful only for stubble. Moreover, country regulations are increasingly controlling agri-food waste incineration in order to prevent fires and unhealthy emissions. Thus, other solutions to manage such agri-food must be explored. In the case of barley straw there have been intents to use such waste as biofuel source [6–9] with successful results. Nonetheless, the use of this waste as biofuel source is only possible if a treatment plant is near enough in terms of transport costs. There is also literature dealing with the use of barley straws in the paper and board industry [10,11]. Other researchers have proposed barley straws for algae control purposes [12,13] and to prevent soil erosion on some plantations [14,15]. Thus, barley straws have showed that it is possible to create value from such wastes.

Composite materials reinforcement is a field were the exploitation of lignocellulosic waste has been extensively explored [16,17]. The use of a variety of agri-food waste from annual plants as composite reinforcements has revealed the potential of such fibers as strength and stiffness enhancers [18–20]. Lignocellulosic reinforced materials are intended to be greener than glass fiber reinforced ones, while showing similar mechanical to be commercially competitive. The main obstacles in obtaining comparatively high strengths and stiffness with lignocellulosic fibers are, on the one hand the compatibility between hydrophobic polymer matrices and hydrophilic natural fibers that hinder obtaining strong interfaces [21,22]. On the other hand, the intrinsic properties of natural fibers are lower than those of mineral ones [23,24]. The literature shows how the use of coupling agents allows obtaining strong interfaces for polyolefin-based materials, specifically maleic anhydride-grafted polymers [22,25,26]. Thus, in the case of polyolefin-based composites, a careful dosage of coupling agent solves strong interfaces issues. The intrinsic properties of natural fibers are notably lower than glass fiber. Moreover, the properties of natural fibers show higher scatter than manmade materials. Thus, it is not possible to obtain the same strengths at the same reinforcement contents. Nonetheless, it is possible to add higher amounts of natural fiber to a composite than glass fiber and obtain similar mechanic properties [27,28].

Surprisingly, the literature about barley straw reinforced polymers is scarce. Barley straws are mainly used as concrete or elastomer fillers [29–32]. Hyvärinen and Kärki explored using barley straw instead of wood fibers as polypropylene reinforcement [33]. The researchers found how the mechanical properties of barley straw reinforced materials were lower than wood fiber reinforced ones. Silva-Guzman et al. researched the effect of barley straw on the mechanical properties of a corn starch polymer-based composite [34]. The authors observed a positive effect of the presence of the reinforcements on the strength and stiffness of the materials. Nonetheless, the authors used low reinforcement contents, with a 15% *w/w* highest percentage. Rojas-Leon et al. used barley straw particles with recycled high-density polyethylene (HDPE) to obtain particleboards [35]. In this paper the interface between barley straw and HDPE was weak as the mechanical properties of the materials decreased with the filler contents. Serra-Parareda et al. researched the effect of barley straw content on the tensile strength of mold injected composites [36]. In this paper the authors found that adding a 6% of coupling agent returned the highest tensile strength values. The authors also obtained the intrinsic tensile strength and Young's modulus of the reinforcements. To the extent of authors' knowledge there is no literature on the flexural strength of barley straw reinforced polyolefin composites.

Knowing the flexural properties of a material is of great importance for engineers. Moreover, when the material is clearly anisotropic, as semi-oriented short fiber reinforced composites [37–39]. Usually, products and components are used under bending conditions and purely tensile cases are scarce in comparison. Thus, designers are interested in previewing the behavior of such components under flexural loads [40,41]. Additionally, the intrinsic flexural strength of barley straw is unknown in the literature. Knowing such value can be used to model the behavior of composites at different reinforcement contents.

In the current investigation barley straw fibers were submitted to elevated temperatures and then defibrated to obtain single fibers. Fully biobased composites were prepared based on a biobased polyethylene matrix reinforced with 15, 30 and 45 wt.% of barley fibers. A coupling agent was added to the formulation to enhance the interfacial adhesion. The materials were injection-molded and subjected under three-point bending test to evaluate the flexural properties. The properties were studied from a macro and micromechanical viewpoint, where the intrinsic flexural strength of the fibers, the coupling factors, and the contribution of the reinforcements to the flexural strength of the composite were assessed as main important outcomes. Overall, the current investigation explores the potential of barley straw residues in added value applications by its incorporation in a fully biobased matrix, contributing to global sustainable development.

2. Results

2.1. Fibers Characterization

Barley straws were submitted to steam-water treatment with further defibration by means of Sprout Waldron equipment, obtaining barley thermomechanical (TM) fibers. The chemical composition and morphology of the fibers was examined as two main important factors affecting composite's properties. On the one hand, the chemical composition of the fibers plays a key role in establishing the extend of interaction between the fibers and the matrix, assisted by the coupling agent. This phenomenon will affect the stress-transfer between the phases inside the composite [42,43]. On the other hand, a definite fiber aspect ratio is required for the effective stress-transfer between the phases. In this way, when the stress concentration at the fiber ends, this leads to the matrix cracking. Thereby, shorter aspect ratios will bring to more fiber ends, acting as stress concentration points with failure potential [44].

Hence, the initial evaluation of the chemical and morphological composition is needed. Table 1 presents the chemical constituents and the mean fiber length and diameter of the original barley straw and the thermomechanical fibers. For readers' convenience, illustrations of untreated barley straw and thermomechanically treated barley fibers are presented in Figure 1.

Table 1. Chemical and morphological composition of barley straw and barley thermomechanical (TMP) fibers.

Composition/Morphology	Barley Straw	Barley TMP Fibers
Holocellulose (wt.%)	70.12 ± 0.54	77.67 ± 0.61
Klason lignin (wt.%)	16.45 ± 0.34	15.30 ± 0.46
Extractives (wt.%)	5.90 ± 0.76	2.73 ± 0.12
Ashes (wt.%)	7.1 ± 0.2	4.3 ± 0.3
Length [1] (μm)	-	745 ± 21
Diameter (μm)	-	19.6 ± 0.6
Aspect ratio (length/diameter)	-	38.0

[1] Length weighted in length.

Figure 1. Barley straw images (**a**) before being treated and (**b**) after the thermomechanical process.

From Table 1, barley straw is rich in holocellulose with a relatively small portion of lignin in comparison with other sources of natural fibers. For example, wood fibers possess higher lignin content, with minor amount of holocellulose. This is explained by the fact that in wood fibers lignin is needed to ensure the maintenance of the fiber cell wall structure [45,46]. The thermomechanical treatment removed part of the lignin, some of the extractives and ashes. As expected, an increase in the carbohydrate content (holocellulose) was experimented owing to changes of the lignin, extractives, and ashes content. The thermomechanical treatment also promoted the release of fiber elements with high aspect ratio (38.0). The weighted fiber length is here considered.

By treating the fibers at high temperatures, the lignin is softened, and fibers breakage is more likely to occur at the outsider layers of the fiber cell wall, between the primary wall and middle lamella. Here is where the largest concentration of lignin (~70 wt.%) is found, attaching the individual fibers together, with minor amounts of cellulose (~10 wt.%) and hemicellulose (~20 wt.%) [45].

During the thermomechanical treatment, part of the lignin can be dissolved in the hot water and released from the fiber cell wall during the mechanical defibration. Lignin is bonded to the surface of carbohydrates (Figure 2), therefore its removal can finally lead to the release of hemicelluloses, extractives, and inorganic matter. As a result, the global yield in thermomechanical processes renders values between the 85% and 95% depending on the severity of the treatment, indicating the loss of the chemical constituents throughout the process [47,48].

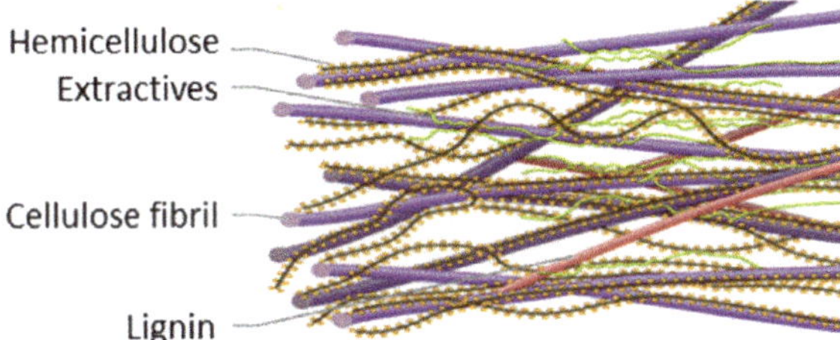

Figure 2. Illustration of the distribution of the lignin and carbohydrates in the fiber surface.

Overall, the thermomechanical fibers produced from barley straw show high amount of holocellulose fibers with relatively high aspect ratio. Therefore, in regions with big availability of this biomass, deforestation can be prevented. These fibers show to be good candidates as reinforcing fibers in composite materials.

2.2. Optimization of the Coupling Agent

The flexural properties in composite materials depend on the type and amount of reinforcement, orientation and morphology of the fibers, the dispersion of the reinforcement inside the matrix, and largely on the quality at the interphase [5,37,49]. However, the different nature of natural fibers and thermoplastics hinders the spontaneous interactions between both materials. The lack of compatibility is explained by the different chemical structure of thermoplastics and natural fibers driving to different polarities. The hydroxyl groups in the fiber surface gives them and hydrophilic nature, whereas the hydrocarbon structure of thermoplastics confers them hydrophobicity.

As a result, the poor compatibility hinders the stress-transfer capacity and makes difficult the increment of the strength by the addition of the lignocellulosic reinforcement. To enhance the interfacial adhesion, coupling agents have proved to work efficiently in this purpose. More specifically, maleic anhydride polyethylene (MAPE) can be used to increase the interactions between both phases. In this context, the coupling agent form linkages with the hydroxyl groups in the fibers' surface by means of hydrogen bonds and covalent interaction with the maleic groups, and by chain entangling with the unmodified BioPE chains, as illustrated in Figure 3.

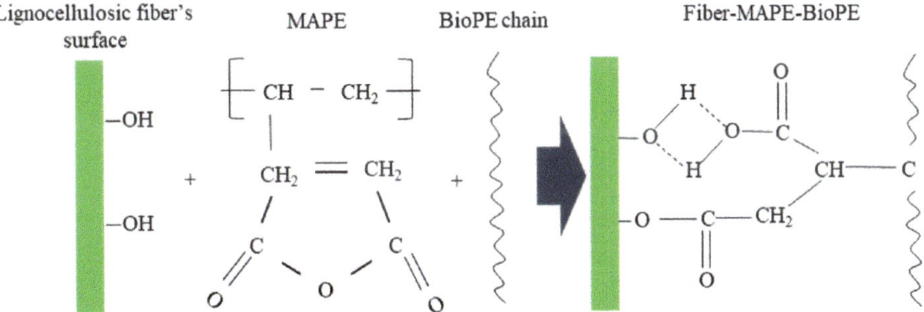

Figure 3. Illustration of maleic anhydride polyethylene (MAPE) interaction between the fiber and the matrix.

The efficiency of the coupling agent depends largely on the amount of bonding and the interaction quality with the natural fibers [50,51]. The optimal content of MAPE in natural fiber composites has been found to be between 4 and 8 wt.% with respect to fiber content [42,52,53]. The amount of MAPE added will depend on the fiber content, thus, the optimal amount of MAPE needed to enhance the interfacial bonding will be investigated in view of the fiber loading.

To investigate how the content MAPE affected the interfacial adhesion, varying amounts of MAPE (0, 2, 4, 6, 8, and 10 wt.%) with respect to fiber content were added to composites reinforced with 30 wt.% of barley fibers. The coupling agent was optimized to achieve the highest flexural strength, indicative of an optimal fiber-to-matrix interfacial union. When the amount of coupling agent was optimized, the same MAPE percentage was then applied to the rest of the composites with different fiber loadings. These results are shown in Figure 4.

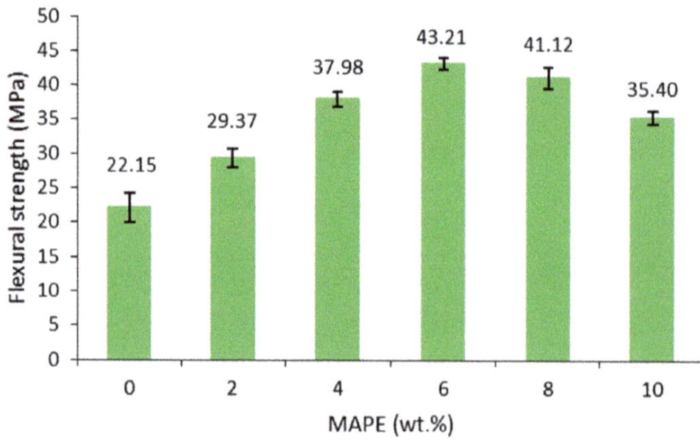

Figure 4. Flexural strength of composites at 30 wt.% and different MAPE content.

The composite material without MAPE showed a similar flexural strength than the neat matrix (21.25 MPa), evidencing scarce compatibility between composite phases. Still, however, the addition of barley TMP fibers into the polymer did not decrease the flexural strength. However, by adding the coupling agent the flexural increases, reaching a maximum value at 6 wt.% of MAPE. For lower amounts of coupling agent, little improvement was observed, whereas much high amounts of coupling agent the gaining in property was again reduced. The reduction of the flexural strength at too high amounts of coupling agent can be attributed to the much shorter polymer lengths of MAPE polymer, as compared to the polymer itself; the benefits of the coupling agent were less compared to the effect of shorter polymer chains in the formulation.

Once the content of MAPE was optimized, the flexural properties of the composite materials at other formulations were examined.

2.3. Flexural Properties of Barley Fiber Composites

The barley fibers were incorporated to a biobased polyethylene, and the flexural properties measured. The results of the bending test as function of the fiber loading are presented in Table 2, where V^f is the reinforcement volume fraction, σ_f^c is the flexural strength of the composite, ε_f^c is the deformation at the maximum flexural strength value, and σ_f^{m*} is the contribution of the matrix to the tensile strength.

Table 2. Flexural properties of BioPE composites reinforced with barley fibers.

Sample	Reinforcement (wt.%)	V^f	σ_f^c (MPa)	ε_f^c (%)	σ_f^{m*} (MPa)
BioPE	0	0	21.25 ± 0.95	7.18 ± 0.41	21.25
	15	0.111	30.21 ± 1.23	4.03 ± 0.28	18.21
BioPE/Barley fibers	30	0.233	43.21 ± 0.89	3.52 ± 0.31	16.98
	45	0.367	52.45 ± 1.45	2.85 ± 0.19	15.14

The values of σ_f^{m*} were obtained from the stress-strain curves of the neat matrix by computing the stress of the matrix at the deformation where the maximum stress of the composite was produced (Figure 5).

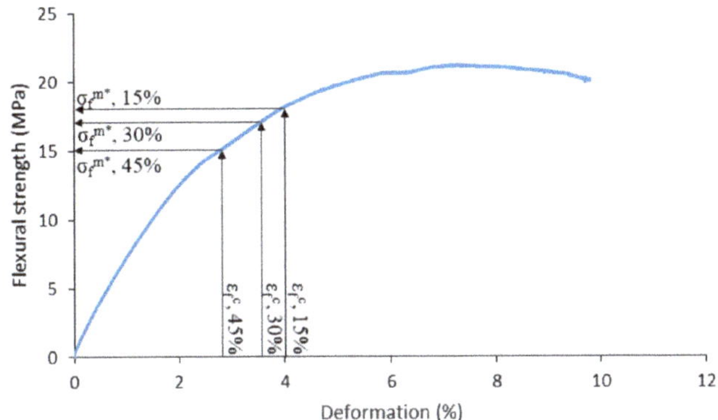

Figure 5. Flexural stress-strain curve of BioPE. Evaluation of the matrix contribution to the flexural strength of the composite.

The flexural strength of the composites followed a linear evolution with the fiber volume fraction. This indicated a proper stress transfer between the phases and a good dispersion of the reinforcement inside the plastic matrix. The addition of the fibers produced an enhancement in the flexural strength about the 42%, 103% and 147% in the composite reinforced with the 15, 30 and 45 wt.%, with respect to the neat matrix.

These are remarkable increments considering the type of raw biomass used, which is an agricultural residue. In fact, barley composites exhibited comparative flexural properties than other natural fiber composites by using wood fibers, such as spruce, and higher than other agricultural residues [54–56]. This performance can be attributed to the chemical composition of barley fibers.

Cellulose is the major crystalline compound and its aligned structure confers the strength and stiffness to the fiber cell wall structure. As a result, one can expect a higher contribution to the flexural strength of the composite when the reinforcement possesses higher amounts of holocellulose [57]. Besides, lignin is an amorphous polymer with a certain degree of hydrophobicity, which does not significantly contribute to the mechanical properties of the fibers, though, the compound plays a major role in binding the cellulosic chains and favoring the stress-transfer within the fibers and with the matrix [58] (Figure 6). According to Bledzki et al. [59,60], an increment on the composite's strength can be attributed to higher cellulose and lignin content, as well as to an optimal dispersion and interfacial adhesion of the reinforcement with the matrix. Moreover, Shebani et al. [61] stated that optimal amounts of lignin can act as binding between the cellulose fibrils, granting to the stress transfer between the fibrils. This statement is in accordance with previous investigation of the research group dealing with the influence of lignin in natural fiber composites [45].

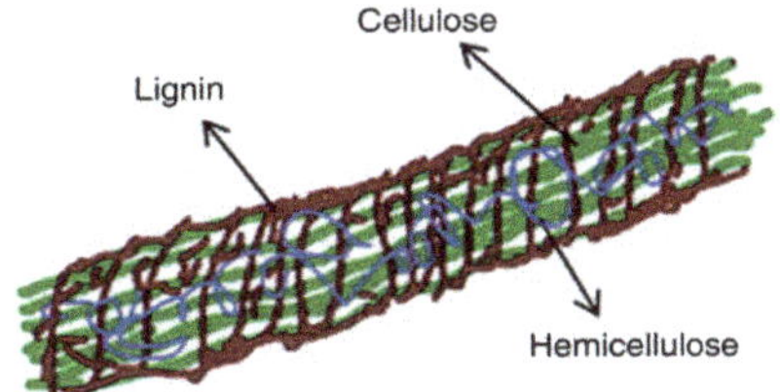

Figure 6. Hierarchical structure of lignocellulosic biomass.

In the present case, fibers' breakage occurred at the outer layers of the fiber cell wall during the thermomechanical treatment, where the major concentration of lignin is placed. The outer layers are covered in its surface by lignin, and it is there where one can hope for an optimal compatibility fiber-to-fiber and fiber-to-matrix, favoring the stress-transfer throughout the fibers. The fact that the BioPE can be reinforced up to a 45 wt.% of these fibers is explained by the good compatibility given by the chemical composition of the fibers.

Apart from its chemical composition, the high aspect ratios of the fibers also confer the material a larger capacity of transferring the stress through the fibers and incrementing the final strength of the material.

The deformation of the materials was significantly affected by the addition of a more rigid phase. This fact is attributed to the increased adhesion between the phases and the greater rigidity of barley fibers in comparison with the soft BioPE [41,62]. This reduced the deformation ability of the material. A micro-mechanical analysis was also performed to better understand the behavior of the composites.

2.4. Intrinsic Flexural Strength Properties

The strength of natural fiber composites is a combination of the strength supported by the polymeric phase and the stress effectively transferred to the reinforcing fibers. As abovementioned, the stress supported by the polymeric phase is obtained from the stress-strain curve of the neat matrix. Thereby, the difference between the strength of the composite and the stress supported by the plastic matrix is attributed to the stress transferred to the reinforcement. Thereafter, it is possible to quantify the effectiveness of the fibers inside the composite, as well as its intrinsic mechanical properties.

One of the simplest methods used to express the contribution of the phases to the material's strength is by using the modified Rule of Mixtures (mRoM) [63,64]. The model was initially developed to be applied to tensile properties, though, it can also be extended to flexural ones. The mRoM for tensile and flexural properties are shown in Equations (1) and (2), respectively.

$$\text{Tensile mRoM} \quad \sigma_t^c = f_{c,t} \times \sigma_t^F \times V^f + \sigma_t^{m^*} \times \left(1 - V^f\right) \tag{1}$$

$$\text{Flexural mRoM} \quad \sigma_f^c = f_{c,f} \times \sigma_f^F \times V^f + \sigma_f^{m^*} \times \left(1 - V^f\right) \tag{2}$$

where σ_t^F and σ_f^F are the intrinsic tensile and flexural strength of the fibers, and $f_{c,t}$ and $f_{c,f}$ are the tensile and flexural coupling factors. Generally, in short semi-aligned fiber composites with strong interfacial adhesion, the coupling factor tends to a value between 0.18 and 0.20. In its current shape, the mRoM contain two incognita, which are the intrinsic strength and the coupling factor.

The value of the intrinsic tensile strength of the fibers was calculated in previous works [36] by using the Kelly and Tyson modified equation and its solution, provided by Bowyer and Bader [65,66]. In that work, a pre-evaluation of the tensile properties in view of the fiber orientation, fiber morphology and interfacial adhesion was carried out. The investigation allowed the acquisition of the orientation factor (0.309) and interfacial shear strength (10.49), as important outcomes. At a 6 wt.% of MAPE, the intrinsic tensile strength of barley fibers at a 30% of reinforcement was 521.2 MPa. Though, the current investigation incorporates the tensile properties of composites reinforced with a 15 and 45 wt.%. By following the same methodology, the intrinsic tensile strength of the fibers was obtained, with values of 532.9 and 500.5 MPa, at a 15 and 45 wt.%, respectively. Once computed the intrinsic tensile strengths, one can calculate the tensile coupling factors from Equation (1) at each fiber loading.

Nonetheless, the calculus of the intrinsic flexural strength is not as straightforward as one could expect. For example, Hashemi [67] proposed a correlation between the composite's and fiber's tensile and flexural strength, defined by $\sigma_f^F = \left(\sigma_f^c / \sigma_t^c\right) \times \sigma_t^F$ However, as reported by the same author, this assumption may not be necessarily correct.

Recent work methodologies suggested to only account for the fiber contribution to the composite strength. A correlation was established between the contribution of the fibers to both the tensile and flexural strength of the composite, and the intrinsic flexural and tensile strength of the reinforcement.

This assumption is made upon the fact that the tensile and flexural coupling factors are in the same order of magnitude, since the factor is not dependent on the type of test conducted, either flexural or tensile. Additionally, the tensile coupling factor (fc,t) and the flexural coupling factor (fc,t), which largely depend on the quality at the interphase, fiber's morphology and dispersion of the fibers inside the matrix, should acquire alike values in both tests. Assuming this hypothesis, the net contribution of the fibers to the tensile ($f_{c,t} \times \sigma_t^F \times V^f$) and flexural ($f_{c,f} \times \sigma_f^F \times V^f$) strength of the composite should be directly correlated to the intrinsic tensile strength (σ_t^F) and intrinsic flexural strength (σ_f^F) of the fibers [38,39,68].

The global contribution of the fibers to the tensile and flexural strength of the composite can be obtained by reorganizing the mRoM. Thereby, it is possible to isolate the net contribution of the fibers to the strength of the composite with the fiber volume fraction. Afterwards, if the net contribution is plotted versus the volume fraction in each of the composites, the fiber flexural strength factor (FFSF) (Equation (3)) and the fiber tensile strength factor (FTSF) (Equation (4)) is obtained from the slope of the line [69].

$$\text{FFSF} \quad f_{c,\,f} \times \sigma_f^F \;=\; \left(\frac{\sigma_f^c - \sigma_f^{m*} \times (1 - V^f)}{V^f} \right) \tag{3}$$

$$\text{FTSF} \quad f_{c,\,t} \times \sigma_t^F \;=\; \left(\frac{\sigma_t^c - \sigma_t^{m*} \times (1 - V^f)}{V^f} \right) \tag{4}$$

Knowing the intrinsic tensile strength, and the global contribution of the fibers to the tensile and flexural strength of the composite, it is possible to calculate the intrinsic flexural strength of the fibers following Equation (5).

$$\frac{\sigma_f^F}{\sigma_t^F} \;=\; \frac{\text{FFSF}}{\text{FTSF}} \tag{5}$$

To compute the contribution of the fibers to the tensile strength of the composite, the tensile properties are needed (Table 3). The properties were extracted from the previous work dealing with tensile properties [36].

Table 3. Tensile properties of BioPE composites reinforced with barley fibers.

Sample	Reinforcement (wt.%)	V^f	σ_t^c (MPa)	ε_t^c (%)	σ_t^{m*} (MPa)
BioPE	0	0	18.05 ± 0.74	12.18 ± 0.34	18.05
BioPE/Barley fibers	15	0.111	25.21 ± 0.64	7.65 ± 0.24	16.37
	30	0.233	34.70 ± 0.90	6.45 ± 0.31	16.76
	45	0.367	43.10 ± 0.57	4.69 ± 0.33	15.86

Briefly, the tensile strength followed a linear evolution with the fiber content. Increments in the tensile strength parameter were obtained about the 40%, 92% and 139%. The global contribution of the fibers to the composite strength computed by means of the FTSF and FFSF are presented in Figure 7.

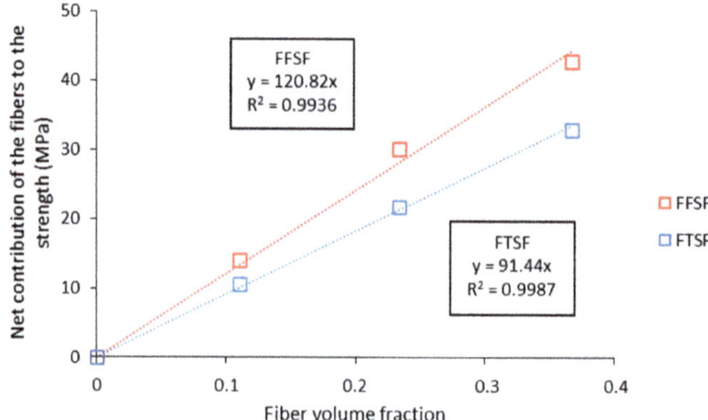

Figure 7. Fiber tensile strength factor (FTSF) and fiber flexural strength factor (FFSF).

The contribution of the fibers to the flexural strength (FFSF = 120.8) was significantly higher than in the tensile one (FTSF = 91.44). This is attributed to the fact that composites subjected to flexural loads support a combination of compressive and tensile forces at the cross-sectional area of the specimens (Figure 8).

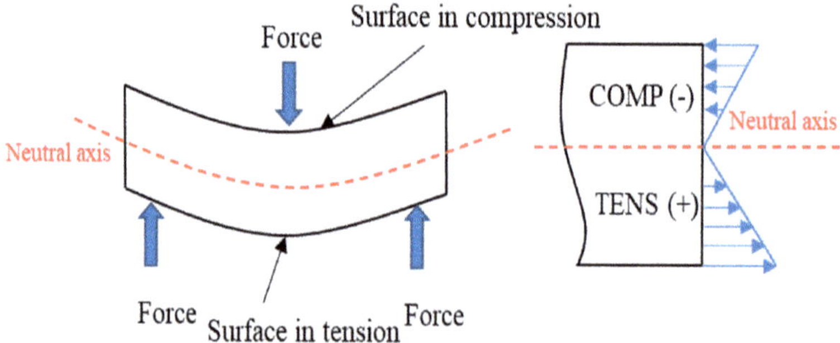

Figure 8. Combination of compression and tension forces during the flexural test.

Some authors explain that while composites subjected to tensile test are fully loaded under tensile stresses, flexural specimens are loaded under compressive and tensile forces at the same time. Since most of the thermoplastics have a larger capacity to withstand the load under compression rather than tensile, the part of the specimen subjected to compression is expected to contribute more than the one submitted to tensile stress. As a result, flexural specimens will support higher stresses than tensile ones. Other authors state that the anisotropy of the fibers and their semi-alignment inside the plastic can contribute more extensively to the flexural strength [56].

Overall, the FFSF was found to be higher than in other composites reinforced with different sources of agricultural residues, reflecting the potential of barley straws in composites field. In comparison with wood fiber reinforced composites, the FFSF did not differ much, though, larger discrepancies could be observed with the FTSF. Nonetheless, this could be an advantage for composite materials subjected to flexural loads since the replacement of agricultural residues for wood fibers would be an attractive alternative.

Considering the relationship between the contribution of the fibers to the flexural and tensile strength of the composite (FFSF/FTSF), and with knowledge of the intrinsic tensile strength of the fibers, it is therefore possible to determine the intrinsic flexural strength according to Equation (5).

Then, by using the mRoM for both the tensile and flexural properties, the respective coupling factors can be obtained and compared (Table 4).

Table 4. Intrinsic flexural (σ_f^F) and tensile strength (σ_t^F) of the fibers, and flexural (fc,f) and tensile (fc,t) coupling factors.

Sample	Reinforcement (wt.%)	$\frac{FFSF}{FTSF}$	Tensile		Flexural	
			σ_t^F (MPa)	fc,t	σ_f^F (MPa)	fc,f
BioPE + barley	15		532.9	0.18	703.4	0.18
	30	1.32	521.2	0.18	688.0	0.19
	45		500.5	0.18	660.7	0.18

The intrinsic flexural strength increased to 703.4 MPa at a 15 wt.% of reinforcement, being lower at the 45 wt.% (660.7 MPa). The followed methodology was proved to work efficiently owing to the great similarities between the tensile and flexural coupling factors. As previously mentioned, the coupling factor in natural fiber composites with optimal interfaces is between 0.18 and 0.20, proving the good interface in barley composites.

3. Materials and Methods

3.1. Materials

Composite materials were prepared using biobased polyethylene (BioPE) as polymer matrix and barley straw residues as reinforcement. BioPE was kindly supplied by Braskem (Sao Paulo, Brazil). BioPE is obtained from bioethanol coming from sugarcane feedstocks. Thereby, the polymer is completely biobased and 100% recyclable in the same chain established for the conventional fossil-based polyethylene. The melt flow index of the polymer is 20 g/10 for hammer weight of 2.16 kg, with a density of 0.955 g/cm^3. Maleic anhydride polyethylene was added as coupling agent to enhance the interfacial adhesion between the matrix and the reinforcement. The coupling agent (Fusabond MB100D) was supplied by DuPont (Wilmington, DE, USA). Barley straws residues were kindly provided by Mas Clarà S.A. (Girona, Spain). The length of a single barley straw ranged from 5 to 50 cm, with diameters between 0.1 and 0.6 cm.

Ethanol (95 wt.%), toluene (99.5 wt.%) and sulfuric acid (72 wt.%) were employed for the chemical characterization of the fibers. All reagents used in the present investigation were supplied by Sigma-Aldrich and used as received.

3.2. Methods

3.2.1. Thermomechanical (TM) Barley Straw Fiber Production and Characterization

Barley straw was chopped by means of a blade mill with a 3 mm mesh. Straw particles were then subjected to a thermomechanical treatment for the extraction of single fibers (TMP fibers). For this, the lignocellulosic material was submitted to steam-water treatment in a pressurized reactor at 160 °C temperature and solid to liquid ratio of 1:6 for 15 min. Afterwards, the obtained suspension was filtered and washed thoroughly with distilled water. The obtained pulp was mechanically defibrated by using Sprout Waldon equipment, responsible of the fiber defibering. Finally, fibers were oven-dried at 80 °C until constant weight.

The chemical composition and morphology of the fibers was examined. The size distribution analysis was carried out using MORFI equipment (TechPAP, Gières, France). A minimum of 4 samples were analyzed, taking 30,000 images of fibers in each analysis. The analysis of the chemical constituents was carried out from the analysis of the ethanol soluble extractives (TAPPI T204 cm-07), ashes (ISO 2144:2019) and lignin (ISO/DIS 21436). The holocellulose content (cellulose + hemicelluloses) was measured by difference.

3.2.2. Composites Preparation and Sample Obtaining

BioPE and barley TMP fibers were blended at weight ratios of 85/15, 70/30 and 55/45 (matrix/reinforcement) by means of an intensive Gelimat kinetic mixer. Initially, the fibers were introduced in the mixer at a speed of 300 rpm. The polymer and the coupling agent were then added to the mixer chamber maintaining constant speed. The speed was then increased up to 2500 rpm until the polymer was completely melted. The composite is then after discharged and cooled down and pelletized using a blade mill equipped with a 5 mm mesh. The material was oven-dried until constant weight.

The specimens for the flexural test were produced with a steel mold in an injection molding machine Aurburg 220 M 350-90U (Aurburg, Loßburg, Germany). Tensile specimens were also acquired for the determination of the tensile properties of the composites.

3.2.3. Mechanical Test

Prior to testing, specimens were placed in a conditioning chamber (Dycometal, Sant Boi de Llobregat, Spain) at 23 °C and 50% relative humidity for 48 h, according to ASTM D618 standard. Flexural properties of the specimens were determined by means of an INSTRON universal testing machine equipped with a 5 kN load cell. The flexural test was performed following ASTM D790. Tensile properties were also measured following ASTM D638 standard. At least five specimens of each composite formulation were tested.

Figure 9 presents a schematic flowchart of the experimental procedure, including composite's preparation and the analysis of its properties.

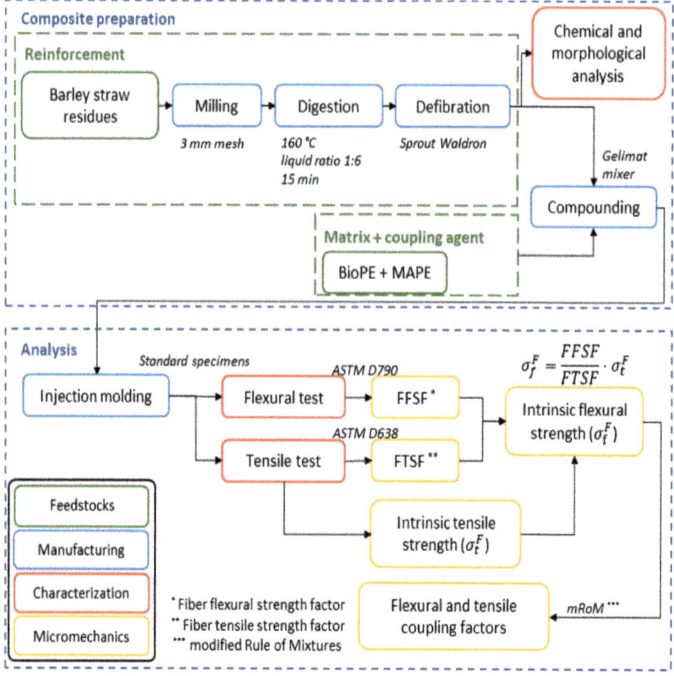

Figure 9. Flowchart of the current investigation.

4. Conclusions

The present work evaluates the feasibility of incorporating barley straw fibers as reinforcement in a biobased polyethylene to develop a fully biobased and 100% recyclable material. Barley straw was treated by means of a thermomechanical process and the resulting fibers were evaluated in terms of its chemical composition and its morphology. The efficiency of barley fibers was enhanced by the addition of anhydride maleic polyethylene as coupling agent. The flexural behavior of the material was investigated as important property determining the suitability of the material for several applications. The addition of barley straw fibers caused enlargement in the flexural strength about the 42%, 103% and 147% at 15, 30 and 45 wt.% fiber content, respectively.

A methodology was followed to determine the intrinsic flexural strength of the fibers. The methodology assumes that the flexural and tensile coupling factors are in the same order of magnitude. The coupling factors were found to be in the range from 0.18 to 0.20, an indication of the existence of strong interfaces for semi-aligned short fiber reinforced composites. The intrinsic flexural strength of barley straw changed with the amount of reinforcement, showing values ranging from 700 MPa at a 15 wt.% to 660 MPa at a 45 wt.% reinforcement content. The results from the study show the suitability of barley straw biobased composites for semi-structural and engineering purposes.

Author Contributions: F.S.-P. and F.J., investigation; E.E., validation; A.R. and F.V., writing—revision; F.X.E. and F.V., supervision. All authors have read and agreed to the published version of the manuscript.

Funding: This research received no external funding.

Acknowledgments: The authors are grateful to Spain's DGICyT, MICINN for funding this research within the framework of the Projects CTQ2016-78729-R and supported by the Spanish Ministry of Science and Education through the National Program FPU (Grant Number FPU14/02278), and to the staff of the Central Service for Research Support (SCAI) at the University of Córdoba. The authors wish to acknowledge the financial support of the Càtedra de Processos Industrials Sostenibles of the University of Girona.

Conflicts of Interest: The authors declare no conflict of interest.

References

1. United Nations, Department of Economic and Social Affairs. *World Population Prospects 2019: Highlights*; United Nations: New York. NY, USA, 2019.

2. González, Z.; Rodríguez, A.; Vargas, F.; Jiménez, L. Influence of the operational variables on the pulping and beating of the orange tree pruning. *Ind. Crops Prod.* **2013**, *49*, 785–789. [CrossRef]

3. González, Z.; Vargas, F.; Jiménez, L.; Rodríguez, A. Orange tree prunings as raw material for cellulose production by the Kraft process. *Cell. Chem. Technol.* **2013**, *47*, 603–611.

4. Anastas, P.T.; Warner, J.C. Principles of green chemistry. In *Green Chemistry: Theory and Practice*; Oxford University: New York, NY, USA, 1998.

5. Espinach, F.X.; Julian, F.; Verdaguer, N.; Torres, L.; Pelach, M.A.; Vilaseca, F.; Mutjé, P. Analysis of tensile and flexural modulus in hemp strands/polypropylene composites. *Compos. Part B Eng.* **2013**, *47*, 339–343. [CrossRef]

6. Petts, J. *Waste Incineration and The Environment*; Royal Society of Chemistry: Cambridge, UK, 1994.

7. Han, M.; Kang, K.E.; Kim, Y.; Choi, G.W. High efficiency bioethanol production from barley straw using a continuous pretreatment reactor. *Process Biochem.* **2013**, *48*, 488–495. [CrossRef]

8. Mostafaeipour, A.; Sedaghat, A.; Hedayatpour, M.; Jahangiri, M. Location planning for production of bioethanol fuel from agricultural residues in the south of Caspian Sea. *Environ. Dev.* **2020**, *33*, 100500. [CrossRef]

9. Rezania, S.; Oryani, B.; Cho, J.; Talaiekhozani, A.; Sabbagh, F.; Hashemi, B.; Rupani, P.F.; Mohammadi, A.A. Different pretreatment technologies of lignocellulosic biomass for bioethanol production: An overview. *Energy* **2020**, *199*, 117457. [CrossRef]

10. Vargas, F.; González, Z.; Rojas, O.; Garrote, G.; Rodríguez, A. Barley Straw (Hordeum vulgare) as a Supplementary Raw Material for *Eucalyptus camaldulensis* and *Pinus sylvestris* Kraft Pulp in the Paper Industry. *BioResources* **2015**, *10*, 3682–3693. [CrossRef]

11. Vargas, F.; González, F.; González, Z.; Sánchez, R.; Jiménez, L.; Rodríguez, A. Cellulosic pulps of cereal straws as raw material for the manufacture of ecological packaging. *BioResources* **2012**, *7*, 4161–4170.

12. Everall, N.C.; Lees, D.R. The use of barley-straw to control general and blue-green algal growth in a Derbyshire reservoir. *Water Res.* **1996**, *30*, 269–276. [CrossRef]

13. Spence, D.; Lembi, C. Evaluation of barley straw as an alternative algal control method in Northern California rice fields. *J. Aquat. Plant. Manag.* **2007**, *45*, 84–90.

14. Cerdà, A.; González-Pelayo, Ó.; Giménez-Morera, A.; Jordán, A.; Pereira, P.; Novara, A.; Brevik, E.C.; Prosdocimi, M.; Mahmoodadbadi, M.; Keesstra, S.B.; et al. Use of barley straw residues to avoid high erosion and runoff rates on persimmon plantations in Eastern Spain under low frequency–high magnitude simulated rainfall events. *Soil Res.* **2016**, *54*, 154–165. [CrossRef]

15. Prosdocimi, M.; Jordán, A.; Tarolli, P.; Keesstra, S.; Novara, A.; Cerdà, A. The immediate effectiveness of barley straw mulch in reducing soil erodibility and surface runoff generation in Mediterranean vineyards. *Sci. Total Environ.* **2016**, *547*, 323–330. [CrossRef] [PubMed]

16. Yan, L.; Kasal, B.; Huang, L. A review of recent research on the use of cellulosic fibres, their fibre fabric reinforced cementitious, geo-polymer and polymer composites in civil engineering. *Compos. Part B Eng.* **2016**, *92*, 94–132. [CrossRef]

17. Pickering, K.L.; Efendy, M.G.A.; Le, T.M. A review of recent developments in natural fibre composites and their mechanical performance. *Compos. Part A Appl. Sci. Manuf.* **2016**, *83*, 98–112. [CrossRef]

18. Wirawan, R.; Sapuan, S.M.; Abdan, K.; Yunus, R.B. Tensile and impact properties of sugarcane bagasse/poly (vinyl chloride) composites. *Key Eng. Mater.* **2011**, *471*, 167–172. [CrossRef]

19. Zabihzadeh, S.M. Influence of Plastic Type and Compatibilizer on Thermal Properties of Wheat Straw Flour/Thermoplastic Composites. *J. Thermoplast. Compos. Mater.* **2010**, *23*, 817–826. [CrossRef]

20. Wang, W.; Yuan, S.; Bu, F.; Li, G.; Wang, Q. Wheat-Straw-HDPE Composite Produced by the Hot-pressing Method. *J. Thermoplast. Compos. Mater.* **2011**, *24*, 251–261. [CrossRef]

21. Zhou, Y.; Fan, M.; Chen, L. Interface and bonding mechanisms of plant fibre composites: An overview. *Compos. Part B Eng.* **2016**, *101*, 31–45. [CrossRef]

22. Tarrés, Q.; Vilaseca, F.; Herrera-Franco, P.J.; Espinach, F.X.; Delgado-Aguilar, M.; Mutjé, P. Interface and micromechanical characterization of tensile strength of bio-based composites from polypropylene and henequen strands. *Ind. Crops Prod.* **2019**, *132*, 319–326. [CrossRef]

23. Scarponi, C.; Messano, M. Comparative evaluation between E-Glass and hemp fiber composites application in rotorcraft interiors. *Compos. Part B Eng.* **2015**, *69*, 542–549. [CrossRef]

24. López, J.P.; Méndez, J.A.; Mansouri, N.E.E.; Mutjé, P.; Vilaseca, F. Mean intrinsic tensile properties of stone groundwood fibers from softwood. *BioResources* **2011**, *6*, 5037–5049. [CrossRef]

25. Salem, S.; Oliver-Ortega, H.; Espinach, F.X.; Hamed, K.B.; Nasri, N.; Alcalà, M.; Mutjé, P. Study on the Tensile Strength and Micromechanical Analysis of Alfa Fibers Reinforced High Density Polyethylene Composites. *Fibers Polym.* **2019**, *20*, 602–610. [CrossRef]

26. Sullins, T.; Pillay, S.; Komus, A.; Ning, H. Hemp fiber reinforced polypropylene composites: The effects of material treatments. *Compos. Part B Eng.* **2017**, *114*, 15–22. [CrossRef]

27. Kumar, R.; Ul-Haq, M.I.; Raina, A.; Anand, A. Industrial applications of natural fibre-reinforced polymer composites–challenges and opportunities. *Int. J. Sustain. Eng.* **2019**, *12*, 212–220. [CrossRef]

28. Vallejos, M.E.; Espinach, F.X.; Julián, F.; Torres, L.; Vilaseca, F.; Mutjé, P. Micromechanics of hemp strands in polypropylene composites. *Compos. Sci. Technol.* **2012**, *72*, 1209–1213. [CrossRef]

29. Masłowski, M.; Miedzianowska, J.; Strzelec, K. Natural rubber biocomposites containing corn, barley and wheat straw. *Polym. Test.* **2017**, *63*, 84–91. [CrossRef]

30. Bouasker, M.; Belayachi, N.; Hoxha, D.; Al-Mukhtar, M. Physical Characterization of Natural Straw Fibers as Aggregates for Construction Materials Applications. *Materials* **2014**, *7*, 3034–3048. [CrossRef]

31. Bederina, M.; Belhadj, B.; Ammari, M.S.; Gouilleux, A.; Makhloufi, Z.; Montrelay, N.; Quéneudéc, M. Improvement of the properties of a sand concrete containing barley straws - Treatment of the barley straws. *Constr. Build. Mater.* **2016**, *115*, 464–477. [CrossRef]

32. Belhadj, B.; Bederina, M.; Makhloufi, Z.; Dheilly, R.M.; Montrelay, N.; Quéneudéc, M. Contribution to the development of a sand concrete lightened by the addition of barley straws. *Constr. Build. Mater.* **2016**, *113*, 513–522. [CrossRef]

33. Hyvärinen, M.; Kärki, T. The Effects of the Substitution of Wood Fiberwith Agro-based Fiber (Barley Straw) on the Properties of Natural Fiber/Polypropylene Composites. *MATEC Web Conf.* **2015**, *30*, 01014. [CrossRef]
34. Silva-Guzmán, J.A.; Anda, R.R.; Fuentes-Talavera, F.J.; Manríquez-González, R.; Lomelí-Ramírez, M.G. Properties of Thermoplastic Corn Starch Based Green Composites Reinforced with Barley (Hordeum vulgare L.) Straw Particles Obtained by Thermal Compression. *Fibers Polym.* **2018**, *19*, 1970–1979. [CrossRef]
35. Rojas-Leon, A.; Guzmán-Ortiz, F.A.; Bolarín-Miró, A.M.; Otazo-Sánchez, E.M.; Prieto-García, F.; Fuentes-Talavera, F.J.; Román-Gutierrez, A.D. Eco-innovation of barley and HDPE wastes: A proposal of sustainable particleboards. *Rev. Mex. Ing. Química* **2019**, *18*, 57–68. [CrossRef]
36. Serra-Parareda, F.; Tarrés, Q.; Delgado-Aguilar, M.; Espinach, F.X.; Mutjé, P.; Vilaseca, F. Biobased Composites from Biobased-Polyethylene and Barley Thermomechanical Fibers: Micromechanics of Composites. *Materials* **2019**, *12*, 4182. [CrossRef] [PubMed]
37. Tarrés, Q.; Oliver-Ortega, H.; Espinach, F.X.; Mutjé, P.; Delgado-Aguilar, M.; Méndez, J.A. Determination of Mean Intrinsic Flexural Strength and Coupling Factor of Natural Fiber Reinforcement in Polylactic Acid Biocomposites. *Polymers* **2019**, *11*, 1736. [CrossRef]
38. Espinach, F.X.; Méndez, J.A.; Granda, L.A.; Pelach, M.A.; Delgado-Aguilar, M.; Mutjé, P. Bleached kraft softwood fibers reinforced polylactic acid composites, tensile and flexural strengths. *Nat. Fiber-Reinforced Biodegrad. Bioresorbable Polym. Compos.* **2017**, 73–90. [CrossRef]
39. Gironès, J.; Lopez, J.P.; Vilaseca, F.; Bayer, R.; Herrera-Franco, P.J.; Mutjé, P. Biocomposites from Musa textilis and polypropylene: Evaluation of flexural properties and impact strength. *Compos. Sci. Technol.* **2011**, *71*, 122–128. [CrossRef]
40. Oliver-Ortega, H.; Julian, F.; Espinach, F.X.; Tarrés, Q.; Ardanuy, M.; Mutjé, P. Research on the use of lignocellulosic fibers reinforced bio-polyamide 11 with composites for automotive parts: Car door handle case study. *J. Clean. Prod.* **2019**, *226*, 64–73. [CrossRef]
41. Serrano, A.; Espinach, F.X.; Tresserras, J.; Pellicer, N.; Alcala, M.; Mutje, P. Study on the technical feasibility of replacing glass fibers by old newspaper recycled fibers as polypropylene reinforcement. *J. Clean. Prod.* **2014**, *65*, 489–496. [CrossRef]
42. Doan, T.T.L.; Gao, S.L.; Mäder, E. Jute/polypropylene composites I. Effect of matrix modification. *Compos. Sci. Technol.* **2006**, *66*, 952–963. [CrossRef]
43. Granda, L.A.; Espinach, F.X.; Tarrés, Q.; Méndez, J.A.; Delgado-Aguilar, M.; Mutjé, P. Towards a good interphase between bleached kraft softwood fibers and poly(lactic) acid. *Compos. Part B Eng.* **2016**, *99*, 514–520. [CrossRef]
44. Amuthakkannan, P.; Manikandan, V.; Winowlin-Jappes, J.T.; Uthayakumar, M. Effect of fibre length and fibre content on mechanical properties of short basalt fibre reinforced polymer matrix composites. *Mater. Phys. Mech.* **2016**, *16*, 107–117.
45. Serra-Parareda, F.; Tarrés, Q.; Espinach, F.X.; Vilaseca, F.; Mutjé, P.; Delgado-Aguilar, M. Influence of lignin content on the intrinsic modulus of natural fibers and on the stiffness of composite materials. *Int. J. Biol. Macromol.* **2020**, *115*, 81–90. [CrossRef] [PubMed]
46. Ververis, C.; Georghiou, K.; Christodoulakis, N.; Santas, P.; Santas, R. Fiber dimensions, lignin and cellulose content of various plant materials and their suitability for paper production. *Ind. Crops Prod.* **2004**, *19*, 245–254. [CrossRef]
47. Reixach, R.; Franco-Marquès, E.; El Mansouri, N.E.; Ramirez de Cartagena, F.; Arbat, G.; Espinach, F.X.; Mutjé, P. Micromechanics of Mechanical, Thermomechanical, and Chemi-Thermomechanical Pulp from Orange Tree Pruning as Polypropylene Reinforcement: A Comparative Study. *BioResources* **2013**, *8*, 3231–3246. [CrossRef]
48. Theng, D.; Arbat, G.; Delgado-Aguilar, M.; Vilaseca, F.; Ngo, B.; Mutjé, P. All-lignocellulosic fiberboard from corn biomass and cellulose nanofibers. *Ind. Crops Prod.* **2015**, *76*, 166–173. [CrossRef]
49. Tarrés, Q.; Soler, J.; Rojas-Sola, J.I.; Oliver-Ortega, H.; Julián, F.; Espinach, F.X.; Mutjé, P.; Delgado-Aguilar, M. Flexural Properties and Mean Intrinsic Flexural Strength of Old Newspaper Reinforced Polypropylene Composites. *Polymers* **2019**, *11*, 1244. [CrossRef]
50. Mutjé, P.; Vallejos, M.E.; Gironès, J.; Vilaseca, F.; López, A.; López, J.P.; Méndez, J.A. Effect of maleated polypropylene as coupling agent for polypropylene composites reinforced with hemp strands. *J. Appl. Polym. Sci.* **2006**, *102*, 833–840. [CrossRef]

51. Franco-Marquès, E.; Méndez, J.A.; Pèlach, M.A.; Vilaseca, F.; Bayer, J.; Mutjé, P. Influence of coupling agents in the preparation of polypropylene composites reinforced with recycled fibers. *Chem. Eng. J.* **2011**, *166*, 1170–1178. [CrossRef]

52. Granda, L.A.; Espinach, F.X.; López, F.; García, J.C.; Delgado-Aguilar, M.; Mutjé, P. Semichemical fibres of Leucaena collinsii reinforced polypropylene: Macromechanical and micromechanical analysis. *Compos. Part B Eng.* **2016**, *91*, 384–391. [CrossRef]

53. Faruk, O.; Bledzki, A.K.; Fink, H.P.; Sain, M. Biocomposites reinforced with natural fibers: 2000–2010. *Prog Polym. Sci.* **2012**, *37*, 1552–1596. [CrossRef]

54. Reis, P.N.B.; Ferreira, J.A.M.; Silva, P.A.A. Mechanical behaviour of composites filled by agro-waste materials. *Fibers Polym.* **2011**, *12*, 240–246. [CrossRef]

55. Mirmehdi, S.M.; Zeinaly, F.; Dabbagh, F. Date palm wood flour as filler of linear low-density polyethylene. *Compos. Part B Eng.* **2014**, *56*, 137–141. [CrossRef]

56. López, J.P.; Gironès, J.; Mendez, J.A.; Pèlach, M.A.; Vilaseca, F.; Mutjé, P. Impact and flexural properties of stone-ground wood pulp-reinforced polypropylene composites. *Polym. Compos.* **2013**, *34*, 842–848. [CrossRef]

57. Turker, D.; Nadir, A.; Büyüksari, U. Utilization of waste pine cone in manufacture of wood/plastic composite. In Proceedings of the Second International Conference on Sustainable Construction Materials and Technologies, Ancona, Italy, 28–30 June 2010; Volume 3, pp. 1517–1528.

58. Zhang, K.; Barhoum, A.; Chen, X.; Li, H.; Samyn, P. Cellulose Nanofibers: Fabrication and Surface Functionalization Techniques. In *Handbook of Nanofibers*; Springer Nature Switzerland AG: Cham, Switzerland, 2019. [CrossRef]

59. Bledzki, A.K.; Gassan, J. Composites reinforced with cellulose based fibres. *Prog. Polym. Sci.* **1999**, *24*, 221–274. [CrossRef]

60. Bledzki, A.K.; Gassan, J.; Theis, S. Wood-filled thermoplastic composites. *Mech. Compos. Mater.* **1998**, *34*, 563–568. [CrossRef]

61. Shebani, A.N.; Van Reenen, A.J.; Meincken, M. The Effect of Wood Species on the Mechanical and Thermal Properties of Wood—LLDPE Composites. *J. Compos. Mater.* **2009**, *43*, 1305–1318. [CrossRef]

62. Vilaseca, F.; Valadez-Gonzalez, A.; Herrera-Franco, P.J.; Pèlach, M.À.; López, J.P.; Mutjé, P. Biocomposites from abaca strands and polypropylene. Part I: Evaluation of the tensile properties. *Bioresour. Technol.* **2010**, *101*, 387–395. [CrossRef]

63. Alcalá, M.; González, I.; Boufi, S.; Vilaseca, F.; Mutjé, P. All-cellulose composites from unbleached hardwood kraft pulp reinforced with nanofibrillated cellulose. *Cellulose* **2013**, *20*, 2909–2921. [CrossRef]

64. Serra, A.; Tarrés, Q.; Chamorro, M.À.; Soler, J.; Mutjé, P.; Espinach, F.X.; Vilaseca, F. Modeling the Stiffness of Coupled and Uncoupled Recycled Cotton Fibers Reinforced Polypropylene Composites. *Polymers* **2019**, *11*, 1725. [CrossRef]

65. Kelly, A.; Tyson, W.R. Tensile properties of fibre-reinforced metals: Copper/tungsten and copper/molybdenum. *J. Mech. Phys. Solids* **1965**, *13*, 329–338. [CrossRef]

66. Bowyer, W.H.; Bader, M.G. On the re-inforcement of thermoplastics by imperfectly aligned discontinuous fibres. *J. Mater. Sci.* **1972**, *7*, 1315–1321. [CrossRef]

67. Hashemi, S. Hybridisation effect on flexural properties of single- and double-gated injection moulded acrylonitrile butadiene styrene (ABS) filled with short glass fibres and glass beads particles. *J. Mater. Sci.* **2008**, *43*, 4811–4819. [CrossRef]

68. Pimenta, M.T.B.; Carvalho, A.J.F.; Vilaseca, F.; Girones, J.; López, J.P.; Mutjé, P.; Curvelo, A.A.S. Soda-treated sisal/polypropylene composites. *J. Polym. Environ.* **2008**, *16*, 35–39. [CrossRef]

69. Thomason, J. Interfacial strength in thermoplastic composites–at last an industry friendly measurement method? *Compos. Part A Appl. Sci. Manuf.* **2002**, *33*, 1283–1288. [CrossRef]

Sample Availability: Samples of the compounds are not available from the authors.

Article

Effect of the Micronization of Pulp Fibers on the Properties of Green Composites

Bruno F. A. Valente [1], **Armando J. D. Silvestre** [1], **Carlos Pascoal Neto** [2], **Carla Vilela** [1] and **Carmen S. R. Freire** [1,*]

1   CICECO—Aveiro Institute of Materials, Department of Chemistry, University of Aveiro, Campus Universitário de Santiago, 3810-193 Aveiro, Portugal; bfav@ua.pt (B.F.A.V.); armsil@ua.pt (A.J.D.S.); cvilela@ua.pt (C.V.)
2   RAIZ—Research Institute of Forest and Paper, The Navigator Company, Rua José Estevão, 3800-783 Eixo, Portugal; Carlos.Neto@thenavigatorcompany.com
*   Correspondence: cfreire@ua.pt

Abstract: Green composites, composed of bio-based matrices and natural fibers, are a sustainable alternative for composites based on conventional thermoplastics and glass fibers. In this work, micronized bleached Eucalyptus kraft pulp (BEKP) fibers were used as reinforcement in biopolymeric matrices, namely poly(lactic acid) (PLA) and poly(hydroxybutyrate) (PHB). The influence of the load and aspect ratio of the mechanically treated microfibers on the morphology, water uptake, melt flowability, and mechanical and thermal properties of the green composites were investigated. Increasing fiber loads raised the tensile and flexural moduli as well as the tensile strength of the composites, while decreasing their elongation at the break and melt flow rate. The reduced aspect ratio of the micronized fibers (in the range from 11.0 to 28.9) improved their embedment in the matrices, particularly for PHB, leading to superior mechanical performance and lower water uptake when compared with the composites with non-micronized pulp fibers. The overall results show that micronization is a simple and sustainable alternative for conventional chemical treatments in the manufacturing of entirely bio-based composites.

Keywords: poly(lactic acid); poly(hydroxybutyrate); cellulose fibers; micronization; green composites

Citation: Valente, B.F.A.; Silvestre, A.J.D.; Neto, C.P.; Vilela, C.; Freire, C.S.R. Effect of the Micronization of Pulp Fibers on the Properties of Green Composites. *Molecules* **2021**, *26*, 5594. https://doi.org/10.3390/molecules26185594

Academic Editors: Alejandro Rodriguez Pascual, Eduardo Espinosa Víctor and Carlos Martín

Received: 6 August 2021
Accepted: 9 September 2021
Published: 15 September 2021

Publisher's Note: MDPI stays neutral with regard to jurisdictional claims in published maps and institutional affiliations.

1. Introduction

The increasing demand for eco-friendly materials associated with the implementation of legislation and policies towards a more sustainable society has triggered the replacement of synthetic and petrochemical-based materials with bio-based ones [1,2]. In the field of composite materials, as far as reinforcements are concerned, a notorious increase in the use of natural-based fibers in replacement of synthetic counterparts, such as glass or aramid, has been witnessed in the last decade. Several natural fibers such as flax, hemp, jute, kenaf, wood flour, or pulp have been thoroughly investigated [3,4]. In fact, the market of natural fiber-based composites, also commonly referred to as biocomposites, is already established (USD 22.3 billion in 2019) and some large companies, such as Stora Enso, UPM, and Sappi, have launched over the years a range of products composed of conventional fossil-based and non-biodegradable thermoplastics, such as polypropylene (PP) and polyethylene (PE), reinforced with cellulosic fibers [5–8].

PP and PE, together with other polymeric matrices, such as poly(vinyl chloride) (PVC), polystyrene (PS), and acrylonitrile butadiene styrene (ABS), are still the main thermoplastics used in the biocomposite industry [9]. However, and despite the clear environmental benefits over composites reinforced with synthetic fibers, biocomposites whose matrices are derived from fossil resources still pose some environmental threats. Specifically, the non-renewability and non-biodegradability of the matrices as well as the unfeasibility to recycle the composites are still their major drawbacks [10,11]. In this regard, the logical alternative is to manufacture fully bio-based composites by replacing the non-biodegradable petrochemical-derived matrices with bio-based polymers, the so-called bioplastics [12].

Poly(lactic acid) (PLA) and poly(hydroxybutyrates) (PHB) are among the few bioplastics currently produced at a commercial scale. They have comparable properties to some commodity plastics, can be processed with technologies applied to conventional thermoplastics, and, because of the increasing demand for bioplastics and maturing of production technologies, their prices are becoming more affordable [10,13,14]. The ever-increasing number of studies regarding the use of bioplastics in the composite field reflect the growing interest on these sustainable polymers [4]. However, these so-called green composites, for which both the matrix and reinforcement are bio-based, face some of the same challenges of their petroleum-based counterparts. Although the interfacial adhesion between for instance PLA and cellulose fibers is naturally stronger than for many other thermoplastic polymers, the lack of compatibility between the hydrophilic cellulose fibers and the hydrophobic matrices is still an issue [11,15]. The intrinsic hydrophilicity and high aspect ratio of the cellulosic materials often lead to agglomeration and poor dispersion of the fibers in the polymeric matrices [16]. Therefore, the visual aspect of the composites is inevitably impaired, as well as their mechanical performance. The strategies commonly used for composites with petrochemically based matrices have also been investigated for PLA and PHB-based counterparts to overcome these challenges and further increase the overall performance of such materials. Pre-treatments of the fibers, such as alkali treatments [17] or surface modifications including, for example, acetylation [18] or silylation [19], are among the most common. Other compatibilization strategies, such as the use of coupling agents [20–22], which can be done prior to or during melt-mixing, have also been tested. From the industrial point of view, however, the manufacturing process should be as simple, inexpensive, and efficient as possible. Thus, as an alternative to the aforementioned chemical methods, mechanical procedures may also be efficient to overcome some drawbacks. In this sense, size reduction by milling processes, such as pan milling [23], ball milling [24,25], or shear and cooling milling [26], may be used to decrease the length and width of the fibers to the micro or nano range. For instance, ball milling has been already used to reduce the size of bleached pine kraft pulp to particles sizes inferior to 120 μm [24,25]. The results showed that size reduction was an efficient method to increase the dispersion of cellulose materials in PLA or PLA/poly(3-hydroxybutyrate-*co*-3-hydroxyvalerate) (PHBV) based composites. In addition, the size reduced particles still had a reinforcing effect as proved by the increased mechanical properties. However, upon these mechanical treatments, the fibers lost their fibrillar morphology, becoming irregular particles instead. Additionally, the aspect ratio and crystallinity index (CI) of the fibers drastically decreased [24,25].

In this context, the present study aims at manufacturing fully green composites using bio-based matrices, namely PLA and PHB, reinforced with micronized bleached Eucalyptus kraft pulp (BEKP). Micronization is proposed as a mechanical treatment to reduce both the length and width of the fibers to the micrometric range, without compromising their fibrillar morphology and crystallinity but rather improving their dispersion within the polymeric matrices. This strategy has the additional advantage of being practical and free of any solvents or chemicals. The influence of the fiber load and effect of their aspect ratio were studied. The composites were evaluated regarding their interfacial morphology, mechanical performance, water uptake capacity, melt flowability, and thermal properties.

2. Results and Discussion

In the present work, the impact of the micronized fiber load and the effect of their aspect ratio on the properties of green composites with two distinct grades of both PLA and PHB matrices were evaluated. The four bio-based polymeric matrices (i.e., PLA 3D860, PLA 3100HP, PHB P209E, and PHB P226) were selected according to their mechanical properties, melt flow index, and recommended uses [27,28]. To assess the impact of the micronized fibers' load, Cel355 fibers with an intermediate aspect ratio (26.6) were melt-mixed with the thermoplastic matrices. To study the effect of the aspect ratio of the fibers, four micronized fibers with distinct aspect ratios and BEKP (Table 1) were also melt-mixed with all four thermoplastic matrices for a fixed fiber load of 40 wt.% (Figure 1).

Table 1. Average dimensions and crystallinity indexes of the cellulose fibers used in this study.

Fiber Sample	Length (μm)	Width (μm)	Aspect Ratio	Crystallinity Index (%)
Cel60	149 ± 129	13.6 ± 5.4	11.0	54.1
Cel200	257 ± 170	14.9 ± 4.6	17.2	65.4
Cel355	332 ± 211	12.5 ± 5.4	26.6	64.6
Cel500	405 ± 203	14.4 ± 4.5	28.9	68.4
BEKP	770 ± 0.006	18.2 ± 0.1	42.3	70.7

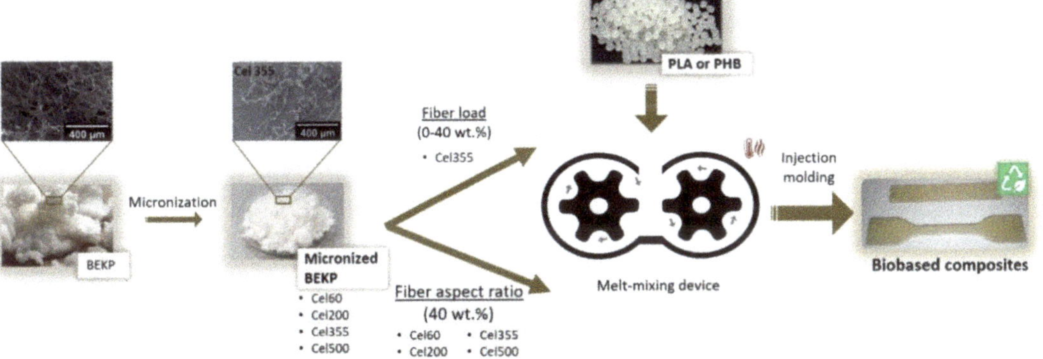

Figure 1. Schematic illustration of the experimental procedure used in the present study.

The macroscopic aspect of the composites specimens after the injection molding is shown in Figure 2. The composites with different loads of Cel355 (Figure 2A) are increasingly darker with the increment of the fiber load, particularly for the composites with PLA. This behavior can be attributed to the high temperatures used during the processing of these composites, particularly during injection molding (195 °C), which led to some thermal degradation of cellulose. A similar observation was reported by Ozyhar et al. [29] for composites of PLA reinforced with 40 wt.% of wood fibers, where the increase in the composites color intensity was also associated to some thermal degradation of the polysaccharides during processing. Besides the increased color intensity, no visible agglomerates of the fibers could be perceived for composites with different reinforcement loads or for those with 40 wt.% micronized fibers with distinct aspect ratios (Figure 2B), which might indicate that the fibers were homogeneously dispersed in the matrices. The density values of the polymeric matrices agreed with the specification of the products [27,28] (Table S1). As for the composites, as expected, an increase on the fiber load raised the density of the composite, given the superior density of the fibers [30] in comparison with the thermoplastic polymers used in the present work. Conversely, the density of the composites with fibers having different aspect ratios remained relatively unchanged (Table S1). Then, the morphology, mechanical properties, water-uptake capacity, flowability, and thermal stability of all the composites were evaluated.

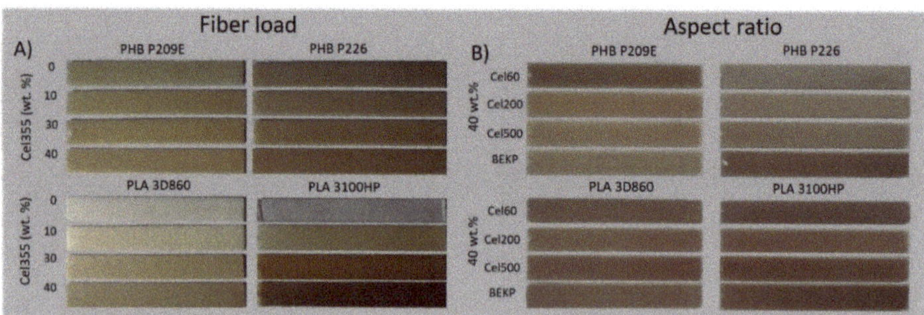

Figure 2. Digital photographs of the PHB- and PLA-based composites with: (**A**) different Cel355 loads (0, 10, 30 and 40 wt.%), and (**B**) with micronized fibers with distinct aspect ratios and non-micronized BEKP fibers.

2.1. Morphology

The mechanical properties of composites are highly dependent on the ability to efficiently transfer energy from the polymeric matrix to the fibers that in turn depends on the dispersion of the fibers and on the interfacial adhesion between them and the matrix [31]. The cross-section fracture surfaces of the neat matrices, obtained after tensile tests, and of the composites with different loads of Cel355 are displayed in Figure 3, while those of the composites with micronized fibers with different aspect ratios are shown in Figure 4. Despite the strong tendency of the hydrophilic cellulose fibers to stack together when compounded with hydrophobic thermoplastic matrices [31], no visible agglomerates or bundles of micronized fibers can be observed on the micrographs of the composites with different contents of Cel355 (Figure 3) or with fibers with different aspects ratios (Figure 4). This is a confirmation of the good dispersion of the micronized fibers in the PLA and PHB matrices, which agrees with the visual observation of the test specimens (Figure 2).

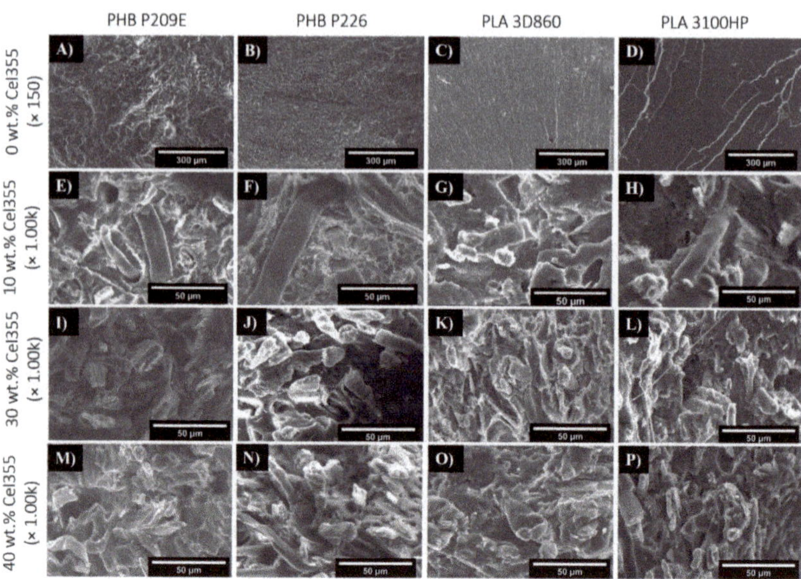

Figure 3. SEM micrographs of the fractured surfaces of (**A–D**) the neat PHB and PLA matrices and of (**E–P**) the corresponding composites reinforced with different Cel355 loads: (**E–H**) 10 wt.%, (**I–L**) 30 wt.% and (**M–P**) 40 wt.%.

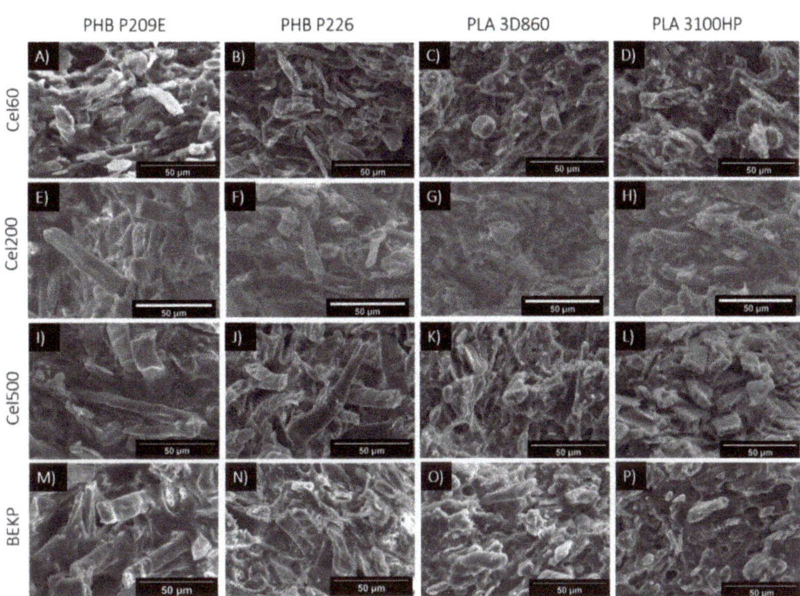

Figure 4. SEM micrographs of the fractured surfaces of composites with fibers having different aspect ratios: (**A–D**) Cel60; (**E–H**) Cel200; (**I–L**) Cel500 and (**M–P**) BEKP.

Besides this good dispersion, the fracture surfaces of the composites also revealed the existence of some fibers pull-outs and voids (Figure 3E, top). This phenomenon has been extensively described in the literature and is usually attributed to the different phobic nature of the constituents [32–34]. However, due to the carboxylic and hydroxyl end groups of PHB and PLA, there is a high degree of compatibility between these matrices and cellulose fibers when compared with common thermoplastic matrices, such as PP or PE [15,35]. The fact that fractured fibers are more prevalent on the micrographs than the pulled-out fibers is also evidence of such a compatibility. Additionally, the micrographs point to a slightly better compatibility of the fibers with PLAs than with the PHBs, which may be related to the chemical structural differences between these two polyesters, with PHBs having a longer monomeric chain.

Concerning the composites loaded with different micronized fibers (Figure 4), as the fiber aspect ratio increases, alongside with the length, the fiber pullouts and voids created during the tensile testing are more frequent and the pulled-out fibers are longer, as would be expected. In regard to the interfaces between the fibers and matrices, although only minor differences can be observed on the micrographs, shorter micronized fibers may have improved dispersion and embedment over the longer and higher aspect ratio non-treated BEKP. Such difference of the effect of the fiber dimensions on the interfacial morphology was described by Madyan et al. [36], in which larger fibers led to bigger gaps and cracks, and smaller fibers were better imbedded in the matrix.

2.2. Mechanical Properties

2.2.1. Tensile Properties

The Young's modulus, tensile strength, and elongation at break of the polymeric matrices and of the corresponding composites are presented in Figure 5. The results show, for both PLA- and PHB-based composites, a gradual increment in the Young's modulus with the increasing Cel355 fiber load, which is in agreement with previous findings on the effect of the fiber load, and can be easily explained by the higher stiffness of the cellulose fibers compared to the polymeric matrices [37]. Specifically, the Young's modulus of composites reinforced with 40 wt.% Cel355 increased by 1.7 GPa (2.37 ± 0.06 GPa) and 2.6 GPa (3.83 ± 0.02 GPa) in comparison with the respective PHB matrices (PHB P209E: 0.64 ± 0.02 GPa; PHB P226: 1.25 ± 0.07 GPa), and by 2.8 GPa (5.16 ± 0.05 GPa) and 3.1 GPa (5.83 ± 0.06 GPa) compared to the corresponding PLA matrices (PLA 3D860: 2.40 ± 0.04 GPa; PLA 3100HP: 2.75 ± 0.04 GPa). As the biggest increases are noted for the PLA-based composites, those findings corroborate the relatively better compatibility of the fiber with PLAs than with PHBs, as previously discussed.

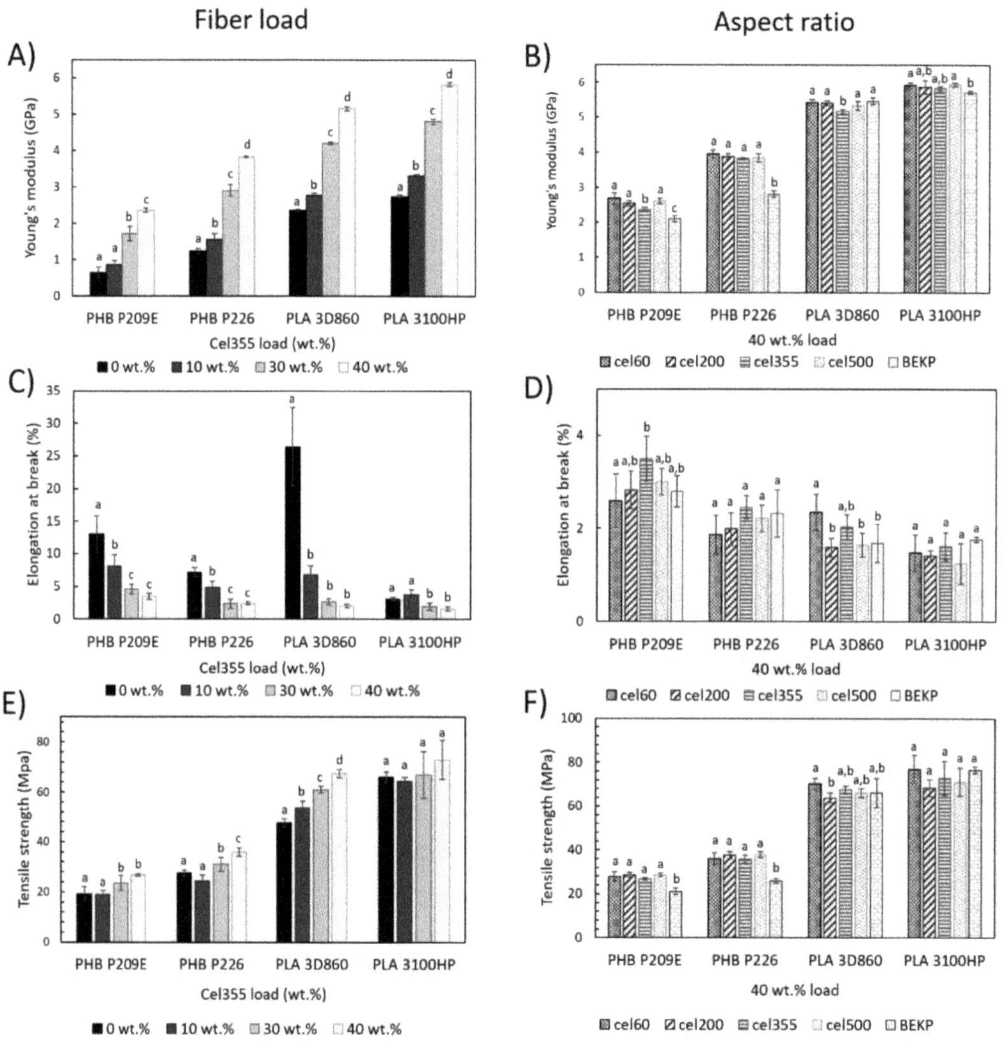

Figure 5. Tensile properties of the PHB and PLA-based composites (**A,C,E**) reinforced with different loads of Cel355 and (**B,D,F**) reinforced with fibers having different aspect ratios for a load of 40 wt.%. Different letters (a,b,c,d) indicate statistically significant differences ($p < 0.05$).

In previous studies of PLA reinforced with a 40 wt.% load of bleached softwood kraft pulp, a similar increase of 3.0 GPa was registered when compared to the matrix [38], which shows that the micronized fibers used in this work, despite their reduced sizes, still have a good reinforcing effect. The effectiveness of such a reinforcing effect is closely related to the fibrillar morphology and crystallinity (Table 1) of the micronized fibers since, unlike other mechanical treatments (e.g., ball milling), the micronization of the fibers still retained their fibrillar morphology and only a slight decrease in the crystallinity was observed [24,25]. Moreover, the Young's modulus of the composites of PHBs reinforced with 40 wt.% Cel355 were within the range of the commercial products based on PP or PE reinforced with 40 wt.% pulp fibers (1.9 to 4.6 GPa) and the Young's modulus of the PLA-based composites were clearly superior to those of the mentioned commercial products [5–7].

Concerning the effect of the aspect ratio on the Young's modulus, for the most part, changes in the micronized fiber aspect ratio (in the range between 11.0 and 28.9) had little influence on this parameter (Figure 5B). Qiang et al. studied the effect of size variations of ball-milled bleached pine kraft pulp in composites with PLA or binary mixtures of PLA and poly(3-hydroxybutyrate-*co*-3-hydroxyvalerate) (PHBV), and also concluded that the loading content contributes more to the variation of the mechanical properties than the fiber aspect ratio [24,25]. Interestingly, PHB-based composites reinforced with micronized fibers had a significantly superior Young's modulus than the composites reinforced with BEKP. In fact, the average Young's modulus of the PHB composites with the micronized fibers is superior to the corresponding composites with BEKP by 26% for PHB P209E and 29% for PHB P226.

With respect to the elongation at break, as expected, a reduction in this parameter was generally observed with the increasing fibers content [10]. All composites with 40 wt.% reinforcement had elongation at break values below 3.5% (Figure 5C), which is inferior to those of the commercial products mentioned before (4.0 to 6.5%) [5–7]. In contrast, with only a few exceptions, the aspect ratio of the fibers did not have a significant effect on the elongation at break of the composites, as portrayed in Figure 5D.

The representation of the tensile strength of the composites as a function of the fiber load (Figure 5E) shows that higher reinforcement contents of micronized fibers generally raised the tensile strength. For example, in composites with 40 wt.% Cel355, an improvement on the tensile strength of 39, 30, 42, and 10% was noted in comparison with the corresponding PHB P209E, PHB P226, PLA 3D860, and PLA 3100HP matrices. The present results contradict some published works in which often the increase in the fiber load tended to decrease the tensile strength, either for PLA or PHB matrices [37,39]. Additionally, the tensile strength values of composites of PLA 3D860 (67.5 ± 1.7 MPa) and PLA 3100HP (72.9 ± 7.7 MPa) reinforced with 40 wt.% Cel355 were superior to those of commercial products of PP reinforced with pulp fibers (47 to 64 MPa) [5–7]. These results are certainly related to the relatively good dispersion of the micronized fibers within the polymeric matrices, as well as to their interfacial adhesion, as previously observed by SEM analysis. The composites with micronized fibers having different aspect ratios had similar tensile strengths (Figure 5F), which has also been previously reported for composites of PLA and mixtures of PLA and PHBV [24,25]. Despite the almost negligible effect of the aspect ratio, the tensile strength of all the PHB composites reinforced with the micronized fibers was far superior to those with BEKP, which agrees with the Young's modulus trend previously described.

2.2.2. Flexural Properties

The effect of the reinforcement of Cel355 fibers on the flexural modulus of composites based on PLA and PHB matrices was in line with the results of the Young's modulus previously discussed, i.e., the increase in the fiber load led to a higher flexural modulus (Figure 6A). This is not surprising, though, given that the flexural modulus often follows the same pattern as the Young's modulus [39,40]. Moreover, the results obtained for PHBs reinforced with 30 wt.% Cel355 (2.5 ± 0.1 GPa for PHB P209E and 3.8 ± 0.1 GPa for PHB P226) are similar or even better to those obtained by Gunning et al., in which PHB matrices were reinforced with 30 wt.% hemp (≈1.6 GPa), Lyocell (≈1.9 GPa) and Jute (≈3.8 GPa) fibers [31]. For PLA-based composites, the flexural modulus of the composites reinforced with 40 wt.% Cel355 (7.3 ± 0.1 GPa for PLA 3D860 and 8.1 ± 0.2 GPa for

PLA 3100HP) was even superior to the values reported for PLA reinforced with 40 wt.% of sisal ($\approx$6.2 GPa) [41], wood ($\approx$4.5 GPa) [37], or with pulp fibers from poplar wood ($\approx$5.7 GPa) [37]. On the contrary to the effect observed for the fiber load, the fiber aspect ratio had little influence on the flexural modulus of the composites. However, PHB-based composites reinforced with BEKP had an inferior modulus than the composites reinforced with micronized fibers (Figure 6B), which is also in agreement with the variation in the Young's modulus. Such results are probably related to the larger dimensions of the BEKP that can lead to larger gaps and cracks, creating weak points for composites to fail [36]. On the contrary, the shorter micronized fibers may have improved distribution and embedment on the matrix, improving its reinforcing effect [42].

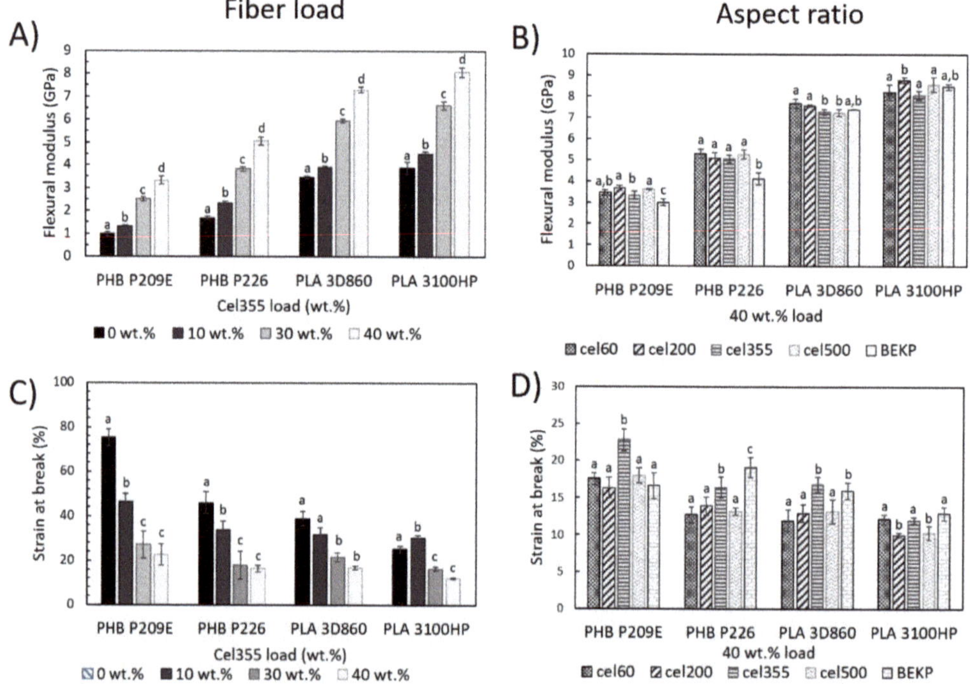

Figure 6. Flexural properties of) the PHB- and PLA-based composites reinforced (**A,C**) with different loads of Cel355 and (**B,D**) with fibers having different aspect ratios for a load of 40 wt.%. Different letters (a,b,c,d) indicate statistically significant differences ($p < 0.05$).

The strain at break, calculated as the ratio between the extension at break and the maximum deflection (20 mm), provides important information about the flexibility of the composites. As observed in Figure 6C, the polymeric matrices are more flexible than the corresponding composites and higher fiber content led to lower flexibility, which can be also credited to the high stiffness of the fibers in comparison with the matrices [10]. For example, the incorporation of only 10 wt.% of Cel355 in the most flexible matrix (PHB P209E) led to a reduction of the strain at break from 75.1 $\pm$ 4.1% to only 46.3 $\pm$ 4.9%. However, only small differences could be observed for the strain at break in the composites reinforced with fibers having different aspect ratios. The results of the strain at break are, in general, in agreement with the elongation at break determined on the tensile tests.

2.2.3. Impact Properties

The impact strength of neat polymeric matrices and composites as a function of the fiber load and aspect ratio were evaluated following the Charpy edgewise impact test (Figure 7). Regarding the polymeric matrices, the impact strength of the unnotched specimens of the PHBs, which are 38.0 ± 8.7 kJ m^{-2} for PHB P209E and 18.2 ± 5.2 kJ m^{-2} for PHB P226, were within the range of impact strengths reported in the literature, which can vary from 5 kJ m^{-2} to over 65 kJ m^{-2} [43,44]. The impact strength of PLA 3100HP (21.5 ± 1.8 kJ m^{-2}) was also near the values reported elsewhere [44,45]. However, the energy required to break an unnotched specimen of PLA 3D860 (121.4 ± 32.4 kJ m^{-2}) was more than three times higher than for any PHB matrix studied and more than five times higher than for PLA 3100HP, which can be justified by the fact that PLA 3D860 is designated by the manufacturers as a high impact polymer [28].

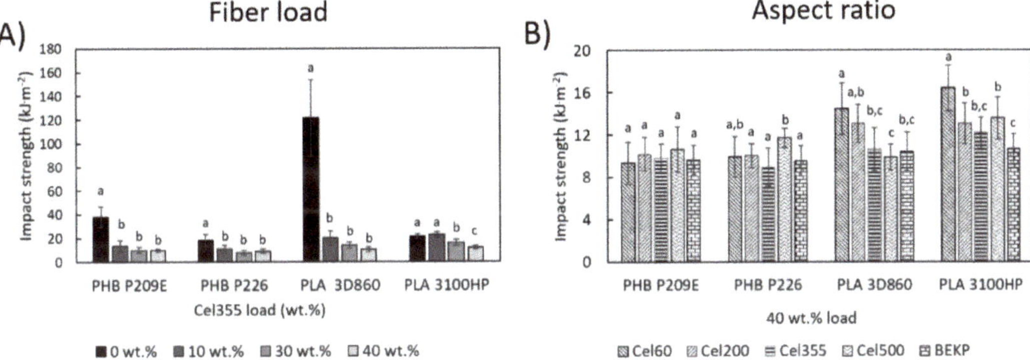

Figure 7. Impact properties of PHB- and PLA-based composites (**A**) reinforced with different loads of Cel355 and (**B**) reinforced with fibers having different aspect ratios for a load of 40 wt.%. Different letters (a,b,c) indicate statistically significant differences ($p < 0.05$).

The incorporation of micronized fibers (Cel355) significantly reduced the ability of the material to absorb the impact (Figure 7A). In fact, the most accentuated decreases were noted for composites with only 10 wt.%. of fibers. For higher cellulose contents, the decreases were not significantly different. In the literature, the effect of fiber incorporation in the impact strength of biocomposites is still unclear and contradictory. Previous studies have shown both increases and decreases in the impact strength upon incorporation of cellulosic fibers [46,47]. It's known that many factors may influence the composite's impact properties, such as the crystallinity and stiffness of the individual components. However, and despite the good homogeneity and dispersion of micronized fibers on the matrices, as observed in the SEM images (Figure 3), some defects on the interface may still be responsible for the deterioration of the impact strength [37,48]. Such defects on the interface, even at lower fiber loads, leads to crack initiation and propagation, which are responsible for the decreased amount of force the material can absorb during impact [32]. Nonetheless, the obtained values are in line with literature data. For instance, Oliver-Ortega et al. recorded an impact strength of 21.7 ± 1.2 kJ m^{-2} in composites of PLA reinforced with 10 wt.% bleached kraft softwood pulp [45], which is similar to the impact strength obtained for PLA 3D860 reinforced with 10 wt.% of Cel355 (20.4 ± 5.6 kJ m^{-2}) and PLA 3100HP (22.8 ± 2.3 kJ m^{-2}).

In Figure 7B, the results indicate that for PHB matrices, the changes in the aspect ratio of the fibers did not translate into different impact strength properties. In contrast, for composites based on PLA, smaller aspect ratio fibers seemed to favor the impact strength. The relatively better interfacial adhesion between the PLA matrices and fibers, as seen by SEM, combined with their smaller aspect ratio, reduced the formation of defects on the composite, consequently lowering sites for initiation and propagation of cracks [36]. In comparison with other materials, the obtained impact properties of composites with a fiber load of 40 wt.%, ranging from 9.0 ± 1.8 kJ m^{-2} to 16.4 ± 2.2 kJ m^{-2}, are inferior to those of the commercial products of PP or PE reinforced with 40 wt.% pulp fibers (33 to 42 kJ m^{-2}) [5–7], which still presents a challenge.

2.3. Water Uptake Capacity

The evaluation of the water absorption of the composites reinforced with natural fibers is of enormous importance because it is normally associated with dimensional stability issues and a decrease of the mechanical properties [49]. The water absorption of the neat PLA and PHB matrices used in this study were under 1.3 ± 0.1%, after 31 days of immersion (Figure 8), which is in accordance with the values reported in the literature for PLA [40] and PHB [50] matrices. Since cellulose has high affinity for water [51,52], all composites showed higher water absorptions that increased with the growing fiber contents, which is not surprising given there is unanimity in the fact that increasing the content of the hydrophilic portion of the composites leads to an increased water absorption [22,45]. Moreover, all composites followed the same uptake pattern: a rapid increase during the first few days of immersion and stabilization after reaching saturation [53]. It is also noticeable that composites with PHBs had higher water-uptakes than composites with PLAs. This observation can also be related to the apparent inferior compatibility between the fibers and PHBs, as previously discussed (Figure 3), which leads to enhanced water penetration thought the composite material [45,51]. These current findings contradict the observations made by Yatigala et al., in which wood fiber-reinforced PLA composites had higher-water uptakes (11.9%) than the PHB counterpart (10.2%) [22]. However, and more importantly, the composites prepared in this work with 30 wt.% of the micronized fiber (Cel355) had significantly inferior water-uptakes (6.4 ± 0.1% to 6.6 ± 0.1% for PHBs and 4.6 ± 0.1% to 4.9 ± 0.1% for PLAs) than the ones previously reported by the authors for the same reinforcement percentage [22].

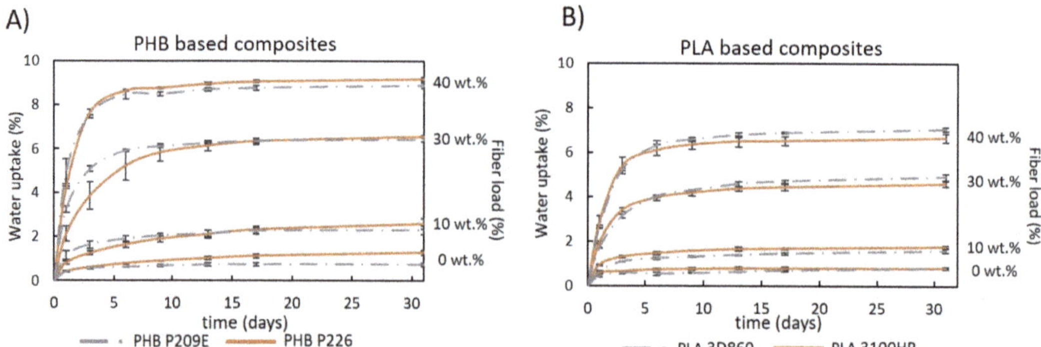

Figure 8. Water uptake as function of time for: (**A**) PHB- and (**B**) PLA-based composites reinforced with Cel355.

Interestingly, composites reinforced with micronized fibers have a clear advantage over the ones reinforced with non-micronized BEKP (Figure 9). Apart from those of PLA 3100HP, composites with BEKP not only have higher water-uptake values (up to 11.1 ± 0.2%) but also higher water absorption rates during the first days, mainly for composites with PHBs. The positive effect of micronization on the water-uptake behavior is in line with the results of the tensile and flexural moduli, which strengthens the idea that the reduced size of micronized fibers favors their embedment in the matrix, preventing water from permeating easily into the composite [51]. In reference to the aspect ratio of the micronized fibers (11.0 to 28.9), with the exception of composites based on PLA 3100HP, all composites with different micronized fibers had similar water-uptake values.

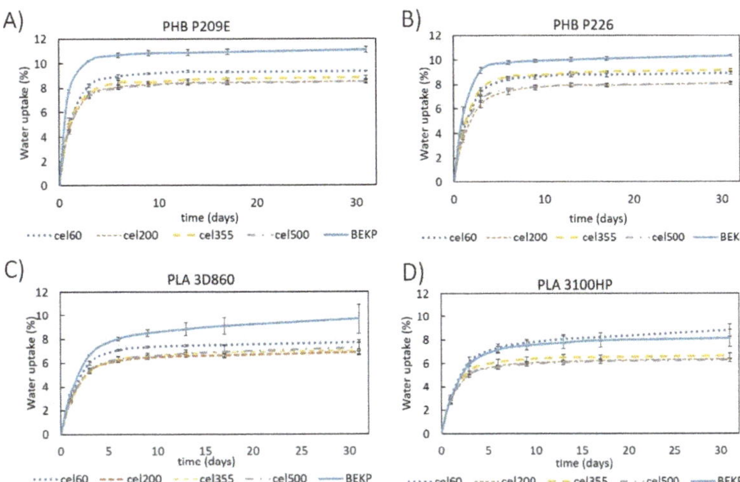

Figure 9. Water uptake as function of time for composites prepared with different micronized cellulose samples for a fiber load of 40 wt.% in matrices of: (**A**) PHB P209E; (**B**) PHB P226; (**C**) PLA 3D860 and (**D**) PLA 3100HP.

2.4. Melt Flow Rate

The melt flow rate measures the ease of a molten thermoplastic material to flow under very specific conditions, such as the diameter and length of the die and the cylinder [54]. The outcome is expressed in grams of the material that flows over the course of ten minutes when standard weights are applied at a predetermined temperature [54]. Figure 10A,C represent the melt flow rate of the PLA and PHB matrices prior to and after melt-mixing, along with the melt flow rate of the composites with different fiber loadings, while Figure 10B,D show the melt flow of the composites having fibers with different aspect ratios.

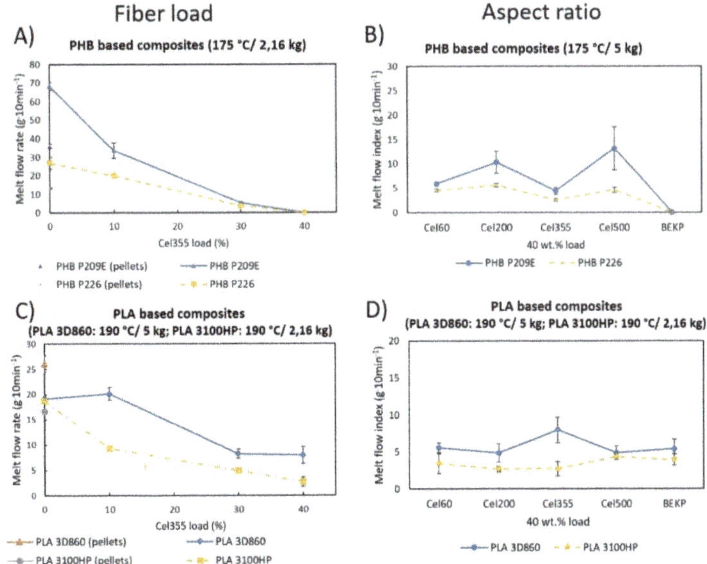

Figure 10. MFR of the polymeric matrices before and after melt-mixing: (**A,C**) their composites with Cel355 and (**B,D**) of the composites with fibers having different aspect ratios.

According to Figure 10A,C, the melt flow rate of the polymeric matrices increased after being submitted to the melt-mixing procedure, except for PLA 3D860, probably due to the presence of additives in this PLA sample. The effect is more pronounced for the PHBs, with increments of 90% for PHB P209E and a 103% increase for PHB P226. Giving that there is an inversely proportional relationship between the MFR and the molecular weight of the thermoplastic polymers, the increase in the MFR is most likely due to some thermal degradation during the melt-mixing procedure, which certainly leads to the decrease in the molecular weight of the polymeric matrices [55]. This is supported by previous studies as Carrasco F. et al., who studied the influence of melt processing on the molecular weight and melt flowability of PLA, concluding that the thermal degradation caused by melt processing decreased the molecular weight of PLA from 212.3 kDa to 162.5 kDa. As a direct consequence, the melt flow rate increased from 7.0 g·10 min^{-1} to 10.7 g·10 min^{-1} [56]. A similar outcome was verified for PHB matrices where melt processing decreased the molecular weight from 535×10^3 g·mol^{-1} to 208×10^3 g·mol^{-1}, leading the MFR to raise from 19 g·10 min^{-1} to 26 g·10 min^{-1} [57].

When the micronized fibers (Cel355) were added to the matrices, the MFR gradually decreased. Similar observations have been reported for composites of PLA or PHB reinforced with different natural fibers (e.g., jute and hemp) [31,58] and the decrease was mainly attributed to the bad dispersion of the fibers within the matrices and to the fiber/fiber and fiber/matrix frictions [31,58]. However, since the composites in the present work show a good dispersion of the micronized fibers, as previously discussed, the decrease in the MFR probably has to do with the increase in the friction between fibers and between fibers with the matrices, which its turn decreases the ease of flow [58]. Moreover, the most pronounced decreases in the MFR were noted for PHB-based composites, which, according to the SEM micrographs, were slightly less compatible with the fibers than the PLA matrices. The poorer interfacial adhesion may increase the friction between the fibers and the matrices, leading in turn to a decreased flowability of the composite material [59].

The fiber aspect ratio had little effect on the MFR, especially for PLA-based composites. However, for the composites based on PHBs, the micronization had a positive effect on the MFR. All the composites reinforced with micronized fibers showed MFRs between 4.5 ± 0.3 g·10 min^{-1} and 13.2 ± 4.5 g·10 min^{-1}, while those reinforced with BEKP did not flow at all. In composites with PHBs, where the compatibility with fibers is slightly inferior than with PLAs, as previously observed by SEM (Figures 3 and 4), the BEKP fibers, which are considerably longer and wider than the micronized fibers, may lead to a higher fiber entanglement, preventing the composite material from flowing [31,60]. In contrast, the shorter micronized fibers with improved dispersion and embedment may have prevented fiber entanglement. Such an observation emphasizes that the micronization of the BEKP fibers is a useful strategy to improve the flowability of the composites, which is of great importance for injection molding applications to ensure proper mold-filling and good quality of the materials [31,61].

2.5. Thermal Analysis

The thermal stability of all the thermoplastic polymers and composites were studied by thermogravimetric analysis (TGA). The thermogravimetric curves of PLA and PHB matrices, as well as of the corresponding composites with different fiber loads, are shown in Figure 11 and those of the composites with fibers having different aspect ratios are represented in Figure S5. The neat polymeric matrices showed a degradation profile with one major weight loss at 236 °C and 281 °C for PHB P209E and PHB P226, respectively, and 336 °C and 364 °C for PLA 3D860 and PLA 3100HP, respectively. These main weight losses, credited to the degradation of the polymeric backbone, are in agreement with the maximum degradation temperatures of other PHB [62,63] and PLA [34,64] grades reported in the literature. Apart from PLA 3100HP, the other matrices showed a small degradation step at 338 °C, 395 °C, and 456 °C for PHB P209E, PHB P226, and PLA 3D860, respectively, which were likely due to the degradation of the additives.

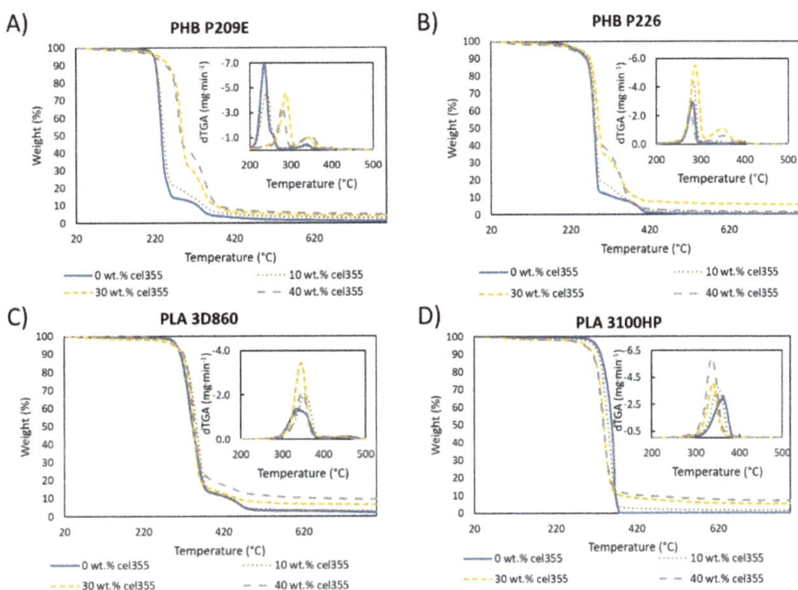

Figure 11. Thermogravimetric and derivative (dTGA) curves of the pure polymeric matrices and composites with different loads of Cel355 (10, 30, and 40 wt.%) in matrices of: (**A**) PHB P209E; (**B**) PHB P226; (**C**) PLA 3D860 and (**D**) PLA 3100HP.

PHB-based composites have three different degradation steps: (i) a small weight loss attributed to water vaporization starting at approximately 100 °C, associated with the humidity of the fibers; (ii) a main weight loss step between 236 and 287 °C, related to the degradation of PHBs' main chain; and (iii) the step related to the thermal degradation of cellulose between 335 and 350 °C (Figure S4c) [22,65]. Additionally, the weight losses associated with this last step match the different reinforcement loads of the fibers. These ternary degradation profiles have also been reported by other authors for PHB-based composites loaded with other natural fibers (e.g., wood fibers and flax) [22,66]. The maximum degradation temperatures of PHB P226-based composites were similar to those of the pristine matrix (around 280 °C), but for composites with PHB P209E (Figure 11A), the incorporation of high contents of cellulose fibers, e.g., 40 wt.% load, increased the maximum thermal degradation from 236 °C to over 280 °C. This increase proves that, although the compatibility of PHB with the micronized fibers is not excellent, as seen in the SEM micrographs, there is some degree of interfacial adhesion between them [35]. Even if increases in the thermal stability have been reported in other studies, as is the example of PHB reinforced with agave fibers [39], the effects of the fiber incorporation on the thermal stability of PHB-based composites are debatable, with most studies showing decreases on the stability upon incorporation of natural fibers such as piassava [67], flax [66] almond shell, or rice husk [68].

For composites with PLA, however, given the relatively good interfacial adhesion between the micronized fibers and these thermoplastic polymers, as well as the similar thermal stabilities of PLAs and micronized cellulose fibers, only a main weight loss can be observed on the thermograms in Figure 11. Regarding the composites with PLA 3D860 as the matrix, the incorporation of fibers led to an increase in the maximum degradation temperature of the material (Figure 11C). For instance, the addition of 40 wt.% of Cel355 raised the maximum degradation temperature by 11 °C from 336 °C to 347 °C, which can be justified by the lower maximum degradation temperature of PLA 3D860 (336 °C) in comparison with Cel355 (350 °C) (Figure S4c). On the contrary, because the maximum degradation temperature of PLA 3100HP (364 °C) was superior to that of Cel355, the gradual increase on the fiber load led to a slight reduction on the maximum degradation temperature of the composite (Figure 11D). In a similar research study, Espinach et al. [64] investigated the thermal properties of composites made of PLA reinforced with bleached kraft softwood and also concluded that the decrease in the thermal stability of the composites was due to the presence of a cellulosic filler with a lower thermal stability than the PLA. The existence of only one peak also strengthens the relatively good compatibility between PLA and cellulosic fibers.

Analogous to the composites with different fiber loads, composites based on PHBs reinforced with 40 wt.% of micronized fibers with different aspect ratios displayed the same ternary degradation profile and PLA-based composites only exhibited one major single degradation step. The similar degradation patterns and degradation temperatures observed in Figure S5 allow us to withdraw the conclusion that the aspect ratio of the micronized fibers does not influence the thermal stability of the composites. This conclusion corroborates the results obtained for the tensile, flexural, and impact mechanical properties, as well as the results from MFR in the sense that different aspect ratios of the micronized fibers gave origin to composites with identical properties.

3. Materials and Methods

3.1. Materials

Two poly(lactic acid) (PLA) pellet samples: Ingeo™ Biopolymer 3D860, with a melt flow rate (MFR) of 5–7 g·10 min^{-1} (210 °C, 2.16 kg), density of 1.22 g·cm^{-3}, and relative viscosity of 4.0, and Ingeo™ Biopolymer 3100HP with an MFR of 24 g·10 min^{-1} (210 °C, 2.16 kg), density of 1.24, relative viscosity of 3.1, and molecular weight of 148 kDa [69], were supplied by NatureWorks (Minnnetonka, Minnesota, USA). Two commercial pellet grades of poly(hydroxybutyrate) (PHB): Biomer® P209E, with an MFR of 10 g·10 min^{-1} (180 °C, 2.16 kg), density of 1.20 g·cm^{-3}, and molecular weight of 0.6 × 10^6 [70], and Biomer® P226 with an MFR of 10 g·10 min^{-1} (180 °C, 5 kg), density of 1.25 g·cm^{-3}, and number-average molecular weight of 22,200 ± 4500 [71], were purchased from Biomer Biopolyesters (Schwalbach, Germany). PHBs samples had 0–40 wt.% of a plasticizer and an unreported amount of a nucleating agent [27]. The chemical and crystalline structures of the thermoplastic matrices were confirmed through Fourier-transform infrared spectroscopy (FTIR) and X-ray diffraction, as shown in Figures S1 and S2, respectively.

Bleached Eucalyptus kraft pulp (BEKP) with a crystallinity index (CI) of 70.1% was kindly provided by The Navigator Company (Aveiro, Portugal). The micronized fibers were obtained from the same BEKP on a drum milling equipment (Pallman Fine grinding PS, Zweibrücken, Germany) armed with knives. The BEKP was fed into the cutting chamber and cut repeatedly between the rotor knifes and stator knifes against each other until the material could pass through the screen insert. Different sieve meshes were used to produce fibers with different aspect ratios, namely Cel60 (11.0), Cel200 (17.2), Cel355 (26.6), and Cel500 (28.9). The micrographs showing their fibrillar morphology as well as the length and width distributions are presented in Figure S3, and the average length, width, aspect ratio, and crystallinity indexes are shown in Table 1. The CI of BEPK and of all the microfibers (Figure S4b) were determined by X-ray diffraction based on the peak height method. A Phillips X'pert MDP diffractometer (PANalytical, The Netherlands) using CuKα radiation (λ = 1.541 Å) with a scan rate of 0.05° s^{-1} was used for this analysis [72].

3.2. Compounding and Processing of the Biocomposites

Composites with different percentage loads of Cel355 (micronized fibers with an intermediate aspect ratio) ranging from 10 to 40 wt.%, relative to the total weight of the composite, were compounded with the four distinct thermoplastic polymers: PHB P209E, PHB P226, PLA3D860, and PLA3100HP. Additionally, to evaluate the influence of the fibers aspect ratio, the four thermoplastic matrices were also compounded with the four micronized pulp fibers and the non-treated BEKP for a fixed reinforcement load of 40 wt.%.

All composites were manufactured by melt-mixing in a Brabender W 30 EHT Plastograph EC mixer (Duisburg, Germany) with a total volume capacity of 30 cm^3. The thermoplastic polymeric matrices and cellulose fibers were mixed for 15 min at a screw speed of 50 rpm at a temperature of 180 °C for PLA 3100HP and 170 °C for the remaining polymers.

Test specimens for the mechanical and water-uptake assays were prepared by injection molding in a Thermo Scientific Haake Minijet II (Waltham, MA, USA). For the PHB-based composites, the injection temperature was set at 175 °C, the mold temperature at 60–65 °C, injection pressure of 400 bar for 20 s, and post-injection pressure of 200 bar for 5 s. For the PLA-based composites, the injection temperature was 195 °C, with a mold temperature of 100–130 °C and injection pressure of 800 bar.

3.3. Characterization

The density of the polymeric matrices and composites were calculated by dividing the weight of the test specimens by their volume. At least five specimens ($80 \times 10 \times 4$ mm^3) with a volume of 3.2 cm^3 were weighted for each sample and both the mean and standard deviation calculated.

Scanning electron microscopy (SEM) analysis were performed on a FE-SEM Hitachi SU70-47 microscope (Hitachi High-Technologies Corporation, Tokyo, Japan), operated at 15.0 kV. The cross-section micrographs were obtained from the tensile test specimens after breaking. Prior to the analysis, all samples were coated with a carbon film.

Tensile and flexural assays were carried out on the universal testing machine Instron 5564 (Instron Corporation, Norwood, MA, USA). The tensile properties of at least six specimens were tested in accordance with the ISO-527-2 procedure (bar type3) at a velocity of 5 mm·min^{-1} using a 10 kN static load cell. For the flexural modulus, the three-point loading model was used according to ISO 178. Five specimens ($80 \times 10 \times 4$ mm^3) were tested at a crosshead velocity of 5 mm·min^{-1} and at a length of span between supports of 64 mm using a 500 N static load cell. To measure the unnotched Charpy (edgewise) impact strength, a Ray Ran Universal Pendulum impact system (Ray-Ran Test Equipment Ltd., Nuneaton, UK), operating a pendulum of 4 J, was used according to ISO 179/1eU. The support span was set at 62 mm. Ten specimens with dimensions of $80 \times 10 \times 4$ mm^3 were tested for each sample and the average values were calculated.

The water-uptake capacity was assessed by immersing composite specimens ($60 \times 10 \times 1$ mm^3) in water at room temperature during a period of 31 days. The weight of the samples was periodically assessed after removing the excess water with tissue paper. The water uptake (%) at time t was calculated according to Equation (1):

$$Water\ uptake\ (\%) = \frac{(W_t - W_0)}{W_0} \times 100 \tag{1}$$

where W_0 is the specimen's initial weight and W_t is the weight of the specimens after the immersion time in grams. The mean and standard deviation were calculated for three replicates.

The melt flow rate of the different samples was evaluated using the Melt Flow Indexer Davenport (MFR-9) (Ametek, Denmark) operated at 190 °C for the PLA-based composites and at 175 °C for the PHB-based composites. The cut-off intervals were chosen according to the ASTM D1238 standard. At least five cut-offs for each sample were weighted and the melt flow rate was calculated as follows:

$$MFR\ \left(g \cdot 10\ min^{-1}\right) = \frac{600 \times m}{t} \tag{2}$$

where m is the average mass of the cut-offs in grams and t is the cut-off time interval in seconds.

The thermal stability of the composites was evaluated with a SETSYS Setaram TGA analyzer (SETARAM Instrumentation, Lyon, France) equipped with a platinum cell.

Approximately 10 mg of each sample was heated from room temperature to 800 °C at a constant rate of 10 °C·m^{-1} under a nitrogen atmosphere.

Statistical analysis of all the mechanical properties data was performed using the analysis of variance (ANOVA) and Tukey's mean comparison test (OriginPro 9.6.5, Origin-Lab Corporation, Northampton, MA, USA) with the statistical significance established at $p < 0.05$.

4. Conclusions

Micronized cellulose fibers obtained from bleached Eucalyptus kraft pulp were investigated as reinforcements in composites with bio-based matrices (PLA and PHB). The influence of the fiber load and fiber aspect ratio on the performance of the composites was thoroughly studied.

The composites with micronized fibers displayed overall good homogeneity with no visible fiber agglomerates. Reinforcing the composites with increasing contents of micronized fibers significantly improved the Young's modulus, tensile strength, and flexural modulus, while decreasing the elongation at break, strain at break, and the impact strength. The increase in the water uptake and the decrease of the melt flow rate were also consequences of the increasing reinforcement content. For the most part, the maximum degradation temperatures of the composites were slightly raised with the incorporation of the fibers, especially in composites based on PHB P209E and PLA 3D860.

For composites with PHBs, the water-uptake resistance, melt flowability, and mechanical performance, namely the tensile strength and both the Young's and flexural moduli, were significantly improved when using micronized fibers rather than non-treated BEKP. However, the aspect ratio of the micronized fibers had little influence on the tensile, flexural, and impact properties, as well as on the thermal stability of the composites. In short, the overall results lead us to conclude that the micronization of the BEKP was an efficient and convenient method to further improve the performance of the composites, without the use of any hazardous solvents or chemicals. The mechanical and flow properties show the potential of the present fully sustainable composites as an alternative to the existing partially bio-based ones for injection molding applications intended for electronics, furniture, house appliances, and other applications.

Supplementary Materials: The following are available online: FTIR-ATR spectra and X-ray diffractograms of the thermoplastic matrices; SEM micrographs and size (width and length) histograms of the micronized fibers; FTIR-ATR spectra, X-ray diffractograms, and TGA thermograms of the micronized fibers; density of the matrices and composites, and TGA thermograms of the composites reinforced with 40 wt.% fibers having different aspect ratios.

Author Contributions: Conceptualization, C.P.N., C.V. and C.S.R.F.; methodology, B.F.A.V.; investigation, B.F.A.V. and C.V.; resources, A.J.D.S., C.P.N. and C.S.R.F.; data curation, B.F.A.V.; writing—original draft preparation, B.F.A.V.; writing—review and editing, B.F.A.V., A.J.D.S., C.P.N., C.V. and C.S.R.F.; supervision, C.P.N., C.V. and C.S.R.F.; project administration, A.J.D.S., C.P.N. and C.S.R.F. All authors have read and agreed to the published version of the manuscript.

Funding: This work was carried out under the project Inpactus: innovative products and technologies from eucalyptus, project number 21874, funded by Portugal 2020 through the European Regional Development Fund (ERDF) in the frame of COMPETE 2020 n°246/AXIS II/2017, and under the project CICECO-Aveiro Institute of Materials, UIDB/50011/2020 and UIDP/50011/2020, financed by national funds through the Portuguese Foundation for Science and Technology (FCT)/MCTES. FCT is also acknowledged for the research contracts under the Scientific Employment Stimulus to C.V. (CEECIND/00263/2018) and C.S.R.F. (CEECIND/00464/2017).

Institutional Review Board Statement: Not applicable. This study did not involve humans or animals.

Informed Consent Statement: Not applicable.

Data Availability Statement: Not applicable.

Conflicts of Interest: The authors declare no conflict of interest.

Sample Availability: Samples of the compounds are not available from the authors.

References

1. Ramamoorthy, S.K.; Skrifvars, M.; Persson, A. A review of natural fibers used in biocomposites: Plant, animal and regenerated cellulose fibers. *Polym. Rev.* **2015**, *55*, 107–162. [CrossRef]
2. Khan, M.Z.; Srivastava, S.K.; Gupta, M. Tensile and flexural properties of natural fiber reinforced polymer composites: A review. *J. Reinf. Plast. Compos.* **2018**, *37*, 1435–1455. [CrossRef]
3. Wang, J.; Liu, X.; Jin, T.; He, H.; Liu, L. Preparation of nanocellulose and its potential in reinforced composites: A review. *J. Biomater. Sci. Polym. Ed.* **2019**, *30*, 919–946. [CrossRef]
4. Gholampour, A.; Ozbakkaloglu, T. A review of natural fiber composites: Properties, modification and processing techniques, characterization, applications. *J. Mater. Sci.* **2020**, *55*, 829–892. [CrossRef]
5. Upmformi. Available online: www.upmformi.com (accessed on 22 April 2021).
6. Aqvacomp Composites. Available online: https://www.aqvacomp.fi/en/materials/ (accessed on 22 April 2021).
7. SAPPI Symbio. Available online: https://www.sappi.com/symbio (accessed on 22 April 2021).
8. Storaenso. Available online: https://www.storaenso.com/en/products/biocomposites (accessed on 22 April 2021).
9. Jaafar, J.; Siregar, J.P.; Mohd Salleh, S.; Mohd Hamdan, M.H.; Cionita, T.; Rihayat, T. Important Considerations in Manufacturing of Natural Fiber Composites: A Review. *Int. J. Precis. Eng. Manuf. Green Technol.* **2019**, *6*, 647–664. [CrossRef]
10. Granda, L.A.; Espinach, F.X.; Tarrés, Q.; Méndez, J.A.; Delgado-Aguilar, M.; Mutjé, P. Towards a good interphase between bleached kraft softwood fibers and poly(lactic) acid. *Compos. Part B Eng.* **2016**, *99*, 514–520. [CrossRef]
11. Peltola, H.; Immonen, K.; Johansson, L.; Virkajärvi, J.; Sandquist, D. Influence of pulp bleaching and compatibilizer selection on performance of pulp fiber reinforced PLA biocomposites. *J. Appl. Polym. Sci.* **2019**, *136*, 47955. [CrossRef]
12. Zini, E.; Scandola, M. Green composites: An overview. *Polym. Compos.* **2011**, *32*, 1905–1915. [CrossRef]
13. Plackett, D.; Södergård, A. Polylactide-based biocomposites. In *Natural Fibers, Biopolymers, and Biocomposites*; Mohanty, A.K., Mirsa, M., Drzal, L., Eds.; Taylor & Francis: Abingdon, UK, 2005; pp. 583–600.
14. Misra, M.; Pandey, J.K.; Mohanty, A.K. *Biocomposites: Design and Mechanical Performance*, 1st ed.; Woodhead Publishing: Sawston, UK, 2015; ISBN 9781782423942.
15. Faludi, G.; Dora, G.; Imre, B.; Renner, K.; Mōczō, J.; Pukánszky, B. PLA/lignocellulosic fiber composites: Particle characteristics, interfacial adhesion, and failure mechanism. *J. Appl. Polym. Sci.* **2014**, *131*, 39902. [CrossRef]
16. Satyanarayana, K.G.; Arizaga, G.G.C.; Wypych, F. Biodegradable composites based on lignocellulosic fibers-An overview. *Prog. Polym. Sci.* **2009**, *34*, 982–1021. [CrossRef]
17. Mazzanti, V.; Pariante, R.; Bonanno, A.; Ruiz de Ballesteros, O.; Mollica, F.; Filippone, G. Reinforcing mechanisms of natural fibers in green composites: Role of fibers morphology in a PLA/hemp model system. *Compos. Sci. Technol.* **2019**, *180*, 51–59. [CrossRef]
18. Mukherjee, T.; Sani, M.; Kao, N.; Gupta, R.K.; Quazi, N.; Bhattacharya, S. Improved dispersion of cellulose microcrystals in polylactic acid (PLA) based composites applying surface acetylation. *Chem. Eng. Sci.* **2013**, *101*, 655–662. [CrossRef]
19. Srubar, W.V.; Pilla, S.; Wright, Z.C.; Ryan, C.A.; Greene, J.P.; Frank, C.W.; Billington, S.L. Mechanisms and impact of fiber-matrix compatibilization techniques on the material characterization of PHBV/oak wood flour engineered biobased composites. *Compos. Sci. Technol.* **2012**, *72*, 708–715. [CrossRef]
20. Yu, T.; Hu, C.; Chen, X.; Li, Y. Effect of diisocyanates as compatibilizer on the properties of ramie/poly(lactic acid) (PLA) composites. *Compos. Part A Appl. Sci. Manuf.* **2015**, *76*, 20–27. [CrossRef]
21. Solle, M.A.; Arroyo, J.; Burgess, M.H.; Warnat, S.; Ryan, C.A. Value-added composite bioproducts reinforced with regionally significant agricultural residues. *Compos. Part A Appl. Sci. Manuf.* **2019**, *124*, 105441. [CrossRef]
22. Yatigala, N.S.; Bajwa, D.S.; Bajwa, S.G. Compatibilization improves physico-mechanical properties of biodegradable biobased polymer composites. *Compos. Part A Appl. Sci. Manuf.* **2018**, *107*, 315–325. [CrossRef]
23. Niu, Z.; Chen, Y.; Feng, J. Preparation, structure, and property of wood flour incorporated polypropylene composites prepared by a solid-state mechanochemical method. *J. Appl. Polym. Sci.* **2016**, *133*, 43108. [CrossRef]
24. Qiang, T.; Wang, J.; Wolcott, M.; Qiang, T.; Wang, J.; Wolcott, M.P. Facile Fabrication of 100% Bio-based and Degradable Ternary Cellulose/PHBV/PLA Composites. *Materials* **2018**, *11*, 330. [CrossRef] [PubMed]
25. Qiang, T.; Wang, J.; Wolcott, M.P. Facile Preparation of Cellulose/Polylactide Composite Materials with Tunable Mechanical Properties. *Polym. Plast. Technol. Eng.* **2018**, *57*, 1288–1295. [CrossRef]
26. Yang, S.; Bai, S.; Wang, Q. Sustainable packaging biocomposites from polylactic acid and wheat straw: Enhanced physical performance by solid state shear milling process. *Compos. Sci. Technol.* **2018**, *158*, 34–42. [CrossRef]
27. Biomer Biopolyesters. Available online: www.biomer.de (accessed on 23 April 2021).
28. NatureWorks. Available online: https://www.natureworksllc.com/Products/3-series-for-injection-molding (accessed on 22 April 2021).
29. Ozyhar, T.; Baradel, F.; Zoppe, J. Effect of functional mineral additive on processability and material properties of wood-fiber reinforced poly(lactic acid) (PLA) composites. *Compos. Part A Appl. Sci. Manuf.* **2020**, *132*, 105827. [CrossRef]
30. Khouaja, A.; Koubaa, A.; Ben Daly, H. Dielectric properties and thermal stability of cellulose high-density polyethylene bio-based composites. *Ind. Crops Prod.* **2021**, *171*, 113928. [CrossRef]

31. Gunning, M.A.; Geever, L.M.; Killion, J.A.; Lyons, J.G.; Higginbotham, C.L. Mechanical and biodegradation performance of short natural fibre polyhydroxybutyrate composites. *Polym. Test.* **2013**, *32*, 1603–1611. [CrossRef]
32. Faruk, O.; Bledzki, A.K.; Fink, H.-P.; Sain, M. Biocomposites reinforced with natural fibers: 2000–2010. *Prog. Polym. Sci.* **2012**, *37*, 1552–1596. [CrossRef]
33. Arbelaiz, A.; Txueka, U.; Mezo, I.; Orue, A. Biocomposites Based on Poly(Lactic Acid) Matrix and Reinforced with Lignocellulosic Fibers: The Effect of Fiber Type and Matrix Modification. *J. Nat. Fibers* **2020**, *17*, 26. [CrossRef]
34. Lee, M.C.; Koay, S.C.; Chan, M.Y.; Choo, H.L.; Pang, M.M.; Chou, P.M.; Tshai, K.Y. Properties of poly(lactic acid)/durian husk fiber biocomposites: Effects of fiber content and processing aid. *J. Thermoplast. Compos. Mater.* **2020**, *33*, 1518–1532. [CrossRef]
35. Robledo-Ortíz, J.R.; González-López, M.E.; Martín del Campo, A.S.; Pérez-Fonseca, A.A. Lignocellulosic Materials as Reinforcement of Polyhydroxybutyrate and its Copolymer with Hydroxyvalerate: A Review. *J. Polym. Environ.* **2021**, *29*, 1350–1364. [CrossRef]
36. Madyan, O.A.; Wang, Y.; Corker, J.; Zhou, Y.; Du, G.; Fan, M. Classification of wood fibre geometry and its behaviour in wood poly(lactic acid) composites. *Compos. Part A Appl. Sci. Manuf.* **2020**, *133*, 105871. [CrossRef]
37. Yang, Z.; Feng, X.; Bi, Y.; Zhou, Z.; Yue, J.; Xu, M. Bleached extruder chemi-mechanical pulp fiber-PLA composites: Comparison of mechanical, thermal, and rheological properties with those of wood flour-PLA bio-composites. *J. Appl. Polym. Sci.* **2016**, *133*, 44241. [CrossRef]
38. Immonen, K.; Anttila, U.; Wikström, L. Coupling of PLA and bleached softwood kraft pulp (BSKP) for enhanced properties of biocomposites. *J. Thermoplast. Compos. Mater.* **2019**, *32*, 328–341. [CrossRef]
39. Smith, M.K.M.; Paleri, D.M.; Abdelwahab, M.; Mielewski, D.F.; Misra, M.; Mohanty, A.K. Sustainable composites from poly(3-hydroxybutyrate) (PHB) bioplastic and agave natural fibre. *Green Chem.* **2020**, *22*, 3906–3916. [CrossRef]
40. Baghaei, B.; Skrifvars, M.; Rissanen, M.; Ramamoorthy, S.K. Mechanical and thermal characterization of compression moulded polylactic acid natural fiber composites reinforced with hemp and lyocell fibers. *J. Appl. Polym. Sci.* **2014**, *131*, 40534. [CrossRef]
41. Liang, Z.; Wu, H.; Liu, R.; Wu, C. Preparation of long sisal fiber-reinforced polylactic acid biocomposites with highly improved mechanical performance. *Polymers* **2021**, *13*, 1124. [CrossRef] [PubMed]
42. Mendes, J.F.; Castro, L.S.; Corrêa, A.C.; Marconcini, J.M.; Mattoso, L.H.C.; Mendes, R.F. Effects of short fibers and processing additives on HDPE composites properties reinforced with Pinus and Eucalyptus fibers. *J. Appl. Polym. Sci.* **2021**, *138*, 50178. [CrossRef]
43. Ren, H.; Zhang, Y.; Zhai, H.; Chen, J. Production and evaluation of biodegradable composites based on polyhydroxybutyrate and polylactic acid reinforced with short and long pulp fibers. *Cellul. Chem. Technol.* **2015**, *49*, 641–652.
44. Graupner, N.; Müssig, J. A comparison of the mechanical characteristics of kenaf and lyocell fibre reinforced poly(lactic acid) (PLA) and poly(3-hydroxybutyrate) (PHB) composites. *Compos. Part A Appl. Sci. Manuf.* **2011**, *42*, 2010–2019. [CrossRef]
45. Oliver-Ortega, H.; Tarrés, Q.; Mutjé, P.; Delgado-Aguilar, M.; Méndez, J.A.; Espinach, F.X. Impact strength and water uptake behavior of bleached kraft softwood-reinforced PLA composites as alternative to PP-based materials. *Polymers* **2020**, *12*, 2144. [CrossRef]
46. Virtanen, S.; Wikström, L.; Immonen, K.; Anttila, U.; Retulainen, E. Cellulose kraft pulp reinforced polylactic acid (PLA) composites: Effect of fibre moisture content. *AIMS Mater. Sci.* **2016**, *3*, 756–769. [CrossRef]
47. Peltola, H.; Pääkkönen, E.; Jetsu, P.; Heinemann, S. Wood based PLA and PP composites: Effect of fibre type and matrix polymer on fibre morphology, dispersion and composite properties. *Compos. Part A Appl. Sci. Manuf.* **2014**, *61*, 13–22. [CrossRef]
48. Singh, S.; Mohanty, A.K. Wood fiber reinforced bacterial bioplastic composites: Fabrication and performance evaluation. *Compos. Sci. Technol.* **2007**, *67*, 1753–1763. [CrossRef]
49. Espert, A.; Vilaplana, F.; Karlsson, S. Comparison of water absorption in natural cellulosic fibres from wood and one-year crops in polypropylene composites and its influence on their mechanical properties. *Compos. Part A Appl. Sci. Manuf.* **2004**, *35*, 1267–1276. [CrossRef]
50. Gallardo-Cervantes, M.; González-García, Y.; Pérez-Fonseca, A.A.; González-López, M.E.; Manríquez-González, R.; Rodrigue, D.; Robledo-Ortíz, J.R. Biodegradability and improved mechanical performance of polyhydroxyalkanoates/agave fiber biocomposites compatibilized by different strategies. *J. Appl. Polym. Sci.* **2021**, *138*, 50182. [CrossRef]
51. Trinh, B.M.; Ogunsona, E.O.; Mekonnen, T.H. Thin-structured and compostable wood fiber-polymer biocomposites: Fabrication and performance evaluation. *Compos. Part A Appl. Sci. Manuf.* **2021**, *140*, 106150. [CrossRef]
52. Huerta-Cardoso, O.; Durazo-Cardenas, I.; Longhurst, P.; Simms, N.J.; Encinas-Oropesa, A. Fabrication of agave tequilana bagasse/PLA composite and preliminary mechanical properties assessment. *Ind. Crops Prod.* **2020**, *152*, 112523. [CrossRef]
53. Vilela, C.; Engström, J.; Valente, B.F.A.; Jawerth, M.; Carlmark, A.; Freire, C.S.R. Exploiting poly(ε-caprolactone) and cellulose nanofibrils modified with latex nanoparticles for the development of biodegradable nanocomposites. *Polym. Compos.* **2019**, *40*, 1342–1353. [CrossRef]
54. Singh, R.; Kumar, R.; Hashmi, M.S.J. Friction Welding of Dissimilar Plastic-Based Material by Metal Powder Reinforcement. *Ref. Modul. Mater. Sci. Mater. Eng.* **2017**, *101*, 77. [CrossRef]
55. Nagarajan, V.; Misra, M.; Mohanty, A.K. New engineered biocomposites from poly(3-hydroxybutyrate-co-3-hydroxyvalerate) (PHBV)/poly(butylene adipate-co-terephthalate) (PBAT) blends and switchgrass: Fabrication and performance evaluation. *Ind. Crops Prod.* **2013**, *42*, 461–468. [CrossRef]

56. Carrasco, F.; Pagès, P.; Gámez-Pérez, J.; Santana, O.O.; Maspoch, M.L. Processing of poly(lactic acid): Characterization of chemical structure, thermal stability and mechanical properties. *Polym. Degrad. Stab.* **2010**, *95*, 116–125. [CrossRef]
57. Pachekoski, W.M.; Dalmolin, C.; Agnelli, J.A.M. The influence of the industrial processing on the degradation of poly(hidroxybutyrate)-PHB. *Mater. Res.* **2013**, *16*, 327–332. [CrossRef]
58. Long, H.; Wu, Z.; Dong, Q.; Shen, Y.; Zhou, W.; Luo, Y.; Zhang, C.; Dong, X. Effect of polyethylene glycol on mechanical properties of bamboo fiber-reinforced polylactic acid composites. *J. Appl. Polym. Sci.* **2019**, *136*, 47709. [CrossRef]
59. Lee, C.H.; Khalina, A.; Lee, S.H. Importance of interfacial adhesion condition on characterization of plant-fiber-reinforced polymer composites: A review. *Polymers* **2021**, *13*, 438. [CrossRef]
60. Jaszkiewicz, A.; Meljon, A.; Bledzki, A.K.; Radwanski, M. Gaining knowledge on the processability of PLA-based short-fibre compounds-A comprehensive comparison with their PP counterparts. *Compos. Part A Appl. Sci. Manuf.* **2016**, *83*, 140–151. [CrossRef]
61. Spear, M.J.; Eder, A.; Carus, M. *Wood polymer Composites*; Woodhead Publishing: Sawston, UK, 2015; ISBN 9781782424772.
62. Reis, K.C.; Pereira, L.; Melo, I.C.N.A.; Marconcini, J.M.; Trugilho, P.F.; Tonoli, G.H.D. Particles of coffee wastes as reinforcement in polyhydroxybutyrate (PHB) based composites. *Mater. Res.* **2015**, *18*, 546–552. [CrossRef]
63. Aydemir, D.; Gardner, D.J. The effects of cellulosic fillers on the mechanical, morphological, thermal, viscoelastic, and rheological properties of polyhydroxybutyrate biopolymers. *Polym. Compos.* **2020**, *41*, 3842–3856. [CrossRef]
64. Espinach, F.X.; Boufi, S.; Delgado-Aguilar, M.; Julián, F.; Mutjé, P.; Méndez, J.A. Composites from poly(lactic acid) and bleached chemical fibres: Thermal properties. *Compos. Part B Eng.* **2018**, *134*, 169–176. [CrossRef]
65. Yeng, L.C.; Wahit, M.U.; Othman, N. Thermal and flexural properties of regenerated cellulose(RC)/poly(3-hydroxybutyrate)(PHB) biocomposites. *J. Teknol.* **2015**, *75*, 107–112. [CrossRef]
66. Ventura, H.; Claramunt, J.; Rodríguez-Pérez, M.A.; Ardanuy, M. Effects of hydrothermal aging on the water uptake and tensile properties of PHB/flax fabric biocomposites. *Polym. Degrad. Stab.* **2017**, *142*, 129–138. [CrossRef]
67. Santos, E.B.C.; Barros, J.J.P.; Moura, D.A.D.; Moreno, C.G.; Fim, F.D.C.; Silva, L.B.D. Rheological and thermal behavior of PHB/piassava fiber residue-based green composites modified with warm water. *J. Mater. Res. Technol.* **2019**, *8*, 531–540. [CrossRef]
68. Sánchez-Safont, E.L.; Aldureid, A.; Lagarón, J.M.; Gámez-Pérez, J.; Cabedo, L. Biocomposites of different lignocellulosic wastes for sustainable food packaging applications. *Compos. Part B Eng.* **2018**, *145*, 215–225. [CrossRef]
69. Mysiukiewicz, O.; Barczewski, M.; Skórczewska, K.; Matykiewicz, D. Correlation between processing parameters and degradation of different polylactide grades during twin-screw extrusion. *Polymers* **2020**, *12*, 1333. [CrossRef]
70. Taguchi, S.; Iwata, T.; Abe, H.; Doi, Y.; Aqida, S.N. Poly(hydroxyalkanoate)s. In *Reference Module in Materials Science and Materials Engineering*; Elsevier: Amsterdam, The Netherlands, 2016.
71. Srubar, W.V.; Wright, Z.C.; Tsui, A.; Michel, A.T.; Billington, S.L.; Frank, C.W. Characterizing the effects of ambient aging on the mechanical and physical properties of two commercially available bacterial thermoplastics. *Polym. Degrad. Stab.* **2012**, *97*, 1922–1929. [CrossRef]
72. Park, S.; Baker, J.O.; Himmel, M.E.; Parilla, P.A.; Johnson, D.K. Cellulose crystallinity index: Measurement techniques and their impact on interpreting cellulase performance. *Biotechnol. Biofuels* **2010**, *3*, 10. [CrossRef] [PubMed]

 *molecules*

Article

Micro-Fibrillated Cellulose in Adhesive Systems for the Production of Wood-Based Panels

Emmanouil Karagiannidis *, Charles Markessini and Eleftheria Athanassiadou

CHIMAR HELLAS S.A.,15 km National Road, Thessaloniki–Polygyros, 57001 Thessaloniki, Greece; CharlesM@ari.gr (C.M.); eathan@ari.gr (E.A.)
* Correspondence: manos.karag@ari.gr; Tel.: +30-2310-424167

Academic Editors: Alejandro Rodríguez, Eduardo Espinosa and Fabrizio Sarasini
Received: 31 July 2020; Accepted: 18 October 2020; Published: 21 October 2020

Abstract: Micro-Fibrillated Cellulose (MFC) is a new type of bio-based additive, coming from wood cellulose. It can compete and substitute oil derived chemicals in several application fields. In the present work, the use of micro-fibrillated cellulose, in waterborne adhesive systems applied in the manufacture of composite wood-based panels was evaluated. Research was conducted to test the potential of improving the performance of wood-based panel types such as particleboard, waferboard or randomly-oriented strand board and plywood, by the application of MFC and the substitution of conventional and non-renewable chemical compounds. The approaches followed to introduce MFC into the adhesive systems were three, i.e., MFC 2% suspension added during the adhesive resin synthesis, MFC 10% paste admixed with the already prepared adhesive resin and MFC 2% suspension admixed with the already prepared resin. It was found that MFC improves not only the performance of the final wood panel products but also the behaviour of the applied adhesive polymer colloids (e.g., rheology improvement), especially when admixed with the already prepared resins. Moreover, it was proven that when MFC is introduced into the adhesive resin system, there is a possibility of decreasing the resin consumption, by maintaining the board performance. MFC's robustness to pH, shear and temperature makes it a highly interesting new additive for adhesive producers. In addition, its natural origin can give adhesive producers the opportunity to move over to more environmentally friendly product solutions.

Keywords: micro-fibrillated cellulose; formaldehyde adhesives; wood-based panels

1. Introduction

Cellulose is the most abundant, inexpensive and readily available biopolymer found in nature. It is contained at very high levels in cotton (~94%) and wood (~50%), which in turn are the major sources for cellulose products (paper, textiles, cosmetics, hygiene products, etc.) [1]. Traditionally, cellulose comes from vegetal resources and their waste.

Composite or engineered wood products such as particleboard (PB), plywood, waferboard or oriented strand board (OSB), medium density fibreboard (MDF) and the like are widely used in furniture manufacturing to replace the more expensive and scarcer natural wood (solid wood). For the manufacturing of the aforementioned wood products, it is necessary to mix or coat wood, coming from various forest species, in the form of particles, fibres, veneers or flakes, with special gluing systems, comprising adhesives (thermosetting polymer resins) and several chemical additives. Heat and pressure are then applied to form the final polymer network and bind the wood elements together and form the final wood panel product. The application of the adhesive system on the specially prepared wood parts is carried out by spraying the gluing system in the form of a mist through suitable spraying nozzles (PB, OSB, MDF), or spreading the adhesive system on the veneers (plywood) using, for example, a roller-coater to adequately distribute it on the wood surface. In the framework of

EXILVA EU -funded -project, the objective of participant R&D company CHIMAR was to develop a new technology for the synthesis or the reinforcement of polymeric wood adhesives, by using EXILVA Micro-Fibrillated Cellulose (MFC), a product developed by Borregaard, based on an innovative technology for its production at commercial scale.

Micro-Fibrillated Cellulose (MFC) is characterized by its extended surface area with an expanded number of functional hydroxyl (-OH) groups. These characteristics are common for all MFC products. Moreover, MFC possesses several interesting properties, such as very high aspect ratios of the fibrillated fibres [2]. A number of reviews have shown mechanical improvement offered by MFC [3]. Therefore, it has a great potential to reinforce synthetic, petroleum-based resins that are already used in the market. In addition, positive effects of nanocellulose addition to amino-plastic adhesives in the bonding of particleboard and oriented strand panels have been found [4]. Particleboard panels have been manufactured using CNF (Cellulose Nano-Fibres, similar material to MFC) as a sole binder and met the industry requirements in terms of mechanical properties [5]. Another study revealed that MFC can be utilized as a rheology modifier, giving a more viscous urea-formaldehyde (UF) resin compared to the unmodified resin, in the manufacture of particleboard panels [6]. In the same study, a higher fraction of adhesive was proven to be available for bond-line formation and a larger part of the wooden particles was covered with adhesive, when MFC was used as a reinforcing agent of UF resin. Changed adhesive distribution together with improved adhesive toughness are proposed to contribute to improved board strength [6]. Nanocellulose reinforced UF resins have displayed improvement of storage modulus values and the wood composites manufactured with these reinforced UF resins showed enhanced performance [7]. A study on the effect of the length and content of cellulose fibers in phenol-formaldehyde (PF) resin showed an improvement in the mechanical properties of the wood composites manufactured with this resin, compared to the reference composites, manufactured with the unmodified PF resin [8].

In the framework of the present study, MFC was used in combination with currently used petrochemical adhesives, i.e., amino resins like urea-formaldehyde (UF) and melamine-urea-formaldehyde (MUF), and phenolic resins like phenol-formaldehyde (PF). The MFC was introduced either during the polycondensation stage of the resin synthesis or as an additive admixed together with the already prepared resin and other chemicals in a glue mixture, which eventually was sprayed onto wood particles or spread onto wood veneers, depending on the panel type. Three types of panels were manufactured at CHIMAR pilot plant: particleboard (PB), waferboard and plywood panels.

2. Materials and Methods

2.1. Introduction of MFC

Three different approaches to introduce MFC into adhesive systems were followed (Figure 1):

Approach 1: Introduction of MFC 2% suspension in the resin as a raw material, during its synthesis.

Approach 2: Admixing MFC 10% paste with the resin after its synthesis (addition level: 0.25–1.0% *w/w*, dry/dry resin).

Approach 3: Admixing MFC 2% suspension with the resin after its synthesis (addition level: 0.01–0.03% *w/w*, dry/dry resin).

Each approach was specially chosen for each adhesive type and the respective application (UF/MUF for particleboard panels, MUF for waferboard panels, PF for plywood panels), according to the best manufacturing practices of the wood-based panel sector.

The scope of following this path was to investigate whether MFC reacts chemically or physically with the resin as well as to evaluate its contribution to the improvement of the performance of the produced boards. It was assumed that when added during the resin synthesis (Approach 1) and under specific conditions (temperature, pH, etc.), MFC would participate in the polymer network. On the other side, when added into the resin system, after the synthesis of the resin (Approach 2, Approach 3), it would modify the physical properties of the resin, i.e., its rheology. All these different effects were investigated at CHIMAR R&D premises.

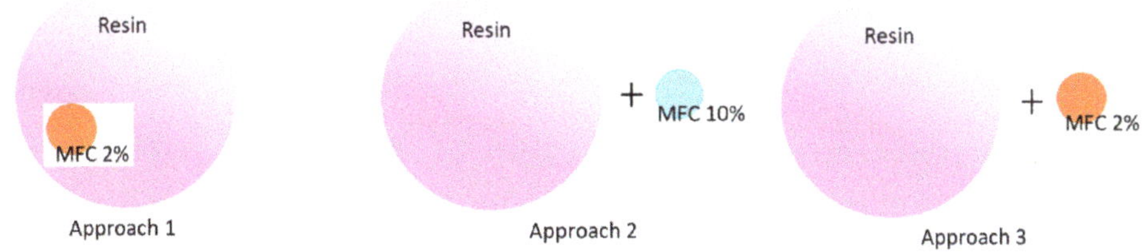

Figure 1. Approaches of introduction of MFC into adhesive systems.

2.2. Materials

Formaldehyde 37% was supplied by PanReac (Barcelona, Spain). Phenol 90% was supplied by Merck (Darmstadt, Germany). The following industrial-grade reagents were supplied by Elton Group S.A. (Thessaloniki, Greece): urea, melamine, formic acid (80%), sodium hydroxide (50%). Ammonium sulphate was purchased by New Trade Fertilizers (Athens, Greece). EXILVA MFC, P-series, at concentrations of 2% (L-grade) and 10% (V-grade) were supplied by Borregaard (Sarpsborg, Norway). Wheat flour was also used as a filler in the production of plywood panels.

UF and MUF resins were applied in the production of particleboard panels, MUF resin was applied in the production of waferboard panels and PF resin was used for the production of plywood panels. All resins were synthesised at the laboratory of CHIMAR.

Ammonium sulphate, in the form of 40% aqueous solution, was used as hardener for the UF and MUF resins during the manufacture of particleboard and waferboard panels. For the production of plywood panels with 3 plies, the phenolic resin was mixed with wheat flour and water, in order to obtain the appropriate viscosity to be spread uniformly onto the wood veneers.

2.3. Synthesis of UF, MUF and PF Resins

The synthesis of UF resins was carried out according to a two-step process, comprising the methylolation and condensation stages. MFC 2% suspension was added at the beginning of the synthesis process, together with the rest of the raw materials (Approach 1). The methylolation step was realized by the addition of formaldehyde to urea to give the so-called methylolureas, followed by the addition of sodium hydroxide to adjust the pH to an alkaline value. The condensation reaction, which followed, was done at elevated temperature, in the acid condensation stage, using formic acid to reduce the pH. When the required viscosity was reached, the pH was adjusted again back to the alkaline region. At the end, a final portion of urea was added in order to reach a final F:U molar ratio in the range of 1.0–1.1. A reference UF resin was synthesised in parallel to the MFC-modified UF resin. The solid content of the resins was in the range of 64–66%. The MFC 2% suspension was added at a level of 0.22% *w/w* dry MFC on liquid resin. All other settings of the resin formulation were kept the same as those of the reference resin.

MUF resins were synthesised in a similar process to the UF resin process, including melamine as well as urea during the methylolation step. A reference MUF resin was synthesised in parallel to the MFC-modified MUF resin. The final F:(U + M) molar ratio was in the range of 1.0–1.1. The melamine content was within the range of 15–30%, depending on the application (particleboard or waferboard). The solid content of the resins was in the range of 64–66%. For Approach 1, MFC 2% suspension was added in the resin during its synthesis, at a level of 0.32% *w/w* dry MFC on liquid resin. All other settings of the resin formulations were kept the same as in the reference resin.

The synthesis of PF resins was realized according to the resole synthesis process comprising the hydroxymethylation and condensation stages. Formaldehyde was added to phenol, using sodium hydroxide as basic catalyst, for the hydroxymethylation to occur. The condensation reaction took place afterwards, at increased temperature. When the desired viscosity was reached, the resin was cooled down to room temperature and stored at 5 °C, as suggested for this type of resin. A reference PF resin was synthesised in parallel to the MFC-modified PF resin. The final F:P molar ratio was in the range of 2.0–2.2 and the resin solid content in the range of 40–45%. For Approach 1, MFC 2% suspension was added in the resin during its synthesis, at a level of 0.22% *w/w* dry MFC on liquid resin. All other settings of the resin formulation were kept the same as in the reference resin.

All resins were delivered in liquid form.

2.4. Resin Characterization

The properties of the produced resins were determined based on standard lab analysis methods. All the produced UF and MUF (amino) resins were characterized for their solid content, pH value, viscosity, gelation time, specific gravity and water tolerance. The solid content was determined by drying 2 g of resin in an oven at 120 °C for 2 h. The pH value of the resins was measured using a digital GLP21 pH-meter from CRISON (Barcelona, Spain) with a single Hamilton glass electrode attached. The viscosity was measured at 25 °C using a Brookfield rotational viscometer, with a small sample adapter (SC4-45Y), which requires a sample chamber SC4-13R and a spindle SC4-18. The gelation time of the resins was determined by measuring the time needed for gelation in boiling water after addition of 3.5% ammonium sulphate hardener (*w/w* dry/dry resin). For the determination of the water tolerance of the resins at 25 °C, a 10 mL sample of resin was added in a volumetric cylinder and the quantity of water needed until the sample started to coagulate was measured. The specific gravity was determined using suitable glass hydrometers.

Additionally, Differential Scanning Calorimetry (DSC) analysis was performed for specific experiments with UF resins, which were used for particleboard production, to evaluate the curing behavior of the UF resins with and without the addition of MFC. The testing was done in a Shimadzu DSC-50 device, using Aluminum High Pressure Cells (222-01701-91). In all measurements, the resin was catalysed with 3,5% *w/w* ammonium sulphate (in the form of 40% *w/w* aqueous solution) and all formulations were adjusted to the same solid content, by addition of water where needed. Test samples of 14–16 mg for each variable were placed in the aluminum cells, sealed and measured immediately. All samples were heated up to 200 °C at a heating rate of 10 °C/min under nitrogen atmosphere. For each test sample, the peak temperature, the onset temperature and the heat release were recorded.

For the PF resins' characterization, the solid content, pH value, viscosity, gelation time and alkalinity were determined. The pH value, the viscosity, the gelation time and the solid content measurements were in accordance with the practice followed for the UF and MUF resins. The alkalinity measurement is a titration with HCl acid, to calculate the remaining NaOH in the resin.

2.5. Panel Manufacturing Process

Three types of panels were manufactured at CHIMAR pilot plant: particleboard (PB), waferboard and plywood panels. The panel production process followed the industrial manufacturing practice. The mechanical and wet properties of the boards produced were determined in accordance with the European standards (EN) in force, i.e., density (EN 323), tensile strength/Internal Bond (IB) (EN 310), bending strength/Modulus Of Rupture (MOR)/Modulus Of Elasticity (MOE) (EN 319), Thickness Swelling (TS) (EN 317), as well as the formaldehyde emission potential, were measured by determination of the Formaldehyde content according to the Perforator Method (International Standard ISO 12460-5). The properties of the panels should satisfy the specifications as stated in product standards EN 312 (PB), EN-300 (Waferboard/OSB) and EN-314.02 (plywood).

2.5.1. Particleboard

Particleboard panels were manufactured at CHIMAR wood-based panels pilot production plant, according to industrial practice. The gluing mixture, consisting of resin, hardener and water, was sprayed on the wood particles, consisting mainly of a mixture of pine with 20%–30% poplar wood, and having a moisture content of ca. 3%. The particles were blended with the gluing mixture in a Lödige FM130D blender equipped with a suitable spraying system attached on the top side, for about 3 min. The resinated wood particles were then formed in a mattress (mat) with dimensions of 44 × 44 cm. The mat was manually pre-pressed, and it was then led to the hydraulic press and pressed at high temperature for a sufficient time to achieve the cross-linking and total curing of the resin. The pressing temperature and time were defined by previous experience with similar experiments and were the same for all particleboard panels produced. The target density of the boards was 650 kg/m^3 and the target thickness was 15 mm. After they were given time to cool down, the produced boards were sanded and proper specimens were taken from them, for testing the board mechanical and wet properties, as well as their formaldehyde content, according to the perforator method (ISO 12460-5). Each resin system formulation was applied in duplicates of panels.

2.5.2. Waferboard/Randomly-Oriented Strand Board

The production of waferboard panels was carried out following a very similar procedure to that of the particleboard, but using different blending and spaying devices. A cement mixer was used as a blending device, in order to avoid breaking down the wood strands. The glue mixture was sprayed through a high-pressure handheld spraying gun. The mat forming and the pressing process were the same as in the case of particleboards production. The pressing temperature and time were defined by previous experience with similar experiments and were the same for all the panels produced. The target density of the boards was 600 kg/m^3 and the target thickness was 15 mm. Each formulation consisted of one panel.

2.5.3. Plywood

For the plywood manufacture, a glue mixture consisting of resin, wheat flour and water was prepared and spread on the wood veneers (50 × 50 cm × 1.5 mm), with the assistance of a spatula and a roller. The wood species that was used for the present study was beech. The glued veneers were stacked and pre-pressed. Afterwards, they were pressed in the hydraulic press at high temperature for a sufficient time to totally cure the resin. The pressing temperature and time were defined by previous experience with similar experiments and were common for all plywood panels. After having been cooled down, the produced panels were cut in specimens and their mechanical properties were determined. Each resin system formulation was applied in duplicates of panels.

3. Results

During the experimental part, a lot of different cases were investigated, concerning the MFC concentration, the MFC addition levels (dry/liquid resin), the resin loading levels, the resin and panel type, etc., and they are all presented in the following sections, divided according to the panel type.

3.1. Particleboard Production

As mentioned earlier, the resins that were utilized for the particleboard manufacture were of UF and MUF type. Table 1 below presents the analysis results from the characterization of a conventional and an MFC-reinforced resin of both types (the latter synthesized according to Approach 1).

Table 1. Urea-Formaldehyde (UF) and Melamine-Urea-Formaldehyde (MUF) resins analysis results, Particleboard.

Resin Type	UF Reference	MFC-Reinforced UF	MUF Reference	MFC-Reinforced MUF
% MFC, dry/liquid resin	0.00	0.22	0.00	0.32
Solids (%)	65.8	67.0	63.6	63.6
pH	8.5	7.9	9.2	9.6
Viscosity (cP)	315	370	195	450
Gel time (s)	60	56	73	95
Water tolerance (mL:mL)	1/1.4	1/0.6	1/0.5	1/0.6
Specific gravity	1.279	1.295	1.265	1.233

The analysis results indicate that there is no significant difference between the properties of the reference UF resin and the ones of the MFC-reinforced UF resin. When comparing the reference MUF resin with the MFC-reinforced one, there is a notably shorter gel time (higher reactivity) of the reference resin, which could contribute to its better performance in particleboard mechanical properties against the MFC reinforced MUF resin, as presented in the next set of experiments.

3.1.1. MFC 2% Suspension Either Added as Raw Material of UF or MUF Resin during Synthesis or Admixed with the Ready UF or MUF Resin (Approach 1 vs. Approach 3)

The results obtained from the analysis of the properties of the panels produced during experiment 3.1.1 are presented in Table 2.

Table 2. Particleboard panels analysis, MFC 2% suspension, Approach 1 vs. Approach 3, Average property values, Standard deviation [1] (in brackets).

Formulation	UF Reference	MFC-Reinforced UF (Approach 1)	UF + 0.22% MFC (Approach 3)	MUF Reference	MFC-Reinforced MUF (Approach 1)	MUF + 0.32% MFC (Approach 3)
% Binder, dry/dry wood	9.0	9.0	9.0	9.0	9.0	9.0
% MFC, dry/liquid resin	0.00	0.22	0.22	0.00	0.32	0.32
Internal bond (MPa)	0.58 (0.04)	0.52 (0.06)	0.63 (0.03)	0.90 (0.09)	0.82 (0.10)	0.95 (0.06)
24 h Thickness Swelling (%)	54.9 (4.4)	54.1 (4.1)	62.6 (4.0)	27.2 (1.4)	26.9 (0.7)	29.3 (1.7)
24 h Water Absorption (%)	100 (1.6)	96 (1.0)	99 (1.0)	72 (2.3)	76 (1.8)	75 (2.2)
Modulus of Rupture (MPa)	13.7	13.5	13.9	17.3	15.8	16.2
Modulus of Elasticity (MPa)	2641	2568	2418	2677	2627	2335
Formaldehyde content (mg/100 g ODB [2])	6.7	5.6	6.7	5.4	6.7	7.4

[1] Standard deviation was added on properties where adequate population data existed. [2] Oven dried board.

It is clear that for both types of resin, UF and MUF, when Approach 1 is followed and MFC 2% is introduced in the resin during its synthesis, the mechanical properties and especially the tensile strength (Internal Bond) are impaired. When Approach 3 is followed and MFC 2% is admixed with the ready resin, the internal bond is improved in both UF and MUF cases. The water resistance of the boards is not significantly influenced in either case. The bending properties (modulus of rupture and modulus of elasticity) follow a trend similar to the tensile strength, which is, however, not considered to be significant. The formaldehyde content of the board seems to be reduced by Approach 1 in the UF resin, while in the MUF case, the formaldehyde content is increased by the addition of MFC. On the other side, when Approach 3 is followed in the UF resin case, the formaldehyde content is not affected, while in the MUF case, it is further increased by the addition of MFC.

3.1.2. MFC 10% Paste at Increasing Addition Levels, Keeping the Resin Loading Level Stable. MFC Admixed with UF Resin (Approach 2)

The results obtained from the analysis of the properties of the panels produced during experiment 3.1.2 are presented in Table 3.

Table 3. Particleboard panels analysis, MFC 10% increasing levels, resin stable, Approach 2, Average property values, Standard deviation in brackets).

Formulation	UF Reference	UF + 0.25% MFC	UF + 0.5% MFC	UF + 0.75% MFC	UF + 1.0% MFC
% Binder, dry/dry wood	7.0	7.0	7.0	7.0	7.0
% MFC, dry/liquid resin	0.00	0.25	0.50	0.75	1.00
Internal bond (MPa)	0.30 (0.04)	0.33 (0.06)	0.36 (0.03)	0.39 (0.05)	0.45 (0.04)
24 h Thickness Swelling (%)	45.3 (2.5)	45.5 (3.1)	42.9 (2.1)	42.8 (2.6)	43.6 (2.1)
24 h Water Absorption (%)	97 (1.9)	96 (1.8)	95 (1.6)	99 (2.1)	99 (2.2)
Modulus of Rupture (MPa)	13.4	13.4	13.5	13.6	13.9
Modulus of Elasticity (MPa)	2389	2402	2412	2445	2455
Formaldehyde content (mg/100 g ODB)	8.7	9.2	9.1	9.0	8.5

When MFC 10% is admixed with the UF resin at elevated levels of addition, it proportionally improves the mechanical properties, with more significant differences appearing on the tensile strength results. Water resistance of the boards does not seem to be importantly affected. The fluctuation of the results obtained for the Formaldehyde content is within the accuracy range, so no conclusions can be drawn.

3.1.3. MFC 10% Paste at Increasing Addition Levels, Decreasing the Resin Loading Level. MFC Admixed with UF Resin (Approach 2)

The results obtained from the analysis of the properties of the panels produced during experiment 3.1.3 are presented in Table 4.

Table 4. Particleboard panels analysis, MFC 10% increasing levels, resin decreasing, Approach 2, Average property values, Standard deviation (in brackets).

Formulation	7.0% UF Reference	6.4% UF + 0.25% MFC	5.8% UF + 0.5% MFC	5.4% UF + 0.75% MFC	5.0% UF + 1.0% MFC	4.7% UF + 1.0% MFC
% Binder, dry/dry wood	7.0	6.4	5.8	5.4	5.0	4.7
% MFC, dry/liquid resin	0.00	0.25	0.50	0.75	1.00	1.00
Internal bond (MPa)	0.36 (0.02)	0.36 (0.02)	0.31 (0.03)	0.29 (0.02)	0.26 (0.02)	0.24 (0.02)
24 h Thickness Swelling (%)	54.1 (1.3)	51.3 (3.8)	59.1 (2.5)	64.8 (2.6)	65.9 (3.2)	72.3 (2.3)
24 h Water Absorption (%)	126 (2.4)	120 (2.3)	127 (4.3)	138 (1.5)	139 (2.8)	146 (2.9)
Modulus of Rupture (MPa)	9.6	11.7	10.5	8.9	8.3	7.3
Modulus of Elasticity (MPa)	1842	1994	1879	1695	1713	1443
Formaldehyde content (mg/100 g ODB)	8.0	7.4	7.3	8.2	8.3	8.3

This experiment proves that when MFC is admixed with the resin up to a specific level (0.25% dry/liquid resin), it allows the reduction in the resin loading level, since all board properties are equal to or even better than in the reference formulation (comparison between the two first formulations). This tendency disappears by further reduction in resin loading, even with an increase in the addition levels of MFC. Beyond that point, it seems that the influence of the reduction in the petrochemical resin is more significant than the introduction of MFC at increasing levels, and this becomes evident from the results of all properties.

3.1.4. MFC 2% Suspension at Increasing Addition Levels, Keeping the Resin Level Stable. MFC Admixed with UF Resin (Approach 3), Average Property Values, Standard Deviation (in Brackets)

The results obtained from the analysis of the properties of the panels produced during experiment 3.1.4 are presented in Table 5.

Table 5. Particleboard panels analysis, MFC 2% increasing levels, resin stable, Approach 3.

Formulation	UF Reference	UF + 0.01% MFC	UF + 0.02% MFC	UF + 0.03% MFC
% Binder, dry/dry wood	9.0	9.0	9.0	9.0
% MFC, dry/liquid resin	0.00	0.01	0.02	0.03
Internal bond (MPa)	0.42 (0.02)	0.59 (0.02)	0.66 (0.03)	0.67 (0.06)
24 h Thickness Swelling (%)	44.2 (2.6)	41.4 (1.5)	39.4 (0.9)	40.0 (2.3)
24 h Water Absorption (%)	92 (3.8)	87 (4.2)	86 (4.1)	83 (5.4)
Modulus of Rupture (MPa)	10.1	11.2	11.7	11.6
Modulus of Elasticity (MPa)	2146	2198	2273	2275
Formaldehyde content (mg/100 g ODB)	7.5	7.0	6.2	6.2

Admixing MFC 2% suspension with the ready resin at low levels 0.01–0.03% dry/liquid resin (Approach 3) shows a significant advantage of increased levels of MFC to the mechanical properties and water resistance of the produced boards. Formaldehyde content was gradually decreased when increasing the MFC levels, which is a positive effect too.

3.1.5. MFC 2% Suspension at Increasing Addition Levels, Decreasing the Resin Loading Level. MFC Admixed with UF Resin (Approach 3), Average Property Values, Standard Deviation (in Brackets)

The results obtained from the analysis of the properties of the panels produced during experiment 3.1.5 are presented in Table 6.

Table 6. Particleboard panels analysis, MFC 2% increasing levels, resin decreasing, Approach 3.

Formulation	9.0% UF Reference	8.5% UF + 0.01% MFC	8.0% UF + 0.02% MFC	7.5% UF + 0.03% MFC
% Binder, dry/dry wood	9.0	8.5	8.0	7.5
% MFC, dry/liquid resin	0.00	0.01	0.02	0.03
Internal bond (MPa)	0.38 (0.03)	0.43 (0.03)	0.43 (0.05)	0.45 (0.04)
24 h Thickness Swelling (%)	58.0 (2.6)	59.0 (2.7)	58.6 (1.7)	61.6 (3.6)
24 h Water Absorption (%)	112 (3.1)	114 (3.3)	118 (2.2)	119 (3.4)
Modulus of Rupture (MPa)	9.7	9.9	9.5	8.8
Modulus of Elasticity (MPa)	2045	2095	2012	1917
Formaldehyde content (mg/100 g ODB)	10.5	9.2	8.6	8.0

When MFC 2% suspension is admixed with the ready resin, at gradually increasing but low dry/liquid resin levels (Approach 3) with parallel reduction in the resin level, the boards' tensile strength is improved even at a 17% decrease in the resin level. The water resistance and bending strength of the boards are slightly impaired, but the fluctuation of the results is within the accuracy range, so no significant differences are noted. Formaldehyde content is, respectively, reduced.

The slight differences observed in the results for the same formulations among the various experiments, derive from the variability of the process and possibly from different prevailing environmental conditions.

3.2. Waferboard Production

For the manufacture of waferboard panels, MUF resin type was used. The results from the analysis of one resin sample of this type are presented in Table 7:

Table 7. MUF resin analysis results, Waferboard.

Resin Type	MUF
Solids (%)	64.2
pH	9.3
Viscosity (cP)	360
Gel time (s)	66
Water tolerance (mL:mL)	1/0.5
Specific gravity	1.279

3.2.1. MFC 10% Paste Admixed with MUF Resin at Increasing Addition Levels, Keeping the Resin Level Stable (Approach 2), Average Property Values, Standard Deviation (in Brackets)

The results obtained from the analysis of the properties of the panels produced during experiment 3.2.1 are presented in Table 8.

Table 8. Waferboard panels analysis, MFC 10% increasing levels, resin stable, Approach 2.

Formulation	MUF Reference	MUF + 0.25% MFC	MUF + 0.50% MFC	MUF + 0.75% MFC	MUF + 1.00% MFC
% Binder, dry/dry wood	12.0	12.0	12.0	12.0	12.0
% MFC, dry/liquid resin	0.00	0.25	0.50	0.75	1.00
Internal bond (MPa)	0.50 (0.06)	0.49 (0.08)	0.51 (0.10)	0.52 (0.03)	0.61 (0.06)
24 h Thickness Swelling (%)	22.1 (2.1)	24.5 (2.8)	22.4 (0.8)	20.8 (0.9)	19.8 (5.3)
24 h Water Absorption (%)	61 (2.2)	68 (3.3)	65 (7.4)	60 (7.2)	61 (6.3)
Modulus of Rupture (MPa)	21.4	20.8	23.1	22.1	24.1
Modulus of Elasticity (MPa)	3395	3257	3679	3820	3789
Formaldehyde content (mg/100 g ODB)	7.4	6.8	7.4	7.8	7.5

The analysis of the produced waferboard panels following Approach 2 (admixing MFC 10% paste with the ready resin) indicated that only adding 1.00% dry MFC on liquid MUF resin significantly improves the Internal Bond and slightly improves the Modulus of Rupture/Modulus of Elasticity and water resistance of the produced boards. The latter effect is shown by the decrease in thickness swelling. The formaldehyde content was not affected. However, the higher the MFC level, the more viscous the glue mixture, almost reaching the limit of being sprayable when that level of MFC is mixed with the resin.

3.2.2. MFC 10% Paste Admixed with MUF Resin at Increasing Addition Levels, Decreasing the Resin Level (Approach 2), Average Property Values, Standard Deviation (in brackets)

The results obtained from the analysis of the properties of the panels produced during experiment 3.2.2 are presented in Table 9.

Table 9. Waferboard panels analysis, MFC 10% increasing levels, resin decreasing, Approach 2.

Formulation	12% MUF Reference	11.3% MUF + 0.25% MFC	10.5% MUF + 0.50% MFC	9.8% MUF + 0.75% MFC	9.0% MUF + 1.00% MFC
% Binder, dry/dry wood	12.0	11.3	10.5	9.8	9.0
% MFC, dry/liquid resin	0.00	0.25	0.50	0.75	1.00
Internal bond (MPa)	0.58 (0.13)	0.65 (0.12)	0.59 (0.09)	0.54 (0.11)	0.57 (0.08)
24 h Thickness Swelling (%)	21.1 (2.1)	19.9 (1.1)	22.8 (2.5)	25.4 (2.5)	29.8 (2.7)
24 h Water Absorption (%)	66 (4.4)	63 (5.2)	65 (3.8)	83 (10.2)	78 (5.5)
Modulus of Rupture (MPa)	18.3	19.5	18.3	15.0	16.0
Modulus of Elasticity (MPa)	2923	3295	3069	2600	2711
Formaldehyde content (mg/100 g ODB)	7.3	7.1	7.0	6.7	7.0

Up to a certain point, 11.3% MUF/0.25% MFC, admixing MFC 10% paste with the resin (Approach 2), even at reduced resin loading level, could be proven advantageous, especially concerning mechanical properties. The water resistance (thickness swelling and water absorption) is also positively affected by the addition of MFC. Beyond that point, all the properties start to be impaired, probably because the influence of the reduction in the petrochemical resin is more significant than the introduction of MFC. The strong hydrophilic character of the MFC, which is due to the high amount of -OH groups on its surface, possibly leads to higher water adsorption, reducing the water resistance of the produced boards. The fluctuation of the results obtained for the formaldehyde content is within the accuracy range, so no conclusions can be drawn.

The slight differences observed in the results for the same formulations among the various experiments derive from the variability of the process and possibly from different prevailing environmental conditions.

3.3. Plywood Production

For the manufacture of plywood panels, PF resins were synthesized and the results from the analysis of a reference PF resin and of an MFC-reinforced one (Approach 1), used during this series of experiments, are presented in Table 10.

Table 10. Phenol-Formaldehyde (PF) resin analysis results, Plywood.

Resin Type	PF Reference	MFC-Reinforced PF
% MFC, dry/liquid resin	0.00	0.22
Solids (%)	40.6	40.9
pH	11.8	11.6
Viscosity (cP)	360	430
Gel time (min)	17	26
Alkalinity (%)	6.8	6.6

A comparison between the reference PF resin and the MFC-reinforced one concerning their analysis results indicates a shorter gel time for the reference PF, and thus a higher reactivity, which, however, could cause premature resin curing reaction. Moreover, the viscosity of the MFC-reinforced PF is higher, which makes sense, since MFC can operate as a thickener and thus it could, as well, prevent the over-penetration of the glue into the wood substrate.

3.3.1. MFC 2% Suspension Added as Raw Material in PF Resin during its Synthesis (Approach 1), Average Property Values, Standard Deviation (in Brackets)

The results obtained from the analysis of the properties of the panels produced during experiment 3.3.1 are presented in Table 11.

Table 11. Plywood panels analysis, MFC 2%, Approach 1.

Formulation	PF Reference	MFC-Reinforced PF (Approach 1)
Binder, g/m^2	58	58
% MFC, dry/liquid resin	0.00	0.22
Shear Strength, N/mm^2	1.2 (0.49)	1.5 (0.41)
Wood failure, %	72 (16)	70 (25)

When MFC 2% suspension was added in the PF resin during its synthesis (Approach 1), it helped to improve the shear strength and achieve equal wood failure as compared to the reference plywood panels. Both characteristics mentioned above, i.e., lower resin reactivity and higher viscosity, could have contributed to the improved shear strength of the panels with the MFC-reinforced formulation, apart from the fact that MFC participates in the formulation as well.

3.3.2. MFC 10% Paste Evaluated as Rheology Modifier (Thickener), Replacing Wheat Flour, Admixed with the Ready PF Resin (Approach 2), Average Property Values, Standard Deviation (in Brackets)

The results obtained from the analysis of the properties of the panels produced during experiment 3.3.2 are presented in Table 12.

Table 12. Plywood panels analysis, MFC 10%, Approach 2.

Formulation	PF Reference	PF + 0.50% MFC (Approach 2)
Binder, g/m^2	58	58
% MFC, dry/liquid resin	0.00	0.50
% Wheat flour, dry/liquid resin	12.0	0.0
% Water, liquid/liquid resin	8.0	0.0
Shear Strength, N/mm^2	1.4 (0.56)	1.5 (0.48)
Wood failure, %	78 (17)	73 (19)

In this case, the addition level of MFC 10% paste was chosen by preparing a glue mixture with similar viscosity to that of the reference glue mixture containing flour. It was observed that the MFC-based formulation was spread more smoothly and evenly than the reference and it prevented early penetration of the glue mixture in the wood substrate, which translates to a rheology improvement of the glue mixture. The panel properties of the two formulations were similar.

3.3.3. MFC 10% Paste Admixed with PF Resin at Elevated Addition Levels, Keeping the Resin Loading Level Stable (Approach 2), Average Property Values, Standard Deviation (in Brackets)

The results obtained from the analysis of the properties of the panels produced during experiment 3.3.3 are presented in Table 13.

Table 13. Plywood panels analysis, MFC 10% increasing levels, Approach 2.

Formulation	PF Reference	PF + 0.23% MFC	PF + 0.46% MFC	PF + 0.68% MFC
Binder, g/m^2	58	58	58	58
% MFC, dry/liquid resin	0.00	0.23	0.46	0.68
Shear Strength, N/mm^2	1.5 (0.61)	1.3 (0.68)	1.5 (0.66)	1.5 (0.58)
Wood failure, %	72 (21)	82 (22)	80 (25)	76 (24)

There was no statistically significant positive effect on the shear strength of plywood panels when the MFC addition levels according to Approach 2 were gradually increased. It was shown that an addition level from 0.23 to 0.46% of dry MFC/liquid resin is enough to slightly improve the % wood failure.

The slight differences observed in the results for the same formulations among the various experiments derive from the variability of the process and possibly from different prevailing environmental conditions.

In total, MFC seems to modify the rheology of the adhesive systems where it is introduced, by increasing their viscosity, while allowing a homogeneous distribution of the resin mixture when applied on the wood. This phenomenon could be attributed to the non-Newtonian character rendered to the resin mixture by MFC, which brings a shear thinning behaviour to it. Such a characteristic is important in the manufacturing process of wood-based panels, since the glue mixture is sprayed on the wood-particles/wood-strands under very high pressure, through spraying nozzles, which create a very fine mist of the mixture. When MFC participates in the mixture, the shear thinning effect facilitates the easy spray of the resin mixture, which is then evenly distributed on the wood particles. The small resin droplets regain their viscosity afterwards, and remain on the wood surface, thus avoiding the resin loss due to absorption of the resin mixture from the wood substrate, which happens at a high extent in the case of the reference, without MFC, formulations. The avoidance of resin loss leads to optimum adhesion performance and improved product properties. In this study, this effect was evident in the case of particleboard and RSB (waferboard), when MFC was admixed with the resin during the preparation of the glue mixture, either at high addition levels (MFC 10% paste/Approach 2) or at low addition levels (MFC 2% suspension/Approach 3).

3.4. Differential Scanning Calorimetry (DSC)

The potential effect of MFC addition on the curing behaviour of the UF adhesive, in some of the experiments described above, was evaluated by Differential Scanning Calorimetry (DSC). In particular, the DSC analysis was performed for the UF adhesive systems for experiments 3.1.1, 3.1.2 and 3.1.4, in order to evaluate the MFC influence when used in combination with UF resins via all three approaches in particleboard production. Since UF resins are thermosetting polymers, their curing reaction is an exothermal one, as depicted by the peaks of the DSC plots, which are presented henceforward. The DSC analysis results are also presented in the tables below:

Experiment 3.1.1. MFC 2% Suspension in UF Resin during Synthesis or Admixed with the Ready UF Resin (Approach 1 vs. Approach 3)

The results obtained from the DSC analysis of the resin systems used during experiment 3.1.1 are presented in Table 14 and the corresponding thermograms are depicted in Figure 2.

Table 14. DSC analysis results, Experiment 3.1.1 (Approach 1 vs. Approach 3).

Sample	Peak (°C)	Onset (°C)	Heat (J/g)
UF Reference	96.95	88.41	45.22
MFC-Reinforced UF (Approach 1)	98.51	88.20	43.94
UF + 0.22% MFC (Approach 3)	98.10	88.15	47.26

The results obtained from the DSC analysis of the first experiment suggest that the reactivity of the UF resins including MFC (either during the resin synthesis or admixed with the resin) was impaired, since higher temperature is necessary for the curing process to take place, according to the respective peak temperatures obtained. It was also observed that the MFC-reinforced UF resin (Approach 1) cures at relatively higher temperature than the UF resin where MFC was admixed (Approach 3). These results are in agreement with the particleboard analysis results, where Approach 3 was proven to be advantageous against Approach 1.

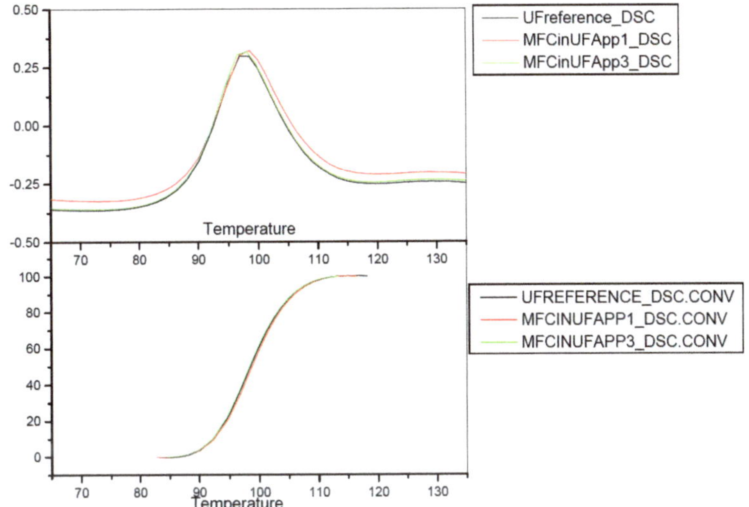

Figure 2. DSC curves, Experiment 3.1.1.

Experiment 3.1.2. MFC 10% Paste at Increasing Addition Levels, Keeping the Resin Loading Level Stable. MFC Admixed with UF Resin (Approach 2)

The results obtained from the DSC analysis of the resin systems used during experiment 3.1.2 are presented in Table 15 and the corresponding thermograms are depicted in Figure 3.

Table 15. DSC analysis results, Experiment 3.1.2 (Approach 2).

Sample	Peak (°C)	Onset (°C)	Heat (J/g)
UF reference	96.81	86.76	46.41
UF + 0.25% MFC	97.04	88.36	45.12
UF + 0.50% MFC	97.76	87.28	45.00
UF + 0.75% MFC	96.98	86.95	49.02
UF + 1.00% MFC	97.38	86.83	47.49

The results obtained in this case suggest that the reference UF resin formulation is more reactive than the formulations where MFC was added. This is not in accordance with the particleboard analysis results. It should be noted, however, that in board production, the curing behaviour of the adhesive system is usually affected by the participation of the wood substrate, due to the pH value and the pH buffering capacity of wood, and this may the reason for the difference between the DSC and the particleboard production results. Furthermore, the improved behaviour of the MFC-containing formulations may be attributed to the rheology effect of the MFC on the resin mixtures during the particleboard manufacturing process, as explained previously. On the other side, the DSC results show that the onset temperature declines with the increase in MFC level in the system, which indicates an increase in reactivity, a result conforming with the particleboard analysis results.

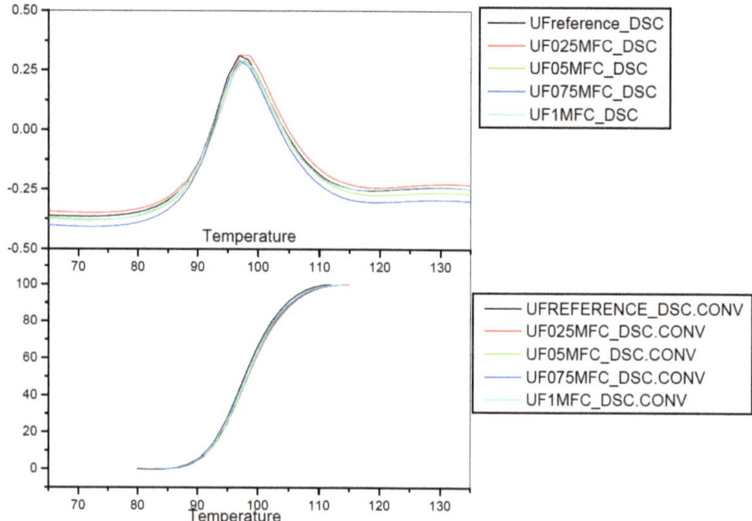

Figure 3. DSC curves, Experiment 3.1.2.

Experiment 3.1.4. MFC 2% Suspension at Increasing Addition Levels, Keeping the Resin Level Stable. MFC Admixed with UF Resin (Approach 3)

The results obtained from the DSC analysis of the resin systems used during experiment 3.1.4 are presented in Table 16 and the corresponding thermograms are depicted in Figure 4.

Table 16. DSC analysis results, Experiment 3.1.4 (Approach 3).

Sample	Peak (°C)	Onset (°C)	Heat (J/g)
UF Reference	97.02	86.35	50.25
UF + 0.01% MFC	98.12	86.06	50.26
UF + 0.02% MFC	96.95	88.41	45.22
UF + 0.03% MFC	98.51	88.20	43.94

The results obtained from this DSC analysis suggest that the most reactive formulation is that with the lowest MFC addition level, followed by the reference UF formulation, especially when observing the onset temperatures. In addition, the highest temperature necessary for the curing reaction to take place was obtained by the analysis of the formulation with the highest MFC addition level. However, the particleboard results indicated that the higher the MFC addition level, the better the board performance. It is hence deduced that here, again, the wood substrate interaction with the resin in combination with the MFC addition may have positively influenced the curing process of the adhesive system and also the board performance.

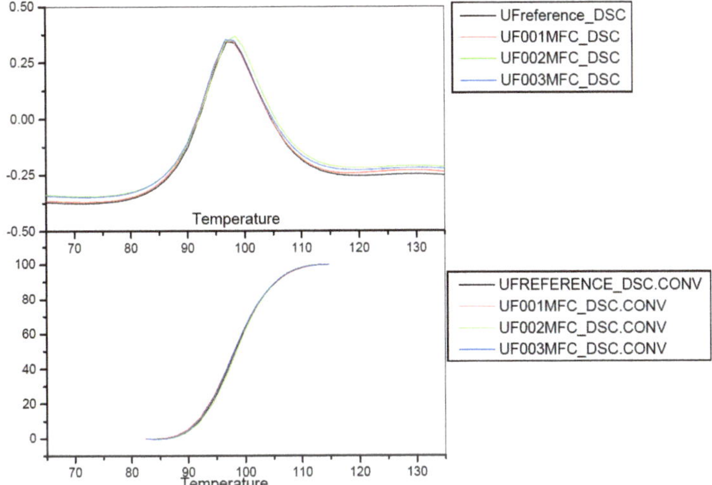

Figure 4. DSC curves, Experiment 3.1.4.

4. Discussion and Conclusions

The conclusions drawn from the entire experimental process are the following:

Particleboard

- The introduction of MFC 2% suspension in the UF and MUF resins during their synthesis (Approach 1), at a range of 0.22% and 0.32% (dry MFC/liquid resin), respectively, deteriorates the performance of particleboard panels, especially the internal bond.
- When MFC 2% suspension is mixed with the ready UF and MUF resins in the glue mixture (Approach 3), at the same range of 0.22% and 0.32% (dry MFC/liquid resin) as in Approach 1, respectively, the performance of particleboard panels improves, especially the internal bond. The other panel properties are not significantly affected.
- Increasing the addition levels of MFC 10% paste in UF resin (Approach 2), at a range of 0.25%–1.00% (dry MFC/liquid resin), improves the tensile strength (internal bond) of the produced particleboards. The other panel properties are not significantly affected.
- The addition of MFC 10% paste in UF resin (Approach 2), up to a level of 0.25% (dry MFC/liquid resin), allows the reduction in UF resin consumption in particleboard panels, by maintaining or even improving the board performance.
- Increasing addition levels of MFC 2% suspension in UF resin (Approach 3), at a range of 0.01%–0.03% (dry MFC/liquid resin), clearly improves the performance of the produced boards.
- The addition of MFC 2% suspension in UF resin (Approach 3), within a range of 0.01%–0.03% (dry MFC/liquid resin), allows the reduction in UF resin consumption in particleboard panels, by improving the internal bond and reducing the formaldehyde content of the boards but slightly impairing the other panel properties.
- MFC acts as a rheology modifier of the resin mixture, improving its distribution on the wood particles, enabling it to cover a larger surface of wood, and preventing its penetration into the wood substrate.

Waferboard

- Increasing the addition levels of MFC 10% paste in MUF resin (Approach 2), within a range of 0.25%–1.00% (dry MFC/liquid resin), improves the mechanical properties of waferboard panels.

- The addition of MFC 10% paste in MUF resin (Approach 2), up to a level of 0.50% (dry MFC/liquid resin), allows the reduction in MUF resin consumption in waferboard panels, maintaining the board properties.
- Similarly to the particleboard case, MFC can act as a rheology modifier of the resin mixture of the RSB application as well, improving its distribution on the wood strands, enabling it to cover a larger surface of wood and preventing its penetration into the wood substrate.

Plywood

- The introduction of MFC 2% suspension in the PF resin during its synthesis (Approach 1), at an addition level of 0.22%, improves the performance of plywood panels.
- Increasing the addition levels of MFC 10% paste in PF resin (Approach 2), within a range of 0.23%–0.68% (dry MFC/liquid resin), did not offer any significant improvement to the board properties. Nevertheless, MFC can be added as a thickener in the glue mixture of the PF resin for plywood production, substituting the conventionally-applied wheat flour, while maintaining the panel performance.

The conclusions discussed above can be better illustrated in Table 17 below:

Table 17. Conclusions and general evaluation.

Application	Approach	MFC Addition Level, % Dry MFC/Liquid Resin	MFC Grade	Property	Outcome	Remarks
Particleboard	1	0.22–0.32%	2% suspension	Mechanical properties (Internal Bond, Modulus of Rupture, Modulus of Elasticity)	Negative	
				Wet properties (Thickness swelling, Water absorption)	Negative	
				Formaldehyde content	Positive	
	3	0.22–0.32%	2% suspension	Mechanical properties (Internal Bond, Modulus of Rupture, Modulus of Elasticity)	Positive	
				Wet properties (Thickness swelling, Water absorption)	Neutral	
				Formaldehyde content	Neutral	
	2	0.25–1.00%	10% paste	Mechanical properties (Internal Bond, Modulus of Rupture, Modulus of Elasticity)	Positive	Allows reduction in resin, maintaining board properties
				Wet properties (Thickness swelling, Water absorption)	Neutral	
				Formaldehyde content	Neutral	
Waferboard	3	0.01–0.03%	2% suspension	Mechanical properties (Internal Bond, Modulus of Rupture, Modulus of Elasticity)	Positive	Allows reduction in resin, maintaining board properties
				Wet properties (Thickness swelling, Water absorption)	Positive	
				Formaldehyde content	Positive	
	2	0.25–1.00%	10% paste	Mechanical properties (Internal Bond, Modulus of Rupture, Modulus of Elasticity)	Positive	Allows reduction in resin, maintaining board properties
				Wet properties (Thickness swelling, Water absorption)	Neutral	
				Formaldehyde content	Neutral	
Plywood	1	0.22%	2% suspension	Shear strength	Positive	
				Wood failure	Neutral	
	2	0.23–0.68%	10% paste	Shear strength	Neutral	Can substitute conventional thickeners, maintaining board properties
				Wood failure	Neutral	

Table 17 indicates that the addition of MFC 2% suspension in UF resin during particleboard production, at addition levels of 0.01–0.03% (dry MFC/liquid resin), following Approach 3, is the most promising path to follow.

Summarising, the use of MFC in adhesive systems for wood-based panels' production has shown positive results at laboratory scale. When introduced at low addition levels as an additive in the glue mixture of the resin, it improves board performance and allows the reduction in petrochemical substances, such as, for instance, the resin itself.

The findings of this research work pave the way for the development and adoption of adhesive formulations with lower CO_2 footprint, rendering a more environmentally friendly character to the final wood-based panel products.

Author Contributions: Conceptualization: E.K. and E.A.; methodology, validation: E.K. and C.M.; formal analysis, investigation, data curation: E.K.; writing—original draft preparation: E.K.; writing—review and editing: E.A.; supervision: C.M.; project administration, funding acquisition: E.A. All authors have read and agreed to the published version of the manuscript.

Funding: This research work was partly funded by the Bio Based Industries Joint Undertaking under the European Union's Horizon 2020 research and innovation programme under grant agreement No 709746.

Acknowledgments: We would like to thank the team of Borregaard working in the EXILVA EU project, grant agreement No 709746, for their support in carrying out and completing the reported research work. We would also like to thank our colleague Sophia Tsiantzi for her help and support in reviewing and editing.

Conflicts of Interest: The authors declare no conflict of interest. The funders had no role in the design of the study; in the collection, analyses, or interpretation of data; in the writing of the manuscript, or in the decision to publish the results.

References

1. Sundarraj, A.A.; Ranganathan, T.V. A review on cellulose and its utilization from agro-industrial waste. *Drug Invent. Today* **2018**, *10*, 89–94.
2. Andresen, M.; Johansson, L.-S.; Tanem, B.S.; Stenius, P. Properties and characterization of hydrophobized microfibrillated cellulose. *Cellulose* **2006**, *13*, 665–677. [CrossRef]
3. Lavoine, N.; Desloges, I.; Dufresne, A.; Bras, J. Microfibrillated cellulose—Its barrier properties and applications in cellulosic materials: A review. *Carbohydr. Polym.* **2012**, *90*, 735–764. [CrossRef] [PubMed]
4. Veigel, S.; Rathke, J.; Weigl, M.; Gindl-Altmutter, W. Particle Board and Oriented Strand Board Prepared with Nanocellulose-Reinforced Adhesive. *J. Nanomater.* **2012**, *2012*, 158503. [CrossRef]
5. Amini, E.; Tajvidi, M.; Gardner, D.J.; Bousfiled, D.W. Utilization of Cellulose Nanofibrils as a binder for Particleboard Manufacture. *BioResources* **2007**, *12*, 4093–4110. [CrossRef]
6. Mahrdt, E.; Pinkl, S.; Schmidberger, C.; Herwijnen, H.W.G.; Veigel, S.; Gindlt-Altmutter, W. Effect of addition of microfibrillated cellulose to urea-formaldehyde on selected adhesive characteristics and distribution in particle board. *Cellulose* **2016**, *23*, 571–580. [CrossRef]
7. Candan, Z.; Akbulut, T.; Gonultas, O. Dynamic mechanical thermal analysis (DMTA) of nanocellulose reinforced urea-formaldehyde resin, COST FP1205 Innovative applications of regenerated wood cellulose fibres. *Res. Gate* **2014**. [CrossRef]
8. Das, M.C.; Athijayamani, A.; Geethan, K.A.V.; Santhosh, D.; Singh, S.P. Effects of length and content of natural cellulose fiber on the mechanical behaviors of phenol formaldehyde composites. *Mater. Today Proc.* **2020**. [CrossRef]

Sample Availability: Samples of the MFC are available from Borregaard. Samples of resins are available from the authors.

 *molecules*

Article

Fully Biobased Epoxy Resins from Fatty Acids and Lignin

Pablo Ortiz [1,2,*] **, Richard Vendamme** [1,2] **and Walter Eevers** [1,3]

[1]   Flemish Institute for Technological Research–VITO, Separation & Conversion Technology, Boeretang 200, 2400 Mol, Belgium; richard.vendamme@vito.be (R.V.); walter.eevers@vito.be (W.E.)

[2]   Biorizon, Auvergnedijk 2, 4612 PZ Bergen op Zoom, The Netherlands

[3]   Department of Chemistry, University of Antwerp, Groenenborgerlaan 171, B-2020 Antwerpen, Belgium

*   Correspondence: pablo.ortiz@vito.be

Academic Editor: Alejandro Rodríguez

Received: 13 February 2020; Accepted: 2 March 2020; Published: 5 March 2020

Abstract: The use of renewable resources for plastic production is an imperious need for the reduction of the carbon footprint and the transition towards a circular economy. With that goal in mind, fully biobased epoxy resins have been designed and prepared by combining epoxidized linseed oil, lignin, and a biobased diamine derived from fatty acid dimers. The aromatic structures in lignin provide hardness and strength to an otherwise flexible and breakable epoxy resin. The curing of the system was investigated by infrared spectroscopy and differential scanning calorimetry (DSC). The influence of the different components on the thermo-mechanical properties of the epoxy resins was analyzed by DSC, thermal gravimetric analysis (TGA), and tensile tests. As the content of lignin in the resin increases, so does the glass transition, the Young's modulus, and the onset of thermal degradation. This correlation is non-linear, and the higher the percentage of lignin, the more pronounced the effect. All the components of the epoxy resin being commodity chemicals, the present system provides a realistic opportunity for the preparation of fully biorenewable resins at an industrial scale.

Keywords: lignin; lignocellulose; aromatics; biobased; epoxy; fatty acid; biopolymers; biobased materials; biorenewable

1. Introduction

We live surrounded by plastics: from packaging to clothing, including construction materials, components of vehicles, furniture, electronics, and an endless list of commodities. In these days of banishment of plastics, their benefits should be kept in mind. For example, the use of plastics for automotive parts results in lighter vehicles that consume less fuel, therefore reducing the greenhouse emissions [1]. However, the overall carbon footprint of most plastics is high because their production relies on fossil fuels. The use of renewable resources for plastic production is, therefore, an imperious need. It is a challenging endeavor because the high performance of current fossil-derived plastics has set a high bar for biobased plastics. The replacement of petrol-derived monomers by their biobased analogs offers, in principle, a win-win situation; that is, the same characteristics and processing as fossil-derived plastics but derived from renewable sources. However, direct replacement is commonly not the most (resource and waste) efficient option because the biomass-derived building blocks are different from the petrol-derived ones [2]. Furthermore, although possible in some cases (as in the case of polyethylene), few of the current industrially produced biobased polymers have a 100% biomass content [3]. Currently, the use of new, biomass-derived building blocks for the production of fully biobased polymers is practically restricted to academia. Unfortunately, the price and/or availability of the monomers employed in these studies are a great obstacle for their industrial application [4].

Here, we report a fully biobased epoxy resin from biorenewable bulk chemicals: lignin and vegetable oils [5]. Currently, the glycidyl ether of bisphenol A is the main epoxy monomer, and it is synthesized from BPA and epichlorohydrin. There have been many attempts to replace BPA with biobased aromatic structures and subsequently cure it [6]. This approach, however, incurs the above-mentioned limitations, namely not being fully biobased (usually due to the curing agent) and the exotic monomers that are used [7]. Some of the few biobased epoxy building blocks that reach industrial-scale production are two epoxidized vegetable oils (EVO), namely epoxidized soybean oil (ESO) and epoxidized linseed oil (ELO) (Scheme 1). The possibility of using epoxidized vegetable oils (EVO) in epoxy polymers has been known for decades, but due to the lack of rigid structures (cycloaliphatic, aromatic) the resulting polymers had a low strength [8].

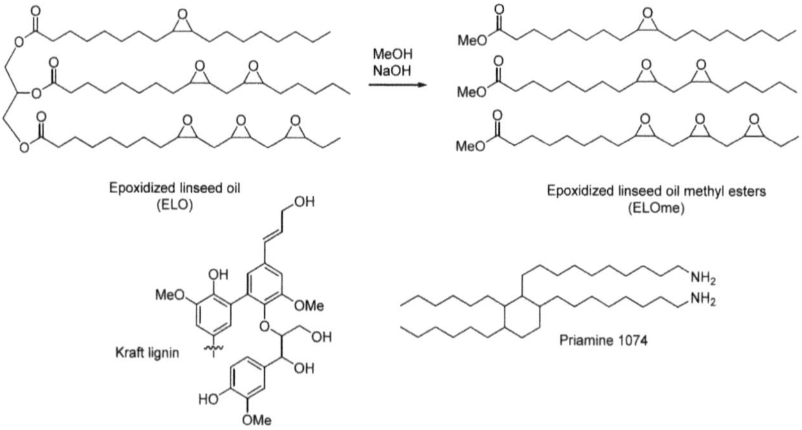

Scheme 1. Biobased building blocks used in this study to produce biobased epoxy resins.

Lignin, with its highly cross-linked phenolic network (Scheme 1), could act as a counterbalance for that deficit of stiffness. Moreover, lignin is currently a waste product from the pulp and paper industry and the only biorenewable source of aromatics that can provide enough feedstock to replace petrol-based aromatics. Technical (industrially produced) lignins are currently the only lignins available at a large scale. Among them, the Kraft process is the dominant in the pulp and paper industry (Scheme 1) [9]. By the use of high temperatures and basic conditions, lignin and hemicellulose are separated from cellulose. High-quality cellulose is obtained, but the resulting lignin is altered. In the pulp and paper industry, this is not a problem because lignin is burned to generate energy and to recover inorganic chemicals. However, the use of this lignin for polymer synthesis represents a challenge due to high polydispersity, limited solubility, and heterogenicity [10]. The fractionation of kraft lignin (KL) [11,12] is an interesting technique that has successfully been applied for epoxy resins synthesis [13]. In epoxy resins, the catalog of hardeners that can be used to cure epoxy resins is broad, the polyamines being the most common choice. Unfortunately, the industrial production of biobased polyamines is still in its infancy [7,14]. One of the few marketed biobased diamines is Priamine® from Croda, derived from dimerized fatty acids (Scheme 1).

2. Results and Discussion

2.1. Design of a Fully Biobased Epoxy Resin

The solubility tests of Lignoboost® KL showed that it was fully soluble in tetrahydrofuran, partially soluble in methanol and acetone but not quite in ethyl acetate. Consequently, methanol and acetone were chosen as fractionation solvents, and the properties of the extracted lignin were analyzed (Table 1 and Supplementary Materials S1–S6).

Table 1. Characterization of lignin fractions.

Lignin Fraction	Yield (%)	Mw (g/mol) [1]	PDI [1]	T_g (°C) [2]	Aliphatic OH (mmol/g) [3]	Aromatic OH (mmol/g) [3]	COOH (mmol/g) [3]
KL	-	5200	3.5	132	2.0	4.3	0.4
KL_{MeOH}	48	1610	2.0	89	1.7	4.5	0.5
$KL_{Acetone}$	61	2540	2.2	114	1.5	4.6	0.5

[1] Determined by GPC. [2] Determined by DSC. [3] Determined by ^{31}P NMR.

The methanol extracted fraction, KL_{MeOH}, with the lowest polydispersity index (PDI), molecular weight (Mw), and glass transition (T_g), was chosen for the biopolymer development. Initial efforts were directed towards using lignin as a curing agent for the epoxy group. The quest for a lignin-epoxy resin goes decades back [15,16]. There is an extensive body of literature showing that the hydroxyl groups in the lignin can react with terminal epoxy groups in molecules such as (polyethelene) glycerol digylicidy ether [17–21]. However, EVO such as ESO or ELO did not yield a fully cured product, probably due to the lower reactivity of internal epoxy groups. Consequently, the addition of a second curing agent, namely a biobased diamine derived from dimer fatty acids (Priamine 1074 ®) was considered and successfully tested. The miscibility of the non-polar oils with polar lignin was not always optimal but could be addressed by transesterifying the EVO with methanol (Scheme 1). Apart from improving the compatibility of the mixture, the use of methyl esters of EVO (ESO_{me} and ELO_{me}) led to samples with higher T_g than those prepared with the EVO. Although the transesterification process was carried out in the lab, it is worth noting that there are commercial suppliers of ESO_{me} and ELO_{me}.

2.2. Possible Curing Mechanism

The reactivity of epoxy groups towards hydroxyl and amine groups is well understood [22], and its curing can be followed using infrared spectroscopy [13]. In our system, besides the epoxy and the amine groups, the hydroxyl groups from lignin were also present. In order to better understand the interaction between the three components, the curing mechanism was studied. More specifically, we wanted to prove that the lignin reacts with the EVO and that it gets incorporated in the structure. Our hypothesis was that the reaction of lignin with an excess of EVO led to an intermediate with epoxy groups still available to be cured upon addition of the diamine (Figure 1 and Scheme 2) [19]. In order to prove that the lignin reacted with the epoxide groups and was not merely embedded in the cured epoxy matrix, the curing of the resin was analyzed by DSC and an exotherm was observed (Supplementary Materials S7). This is in line with previous research [20]. Furthermore, a cured sample was immersed in tetrahydrofuran and kept for 48 h. Afterwards, the epoxy resin was removed, and the liquid was analyzed by ^{1}H NMR. It showed that the leaching corresponded to the diamine and not to lignin (Supplementary Materials S8). This result is not surprising considering that an excess of diamine in relation to epoxides was used. IR showed the disappearance of the stretch corresponding to the epoxy group at 825 cm^{-1} as well as the increase of the broad hydroxyl stretch between 3200 and 3500 cm^{-1} (Figure 2). To our surprise, we noticed that the curing with diamine resulted in the appearance of another peak at around 1650 cm^{-1} (Figure 1). We identified it as that of the C=O stretch of the amide, resulting from the reaction of the ester of EVO with the excess of diamine (Scheme 2).

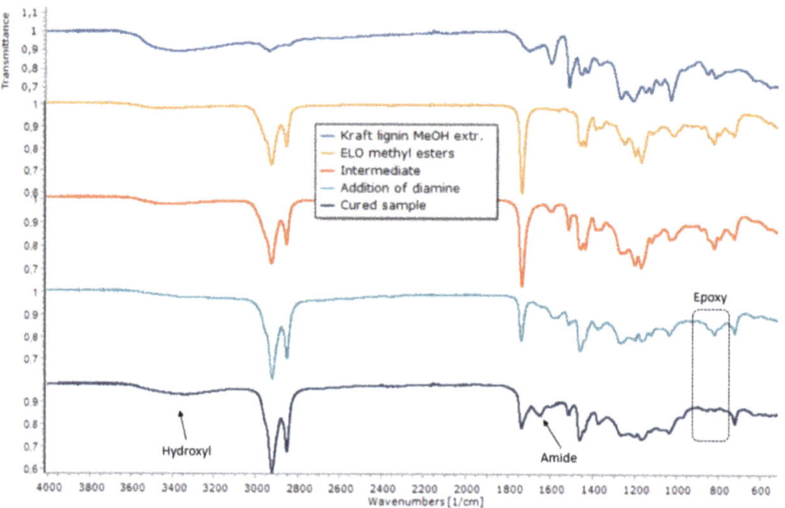

Figure 1. Infrared spectra of the starting materials and the evolution of the mixture during the curing process.

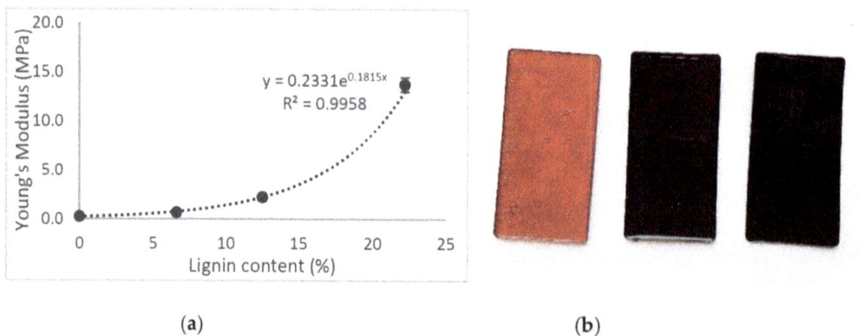

Scheme 2. Proposed mechanism of the curing of the epoxy resins.

(a)

(b)

Figure 2. (**a**) Correlation between the lignin content and Young's modulus of the cured epoxy resins. (**b**) Physical appearance of epoxy resins with 0%, 12.5%, and 22.2% weight of KL$_{MeOH}$.

2.3. Influence of the Dimer Diamine Content

This finding encouraged us to further explore the effect of the amount of the diamine on the properties of the epoxy resins. Fixing the amount of lignin to half the weight of the ELO_{me}, various equivalents of diamine to epoxy equivalents in ELO_{me} were tested (Table 2). The Young's modulus (E_{Young}) was exponentially correlated (Supplementary Materials S9) with the amount of diamine, which is understandable considering its long alkyl chain. Interestingly, other thermomechanical properties did not follow the same trend. Both the tensile strength and glass transition temperature gave similar results for all the samples but one. When one equivalent of diamine was used, the sample had a higher strength and T_g. These results suggest that with one equivalent of diamine, the cross-linking is the highest. Furthermore, they reinforce the proposed curing mechanism. Because some of the OH groups of lignin reacted with the epoxy groups of the EVO, the one equivalent of the diamine was actually in excess to the epoxy groups of the EVO, and the higher cross-linking must have come from the amide formation.

Table 2. The effect of the diamine content on the thermomechanical properties of the epoxy resins.

Diamine Equiv. to ELO_{me}.	E_{Young} (MPa) [1]	Σ_{Break} (MPa) [1]	E_{Break} (%) [1]	T_g (°C) [2]
0.75	22.1 ± 1.6	2.5 ± 0.1	47 ± 2	−2
1	13.8 ± 1.3	4.0 ± 0.7	114 ± 14	0
1.25	7.4 ± 0.4	2.2 ± 0.1	100 ± 33	−5
1.5	4.7 ± 1.1	2.2 ± 0.3	144 ± 15	−6

[1] Calculated by tensile test. Data are the average of three test pieces ± SD. [2] Determined by DSC (S10 in Supplementary Materials).

2.4. Influence of the Lignin Content on the Mechanical Properties

Consequently, we fixed the amount of diamine to one equivalent to the epoxy groups in the ELO_{me}, and we moved on to explore the influence of the KL_{MeOH} content on the thermomechanical properties of the resulting polymers (Table 3). Both Young's modulus and the tensile strength showed an exponential increase with an increasing amount of lignin (Figure 2a). Because of the long aliphatic chains of both ELO_{me} and the diamine, the polymer resulting from their mixture was an elastic although easily breakable material. The incorporation of lignin dramatically changed the properties, and with 12.5% weight of KL_{MeOH} in the composition, the Young's modulus increased ten-fold. Notably, doubling that amount of lignin resulted in a fifty-fold increase in E_{Young} compared with the sample without lignin. Testing the limits, a resin with an even higher percentage of lignin (36.4%) was also tested but resulted in a brittle sample. Besides the mechanical properties, the effect of lignin was also noticeable in the color of the samples, which ranged from dark orange to dark brown (Figure 2b).

Table 3. Influence of the lignin content on the mechanical properties of the cured epoxy resins.

KL_{MeOH} Content (% weight) [1]	E_{Young} (MPa) [2]	Σ_{Break} (MPa) [2]	E_{Break} (%) [2]
0	0.26 ± 0.01	0.4 ± 0.1	197 ± 11
6	0.7 ± 0.1	0.6 ± 0.1	105 ± 15
12.5	2.2 ± 0.4	1.8 ± 0.4	129 ± 16
22.2	13.8 ± 1.3	4.0 ± 0.7	114 ± 14

[1] Lignin content as the % weight of lignin in the resin. [2] Calculated by tensile test. Data are the average of three test pieces ± SD.

2.5. Influence of the Lignin Content on the Thermal Properties

Lignin also had a marked effect on the T_g. The cross-linked epoxy resin resulting from the curing between the ELO_{me} and the diamine showed a glass transition temperature of −20 °C (Figure 3). This result is in line with previous research on epoxy resins derived from EVO and diamines derived from fatty acids [23]. As with the mechanical properties, the effect of KL_{MeOH} on the glass transition followed an exponential increase: 6% of lignin had no influence, but 12.5% of KL_{MeOH} resulted in a T_g of −12 °C, and the resin with 22.2% of lignin had a glass transition of 0 °C (Figure 3). A similar exponential trend had been observed before for an epoxy-lignin-amine system [19]. In that case, the epoxy component was polyethelene glycerol digylicidy ether and the amine triethylene tetramine. Despite the differences with our system, the glass transitions are surprisingly close for a defined amount of lignin. This fact underlines the strong impact lignin has on the thermal properties of epoxy resins.

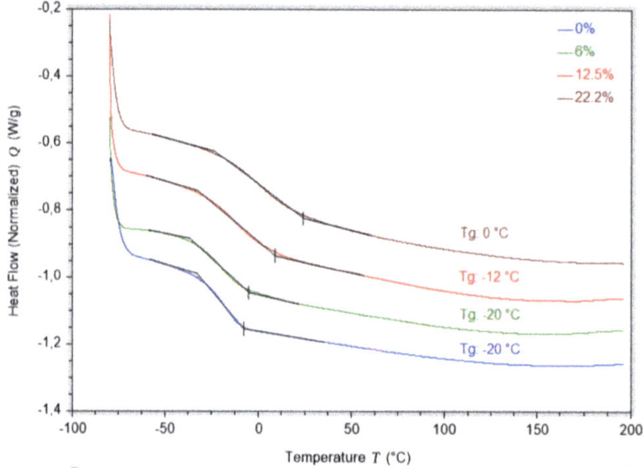

Figure 3. Influence of the KL_{MeOH} content (expressed as weight%) on the T_g.

Likewise, the presence of KL_{MeOH} had a profound impact on the thermal stability of the resins. Although there are some exceptions [24], the incorporation of lignin in a polymeric matrix typically leads to increased thermal stability [25]. TGA experiments (Figure 4 and Table 4) showed increased stability as the content of lignin in the sample rose. The sample without lignin was completely consumed at 500 °C, which is in agreement with previous research on EVO/dimer systems [23]. By contrast, the sample with 12.5% of KL_{MeOH} in weight showed 14% of residual weight. Importantly, the residual weight cannot just be attributed to the remaining lignin, as the sample with 22% weight of KL_{MeOH} retained 40% of weight at 500 °C. Interestingly, the same exponential effect to that of the Young modulus of the resins was also observed here. That is, the effect from doubling the lignin content (from 12.5% to 22% of the total weight) had a greater impact than that from not having lignin at all to having a 12.5% content.

Table 4. Thermal decomposition data from the TGA curves.

Lignin Content (% weight) [1]	2.5% Weight Loss (°C)	5% Weight Loss (°C)	Residual Mass at 800 °C (% weight)
0	249	302	1
12.5	257	292	14
22.2	276	315	40

[1] Lignin content as the % weight of KL_{MeOH} in the resin.

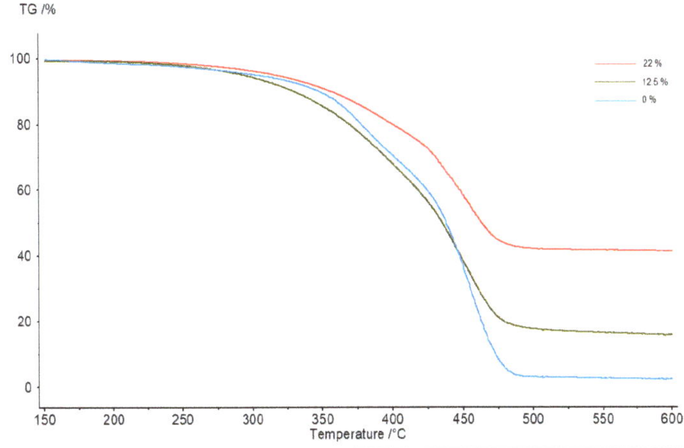

Figure 4. The TGA curves of epoxy resins with an increasing amount of KL_{MeOH} (expressed as weight %).

2.6. Influence of the Type of Lignin on the Thermomechanical Properties

After the observed strong effect of the amount of lignin on the thermomechanical properties of the resins, we wondered how these would be influenced by the characteristics of lignin (Table 5). Different fractions of KL showed a positive correlation between the Mw and the Young's modulus, in line with previous reports [13]. That is, the higher the Mw, the higher the Young's modulus. However, no such trend was observed for the ultimate tensile strength (Σ_{Break}) or elongation at break (E_{Break}). The T_g of the epoxy resins made from different Kraft lignins had a remarkably similar value, and only that of the acetone extracted fraction was slightly higher than the rest. A similar effect had been noted before for lignin-based epoxy resins [13].

Table 5. Effect of the lignin type on the thermomechanical properties of the epoxy resins.

Lignin Type	Lignin Content (% weight) [1]	Mw (g/mol) [2]	Dispersity [2]	E_{Young} (MPa) [3]	Σ_{Break} (MPa) [3]	E_{Break} (%) [3]	T_g (°C) [4]
KL_{MeOH}	22.2	1610	2.0	13.8 ± 1.3	4.0 ± 0.7	114 ± 14	0
$KL_{Acetone}$	22.2	2540	2.2	20.9 ± 1.6	5.7 ± 0.4	121 ± 14	4
KL	22.2	5200	3.5	35.2 ± 9.0	4.1 ± 1.1	41 ± 6	0
West Fraser KL	22.2	4080	3.8	17.6 ± 0.8	6.8 ± 0.7	135 ± 13	0
BCD oil	22.2	330	1.4	31.1 ± 1.3	3.6 ± 0.2	21 ± 3	9
BCD oil	36.4	330	1.4	60.2 ± 9.6	6.5 ± 0.3	21 ± 8	15

[1] Lignin content as the weight% of lignin in the resin. [2] Determined by GPC (Supplementary Materials S4–S6; S11–S12). [3] Calculated by tensile test. Data are the average of three test pieces ± SD. [4] Determined by DSC (Supplementary Materials S13).

One of the features of lignin is the variability and heterogenicity among different samples, and therefore we were interested in comparing a KL from another source. West Fraser KL led to epoxy resins with not striking but noticeable differences. The resin containing West Fraser KL had a slightly lower Young's modulus than Lignoboost KL, which points again to a relation between the Mw and the Young's modulus. However, this seems to not be a linear relation, because West Fraser KL had a Mw of 4080 but a E_{Young} lower than $KL_{Acetone}$, which had a Mw of 2540. Despite its relatively low Young's modulus, the epoxy resin made from West Fraser KL showed the highest Σ_{Break} and E_{Break} of the series.

Fractionation of technical lignins is a good approach to reduce polydispersity and Mw, thus allowing for better miscibility and reactivity in the prepolymer mixture. Even lower molecular weight fractions can be obtained from the depolymerization of lignin. These streams are attractive because they can be in oily form and thus directly blended with the other components without the need for solvents [10]. To that end, solvent-extracted, base-catalyzed depolymerization (BCD) oil from Eucalyptus, with a Mw of 330 and PDI of 1.4, was tested. The polymer resulting from the incorporation of 22% of such lignin showed a T_g of 9 °C, significantly higher than the samples prepared with KL. Moreover, the structural differences of lignin had a great impact on the mechanical properties, leading to harder epoxy resins (higher Young's modulus and lower E_{Break}, Table 5 and Figure 5). Contrary to the case of KL, where reducing the Mw and polydispersity led to samples with a lower E_{Young}, the low Mw of BCD lignin oil did not result in a less stiff material. Moreover, the use of such lignin oil allowed us to increase the content of lignin. A sample with 36.4% of lignin content could be prepared, which displayed a Young's modulus of 60 MPa and a T_g of 15 °C.

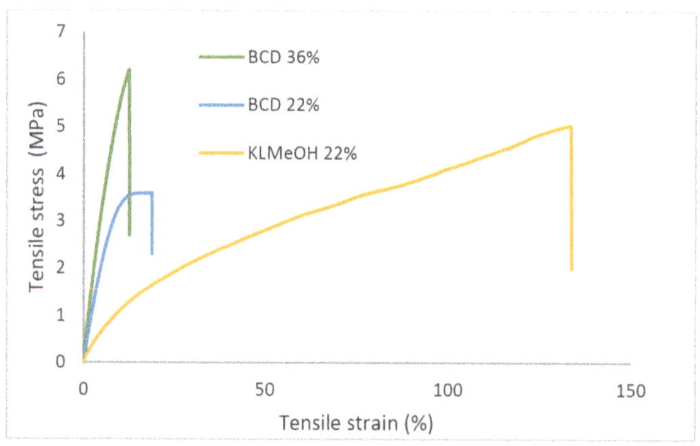

Figure 5. Selected tensile stress-strain curves of different cured epoxy resins.

3. Materials and Methods

3.1. Chemicals

Epoxidized soybean oil (Efka® PL 5382) was kindly provided by BASF; Epoxidized linseed oil (Lankroflex™) was kindly provided by Valtris. Dimer diamine (Priamine 1074®) was kindly provided by Croda (AHEW 138g/equiv.). Lignoboost® Kraft lignin was provided by Innventia Institute. West Fraser Kraft lignin was kindly provided by West Fraser. Base-catalyzed depolymerized (BCD) solvent-extracted eucalyptus lignin oil was kindly provided by Fraunhofer Institute. Tetrahydrofuran was purchased from Fisher Scientific, methanol and ethanol from Acros Organics, and 1-methylimidazol from Alfa Aesar. Sodium hydroxide was purchased from Sigma-Aldrich.

3.2. Transesterification of EVO

The formation of methyl esters of epoxidized soybean and linseed oil was carried out following a modified literature procedure [26]: 2 g of NaOH were dissolved in 50 mL of methanol and transferred to a round bottom flask containing 100 g of epoxidized vegetable oil. 5 mL of dichloromethane were added, and the mixture was stirred for 2 h at 50 °C. It was then let to stand overnight so that the glycerol would settle in the bottom. The upper, liquid phase was transferred to a round bottom flask, and the methanol was removed under reduced pressure. The epoxy equivalent weight of epoxidized soybean oil methyl ester was determined to be 230 g/equiv. and 184 g/equiv. for the epoxidized linseed oil methyl ester.

3.3. Lignin Fractionation and Characterization

Solvent extraction was performed by dissolving 10 g of lignin in 150 mL of a solvent (methanol or acetone), stirring overnight, filtering through a Büchner funnel, followed by concentrating the filtrate and finally drying the lignin at 50 °C under vacuum for 16 h. The lignin hydroxyl content was determined by ^{31}P NMR according to a literature procedure [27].

3.4. Preparation and Curing of Resins

An amount of lignin (LignoBoost Kraft or the solvent-extracted fractions) was dissolved in 2 mL of tetrahydrofuran and mixed with 1 g of epoxidized vegetable oil (linseed, soybean, or their methyl esters). Alternatively, ethanol can be used as a solvent. The mixture was stirred at 70 °C for 1 h in an open flask. Afterwards, an amount of Priamine 1074 was added, and the mixture was stirred for 1 min and poured in a Teflon mold. The Teflon mold was placed in a vacuum oven (850 mbar) at 50 °C for 1 h to allow the solvent to evaporate. Then, the Teflon mold was placed in an oven and cured at 120 °C for 4 h and at 150 °C for 16 h. The curing could also be done at 120 °C for 20 h in the presence of 1-methylimidazole as a catalyst without an influence on the outcome. When BCD Eucalyptus lignin was used, no solvent was required, and thus the required amount of lignin was mixed with 1 g of epoxidized vegetable oil and stirred at 70 °C for 30 min in an open flask. The rest of the protocol was as described above.

3.5. Chemical Analysis

Gel permeation chromatography was performed on a Styragel HR1 column eluting with tetrahydrofuran with RID detection. The hydroxyl content was quantitatively determined using ^{31}P NMR spectroscopy, as described in the literature [27]. The epoxy equivalent weight of methyl esters of the epoxidized vegetable oils was determined by potentiometric titration following ASTM Standard D1652–04.

3.6. Thermal Analysis

Possible thermal phase transitions in all of the cured resin specimens were investigated using differential scanning calorimetry (DSC). The DSC measurements were conducted with TA Discovery DSC 250 by subjecting the cured epoxy resins to a heat–cool–heat cycle from −80 to 200 °C at a 20 °C/min rate under a nitrogen atmosphere. Normalized heat flows were plotted as a function of the temperature, and the plots were analyzed. The glass transition temperature (T_g) was defined as the half value of the heat capacity change (ΔCp/2). Thermal gravimetric analyses (TGA) were performed to gain insight into the degradation temperature of the resins. TGA measurements were carried out in a nitrogen environment (flow rate 20 mL min^{-1}) using a Netzsch STA 449c machine at a 10 °C min^{-1} heating rate from 25 to 800 °C.

3.7. Spectroscopic Analysis

Attenuated total reflectance-Fourier transform infrared (ATR-FTIR) analyses of the epoxy resins were performed using a Nicolet iS10 spectrometer (Thermo Fisher Scientific, Waltham, MA, USA) operating at laboratory atmospheric conditions. The measurements were performed in the attenuated reflectance (ATR) mode using a diamond crystal. The spectral range was 4000 to 500 cm^{-1}. The scans were performed with a 4 cm^{-1} scan resolution and 32 scans per sample. Blank scans were subtracted from the sample scans prior to their analysis.

3.8. Tensile Test

The tensile tests of the cured epoxy resins were conducted according to ASTM D638, using an Instron machine model 2412 (Instron, England). Three replicates were measured for each sample, with the dimensions of ASTM D638 Type V. The tests were performed at 22 ± 2 °C and at 50–60% relative humidity. The crosshead speed was 2.5 mm/min. Young's modulus, ultimate tensile strength, and percentage elongation at break were calculated from the stress-strain curves on the basis of the initial sample dimensions.

4. Conclusions

A series of fully biobased epoxy resins made from renewable resources were prepared, and their thermo-mechanical properties were analyzed. Lignin, rich in aromatic structures, gives stiffness and consistency to the epoxy resins, which are otherwise soft, flexible, and breakable due to long aliphatic chains of the fatty acid derivatives. Solvent extracted Kraft lignin was used due to its more defined profile. At a low lignin content, the effects were barely noticeable, but at a 12.5 weight% of lignin both the glass transition temperature and the Young's modulus increased, and the rise was even more pronounced at a 22 weight%. It is thus not a linear but an exponential correlation. Higher amounts of lignin could not be incorporated if solvent-extracted Kraft lignin was used. Depolymerized and subsequently solvent-extracted lignin, liquid and with low Mw, could be used in higher amounts (36.4%), leading to harder epoxy resins. The fact that all the components of the resin are bulk chemicals opens the door to the production of fully biobased resins on an industrial level.

5. Patents

Ortiz, P.; Vendamme, R. Bio Based Epoxy Resin. PCT/EP2019/075067.

Supplementary Materials: The following are available online, ^{31}P NMR characterization of lignin S1–S3; GPC characterization of lignins S4–S8; Proof of epoxy curing S9–S10; Plot of equiv. of diamine vs. Young's Modulus of cured samples S11; DSC of cured samples S12–S13; TGA of cured samples S14.

Author Contributions: Conceptualization, P.O. and R.V.; methodology, P.O.; investigation, P.O.; writing—original draft preparation, P.O.; writing—review and editing, P.O., R.V. and W.E. All authors have read and agreed to the published version of the manuscript.

Funding: This research was funded by the province of Noord-Brabant (The Netherlands) in the framework of the activities at the Shared Research Center Biorizon.

Acknowledgments: The authors would like to thank BASF for providing epoxidized soybean oil (Efka® PL 5382); Valtris for providing epoxidized linseed oil (Lankroflex™); Croda for providing dimer diamine (Priamine 1074®); Innventia Institute for providing Lignoboost® Kraft lignin. West Fraser for providing West Fraser Kraft lignin; Fraunhofer Institute for providing base-catalyzed depolymerized (BCD) solvent-extracted eucalyptus lignin oil.

Conflicts of Interest: The authors declare no conflict of interest.

References

1. Gupta, A.K. Waste and Biomass to Clean Energy. In *Handbook of Clean Energy Systems*; Wiley: New York, NY, USA, 2015; pp. 1–24.
2. Lie, Y.; Ortiz, P.; Vendamme, R.; Vanbroekhoven, K.; Farmer, T.J. BioLogicTool: A Simple Visual Tool for Assisting in the Logical Selection of Pathways from Biomass to Products. *Ind. Eng. Chem. Res.* **2019**, *58*, 15945–15957. [CrossRef]
3. Aeschelmann, F.; Carus, M. Biobased Building Blocks and Polymers in the World: Capacities, Production, and Applications–Status Quo and Trends Towards 2020. *Ind. Biot.* **2015**, *11*, 154–159. [CrossRef]
4. Bobade, S.K.; Paluvai, N.R.; Mohanty, S.; Nayak, S.K. Bio-Based Thermosetting Resins for Future Generation: A Review. *Polym.-Plast. Technol.* **2016**, *55*, 1863–1896. [CrossRef]
5. Ortiz, P.; Vendamme, R. Bio Based Epoxy Resin. PCT/EP2019/075067.
6. Auvergne, R.; Caillol, S.; David, G.; Boutevin, B.; Pascault, J.-P. Biobased Thermosetting Epoxy: Present and Future. *Chem. Rev.* **2014**, *114*, 1082–1115. [CrossRef] [PubMed]
7. Baroncini, E.A.; Yadav, S.K.; Palmese, G.R.; Stanzione, J.F. Recent Advances in Bio-Based Epoxy Resins and Bio-Based Epoxy Curing Agents. *J. Appl. Polym. Sci.* **2016**, *133*, 44103. [CrossRef]
8. Wang, R.; Schuman, T. Chapter 9-Towards Green: A Review of Recent Developments in Bio-renewable Epoxy Resins from Vegetable Oils. In *Green Materials from Plant Oils*; Liu, Z., Kraus, G., Eds.; The Royal Society of Chemistry: Cambridge, UK, 2015; pp. 202–241.
9. Rinaldi, R.; Jastrzebski, R.; Clough, M.T.; Ralph, J.; Kennema, M.; Bruijnincx, P.C.A.; Weckhuysen, B.M. Paving the Way for Lignin Valorisation: Recent Advances in Bioengineering, Biorefining and Catalysis. *Angew. Chem. Int. Edit.* **2016**, *55*, 8164–8215. [CrossRef]

10. Feghali, E.; Torr, K.M.; van de Pas, D.J.; Ortiz, P.; Vanbroekhoven, K.; Eevers, W.; Vendamme, R. Thermosetting Polymers from Lignin Model Compounds and Depolymerized Lignins. *Top. Curr. Chem. (Z)* **2018**, *376*, 32. [CrossRef]

11. Passoni, V.; Scarica, C.; Levi, M.; Turri, S.; Griffini, G. Fractionation of Industrial Softwood Kraft Lignin: Solvent Selection as a Tool for Tailored Material Properties. *ACS Sustain. Chem. Eng.* **2016**, *4*, 2232–2242. [CrossRef]

12. Duval, A.; Vilaplana, F.; Crestini, C.; Lawoko, M. Solvent Screening for the Fractionation of Industrial Kraft Lignin. *Holzforschung* **2015**, *70*, 11–20. [CrossRef]

13. Gioia, C.; Lo Re, G.; Lawoko, M.; Berglund, L. Tunable Thermosetting Epoxies Based on Fractionated and Well-Characterized Lignins. *J. Am. Chem. Soc.* **2018**, *140*, 4054–4061. [CrossRef]

14. Froidevaux, V.; Negrell, C.; Caillol, S.; Pascault, J.-P.; Boutevin, B. Biobased Amines: From Synthesis to Polymers; Present and Future. *Chem. Rev.* **2016**, *116*, 14181–14224. [CrossRef] [PubMed]

15. Nieh, W.L.-S.; Glasser, W.G. Lignin Epoxide: Synthesis and Characterization. In *Lignin*; Glasser, W.G., Sarkanen, S., Eds.; American Chemical Society: Washington, DC, USA, 1989; Volume 397, pp. 506–514.

16. Simionescu, C.I.; Rusan, V.; Macoveanu, M.M.; Cazacu, G.; Lipsa, R.; Vasile, C.; Stoleriu, A.; Ioanid, A. Lignin/Epoxy Composites. *Compos. Sci. Technol.* **1993**, *48*, 317–323. [CrossRef]

17. Delmas, G.-H.; Benjelloun-Mlayah, B.; Bigot, Y.L.; Delmas, M. Biolignin™ Based Epoxy Resins. *J. Appl. Polym. Sci.* **2013**, *127*, 1863–1872. [CrossRef]

18. Engelmann, G.; Ganster, J. Bio-Based Epoxy Resins with Low Molecular Weight Kraft Lignin and Pyrogallol. *Holzforschung* **2013**, *68*, 435–446. [CrossRef]

19. Nonaka, Y.; Tomita, B.; Hatano, Y. Synthesis of Lignin/Epoxy Resins in Aqueous Systems and Their Properties. *Holzforschung* **2007**, *51*, 183–187. [CrossRef]

20. Li, R.J.; Gutierrez, J.; Chung, Y.-L.; Frank, C.W.; Billington, S.L.; Sattely, E.S. A Lignin-Epoxy Resin Derived from Biomass as an Alternative to Formaldehyde-Based Wood Adhesives. *Green Chem.* **2018**, *20*, 1459–1466. [CrossRef]

21. Ortiz, P.; Wiekamp, M.; Vendamme, R.; Eevers, W. Bio-Based Epoxy Resins from Biorefinery By-Products. *BioResources* **2019**, *14*, 3200–3209.

22. Gibson, G. Chapter 27—Epoxy Resins. In *Brydson's Plastics Materials*, 8th ed.; Gilbert, M., Ed.; Elsevier: Oxford, UK, 2017; pp. 773–797.

23. Stemmelen, M.; Lapinte, V.; Habas, J.-P.; Robin, J.-J. Plant Oil-Based Epoxy Resins from Fatty Diamines and Epoxidized Vegetable Oil. *Eur. Polym. J.* **2015**, *68*, 536–545. [CrossRef]

24. Ismail, T.N.M.T.; Hassan, H.A.; Hirose, S.; Taguchi, Y.; Hatakeyama, T.; Hatakeyama, H. Synthesis and Thermal Properties of Ester-Type Crosslinked Epoxy Resins Derived from Lignosulfonate and Glycerol. *Polym. Int.* **2010**, *59*, 181–186. [CrossRef]

25. Xu, C.; Ferdosian, F. *Conversion of Lignin into Bio-Based Chemicals and Materials*; Green Chemistry and Sustainable Technology; Springer: Berlin/Heidelberg, Germany, 2017.

26. Sahoo, S.K.; Mohanty, S.; Nayak, S.K. Toughened Bio-Based Epoxy Blend Network Modified with Transesterified Epoxidized Soybean Oil: Synthesis and Characterization. *RSC Adv.* **2015**, *5*, 13674–13691. [CrossRef]

27. Pu, Y.; Cao, S.; Ragauskas, A.J. Application of Quantitative 31P NMR in Biomass Lignin and Biofuel Precursors Characterization. *Energy Environ. Sci.* **2011**, *4*, 3154–3166. [CrossRef]

Sample Availability: Samples of the cured epoxy resins are available from the authors.

Article

Horticultural Plant Residues as New Source for Lignocellulose Nanofibers Isolation: Application on the Recycling Paperboard Process

Isabel Bascón-Villegas [1,2,*], Eduardo Espinosa [1,*], Rafael Sánchez [3], Quim Tarrés [4], Fernando Pérez-Rodríguez [2] and Alejandro Rodríguez [1]

[1] Chemical Engineering Department, Bioagres group, Universidad de Córdoba, 14014 Córdoba, Spain; a.rodriguez@uco.es

[2] Department of Food Science and Technology, Universidad de Córdoba, 14014 Córdoba, Spain; b42perof@uco.es

[3] Technological Institute of Packaging, Transport and Logistic (ITENE), 46980 Paterna, Spain; rafael.sanchez@itene.com

[4] Group LEPAMAP, Department of Chemical Engineering, Universidad de Girona, 17071 Girona, Spain; joaquimagusti.tarres@udg.edu

* Correspondence: q12bavii@uco.es (I.B.-V.); eduardo.espinosa@uco.es (E.E.); Tel.: +34-957-21-85-86 (I.B.-V.); +34-957-21-85-86 (E.E.)

Academic Editor: Fabrizio Sarasini
Received: 11 June 2020; Accepted: 14 July 2020; Published: 18 July 2020

Abstract: Horticultural plant residues (tomato, pepper, and eggplant) were identified as new sources for lignocellulose nanofibers (LCNF). Cellulosic pulp was obtained from the different plant residues using an environmentally friendly process, energy-sustainable, simple, and with low-chemical reagent consumption. The chemical composition of the obtained pulps was analyzed in order to study its influence in the nanofibrillation process. Cellulosic fibers were subjected to two different pretreatments, mechanical and TEMPO(2,2,6,6-Tetramethyl-piperidin-1-oxyl)-mediated oxidation, followed by high-pressure homogenization to produce different lignocellulose nanofibers. Then, LCNF were deeply characterized in terms of nanofibrillation yield, cationic demand, carboxyl content, morphology, crystallinity, and thermal stability. The suitability of each raw material to produce lignocellulose nanofibers was analyzed from the point of view of each pretreatment. TEMPO-mediated oxidation was identified as a more effective pretreatment to produce LCNF, however, it produces a decrease in the thermal stability of the LCNF. The different LCNF were added as reinforcing agent on recycled paperboard and compared with the improving produced by the industrial mechanical beating. The analysis of the papersheets' mechanical properties shows that the addition of LCNF as a reinforcing agent in the paperboard recycling process is a viable alternative to mechanical beating, achieving greater reinforcing effect and increasing the products' life cycles.

Keywords: lignocellulose nanofibers; horticultural residues; paperboard; recycling

---

Academic Editor: Fabrizio Sarasini

1. Introduction

The 21st century industrial revolution boosted changes in methods of production and consumption thanks to several factors, such as technological development, globalization of markets and resources, availability of energy, etc. In terms of development and welfare, the linear economic system had been beneficial. However, this model, based on taking, making, and discarding [1], is not compatible with the limited resources and capacity to adapt to environmental impact [2]. As a consequence, nowadays, modern societies are affected by an alarming resource depletion and overproduction of materials which are very difficult to manage. As an alternative, it is proposed that the system takes a regenerative approach, using "the circular economy" concept, with the aim of keeping the resource value, and restricting the raw materials and energy inputs. This is generally known as the "bioeconomy".

The bioeconomy is an economic system in which biomass is converted into value-added materials [3], such as chemicals [4], food and feed, and fuels and energy [5,6], among others. It is based on two assumptions: (i) The biomass is not being totally exploited, its main destination being soil protection, animal feed, burning, composting and silage; and (ii) the use of the biomass could be improved. This can be attained by a product valorization and a more efficient use of the agricultural residues mentioned, extracting more energy and byproducts from them, and decreasing both the cost of agricultural production and the spread of pests and greenhouse gas (GHG) emissions. In addition, the biomass application could be improved, increasing the process yields through innovative technological solutions [7].

The lignocellulosic biomass from agro-industrial activity provides cheap raw materials for the extraction of biopolymers, such as cellulose, hemicellulose, and lignin. Some advantages of these vegetal fibers are their natural abundance, low density, high specific stiffness, and biodegradability [8].

In recent years, with the rise of recycling policies, the paper industry has increased the production of paper from recycled cellulose fiber to produce cardboard packaging. These fibers have poorer physicochemical properties than the original cellulose fiber due to the hornification effect of the fiber. This effect is an alteration of the external layers of the cellulose that occurs in the drying process of paper and during its exposure to the environment, affecting the resistance properties of the paper. Mechanical beating is the most widely used technology at the industrial level, due to its cost and simplicity, to improve the properties of the recycled products that require it. However, even though the mechanical beating increases the specific surface area, swelling capacity, and mechanical properties of the fibers, there is also long-term structural damage and the creation of fines, thus reducing the drainage properties and life span of these products [9]. There are other technologies for dealing with the loss of mechanical properties of recycled products, such as the addition of virgin fiber, the addition of chemical products, or some more innovative ones, such as enzymatic refining or the addition of cellulose nanofibers [10]. Cellulose nanofibers present a high specific surface, compared to the original fiber, which allows a high adhesion capacity with adjacent fibers, acting as a link between fibers and favoring an increase in their mechanical properties [11].

Spain is the largest producer of vegetables in Europe, with a total of 12,629,447 tonnes in 2018, representing over 14% of total European production. In Spain, three of the most representative products are tomatoes, peppers, and eggplants which constitute 37.73%, 10.15%, and 1.89% of the vegetable production, respectively [12]. This production involves the generation of a high amount of waste that needs to be managed properly for its use in the production of high value-added products. The valorization of the lignocellulosic residues from these crops has been studied by several authors as a raw material to produce compost [13], biogas [14], particleboards [15], and cellulosic pulp for paperboard [16].

The aim of this work was to study the suitability of biomass residues from tomato, pepper, and eggplant as new sources for lignocellulosic nanofibers (LCNF) production and their application as a paperboard reinforcing agent. Cellulose pulps were obtained using an environmentally friendly process, energy-sustainable, simple, and with low-chemical reagent consumption. The obtained pulps were physicochemical characterized and used for LCNF production by mechanical and TEMPO-mediated oxidation pretreatment followed by high-pressure homogenization treatment. The LCNF obtained was submitted to a physical-chemical characterization and used to improve the mechanical properties during the paperboard formation and compared with industrial mechanical beating.

2. Results and Discussion

2.1. Chemical Composition of Raw Materials and Obtained Cellulosic Pulps

The first objective of this study is the production of cellulose pulps from horticultural plant residues (tomatoes, peppers, eggplants), which was accomplished by subjecting the raw materials to a soft conditions' soda pulping process. This treatment has previously been successfully applied to produce cellulosic pulps with optimal characteristics for the isolation of lignocellulose nanofibers from herbaceous biomass [17]. The yields from the pulping process were 25.73%, 24.76%, and 20.50% for eggplant, pepper, and tomato plants, respectively.

Figure 1 shows the differences in the chemical composition of the raw materials and that presented by the cellulosic pulps obtained. How the non-structural components (extractables and ashes) are reduced almost entirely after treatment is observed. The cellulose fraction was purified, especially in cellulosic pulps from tomatoes and pepper plants, increasing from 20.9% and 27.91% to 66.4% and 56.77%, respectively. In the case of eggplant, this important purification does not take place, however, and unlike the rest, an increase in the content of hemicellulose up to 30.35% was produced. The hemicellulose content is of special interest in the production of cellulose nanofibers through mechanical treatments acting as a barrier against the aggregation of the fibrillated microfibers. Chaker et al. determined that a hemicellulose content 25% produces a yield twice as high as that presented by fibers with 12% hemicellulose content [18]. The lignin content in the cellulosic pulp is high in comparison with other agri-residues used for lignocellulose nanofiber production [17,19–21]. Lignin acts as binding agent in the lignocellulosic matrix, promoting integrity and impeding its deconstruction. Despite its high lignin content, the pulping process may break ether and ester linkages between lignin and carbohydrates (cellulose and hemicellulose), allowing fiber defibrillation [22]. In addition, the presence of lignin in the fiber can exert an antioxidation action during nanofibrillation, preventing the union of links already broken during nanofibrillation treatment [23].

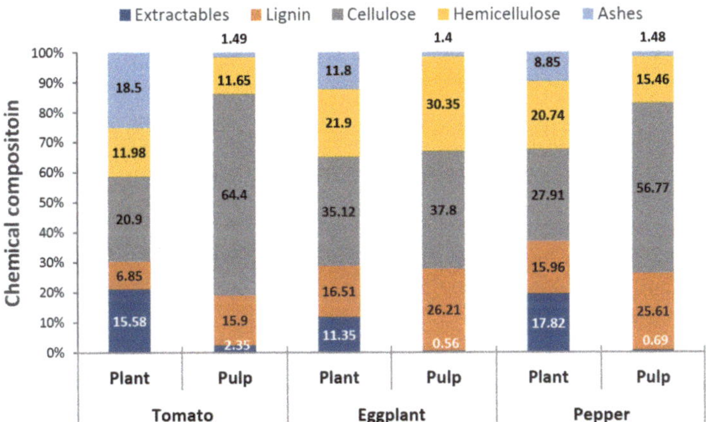

Figure 1. Chemical composition of horticultural plant residues and cellulosic pulps.

The crystallinity structure of the different cellulose pulps was analyzed by the X-ray diffraction technique (Figure 2). It is possible to observe two major reflection peaks at $2\theta = 16.1°$ and $22.5°$, corresponding to 110 and 200 typical reflection planes of cellulose I. The crystallinity index shows values of 31.24%, 44.16%, and 53.21% for eggplant, pepper, and tomato plants, respectively. The crystallinity index (CI) values shown by the cellulosic pulps are low compared to those shown by purified cellulose [20]. This is due to the presence of amorphous components, mainly lignin and hemicellulose, in fiber. These values coincided with the lower values of cellulose in fiber for eggplant and pepper pulp and, therefore, higher contents of hemicellulose and lignin.

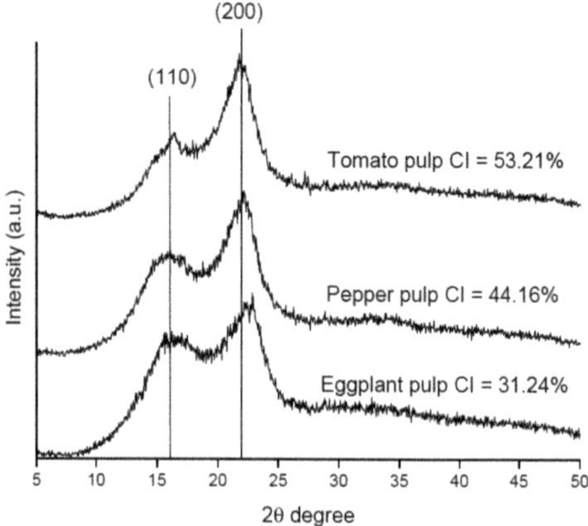

Figure 2. XRD patterns of horticultural residues pulps.

2.2. Lignocellulose Nanofiber Characterization

In order to investigate the effect of the chemical composition of cellulosic pulps and of the different pretreatments on the production of lignocellulose nanofibers, the LCNF obtained were characterized in terms of nanofibrillation yield, cationic demand, carboxyl content, and morphology (Table 1). In general, it is observed that LCNF obtained by TEMPO-mediated oxidation presents a higher nanofibrillation yield (49–70%) compared to those obtained by mechanical pretreatment (18–33%). The nanofibrillation yields of LCNF obtained by mechanical pretreatment are similar to those presented by other cellulose nanofibers produced by the same process [24–28], however, those obtained by TEMPO-mediated oxidation show a slightly lower yield than those shown by other authors, which usually exceed 90% [29–31]. This behavior is also observed in the cationic demand where the TEMPO-oxidized LCNF present considerably higher values of cationic demand. This is mainly due to the greater specific surface (σ_{LCNF}) that these nanofibers present in comparison with those obtained by means of mechanical pretreatment. This greater specific surface results in a greater exposure of the -OH and -COOH groups of the surface of the nanofibers, resulting in higher cationic demand. Regarding the carboxyl content, a slight increase is observed after catalytic oxidation. These values are very low in comparison with other nanofibers obtained by TEMPO-mediated oxidation that present carboxyl content values of 600–1000 µmols/g [26,29–31] This is due to the conversion of primary C6 alcohol groups from the surface of the crystalline regions of the cellulose into carboxyl groups produced during pretreatment. In this case, the low crystallinity, and the high presence of lignin (which may partially or totally consume the NaClO used as a catalytic reaction activator), means that this conversion is not as effective as for bleached pulps.

Regarding the morphology, some differences are observed between the different pretreatments. The diameters obtained vary significantly between the TEMPO-oxidized LCNF and those obtained by mechanical pretreatment. On the one hand, the nanofibers obtained by TEMPO-mediated oxidation show very similar diameter values (12–17 nm) despite the differences shown in the chemical composition. However, in LCNF obtained by mechanical pretreatment, large differences were observed between the different raw materials. It is observed that for LCNF obtained mechanically from eggplant residues, show much smaller diameter than the rest. This is due to their high hemicellulose content, which acts as a key component in the nanofibrillation process [18]. The namometric size of the LCNF was confirmed by direct observation by SEM (Figure 3). It is observed how significant differences exist between the different pretreatments, observing in the case of those obtained mechanically a large proportion of non-nanofibrillated macro/microfibers, as indicated by their low nanofibrillation yield compared to TEMPO-oxidized LCNF. Regarding the length of the lignocellulose nanofibers, a generalized decreased is observed after catalytic oxidation, showing a decrease of 60.07%, 55.94%, and 53.60% for the LCNF obtained from eggplant, pepper, and tomato plants. It is produced by the depolymerization and β-elimination of the cellulose amorphous regions into gluconic acid or cellulose-derived small fragments [32].

Table 1. Characterization of the different lignocellulose nanofibers.

LCNF Sample	η (%)	CD (μeq/g)	CC (μmols/g)	σ_{LCNF} (m^2/g)	Diameter (nm)	Length (nm)
TM-LCNF	17.81 ± 2.59	298.39 ± 48.30	247.35 ± 6.14	24.86	112	5440
PM-LCNF	18.34 ± 3.37	166.46 ± 0.00	148.71 ± 1.66	8.64	278	4317
EM-LCNF	32.61 ± 3.48	248.30± 10.89	127.25 ± 3.99	58.95	42	5132
TT-LCNF	48.77 ± 1.30	707.86 ± 18.54	299.96 ± 48.76	198.65	12	2524
PT-LCNF	69.66 ± 6.11	513.37 ± 37.23	205.81 ± 5.86	148.78	17	1902
ET-LCNF	66.39 ± 1.52	563.38 ± 37.06	186.61 ± 63.78	183.48	14	2049

η: nanofibrillation yield; CD: Cationic demand; CC: Carboxyl content; σ_{LCNF}: Specific surface area of LCNF.

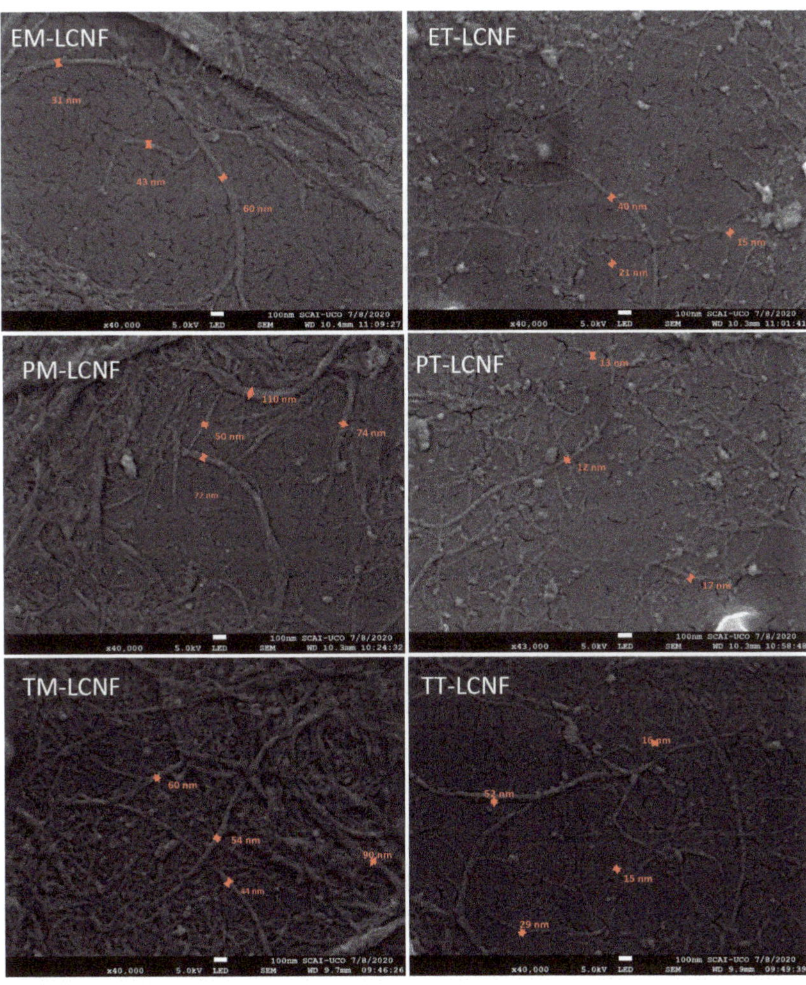

Figure 3. SEM microphotography of the different LCNF.

Figure 4 shows the effect of the different pretreatments on the crystallinity of the lignocellulose nanofibers. In the different patterns crystalline reflection peaks are observed in the planes (110) and (200) as in the case of the cellulose pulps, indicating that LCNF also presents a crystalline structure typical of cellulose I. These observations suggest that the crystalline structure of the fiber is maintained after the different pretreatments and the nanofibrillation treatment. Both pretreatments present similar values for the different lignocellulose nanofibers (56%–60%), all higher than the initial values shown in the cellulose pulps. This may be due to the degradation of the amorphous regions by the action of both pretreatments, maintaining the crystalline regions of cellulose and increasing the crystallinity index of the samples [32].

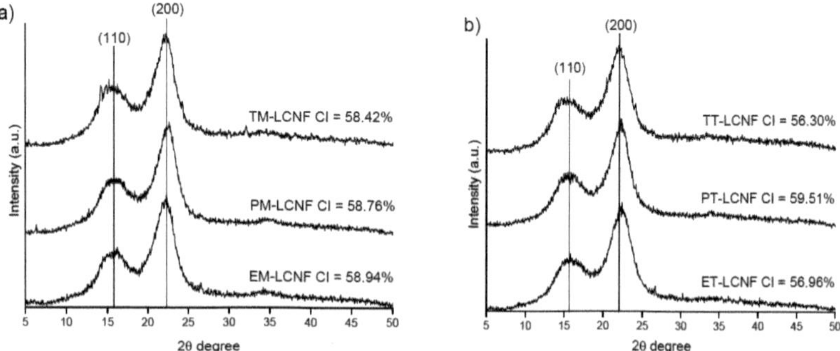

Figure 4. XRD patterns of LCNF obtained by mechanical (**a**) and TEMPO-oxidation (**b**) pretreatments.

The thermal stability of the cellulose nanofibers is also an important parameter to study their suitability in their final application. Figure. 5 shows the thermal behavior of the different cellulose nanofibers. The analysis of the different curves, identifies that the thermal decomposition is carried out in three zones, indicating the presence of distinct components decomposing at different temperatures. The first zone (up to 200 °C), shows a small decrease in the weight of samples related to the evaporation and removal of absorbed and bounded water in fiber [33–35]. The second zone (200–400 °C) corresponds to the active pyrolysis of the lignocellulosic components and is where the main degradation of the samples, including the maximum degradation temperature (T_{max}) shown by the DTG curve. In the third zone (temperature above 400 °C) the passive pyrolysis of the lignocellulosic components takes places, where the low degradation ratio stands out and is identified with the lignin degradation and any carbonaceous matter decomposition [35,36].

The main peak observed in the DTG analysis (Figure 5b,d) shows the temperature where the thermal degradation is maximum, known as T_{max}. The analysis shows that the cellulose nanofibers obtained by mechanical pretreatment show very similar values, being these 356.2 °C, 355.5 °C, and 357.5 °C for eggplant, pepper, and tomato plants. A similar behavior is observed for the lignocellulose nanofibers obtained by TEMPO-mediated oxidation, where the show a very similar value, being 308.8 °C, 317.8 °C, and 306.5 °C for eggplant, pepper and tomato, respectively. The lower stability shown by TEMPO-oxidized nanofibers is due to two main factors: i) The higher specific surface that results in a larger surface area exposed to heat; and ii) the introduction of carboxyl groups on the surface that produces a high number of free ends [24].

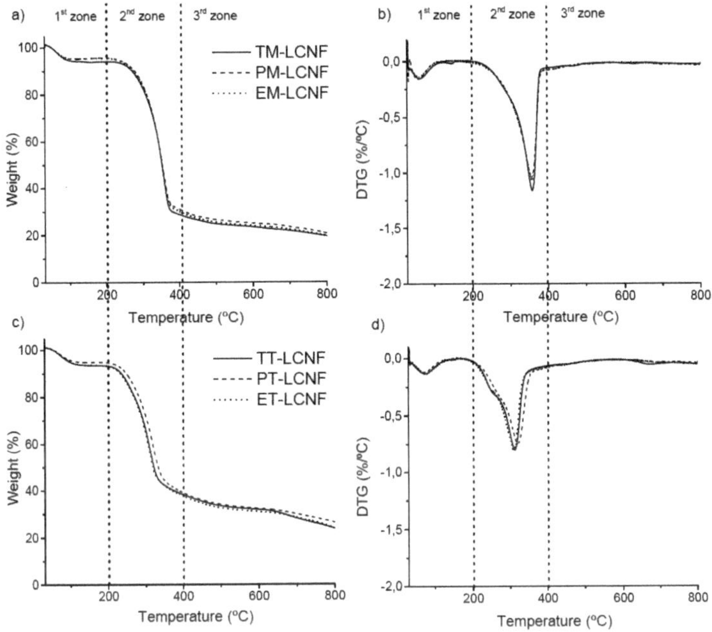

Figure 5. TGA and DTG curves of mechanical (**a,b**) and TEMPO-oxidation (**c,d**) LCNF.

2.3. Lignocellulose Reinforcement on Recycled Paperboard

The recycling process subjects the fibers to dispersion-drying cycles that reduce the binding capacity of the fibers, hornification process and, therefore, the mechanical properties shown by the final products. To correct this decrease, the industry uses different processes in order to increase the union between fibers, such as mechanical beating and the addition of chemical or virgin fiber [9]. The reinforcement effect on the recycled paperboard suspension of the lignocellulose nanofibers obtained in this work was compared with the effect produced by mechanical beating in order to analyze the suitability of the addition of LCNF as an alternative to mechanical beating. Figure 6 shows the evolution of the mechanical properties (breaking length, Young's modulus, tear index, and burst index) of recycled paperboard after the different treatments. In a generalized way, it is observed how both treatments, mechanical beating and LCNF addition, produce an increase in the mechanical properties compared to the values obtained from the original recycled fiber (baseline). The mechanical properties shown by the original recycled fiber were 2726 m for breaking length, 0.73 GPa for the Young's modulus, 26.71 Nm/g for the tear index, and 1.42 KN/g for the burst index. A linear increase is observed as the intensity of the mechanical refining of the amount of LCNF added increases, obtaining the highest values at the most severe conditions (3000 rev and 4.5% LCNF). The addition of LCNF produces a similar increase to the mechanical beating in breaking length and tear index parameters, however, it produces a more pronounced reinforcing effect on the Young's modulus and burst index. The differences in the reinforcing effect were analyzed depending on the raw material used to produce lignocellulose nanofibers. In general terms, it is observed that the addition of LCNF produces a similar effect regardless of the raw material used. Regarding the pretreatment, it is observed that there are no major differences in the reinforcing effect between LCNF obtained by mechanical or TEMPO-mediated oxidation pretreatment. Although all LCNF produce a similar effect, the reinforcement produced by TEMPO-oxidized nanofibers from eggplant residue (ET) stands out. The addition of 4.5% ET produces an increase of 30.63%, 53.42% 19.88%, and 62.68%, compared to 16.18%, 11.00%, 16.21%, and 36.62%

produced by mechanical beating for the breaking length, Young's modulus, tear index, and burst index, respectively.

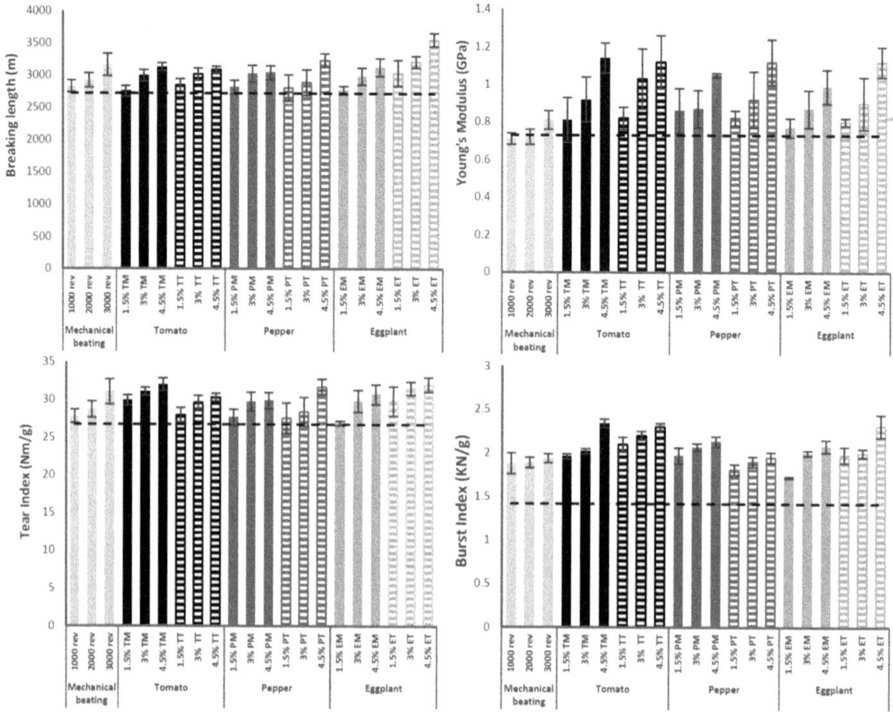

Figure 6. Evolution of the mechanical properties of recycled paperboard after different treatments.

The increase in Elongation at break was also analyzed (Figure 7). For this parameter, as with the other mechanical properties, a linear increase is observed as the treatment becomes more severe, and the reinforcement effect is more effective with the addition of LCNF than with mechanical beating.

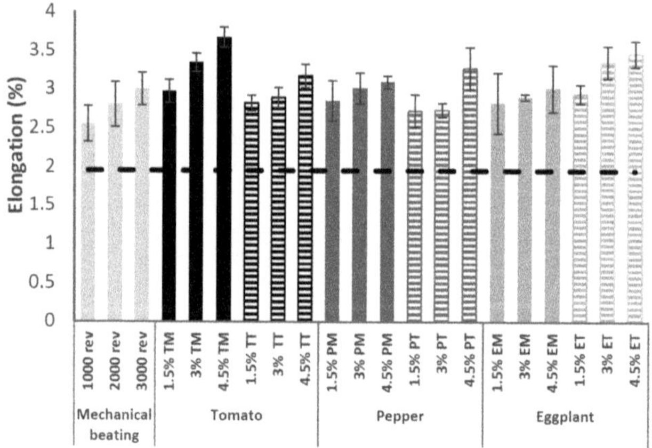

Figure 7. Evolution of the elongation at break of recycled paperboard after different treatments.

The increase in the mechanical properties due to the different treatments is given by: (i) the increase in the specific surface area of fibers which facilitate the bonding with adjacent fibers produced by the mechanical beating, or (ii) the generation of a network embedded between the paperboard substrate fibers and the LCNF added, increasing the bonding capacity and facilitating the interfiber bonding [37].

The industrial paperboard presents a breaking length value of 5656 m. Considering that in the industrial papermaking process formation is produced in an isotropic way, it is necessary to perform a conversion of the isotropic-anisotropic values to compare them with those obtained in an anisotropic laboratory papersheet former. For this, the anisotropic ratio of 1.65 is considered, obtaining that the industrial cardboard shows a breaking length value of 3443 m [38]. According to the data obtained, under the conditions studied in this work, only the addition of 4.5% ET would reach this value (3561 m), fully correcting the loss of the mechanical properties during the recycling process. The use of LCNF, in addition to producing a greater reinforcing effect, would make it possible to increase the recycling cycles of the same fibers from three to 10 or more [37]. The low number of recycling cycles limit when using mechanical beating is due to the structural damages caused by the effect of shearing during the beating. Previous studies have confirmed that the use of this technology can be energy-efficient depending on the nanofibrillation treatment [38]. This phenomenon also explains the increase in density compared to untreated recycled paperboard, as well as the decrease in the porosity.

The influence of the different treatments on the evolution of the physical properties was also analyzed, as shown in Table 2. A decrease in thickness is observed as the severity of the different treatments increases. This is due to the greater bonding strength between the adjacent fibers, resulting in greater compaction of the fibrillar network [39]. These observations are more intense in the addition of LCNF compared to those shown by the mechanical beating treatment. This is due to the fact that in addition to the higher bonding strength, the nanometric size of the LCNF allowing them to fill the gaps between the fiber matrix, increasing the density and decreasing the porosity [39].

Table 2. Evolution of the physical properties of recycled paperboard after different treatments.

Treatment	Sample	Thickness (µm)	Density (g/cm^3)	Porosity (%)
	Recycled paperboard	150.3 ± 2.9	0.36 ± 0.01	75.55 ± 0.79
Mechanical beating	1000 rev	147.8 ± 6.5	0.37 ± 0.01	75.10 ± 1.01
	2000 rev	147.2 ± 8.0	0.38 ± 0.02	74.93 ± 1.49
	3000 rev	146.8 ± 6.6	0.38 ± 0.01	74.75 ± 0.91
Tomato LCNF	1.5% TM	139.9 ± 4.2	0.39 ± 0.01	73.72 ± 0.93
	3% TM	136.6 ± 3.8	0.40 ± 0.02	73.06 ± 1.61
	4.5% TM	134.9 ± 2.7	0.41 ± 0.02	72.72 ± 1.38
	1.5% TT	142.4 ± 5.4	0.39 ± 0.01	74.53 ± 0.57
	3% TT	142.6 ± 1.9	0.39 ± 0.03	74.23 ± 0.74
	4.5% TT	138.4 ± 4.3	0.40 ± 0.02	73.38 ± 1.63
Pepper LCNF	1.5% PM	136.0 ± 4.9	0.40 ± 0.01	72.94 ± 1.34
	3% PM	134.2 ± 7.6	0.41 ± 0.02	72.82 ± 0.69
	4.5% PM	133.1 ± 5.6	0.41 ± 0.02	72.78 ± 0.72
	1.5% PT	147.5 ± 3.0	0.38 ± 0.02	75.08 ± 1.42
	3% PT	147.8 ± 2.1	0.38 ± 0.03	74.86 ± 1.06
	4.5% PT	147.4 ± 5.3	0.39 ± 0.01	74.04 ± 0.36
Eggplant LCNF	1.5% EM	146.0 ± 3.9	0.38 ± 0.01	74.91 ± 0.78
	3% EM	145.3 ± 4.5	0.39 ± 0.02	74.00 ± 1.83
	4.5% EM	138.7 ± 9.2	0.40 ± 0.02	73.43 ± 1.54
	1.5% ET	143.7 ± 8.1	0.38 ± 0.01	74.94 ± 0.83
	3% ET	146.7 ± 4.4	0.38 ± 0.02	74.51 ± 1.23
	4.5% ET	144.5 ± 6.1	0.39 ± 0.01	74.37 ± 0.66

Figure 8 Shows the evolution of the drainage properties of the paperboard slurries after the addition of the different LCNF. As expected, the addition of lignocellulose nanofibers, due to their specific surface area and their hydrophilic nature, results in a high water-holding capacity, thus increasing the viscosity of the suspension and decreasing the drainage capacity of the slurries. The mechanical beating treatment, due to the large generation of fines during the process, also produces a decrease in the drainage capacity showing a drainage degree of 47°SR, 49°SR, and 54°SR for 1000 rev, 2000 rev, and 3000 rev, respectively. These values are lower compared to those obtained by the addition of LCNF; however, this property is not a key parameter when evaluating the suitability of this technology for use in the recycling paperboard industry since this property can be corrected by the combination of nanofibers and electrolytes [40].

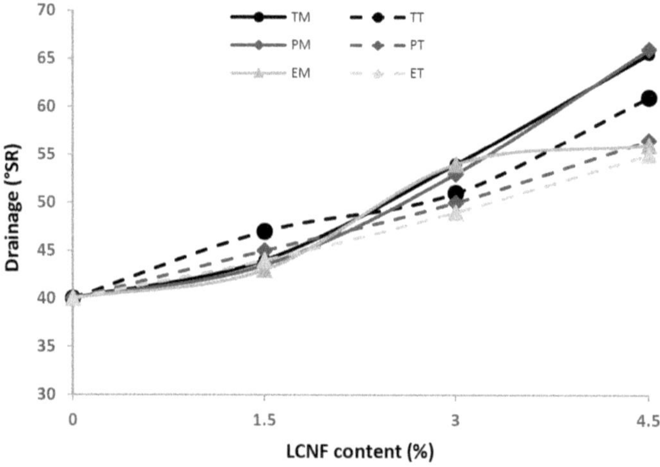

Figure 8. Evolution of drainability properties of cardboard suspension with different LCNF amount.

3. Materials and Methods

3.1. Materials

The greenhouse residues of tomato (*Solanum lycopersicum*), pepper (*Capsicum annuum*), and eggplant (*Solanum melongena*) were used in this work, which corresponded to the entire plants being completely uprooted from the ground (this allows for planting seeds for a new crop), after harvesting the fruits. These greenhouse residues were provided by Aguadulce Cooperative from Aguadulce, Almería (Spain). First, vegetable materials were cleaned and dried at room temperature. Then, they were cut into pieces of 0.1–1 cm and stored in plastic bags for preservation until use. As the paperboard substrate, a suspension of cellulose fiber from recycled paper and cardboard was used. This was not subjected to any mechanical refining, or the addition of chemicals or virgin fiber cellulose. This suspension was kindly supplied by the firm Smurfit Kappa Container 100 S.L. (Mengíbar, Jaén, Spain).

3.2. Soda Pulping

The soda pulping process is carried out to make the cellulose fiber more accessible for the pretreatments performed, also to improve the efficiency of the nanofibrillation process. The raw materials were pulped, according to proper conditions of Specel® process, in a 15 L batch reactor (Metrotec S.A., Lezo, Spain) with 7% NaOH over dried material (o.d.m) at 100 °C for 150 min and a liquid/solid ratio 10/1. The cellulosic pulp was dispersed in a pulp disintegrator (Metrotec S.A., Lezo, Spain) at 1200 rpm for 30 min, and this disintegrated pulp was passed through a Sprout-Bauer beater

(Combustion Engineering, Vienna, Austria). The pulp suspension was separated by sieving through a 0.14 mm mesh to retain uncooked material [17,41]

3.3. Pulp Characterization

The chemical characterization of raw materials and cellulosic pulps was done in terms of their content in ethanol extractables, ash, holocellulose, lignin, and α-cellulose, according to TAPPI standards T-204, T-211, T-222, T-203os61, and T-9m54, respectively. In addition, the pulping yield was calculated according to Equation (1):

$$Yield\ (\%) = \frac{W_1}{W_0} \cdot 100 \tag{1}$$

where W_1 corresponds to the dry weight of the samples after removing the uncooked material and W_0 corresponds to the initial dry weight of raw materials.

3.4. Lignocellulose Nanofibers (LCNF) Production

Lignocellulose nanofibers were produced from the obtained cellulosic pulp by using two different pretreatments: mechanical and TEMPO-mediated oxidation, followed by high-pressure homogenization treatment.

3.4.1. Mechanical Pretreatment

The cellulosic pulp was refined in PFI beater (Metrotec S.A., Lezo, Spain), according to ISO 5264-2:2002, until obtaining the drainage degree (°SR) closest to 90°. For all samples, 30,000 revolutions were required to obtain the above °SR value.

3.4.2. TEMPO-Mediated Oxidation Pretreatment

The cellulosic fibers were subjected to TEMPO-mediated oxidation pretreatment following the methodology described by Besbes et al. [42]. An amount of NaClO suspension (equivalent at 5 mmols/g cellulose) was added with continuous stirring at room temperature. Then, the pH value was maintained at 10.2 with the addition of 0.5 M NaOH until no pH decrease was observed. Finally, the fibers were filtered and washed several times with distilled water.

3.4.3. High-Pressure Homogenization

Both pretreated fibers suspensions (1% concentration) were passed through a high-pressure homogenizer Panda GEA 2K (GEA Niro, Parma, Italy) as nanofibrillation treatment. The high-pressure homogenization was made through 10 cycles (four cycles at 300 bar, three cycles at 600 bar, and three cycles at 900 bar) [24]. This increasing pressure avoids problems of clogging in the equipment because of the initial fiber size. Table 3 shows the codification used for the different samples.

Table 3. Codification of the different lignocellulose nanofibers.

Raw Material	Pretreatment	Treatment	Codification
Tomato	Mechanical	High-pressure homogenization	TM-LCNF
	TEMPO-mediated oxidation		TT-LCNF
Pepper	Mechanical		PM-LCNF
	TEMPO-mediated oxidation		PT-LCNF
Eggplant	Mechanical		EM-LCNF
	TEMPO-mediated oxidation		ET-LCNF

3.5. LCNF Characterization

To determine the nanofibrillation yield, a 0.1% LCNF suspension was centrifuged at 10,000 rpm for 12 min. The precipitated fraction or non-nanofibrillar material was separated from the nanofibrillar material, and it was dried at 100 °C for 24 h [42].

The cationic demand (CD) was determined through the adaptation of the methodology followed by Carrasco et al. [43]. First, 0.20 dried grams of LCNF were diluted in distilled water until 200 g. When the suspension was homogenized, 10 mL of this solution was mixed with 25 mL of a cationic polymer (poly-DADMAC). The suspension was centrifuged at 4000 rpm for 90 min. Then, 10 mL of supernatant was introduced to a Mütek PCD 05 particle charge detector and back-titrated with the anionic polymer (PesNa) until the detector indicated zero conductivity. The CD was determinate according to Equation (2):

$$CD = \frac{(CpolyD. \ VpolyD). \ (CPesNa. \ VPesNa)}{m} \tag{2}$$

where C_{polyD} and V_{polyD} correspond to Poly-DADMAC concentration and volume, respectively; C_{PesNa} and V_{PesNa} correspond to PesNa concentration and volume, respectively; and m is the weight of the dried product (g).

According to the methodology of Besbes et al. [42], the carboxyl content (CC) was determined by conductometric titration. First, 50–100 mg of dry fiber was suspended in 15 mL of HCl (0.01 M) to exchange the Na^+ cations attached to COOH groups by H^+ ions. Then, the suspensions were titrated with NaOH (0.01 M), adding 0.1 mL of NaOH to the sample suspensions and recording the conductivity, observing a reduction, stabilization and increase in the conductivity. The CC was determined according to the Equation (3):

$$CC = \frac{(V_2 - V_1) \cdot [NaOH]}{m} \tag{3}$$

where V_2 and V_1 are the equivalent volumes of added NaOH solution; [NaOH] corresponds to NaOH solution concentration and m stands for the weight of the dried product (g).

Considering the assumptions and methodology described by Carrasco et al. [44], the obtained values of the cationic demand and carboxyl content were used for the theoretical estimation of the specific surface (σ_{LCNF}) and diameter of the LCNF.

The cationic polymer (Poly-DADMAC) used for the cationic demand determination interacts with cellulose fibers through surface adsorption mechanisms. The Poly-DADMAC has a specific area of 4.87×10^{17} $nm^2/\mu eq \cdot g$; based on this value, the specific surfaces of the LCNF obtained were calculated according to Equation (4) [39]:

$$\sigma_{LCNF} = (CD - CC) \cdot \sigma_{Poly-DADMAC} \tag{4}$$

where σ_{LCNF} is the specific surface of LCNF (nm^2/g); CD is the cationic demand ($\mu eq/g$); CC is the carboxyl rate ($\mu mols/g$); and $\sigma_{Poly\text{-}DADMAC}$ corresponds to the specific surface of this cationic polymer ($nm^2/\mu eq$).

Assuming the cylindrical geometry of the LCNF, the value of the specific surface was used to determine its diameter.

All measurements were made in triplicate and mean, and standard deviations were calculated.

3.6. Polymerization Degree and Length

The intrinsic viscosity (η) was obtained according to UNE-57-039-92 and used for the calculation of the degree of polymerization. The degree of polymerization (DP) was calculated using Equations (5) and Equation (6) [45]:

$$DP \ (< 950) : DP = \frac{\eta_s}{0.42} \tag{5}$$

$$DP(> 950) : DP^{0.76} = \frac{\eta_s}{2.28} \tag{6}$$

The length of the nanofibers was calculated from the polymerization degree (DP) using the Equation (7) proposed by Shinoda et al. [46]:

$$Length\ (nm) = 4.286 \cdot DP - 757 \tag{7}$$

3.7. Spectroscopy Analysis

Fourier transform infrared spectroscopy (FTIR) analysis was applied to determine possible changes in the structure, chemical composition, and functional groups during LCNF isolation processes. A Spectrum Two FT-IR Spectrometer (Perkin-Elmer, Massachusetts, USA) was used in the range of 450–4000 cm^{-1} with a resolution of $4\ cm^{-1}$, collecting a total of 40 scans per sample.

3.8. X-ray Diffraction (XRD) Analysis

XRD was used to study the crystal structures of the cellulosic pulps and LCNF. This analysis was performed using a Bruker D8 Discover (Bruker Corporation, Massachusetts, USA) with a monochromatic source $CuK\alpha1$ over an angular range of 5–$50°$ at a scan speed of $1.56°/min$. The Segal method was used to calculate the crystallinity index (CI)(Equation (8)) [47]:

$$CI(\%) = \left(\frac{I_{200} - I_{am}}{I_{200}}\right) \cdot 100 \tag{8}$$

where I_{200} is the diffraction peak intensity at $2\theta = 22.5°$ of the crystalline cellulose regions, and I_{am}, is the intensity minimum between two diffraction peaks ($2\theta = 16.5°$ and $22.5°$) of the amorphous cellulose region.

3.9. Thermogravimetric Analysis

Thermogravimetric analysis (TGA) was performed to evaluate the thermal stability of the LCNF. The derivate thermogravimetric (DTG) was used to analyze the maximum degradation rate (T_{max}). The analysis was carried out using a Mettler Toledo TGA/DSC 1 (Mettler Toledo, Ohio, USA). The samples were taken from room temperature to $800\ °C$ with a heating rate of $10\ °C/min$ and a $50\ mL/min$ nitrogen gas flow rate.

3.10. Reinforcement of Industrial Pulp

LCNF was used as a reinforcing agent on a paperboard industrial pulp (Smurfit Kappa) and compared its effect with that produced by mechanical beating. The industrial cellulose pulp was submitted at different mechanical refining intensities (1000 rev, 2000 rev, and 3000 rev). Lignocellulose nanofibers (mechanical and TEMPO) of tomato, pepper, and eggplant were added at 1.5%, 3%, and 4.5% (o.d.m.) to the industrial cellulose suspension. The process started disintegrating 30 g dry weight pulp for 30 min. Then, LCNF was added in the established proportion and disintegrated for 1 h. After that, 0.5% (*w/w*) cationic starch (Vector SC 20157) and 0.8% (*w/w*) colloidal silica (LUDOX® HS-40) were added to improve the retention of LCNF by improving the bonding of the fibers, based on the dry weight of the pulp and LCNF and kept in constant mechanical agitation. The papersheet formation was carried out in a sheet former ENJO-F-39.71 (Metrotec S.A., Lezo, Spain) according to TAPPI T205ps-95. Before the mechanical testing, the papersheets were conditioned at $25\ °C$ and 50% relative humidity for 48 h. The mechanical characterization was performed using an Instron universal testing machine (Lloyd Instruments, Bognor Regis, United Kingdom) provided with 1 kN load cell, in terms of breaking length and tensile index, elongation, Burst index, and tear index, according to TAPPI standard (T-494-om96, T-494, T-403-om97 and T-414-om98, respectively). The physical properties of the papersheets were analyzed for their thickness, density, and porosity. The thickness was determined according to the standard ISO 534. The density was calculated from the weight of the sheets and their dimensions. The porosity of the sheets was calculated using Equation (9):

$$Porosity\ (\%) = 100 \cdot \left(1 - \frac{\rho_{sample}}{\rho_{cellulose}}\right) \tag{9}$$

where ρ_{sample} is the density of the sheet, and $\rho_{cellulose}$ is the density of cellulose, assumed as 1.5 g/cm^3.

4. Conclusions

Horticultural plant residues, as lignocellulosic source for the isolation of lignocellulose nanofibres, were analyzed. The cellulosic pulps obtained were subjected to two different pretreatments, mechanical and TEMPO-mediated oxidation, and a subsequent high-pressure homogenization process for lignocellulose nanofiber isolation. The different LCNF were added as a reinforcing agent on recycled paperboard and compared with the improvement produced by industrial mechanical beating. The addition of 4.5% TEMPO-oxidized LCNF from eggplant residues produces the greater increase in comparison with the other LCNF. It produces an increase of 30.63%, 53.42%, 19.88%, and 62.68%, compared to 16.18%, 11.00%, 16.21%, and 36.62% produced by mechanical beating for breaking length, Young's modulus, tear index, and burst index, respectively. The use of LCNF produces a decrease in the papersheet thickness and porosity, and an increase in the density. This is due to the greater bonding strength between the adjacent fibers and the gap filling resulting in greater compaction of the fiber matrix. The use of LCNF in the paperboard recycling process as an alternative to mechanical beating produces a greater reinforcing effect and would make it possible to increase the recycling cycles of same fibers from three to 10 or more.

Author Contributions: Conceptualization: A.R.; methodology: I.B.-V. and R.S.; validation: A.R.; investigation: I.B.-V., R.S., and Q.T.; writing—original draft preparation: I.B.-V. and E.E.; writing—review and editing: A.R and E.E.; supervision: A.R. and F.P.-R. All authors have read and agreed to the published version of the manuscript.

Funding: This research was funded by Spain's DGICyT, MICINN within the framework of the Project CTQ2016-78729-R.

Acknowledgments: The authors are grateful to the staff of the Central Service for Research Support (SCAI) at the University of Córdoba

Conflicts of Interest: The authors declare no conflict of interest.

References

1. EEllen MacArthur Fundation, Towards a Circular Economy: Business rationale for an accelerated transition. 2015. Available online: https://www.ellenmacarthurfoundation.org/assets/downloads/TCE_Ellen-MacArthur-Foundation_9-Dec-2015.pdf. (accessed on 13 April 2020).
2. Fundación COTEC, Situación y Evolución De La Economía Circular En España. 2017. Available online: http://cotec.es/media/informe-CotecISBN-1.pdf (accessed on 22 April 2020).
3. European Commission, Innovating for Sustainable Growth: A Bioeconomy for Europe, (2012). Available online: https://ec.europa.eu/research/bioeconomy/index.cfm (accessed on 22 April 2020).
4. Sheldon, R.A. Chemicals from renewable biomass: A renaissance in carbohydrate chemistry. *Green Sustain. Chem.* **2018**, *14*, 89–95. [CrossRef]
5. Teja, K.; Sowlati, T. Biomass logistics: A review of important features, optimization modeling and the new trends. *Renew. Sustain. Energy Rev.* **2018**, *94*, 587–599.
6. Pang, S. Advances in thermochemical conversion of woody biomass to energy, fuels and chemicals. *Biotechnol. Adv.* **2019**, *37*, 589–597. [CrossRef] [PubMed]
7. European Commission. Sustainable Agriculture, Forestry and Fisheries in the Bioeconomy. A Challenge for Europe. 4th SCAR Foresight Exercise. 2015. Available online: http://ec.europa.eu/research/scar/pdf/feg4-draft-15_may_2015.pdf (accessed on 22 April 2020).
8. George, M.; Chae, M.; Bressler, D.C. Composite materials with bast fibres: Structural, technical, and environmental properties. *Prog. Mater. Sci.* **2016**, *83*, 1–23. [CrossRef]

9. Delgado-Aguilar, M.; Tarrés, Q.; Pèlach, À.M.; Mutjé, P.; Fullana-I.-Palmer, P. Are Cellulose Nanofibres a Solution for a More Circular Economy of Paper Products? *Environ. Sci. Technol.* **2015**, *49*, 12206–12213. [CrossRef] [PubMed]

10. Delgado-Aguilar, M. *Nanotecnología en el sector papelero: Mejoras en calidad y permanencia de las fibras de alto rendimiento y secundarias en una economía circular mediante el uso de nanofibras y el refino enzimático*; University of Girona: Girona, Spain, 2015.

11. Delgado-Aguilar, M.; González, I.; Pélach, M.A.; De La Fuente, E.; Negro, C. Improvement of deinked old newspaper/old magazine pulp suspensions by means of nanofibrillated cellulose addition. *Cellulose* **2015**, *22*, 789–802. [CrossRef]

12. Anuario de Estadística. *Ministerio de Agricultura, Pesca y Alimentación*. 2018. Available online: https://www.mapa.gob.es/es/estadistica/temas/publicaciones/anuario-de-estadistica/default.aspx (accessed on 23 April 2020).

13. Toledo, M.; Márquez, P.; Siles, J.A.; Chica, A.F.; Martín, M.A. Co-composting of sewage sludge and eggplant waste at full scale: Feasibility study to valorize eggplant waste and minimize the odoriferous impact of sewage sludge. *J. Environ. Manag.* **2019**, *247*, 205–213. [CrossRef]

14. Hamraoui, K.; Gil, A.; El Bari, H.; Siles, J.A.; Chica, A.F.; Martín, M.A. Evaluation of hydrothermal pretreatment for biological treatment of lignocellulosic feedstock (pepper plant and eggplant). *Waste Manag.* **2020**, *102*, 76–84. [CrossRef]

15. Guntekin, E.; Karakus, B. Feasibility of using eggplant (*Solanum melongena*) stalks in the production of experimental particleboard. *Ind. Crop. Prod.* **2008**, *27*, 354–358. [CrossRef]

16. Covino, C.; Sorrentino, A.; di Pierro, P.; Roscigno, G.; PiaVece, A.; Masi, P. Lignocellulosic fibres from enzyme-treated tomato plantes: Characterisation and application in paperboard manufacturing. *Int. J. Biol. Macromol.* **2020**, *161*, 787–796. [CrossRef]

17. Espinosa, E.; Sánchez, R.; Otero, R.; Domínguez-Robles, J.; Rodríguez, A. A comparative study of the suitability of different cereal straws for lignocellulose nanofibres isolation. *Int. J. Biol. Macromol.* **2017**, *103*, 990–999. [CrossRef] [PubMed]

18. Chaker, A.; Alila, S.; Mutje, P.; Rei, M.; Sami, V. Key role of the hemicellulose content and the cell morphology on the nanofibrillation effectiveness of cellulose pulps. *Cellulose* **2013**, *20*, 2863–2875. [CrossRef]

19. Mandal, A.; Chakrabarty, D. Isolation of nanocellulose from waste sugarcane bagasse (SCB) and its characterization. *Carbohydr. Polym.* **2011**, *86*, 1291–1299. [CrossRef]

20. Alotaibi, M.D.; Alshammari, B.A.; Saba, N.; Alothman, O.Y.; Sanjay, M.R.; Almutairi, Z.; Jawaid, M. Characterization of natural fibre obtained from different parts of date palm tree (Phoenix dactylifera L.). *Int. J. Biol Macromol.* **2019**, *135*, 69–76. [CrossRef]

21. Rajan, K.; Djioleu, A.; Kandhola, G.; Labbé, N.; Sakon, J.; Carrier, D.J.; Kim, J.W. Investigating the effects of hemicellulose pre-extraction on the production and characterization of loblolly pine nanocellulose. *Cellulose* **2020**, *8*, 3693–3706. [CrossRef]

22. el Achaby, M.; el Miri, N.; Aboulkas, A.; Zahouily, M.; Bilal, E.; Barakat, A.; Solhy, A. Processing and properties of eco-friendly bio-nanocomposite films filled with cellulose nanocrystals from sugarcane bagasse. *Int. J. Biol. Macromol.* **2017**, *96*, 340–352. [CrossRef]

23. Rojo, E.; Peresin, M.S.; Sampson, W.W.; Hoeger, I.C.; Vartiainen, J.; Laine, J.; Rojas, O.J. Comprehensive elucidation of the effect of residual lignin on the physical, barrier, mechanical and surface properties of nanocellulose films. *Green Chem.* **2015**, *17*, 1853–1866. [CrossRef]

24. Espinosa, E.; Sánchez, R.; González, Z.; Domínguez-Robles, J.; Ferrari, B.; Rodríguez, A. Rapidly growing vegetables as new sources for lignocellulose nanofibre isolation: Physicochemical, thermal and rheological characterisation. *Carbohydr. Polym.* **2017**, *175*, 27–37. [CrossRef]

25. Tarrés, Q.; Pellicer, N.; Balea, A.; Merayo, N.; Negro, C.; Blanco, A.; Delgado-Aguilar, M.; Mutjé, P. Lignocellulosic micro/nanofibres from wood sawdust applied to recycled fibres for the production of paper bags. *Int. J. Biol. Macromol.* **2017**, *105*, 664–670. [CrossRef]

26. Sánchez, R.; Espinosa, E.; Domínguez-robles, J.; Mauricio, J.; Rodríguez, A. Isolation and characterization of lignocellulose nanofibres from different wheat straw pulps. *Int. J. Biol. Macromol.* **2016**, *92*, 1025–1033. [CrossRef]

27. Tarrés, Q.; Espinosa, E.; Domínguez-Robles, J.; Rodríguez, A.; Mutjé, P.; Delgado-Aguilar, M. The suitability of banana leaf residue as raw material for the production of high lignin content micro/nano fibres: From residue to value-added products. *Ind. Crop. Prod.* **2017**, *99*, 27–33. [CrossRef]

28. Tarrés, Q.; Vanesa, N.; Evangelina, M.; Cristina, M.; Delgado-aguilar, M.; Mutjé, P. Lignocellulosic nanofibres from triticale straw: The influence of hemicelluloses and lignin in their production and properties. *Carbohydr. Polym.* **2017**, *163*, 20–27. [CrossRef] [PubMed]

29. Tarrés, Q.; Delgado-Aguilar, M.; Pèlach, M.A.; González, I.; Boufi, S.; Mutjé, P. Remarkable increase of paper strength by combining enzymatic cellulose nanofibres in bulk and TEMPO-oxidized nanofibres as coating. *Cellulose* **2016**, *23*, 3939–3950. [CrossRef]

30. Serra, A.; González, I.; Oliver-Ortega, H.; Tarrès, Q.; Delgado-Aguilar, M.; Mutjé, P. Reducing the amount of catalyst in TEMPO-oxidized cellulose nanofibres: Effect on properties and cost. *Polymers (Basel)* **2017**, *9*, 557. [CrossRef] [PubMed]

31. Tarrés, Q.; Boufi, S.; Mutjé, P.; Delgado-Aguilar, M. Enzymatically hydrolyzed and TEMPO-oxidized cellulose nanofibres for the production of nanopapers: Morphological, optical, thermal and mechanical properties. *Cellulose* **2017**, *24*, 3943–3954. [CrossRef]

32. Sang, X.; Qin, C.; Tong, Z.; Kong, S.; Jia, Z. Mechanism and kinetics studies of carboxyl group formation on the surface of cellulose fibre in a TEMPO-mediated system. *Cellulose* **2017**, *24*, 2415–2425. [CrossRef]

33. Yousefi, H.; Faezipour, M.; Hedjazi, S.; Mazhari, M. Comparative study of paper and nanopaper properties prepared from bacterial cellulose nanofibres and fibres/ground cellulose nanofibres of canola straw. *Ind. Crop. Prod.* **2013**, *43*, 732–737. [CrossRef]

34. Tao, P.; Zhang, Y.; Wu, Z.; Liao, X.; Nie, S. Enzymatic pretreatment for cellulose nanofibrils isolation from bagasse pulp: Transition of cellulose crystal structure. *Carbohydr. Polym.* **2019**, *214*, 1–7. [CrossRef]

35. Alemdar, A.; Sain, M. Isolation and characterization of nanofibres from agricultural residues—Wheat straw and soy hulls. *Bioresour. Technol.* **2008**, *99*, 1664–1671. [CrossRef]

36. Zhang, H.; Nie, S.; Qin, C.; Wang, S. Removal of hexenuronic acid to reduce AOX formation in hot chlorine dioxide bleaching of bagasse pulp. *Ind. Crops. Prod.* **2019**, *128*, 338–345. [CrossRef]

37. Boufi, S.; González, I.; Delgado-Aguilar, M.; Tarrès, Q.; Pèlach, M.À.; Mutjé, P. Nanofibrillated cellulose as an additive in papermaking process: A review. *Carbohydr. Polym.* **2016**, *154*, 151–166. [CrossRef] [PubMed]

38. Espinosa, E.; Rol, F.; Bras, J.; Rodríguez, A. Production of lignocellulose nanofibres from wheat straw by different fibrillation methods. Comparison of its viability in cardboard recycling process. *J. Clean. Prod.* **2019**, *239*, 118083. [CrossRef]

39. Espinosa, E.; Tarrés, Q.; Delgado-Aguilar, M.; González, I.; Mutjé, P.; Rodríguez, A. Suitability of wheat straw semichemical pulp for the fabrication of lignocellulosic nanofibres and their application to papermaking slurries. *Cellulose* **2015**, *23*, 837–852. [CrossRef]

40. Lourenço, A.F.; Gamelas, J.A.F.; Sarmento, P.; Ferreira, P.J.T. Enzymatic nanocellulose in papermaking—The key role as filler flocculant and strengthening agent. *Carbohydr. Polym.* **2019**, *224*, 115200. [CrossRef]

41. Vargas, F.; González, Z.; Sánchez, R.; Jiménez, L.; Rodríguez, A. Cellulosic pulps of cereal straws as raw material. *BioResources* **2012**, *7*, 4161–4170.

42. Besbes, I.; Alila, S.; Boufi, S. Nanofibrillated cellulose from TEMPO-oxidized eucalyptus fibres: Effect of the carboxyl content. *Carbohydr. Polym.* **2011**, *84*, 975–983. [CrossRef]

43. Carrasco, F.; Mutjé, P.; Pelach, M.A. Control of retention in paper-making by colloid titration and zeta potential techniques. *Wood Sci. Technol.* **1998**, *32*, 145–155. [CrossRef]

44. Carrasco, F.; Mutjé, P.; Pelach, M.A. Refining of bleached cellulosic pulps: Characterization by application of the colloidal titration technique. *Wood Sci. Technol.* **1996**, *30*, 227–236. [CrossRef]

45. Marx-Figini, M. The acid-catalyzed degradation of cellulose linters in distinct ranges of bacterial nano-cellulose reinforced fibre-cement composites. *Constr. Build. Mater.* **1987**, *101*, 958–964.

46. Shinoda, R.; Saito, T.; Okita, Y.; Isogai, A. Relationship between length and degree of polymerization of TEMPO-oxidized cellulose nanofibrils. *Biomacromolecules* **2012**, *13*, 842–849. [CrossRef]

47. Segal, L.; Creely, J.J.; Martin, A.E.; Conrad, C.M. Empirical Method for Estimating the Degree of Crystallinity of Native Cellulose Using the X-Ray Diffractometer. *Text Res. J.* **1959**, *29*, 786–794. [CrossRef]

Sample Availability: Samples of the compounds are not available from the authors.

Article

Acquisition of Torrefied Biomass from Jerusalem Artichoke Grown in a Closed Circular System Using Biogas Plant Waste

Szymon Szufa [1,*], Piotr Piersa [1], Łukasz Adrian [1], Jan Sielski [2], Mieczyslaw Grzesik [3], Zdzisława Romanowska-Duda [4], Krzysztof Piotrowski [4] and Wiktoria Lewandowska [5]

[1] Faculty of Process and Environmental Engineering, Lodz University of Technology, Wolczanska 213, 90-924 Lodz, Poland; piotr.piersa@p.lodz.pl (P.P.); lukasz.adrian@p.lodz.pl (Ł.A.)
[2] Department of Molecular Engineering, Lodz University of Technology, Wolczanska 213, 90-924 Lodz, Poland; jan.sielski@p.lodz.pl
[3] Department of Variety Studies, Nursery and Gene Resources, Research Institute of Horticulture, Str. Konstytucji 3 Maja 1/3, 96-100 Skierniewice, Poland; mieczyslaw.grzesik@inhort.pl
[4] Department of Plant Ecophysiology, Faculty of Biology and Environmental Protection, University of Lodz, Str. Banacha 12/16, 92-237 Lodz, Poland; zdzislawa.romanowska@biol.uni.lodz.pl (Z.R.-D.); k_piotrow@o2.pl (K.P.)
[5] Faculty of Chemistry, University of Lodz, Tamka 12, 91-403 Lodz, Poland; wiktoria.lewandowska.uni.lodz@gmail.com
* Correspondence: szymon.szufa@p.lodz.pl; Tel.: +48-606-134-239

Academic Editors: Alejandro Rodríguez and Eduardo Espinosa
Received: 31 July 2020; Accepted: 23 August 2020; Published: 25 August 2020

Abstract: The aim of the research was to investigate the effect of biogas plant waste on the physiological activity, growth, and yield of Jerusalem artichoke and the energetic usefulness of the biomass obtained in this way after the torrefaction process. The use of waste from corn grain biodigestion to methane as a biofertilizer, used alone or supplemented with Apol-humus and Stymjod, caused increased the physiological activity, growth, and yield of Jerusalem artichoke plants and can limit the application of chemical fertilizers, whose production and use in agriculture is harmful for the environment. The experiment, using different equipment, exhibited the high potential of Jerusalem artichoke fertilized by the methods elaborated as a carbonized solid biofuel after the torrefaction process. The use of a special design of the batch reactor using nitrogen, Thermogravimetric analysis, Differential thermal analysis, and Fourier-transform infrared spectroscopy and combustion of Jerusalem artichoke using TG-MS showed a thermo-chemical conversion mass loss on a level of 30% with energy loss (torgas) on a level of 10%. Compared to research results on other energy crops and straw biomass, the isothermal temperature of 245 °C during torrefaction for the carbonized solid biofuel of Jerusalem artichoke biomass fertilized with biogas plant waste is relativlely low. An SEM-EDS analysis of ash from carbonized Jerusalem artichoke after torrefaction was performed after its combustion.

Keywords: torrefaction; Jerusalem artichoke; biofuel; energy crops; agiculture

1. Introduction

One of the greatest global problems is increasing energy consumption, which, in the face of the need to limit the use of fossil fuels, forces the development of crops that will produce the maximum yield of biomass, which could be converted into energy fuel using modern technologies [1]. For this reason, research is necessary to select plants having a high potential of biomass and energy yield on poor soils and to develop plant cultivation technologies that, in addition to high biomass yield and energy efficiency, will be conducive to the environment by reducing the use of chemistry in agricultural production and will strengthen energy security [2].

Compared to other renewable energy sources, biomass provides continuous electricity generation, and is the only widespread source of renewable heat. Biomass co-firing and biomass combustion will contribute to the reduction of CO_2 and SO_2 emissions, support sustainable development, and increase energy security and regeneration of rural areas, due to the increase of forestry and agricultural activity and the provision of heat and electrical energy production. To increases the biomass share up to a 30% or even 40% caloric value, the biomass particles must be milled down to sizes where high caloric values can be expected. There are many different biomass pre-treatment methods that can be used to convert it into more coal-like matter. There are a number of barriers to overcome in order to expand the exploitation of biomass for heat and eleclricity production. One of them concerns the limitations connected to biomass fuel characteristics [3–5].

When coal is compared with wood biomass, which are both still the dominant solid fuels in heat and electricity production in Poland, the inferior properties of biomass are often revealed. Wood biomass fuel has in most cases a high moisture content, resulting in storage complications, such as self-heating and biological degradation, and lower energy densities. It is also a bulkier fuel (with poorer transportation and handling characteristics), and it is more tenacious (the fibrous nature of biomass makes it difficult to reduce it to small homogeneous particles). The biomass properties mentioned above have negative impacts during energy thermal conversion, such as gasification and lower combustion and co-firing efficiencies [6].

Among the methodologies that can be applied to improve the properties of plant biomass and make it a more coal-like material, torrefaction (biomass carbonization) seems to provide many advantages. Carbonization, or torrefaction, is a thermal degradation of biomass structures, which occurs by heating them without air contact under atmospheric pressure. It removes low-weight organic volatile components and moisture as well as depolymerizes the long polysaccharide chains of biomass. This kind of process of wood carbonization is quite a complex research subject due to the fact that wood contains different fractions. Wood cells are built from microfibrils, bundles of cellulose molecules 'coated' with hemicellulose. Another component of wood biomass is lignin, which is deposited between microfibrils and in some types of biomass in the amorphous regions of the microfibril. All of those three fractions exhibit different thermal behavior [7]. The product of torrefaction is a hydrophobic solid fuel with greatly increased grindability and energy density (on a mass basis). More importantly, the energy requirement for processing the torrefied biomass decreases and it no longer requires additional separate handling facilities when we co-combust new fuel with coal in operating power plants. It is suggested that torrefied biomass can be compacted into high-grade pellets with substantially superior fuel properties compared with standard wood pellets from un-treated biomass. The carbonization process can be combined together with the drying and pelletization process, with both energy end economical benefits. The biomass torrefaction process has proved suitable for feedstock for flow gasification, which has not been considered feasible before for raw biomass. This is due to the fact that carbonized biomass forms more solid fuel spherical-shaped particles during milling or grinding than raw biomass. To produce high caloric value carbonized solid biofuel, which will have a reasonable price (lower torrefaction process costs can be achieved by using superheated steam), with better physical-chemical properties before thermo-chemical conversion, such as a hydrophobic nature, low moisture content, and better grindability, several process conditions have to be optimal. These include a 30% loss of the mass and 10% loss in energy in volatile matter plus a low as possible temperature and residential time in the reactor, which can ensure successful sale on the Polish and European market.

Important parameters in the choice of highly efficient plant species for energy and torrefaction purposes comprise their physiological properties, decisive for high biomass yield and the amount of energy obtained. Jerusalem artichoke (*Helianthus tuberosus* L.) meets these requirements and it is well adapted to the conditions of central Europe. The biomass of this species is an important raw material for the production of bioethanol, for burning to obtain heat energy, and is widely used in the feed, food, and medical industries [8–10].

A serious problem in the production of energy plants is the excessive use of artificial fertilizers and pesticides, which pollute the environment. There is a resulting need to decrease their volume without reducing the yield potential. Replacement of chemical fertilization by natural fertilizers, including waste from biogas plants, seems to be one of the ways to address this problem. Due to the fact that this waste contains the majority of nutrients necessary for plant growth, it seems that their use in agriculture can support soil fertilization. At the same time, this solves the problem of utilization and storage, which is dangerous for the environment. The problem is, however, that the fertilizing value of this waste and its impact on plant physiological activity depends on the type of biodegradable biomass, which requires separate research into the methods of its application and management. These difficulties become serious because biogas plants turn out to be a fast developing branch of energy production and they use of different raw materials. Alburquerque et al. [11], Jasiulewicz and Janiszewska [12], and Łagocka et al. [13] indicated that, given the environmental risks and benefits, the use of this waste in agriculture is most rational, provided that methods for its use are developed. The limitation of environmentally harmful synthetic fertilization by the use of biogas plant waste seems to be very important, as has been similarly demonstrated in the case of microalgae and water plants from the family *Lemnaceae*, which applied to medium enabled a reduction of the recommended artificial fertilizer doses [14–17]. The use of waste from biogas plants for fertilizing purposes is part of the strategy of the circulating production of energy plants in which waste becomes a raw material in the next crop cycle. Another unknown problem is the usefulness of the biomass energy produced from the waste of a biogas plant and the possibility of its torrefaction for energy purposes. The world literature available on these issues, especially concerning agricultural management of waste from the corn grain biodigestion to the methane process together with preparations stimulating growth, their influence on physiological processes that regulate plant energy properties, and the development of torrefaction technologies converting such produced biomass into energy fuel, is hard to find. In the majority of cases, it refers to the waste produced by specific biogas plants and to the raw materials used there [18]. Additionally, the possibilities of torrefaction of the biomass produced on this waste, as well as its energy properties so far are not known.

The purpose of this work was to describe the impact of waste from the corn grain biodigestion to methane process, used as biofertiliser either separately or together with Apol-humus and Stymjod, on the growth, yield, and physiological properties of Jerusalem artichoke biomass and the possibility of converting it into valuable energy fuel using the torrefaction processes.

2. Results

The waste from the corn grain biodigestion to methane process had a beneficial influence on the growth biomass yield and physiological activity of Jerusalem artichoke. All quantities of applied waste accelerated the Jerusalem artichoke growth and biomass yield, with 30–40 m^3 ha^{-1} being the most favorable for plant development. The positive impact of this natural fertilizer on growth was enhanced by the additional application of Apol-humus to soil (10 L ha^{-1}) and, furthermore, to a higher degree by supplementary double plant spraying with Stymjod (5 L ha^{-1}) (Figures 1 and 2).

Biogas plant waste, applied alone or supplemented with Apol-hymus and Stymjod, increased the activity of acid and alkaline phosphorylases, RNase, and dehydrogenases. The activities of these enzymes were closely related to the increasing doses of fertilizers (Table 1).

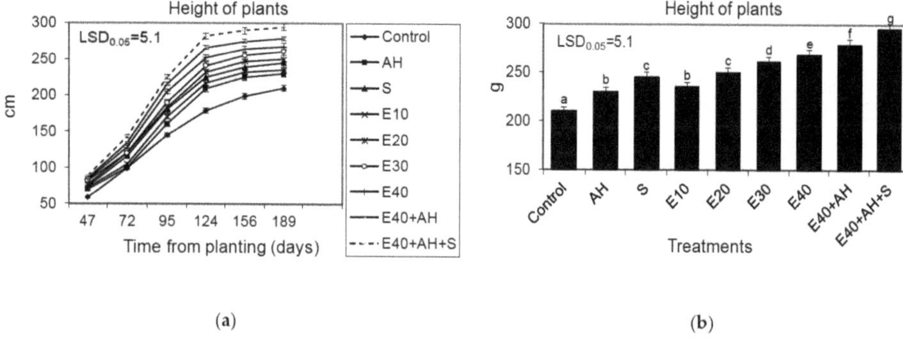

(a) (b)

Figure 1. Kinetics of the growth (**a**) and final height of Jerusalem artichoke plants (**b**) cultivated in a field and fertilized with liquid, non-centrifuged waste from corn grain digestion to methane (E10, 40 m^3 ha^{-1}), Apol-humus (AH; 10 L ha^{-1}), and Stymjod (S; 5 L ha^{-1}). The data marked with the same letters are not significantly different, according to the Newman–Keuls multiple range test at an alpha level of 0.05. The data presented are the average over the years and 10 plants in each repetition of a particular experimental variant.

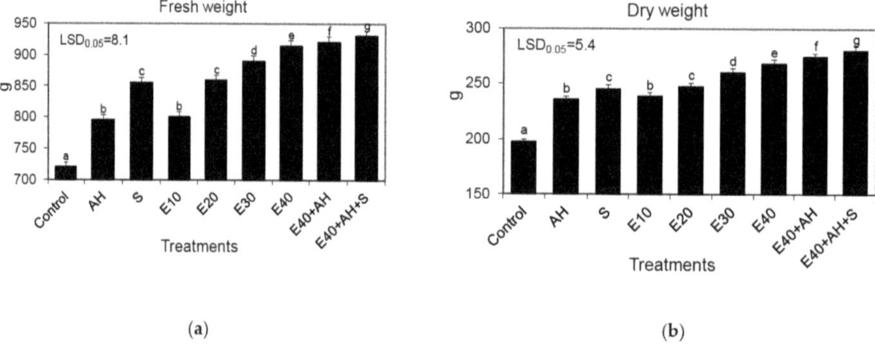

(a) (b)

Figure 2. Fresh (**a**) and dry biomass (**b**) of one Jerusalem artichoke plant cultivated in a field and fertilized with liquid non-centrifuged waste from corn grain digestion to methane (E10, 40 m^3 ha^{-1}), Apol-humus (AH; 10 L ha^{-1}), and Stymjod (S; 5 L ha^{-1}). The data marked with the same letters are not significantly different, according to the Newman–Keuls multiple range test at an alpha level of 0.05. The data presented are the average over the years and five plants in each repetition of a particular experimental variant.

The established correlations between the favorable changes in plant growth and biomass yield and the fertilizer doses studied were also confirmed by the proportionally increased activity of gas exchange and the index of the chlorophyll content in leaves. The growing doses of waste, used alone or supplemented with Apol-humus and Stymjod, also increased the index of the chlorophyll content, net photosynthesis, transpiration, and stomatal conductance, and decreased intercellular CO_2 concentration, inversely proportional to the above-mentioned three parameters of gas exchange. These relationships between the doses used of waste and parameters of gas exchange and the index of the chlorophyll content were similar to those observed between the amount of the fertilizers used and the plant growth and biomass yield (Figures 1–4).

Table 1. Activities of the selected enzymes in leaves of Jerusalem artichoke plants grown in a field and fertilized with liquid non-centrifuged waste from corn grain digestion to methane (E10, 40 m^3 ha^{-1}), Apol-humus (AH; 10 L ha^{-1}), and Stymjod (S; 5 L ha^{-1}).

Waste Doses and Biopreparations Applied to Soil	Phosphorylases		RNase [U g^{-1} f.w.]	Total Dehydrogenases [mg Formazan g Leaf^{-1}]
	(pH = 6.0) [U g^{-1} f.w.]	(pH = 7.5) [U g^{-1} f.w.]		
Control	0.60 a	0.26 a	2.6 a	0.50 a
AH	0.63 b	0.28 b	2.8 b	0.52 b
S	0.66 c	0.31 c	3.2 c	0.66 c
E10	0.63 b	0.28 b	2.9 b	0.52 b
E20	0.67 c	0.31 c	3.3 c	0.66 c
E30	0.71 de	0.33 de	3.6 de	0.69 d
E40	0.73 e	0.34 e	3.7 e	0.72 e
E40 + AH	0.77 f	0.39 f	4.2 f	0.76 f
E40 + AH + S	0.81 g	0.42 g	4.4 g	0.79 g
LSD$_{0.05}$	0.02	0.01	0.1	0.15

The data marked with the same letters within a column are not significantly different, according to a Newman–Keuls multiple range test at an alpha level of 0.05. The data presented are the average over the years and 10 plants in each repetition of a particular experimental variant.

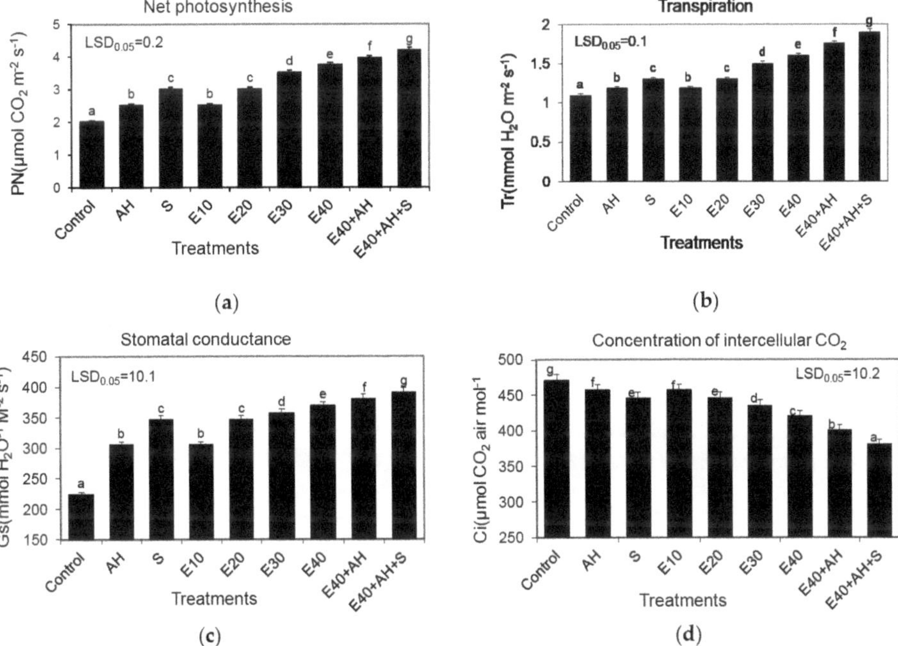

Figure 3. Gas exchange in the leaves of Jerusalem artichoke plants (Net photosynthesis (**a**), Transpiration (**b**), Stomatal conductance (**c**), Concentration of intercellular CO_2 (**d**)) cultivated in a field and fertilized with liquid non-centrifuged waste from corn grain digestion to methane (E10, 40 m^3 ha^{-1}), Apol-humus (AH; 10 L ha^{-1}), and Stymjod (S; 5 L ha^{-1}). The data marked with the same letters are not significantly different, according to a Newman–Keuls multiple range test at an alpha level of 0.05. The data presented are the average over the years and 10 plants in each repetition of a particular experimental variant.

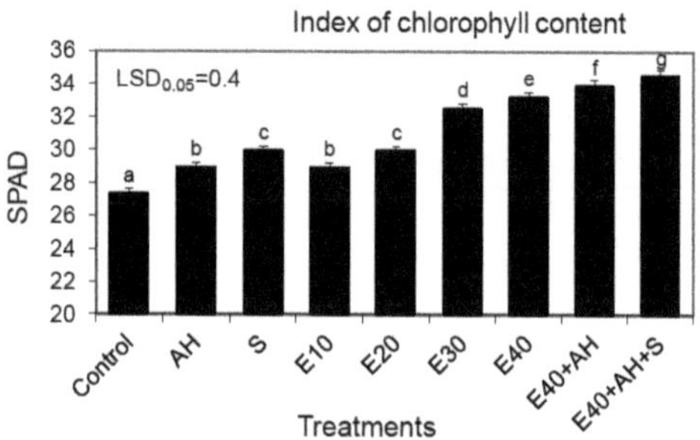

Figure 4. Index of the chlorophyll content in leaves of the Jerusalem artichoke plants cultivated in a field and fertilized with liquid non-centrifuged waste from corn grain digestion to methane (E10, 40 m^3 ha^{-1}), Apol-humus (AH; 10 L ha^{-1}), and Stymjod (S; 5 L ha^{-1}). The data marked with the same letters are not significantly different, according to a Newman–Keuls multiple range test at an alpha level of 0.05. The data presented are the average over the years and 10 plants in each repetition of a particular experimental variant.

The use of biogas plant waste slightly increased the heat of combustion in the analytical state and calorific value in the working state and decreased the ash content in the plants. Table 2 presents the results of the main experiment using a batch reactor for conducting a Jerusalem artichoke torrefaction process in a nitrogen atmosphere. The most important observation is that under 245 °C and a residential time of 13 min, the mass reduction was the closest one of all experimental results to 30%, which is due to the large amount of literature on the most optimal ratio.

Table 2. Experimental research results of the Jerusalem artichoke torrefaction process in nitrogen using a batch reactor.

Sample Number	Mass Reduction, g	Mass Loss, %	Residential Time, min	Torrefaction Temp., °C
1	20/14, 64	26, 80	12	251, 17
2	20/14, 38	28, 10	13	245, 17
3	20/13, 47	32, 65	14	247, 90
4	20/14, 50	27, 50	13	242, 17
5	20/14, 21	28, 95	14	243, 45
6	20/12, 51	37, 45	17	254, 06
7	20/15, 26	23, 70	10	241, 64

In Table 3, the results of a proximate analysis of Jerusalem artichoke before and after the thermo-chemical conversion process are presented. It is quite clear that the C% (weight) in the biomass thermo-chemical process of bio-products increases in tandem with an enhancement in the Jerusalem artichoke torrefaction process temperature. This was contrary to the weight percentages of C_aH and O, which showed a decreasing trend. From the above mechanism, it is clear that dehydration takes place as well as de-carbonization during the Jerusalem artichoke torrefaction process. This clearly shows that the emission of CO_2, CO, or H_2O will result in a decrease in the H and O contents of torrefied biomass. The rising % of the C content was only due to a decrease in the O content.

An FTiR analysis during the thermogravimetric analysis of the Jerusalem artichoke torrefaction process under 245 °C for the production of carbonized solid biofuel shows what kind of volatile matter components are produced, so-called torgas: H_2O, CO, CO_2, CH_4, and C_2 (Figures 5 and 6).

Table 3. Elemental analysis and technical analysis of Jerusalem artichoke before and after the torrefaction process.

Energy Crop	Moisture(%)	C^{ad}, (%)	N^{ad}, (%)	H^{ad}, (%)	S^{ad}, (%)	Cl, (%)	Volatile ^{ad} (%)	Ash (%)	High Heating Value, $(^{MJ}/_{kg})$
Jerusalem artichoke	5.3	48.5	0.27	6.20	0.05	0.115	91.29	2.3	15.82
Torrefied Jerusalem artichoke: (243, 45 °C, 14 min)	2.8	54.37	0.19	5.37	0.05	0.014	73.37	3.94	21.70
(245, 17 °C, 13 min)	2.7	55.04	0.19	5.34	0.05	0.014	72.81	3.84	22.12
(242, 17 °C, 13 min)	2.8	54.79	0.19	5.39	0.05	0.014	72.27	3.71	22.09

^{ad} Add dry basis.

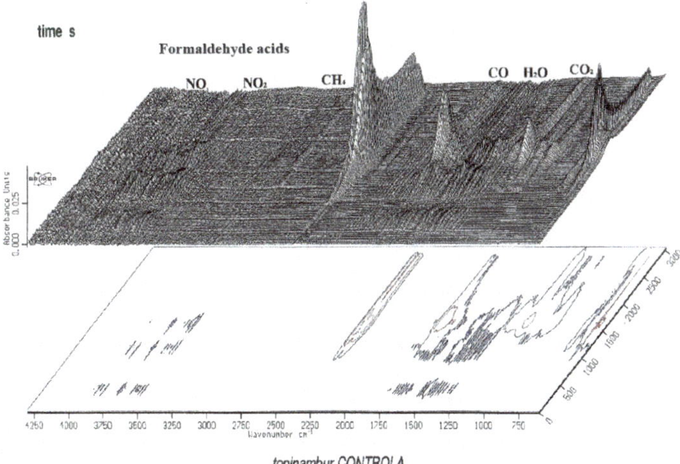

Figure 5. FTiR analysis of Jerusalem artichoke torrefaction by-products: torgas during the torrefaction process under 245 °C.

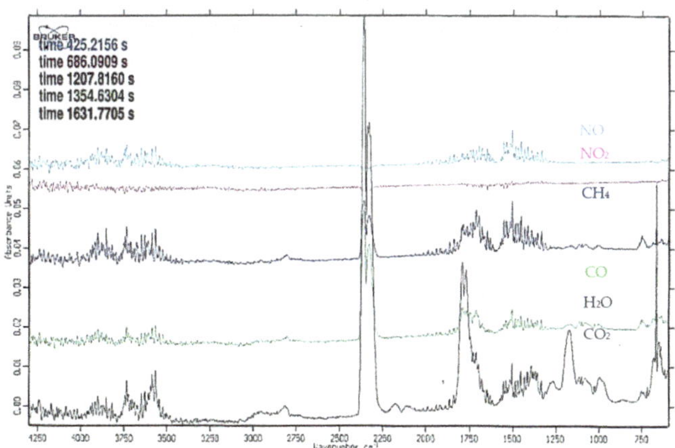

Figure 6. FTiR analysis of Jerusalem artichoke torrefaction by-products: torgas during the torrefaction process under 245 °C.

Figures 7–10 presents the thermogravimetric analysis of the combustion process of torrefied Jerusalem artichoke and TG-MS analysis, which shows what kind of component occurs during combustion. The colored lines represent what kind of volatile components are formulated during the combustion process: H_2O, CO_2, NO, SO_2, NO_2, formaldehyde, and C.

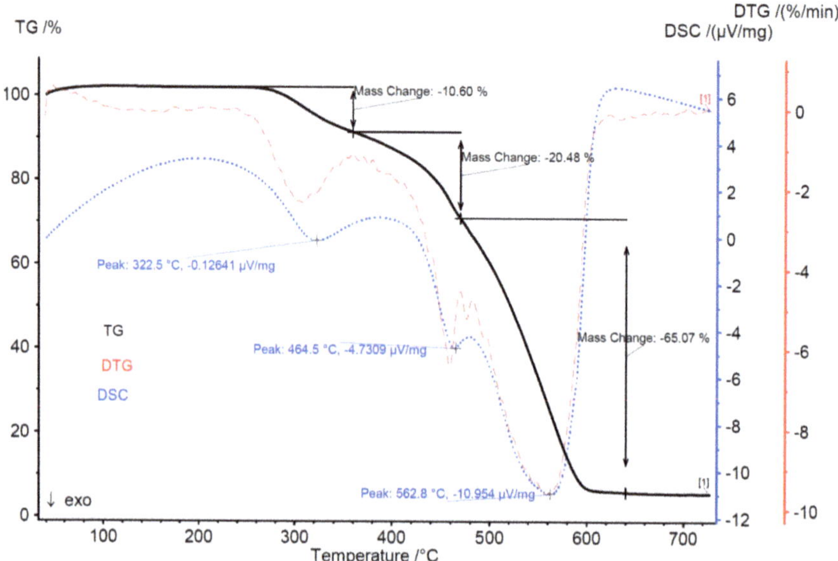

Figure 7. Thermogravimetric analysis of torrefied Jerusalem artichoke combustion process.

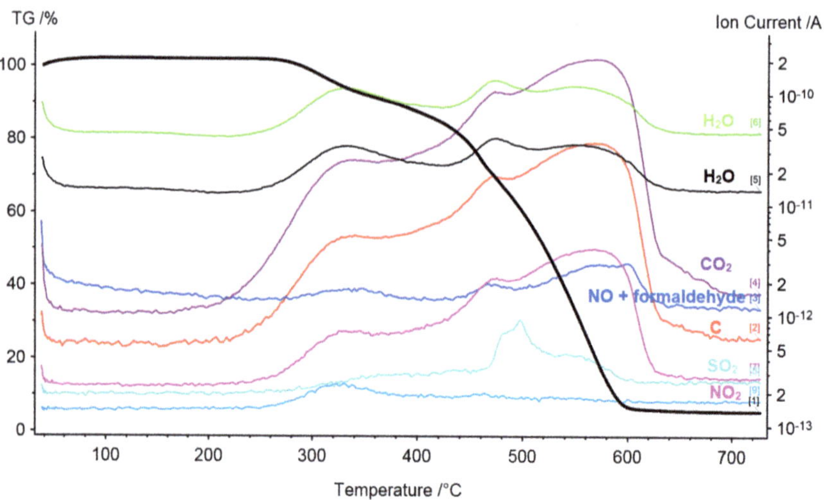

Figure 8. TGA-MS analysis of the Jerusalem artichoke combustion process.

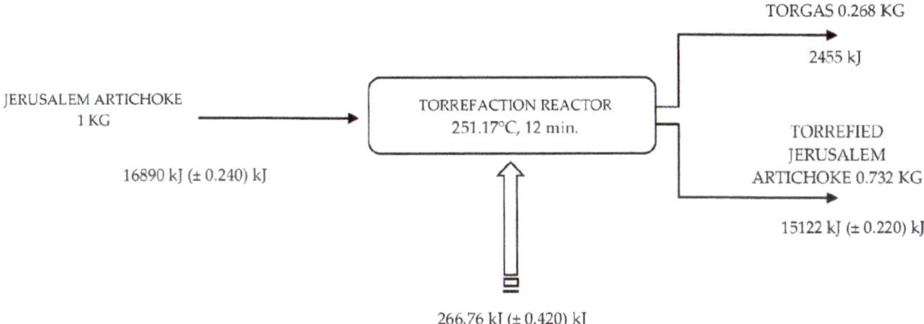

Figure 9. Mass and energy balance of the Jerusalem artichoke torrefaction process (t = 251.17 °C, 12 min).

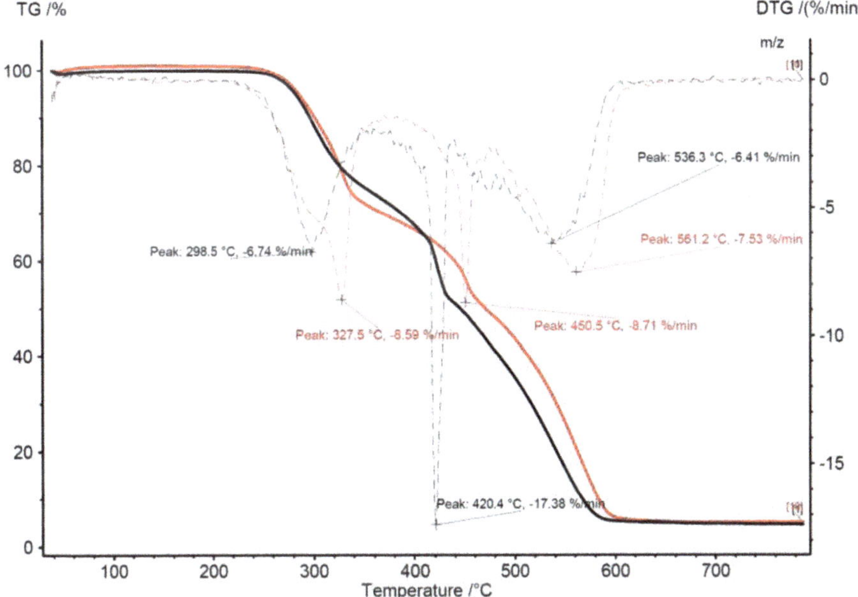

Figure 10. TGA-MS analysis of the torrefied Jerusalem artichoke combustion process.

An elemental analysis shows that during the TG-MS analysis of the torrefaction process of Jerusalem artichoke under 245 °C in torgas, there is 0.1% of CH_4, 0.05% of C_2, 81.2% of CO_2, and 18.6% of CO (Figure 8). Figure 9 represents the mass and energy balance of the Jerusalem artichoke torrefaction process using a batch reactor and in an inert atmosphere of nitrogen. It was calculated that 266.76 (±0.420) kJ is the external energy, which is neccessery to produce carbonized solid biofuel (with a 30% mass loss and 10% energy loss) from Jerusalem artichoke. The novelty of this research is based on the fact that there is a lack of publications describing the use of Jerusalem artichoke growing in Poland on low-class soils with the addition of wastes and biofertilizers as a feedstock for the torrefaction process with a focus on the torrefaction process's optimal parameters for biofuel production and physical and chemical analysis of torrefied biofuels and uses ashes as carriers of C for biofertilisers (Table 4).

Table 4. Content of elements in the ash from untreated Jerusalem artichoke and torrefied Jerusalem artichoke burnt biomass.

Assessed Material	C	O	K	Ca	Mg	Fe	Si	P	S	Cl	Dry Mass
					[Atomic, %]						[%]
Ash composition from torrefied Jerusalem artichoke (average values)	21.75	46.82	23.01	2.63	0.81	0.02	0.34	3.02	0.19	1.41	100.00
Standard deviation, σ	3.00	2.18	2.83	0.51	0.30	0.00	0.10	0.60	0.05	0.40	- *
Ash composition from untreated Jerusalem artichoke (average values)	31.53	36.05	20.09	2.22	0.98	0.11	0.46	1.16	0.19	3.13	100.00
Standard deviation, σ	3.32	4.00	3.88	0.72	0.27	0.01	0.42	0.62	0.05	1.19	- *

* The values of standard deviations do not add up.

Torrefaction is a thermal pre-treatment step for biomass co-firing or biofuel combustion purposes, which takes place in a relatively low temperature range of 225–350 °C to produce a fuel with a bigger energy density mainly by the decomposition of hemicellulose fractions (Figure 11).

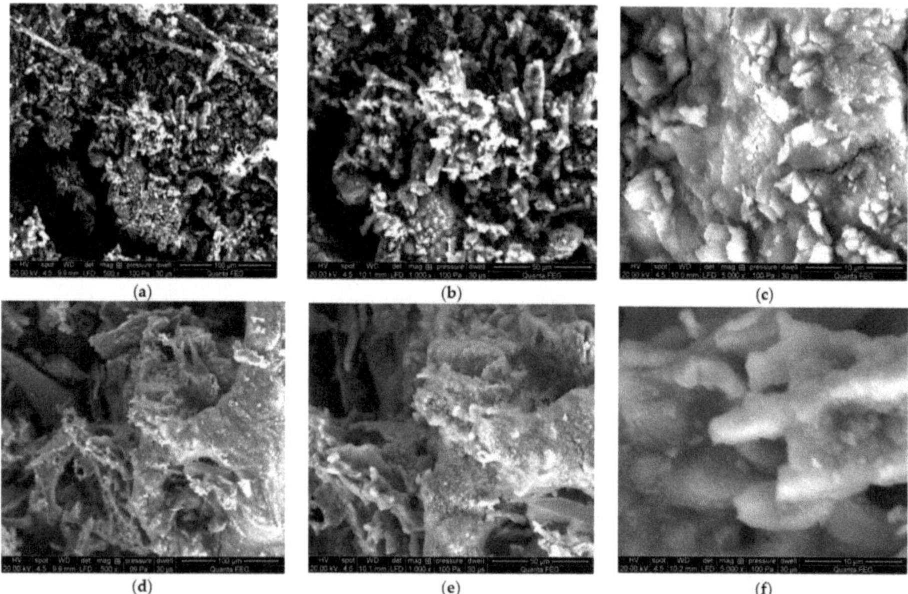

Figure 11. SEM-EDS microscopic images (**a**) 100 µm, (**b**) 50 µm, (**c**) 10 µm of ashes from torrefied Jerusalem artichoke after combustion and (**d**) 100 µm, (**e**) 50 µm, (**f**) 10 µm of ashes from untreated Jerusalem artichoke after combustion.

3. Discussion

Jerusalem artichoke is a very realistic high production potential energy plant that can be used to produce large amounts of biomass (aerial part) and biofuels (using tubers). The amount of yield depends primarily on the plant genotype and soil fertility. Biomass yield can even reach 110 t ha^{-1}, including green mass of 75.6 t ha^{-1}, and tubers of 32.4 t ha^{-1}. The raw material for energy purposes can be tubers that can be used for the production of ethanol or biogas and above-ground parts used for biogas fabrication, burning, or for fuel briquettes and pellets. This plant grows well and produces large amounts of biomass in a wide range of conditions, including moderately compact, well-ventilated, nutrient-rich, and sufficiently moist soils. It can also be cultivated in worse positions for energy needs, especially if it is fertilized with biogas plant waste, as the research presented demonstrates [9].

The main goal of the research was to examine mechanisms controlling the torrefaction process, to determine selected torrefaction parameters of Jerusalem artichoke growing in low-class soils, and to determine the impact of the torrefaction process on the Jerusalem artichoke co-firing with coal process, including the level of pollution reduction. The main thesis of the research is the assumption that the torrefaction process causes an increase in the calorific value of carbonized Jerusalem artichoke, positively affects the improvement fuel values during carbonization, and favorably affects the process of co-firing torrefied biomass with coal, thus limiting a number of operational problems arising as a result of co-firing biomass with hard coal. This research is becoming important as it is commonly estimated that in 2050, about 75% of energy will come from renewable energy sources [19]. Therefore, it is necessary to undertake intensive actions to develop technology for sustainable energy crop production, its conversion into useful biofuel, and to limit the use of synthetic fertilizers, as their production and use in agriculture are energy intensive and cause pollution of the environment. The research performed showed that fertilization with the waste from corn grain biodigestion to methane in a dose of 10–40 m^3 ha^{-1} of Apol-humus and Stymjod increased plant growth, biomass yield, gas exchange, enzymatic activity, and energy value of Jerusalem artichoke biomass proportionally to the fertilizer doses, as was shown in several labour-intensive tests. The correlations found were observed in all tests performed on physiological activity and they were similar regardless of the fact that these plants were cultivated for three subsequent years in podzolic soil. This indicates the proper selection of the plant assessment used tests to show the response of plants to the applied treatments and the possibility of obtaining positive effects from the developed methods of fertilization with waste irrespective of the climatic conditions. The tests implemented also explain the correlation between particular physiological events and their role in regulating growth and yield [20,21]. The changes observed in the kinetics of Jerusalem artichoke growth, biomass yield, gas exchange, and enzyme activity in relation to the fertilization methods demonstrated the positive impact of all used non-centrifuged waste doses, with 30–40 m^3 ha^{-1} being the most favorable for plant growth and physiological activity [19].

The results obtained show that the amounts of nutrients contained in the waste used could influence the photochemical processes of photosynthesis, as was also demonstrated by Kalaji et al. [22] in corn and tomato. They showed that photosynthetic system activity, measured by chlorophyll fluorescence, is related to the nutrient content in plants. The research presented showed also that the increasing doses of waste applied to soil were proportionally related to a higher chlorophyll content in leaves and accelerated growth of plants. Chlorophyll content is the most widely used proxy for N content, as was demonstrated by the studies of Herrmann et al. [23], Homolová et al. [24], and Camino et al. [25]. According to Hamann et al. [26], measurement of the chlorophyll content and nitrogen balance can be a useful non-destructive method to estimate the physiological status of plants, as was found in young apple trees cultivated under water stress conditions.

Research shows that the applied waste from biogas plants influenced the activity of enzymes, which have a key impact on plant growth, biomass yield, physiological processes, and energy properties. Similarly, as in the case of gas exchange, it enhanced the activities of acid and alkaline phosphorylases, RNase, and dehydrogenase in a dose-dependent manner. The alkaline and acid phosphorylases are responsible for the distribution of phosphorus in plants and they catalyze the hydrolysis of organic phosphorus. They also regulate the mineralization potential of organic phosphorus, which can influence the biomass energy value [27]. Stimulation of RNase activity by the waste studied may play an important role in strengthening defense mechanisms in plant tissues, as was also observed in willow and corn plants under the influence of microalgae used as fertilizer. Dehydrogenases play a crucial role in respiration processes important for growth and biomass yield [21]. A close relationship between different fertilization methods, the enzyme activities studied, and plant development was also found in other energy plants, such as Virginia fanpetals, corn and willow treated with algae [17,21,28,29], and in some energy plants fertilized with sewage sludge and ash [30,31].

The studies indicated that enhanced fertility, resulting in higher gas exchange and enzyme activity, also slightly increased the heat of combustion in the analytical state and calorific value in the working state, as well as decreased the ash content in plants proportionally to the applied fertilizer doses. These properties are important when plants are produced for energetic purposes and show that Jerusalem artichoke fertilized with the waste from a biogas plant is suitable for this use. In line with the research of Kordas et al. [32], the results obtained state that increasing the dose of mineral fertilization contributed to the increase in the heat value of plant combustion.

The additional stimulating impact on plant development of the biopreparations added to the waste from the biogas plant in the combined treatments could be caused by humid acids and chitosan polymers contained in Apol-humus and by-nutrients, and humid acids and iodine present in Stymjod. The positive impact of humid acids on plant development has been described in the literature [33]. The favorable impact of chitosan on the development and health of plants under hydrothermal stress was revealed by Górnik et al. [34] in grapevines. The demonstrated stimulatory influence of Stymjod, used with the waste studied, on Jerusalem artichoke plant development could be an effect of macro- and microelements, humid acids, and above all of the presence of iodine [35,36]. Jeznach [36] demonstrated the positive influence of iodine on the cyto-morphological changes in cabbage and tomato, the enlarged diameter of phloem and xylem, and more frequent stomata opening, which resulted in increased gas exchange in the leaves. The application of iodine to cabbage enhanced its resistance to stress and the quantity of several elements in leaves [36]. According to Smoleń et al. [37], iodine application increased the content of phosphorus, potassium, and calcium and decreased the accumulation of iron in stored carrot roots. The intensified physiological processes found in vegetable crops under the influence of iodine could also occur in Jerusalem artichoke and influence its biomass energy properties and sensitivity to growing conditions.

The research presented indicates that biogas plant waste use in Jerusalem artichoke crops under different climate conditions subsequently not only allows a reduction in the doses of synthetic fertilizers that contaminate the surroundings but can also solve the serious problem of utilization and storage, which is expensive and dangerous for the environment [38,39]. Additionally, the use of this waste as a plant fertilizer is safer than sewage sludge, which may often contain harmful compounds that must be removed prior to its use in agriculture [40]. Research indicates that fertilization with the waste from a biogas plant enables a high yield of energy and biomass to be obtained, which could be used in its torrefaction, and may lead to a decrease in the recommended doses of artificial fertilizers, thus limiting environmental pollution. The highest fertilization level of Jerusalem artichoke enables a high quantity of fuels to be obtained as was found in sorghum, which can be a raw material for producing 8455 Nm^3 of biogas ha^{-1} and 200,000 MJ ha^{-1} per year [10].

Regarding research results on the Jerusalem artichoke torrefaction process using a batch reactor, the findings of the torrefaction process temperature are quite important when designing commercial continuous reactors for the production of carbonized solid biofuels. These results indicate that, as is well known in the production of carbonized solid biofuels, in order to obtain the best process conditions of the torrefaction process and a reasonable price of the final product, it is important to achieve a mass loss on a level of 30% and an energy loss (torgas) on a level of 10% in the thermo-chemical conversion. Compared to research results on other energy crops and straw biomass, Jerusalem artichoke's temperature of 245 °C during torrefaction for carbonized solid biofuel production under isothermal conditions is relatively low [41–49]. Research on the Jerusalem artichoke has shown that the amount of ash after the torrefaction process is still at a relatively low level compared to biomass not subjected to the torrefaction process (Jerusalem artichoke unprocessed as a result of the torrefaction process has an ash content of <3%), and solid fossil fuels, such as Polish hard coal, have an ash content of <15%. An SEM-EDS analysis of the ash composition of torrefied Jerusalem artichoke after burning at 700 °C showed a very favorable composition of mineral substances that can be reused as additives to organic fertilizers, a carrier of such elements as K (20.46%) and P (3.36%) and C (22.51%).

An SEM-EDS analysis was performed on fly ash of Jerusalem artichoke sintered at 700 °C.

The method of growing Jerusalem artichoke presented in the experiments carried out is ecological and improves the parameters of the torrefaction process. The Jerusalem artichoke produced as a result of ecological fertilization is characterized by low cultivation costs and the heat energy produced in this carbonized solid biofuel requires about 25% less expenditure than during fertilization with chemical fertilizers.

4. Materials and Methods

4.1. Plants, Waste, and Biopreparations

The bulbs of Jerusalem artichoke used in the experiments were purchased from Chmiel Ecological Farm (Poland). The non-centrifuged liquid waste was obtained from Gamawind Sp. z o.o., Piaszczyna, Poland, a distillery integrated with the biogas plant, which produces alcohol and biogas using corn grain as raw material. Apol-humus, the new generation soil improver, was purchased from the manufacturer Poli-Farm Sp. z o.o., Poland, whereas Stymjod, a nano-organic-mineral fertilizer, was supplied by the producer PHU Jeznach Sp. J., Poland.

4.2. Ultimate Analysis

The ash content of the raw samples of Jerusalem artichoke was measured using an electrical oven using the standard procedure. A main assumption was taken that no ash is lost during the torrefaction stage. Therefore, an ash content value was calculated for each solid residue from the overall mass yields. The C, H, N, and S contents were measured using a Perkin/Elmer Analyser and the elemental analyses procedure was used. All three different samples of Jerusalem artichoke were analyzed: Measurements were repeated in triplicate and the average value, which was corrected for moisture content, is presented in Table 1.

4.3. Plant Treatments and Experimental Design

The research was performed in north Poland in a field where the temperature oscillates from 11 to 21 °C, precipitation is 655 mm, and moist air from the Baltic Sea is noted. The experiments were carried out in podzolic soil, on 7 plots, which were fertilized in April with:
— Non-centrifuged waste from a corn grain biodigestion to methane process in dosages of 0, 10, 20, 30, and 40 m^3 ha^{-1};
— Non-centrifuged waste, 40 m^3 ha^{-1} together with Apol-humus (10 L ha^{-1}); and
— Non-centrifuged waste, 40 m^3 ha^{-1} together with Apol-humus (10 L ha^{-1}) and Stymjod (5 L ha^{-1}).

All the experimental variants and elemental characteristic of wastes and biofertilisers are shown in Figures 1–4 and Tables 1 and 5. Waste and Apol-humus were mixed with the soil after their application while Stymjod was applied twice to leaves in July at a two-week interval. Jerusalem artichoke tubers were planted in the soil (enriched previously in April with waste and Apol-humus) in the first 10 days of May, 10–15 cm deep, 50 cm apart in a row, and 70 cm between rows, as is recommended. Non-fertilized plots/plants served as the control (Figure 12). The applied dosages of waste, Apol-humus, and Stymjod were chosen on the basis of previous research performed in a laboratory, container area, and field [19]. Jerusalem artichoke biomass was collected in November, evaluated for fresh and dry biomass, then chopped in a chopper and torrefied.

Table 5. Content of elements in the non-centrifuged waste from corn grain biodigestion to methane, Stymjod, and Apol-humus.

Assessed Material	pH	N	P	K	Ca	Mg	Fe	Mn	Cu	Zn	B	Dry Mass
		[mg L^{-1}]										[%]
Waste	7.6	2455	278	996	300	115	9.0	0.324	0.175	0.976	3.365	1.4
Stymjod	5.4	1230	6650	62722	945	11574	18.7	885	680	1470	573	-
Apol-humus	12	15.21	15.8	20.2	468	70	140	5.95	0.87	2.40	0.92	-

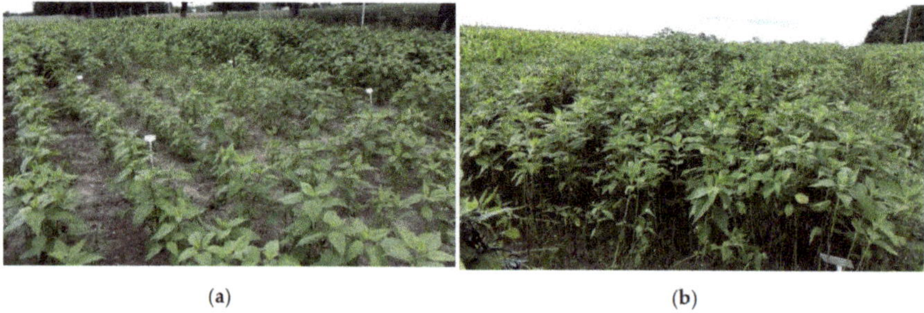

(a)   (b)

Figure 12. Jerusalem artichoke plants fertilized with various doses of biogas plant waste in Piaszczyna, (north Poland). The plot on the left fertilized with the dose of 10 m^3 ha^{-1}, in the middle the control (0 m^3 ha^{-1} and on the right 40 m^3 ha^{-1}. Photo was taken on June 15 (**a**) and 2 of July (**b**).

4.4. Fuel Characteristics

4.4.1. Caloric Value

The caloric value was determined by using two methods. The first one was performed by using a calorimetric bomb and the second one by calculating the CV based on the carbon, hydrogen, and nitrogen content. One type of equation for CV calculation was used to find the value, called the partial least squares regression (PLS) method (Equation (1)), and this gave a higher heating value (HHV) on a dry basis. HHV was calculated using the formula:

$$HHV(PLS) = 5.22C^2 - 319C - 1647H + 38.6CH + 133N + 21028 \tag{1}$$

where C = carbon, H = hydrogen, and N = nitrogen content expressed on a dry mass percentage basis. All caloric values from numerical calculations are the medium values of the determined results and are finally corrected by the dry ash content on a free basis.

The methodology of the experimental analysis of the Jerusalem artichoke torrefaction process for carbonized solid biofuel production comprised several analytical techniques:
— Analysis of the TGA, DTA, TG-FTiR, and TG-MS torrefaction process and biomass co-firing;
— Elemental analysis of biomass torrefaction process products;
— Analysis of the gases formed as a result of torrefaction: FTiR analysis and MS analysis;
— Technical analysis of biomass torrefaction products; and
— SEM-EDS ash analysis of torrefied Jerusalem artichoke after combustion.

Experimental research on the torrefaction process of carbonized energy crops using a specially designed biomass torrefaction installation with a batch reactor was performed in an inert atmosphere, nitrogen (Figure 13).

During biomass decomposition, three zones were distinguished on the weight loss curves of wood during torrefaction using installation with batch reactor Figure 12. The first one corresponded to the most reactive component, hemicellulose, whose decomposition started at 225 °C and finished at 325 °C; the second one was cellulose, whose decomposition temperature rate is from 300 °C up to 375 °C; and the last one was lignin, which represents the widgets' temperature rate of 250–500 °C. The carbonization process of lignocellulosic biomass was described by weight loss kinetics by using different experimental devices. Among those many devices were fluidized bed reactors, thermogarvimetric analyzers, and tube furnaces. In the research presented, a method with a thermogravimetric analyser (TGA) was chosen to determine the weight loss kinetics of energy crop torrefaction. By using this kind of experimental method, we obtained dynamic conditions in which the sample with biomass was placed at a specific heating rate, but it is important to know that experimental heating rates are very often slower than those in real process equipment, such as combustors, reactors, or gasifiers.

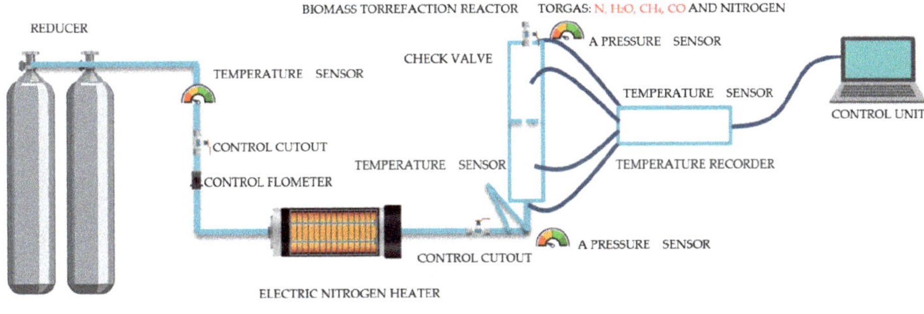

Figure 13. Scheme: installation with a batch reactor for the Jerusalem artichoke torrefaction process using nitrogen.

4.4.2. Ash Analysis of Torrefied Jerusalem Artichoke

The samples were investigated by scanning electron microscopy (SEM) and energy-dispersive X-ray spectroscopy (EDX) using an SEM FEI Quanta 200FEG microscope equipped with an EDX Oxford X-Max spectrometer. Measurements by the EDX technique were performed in at least 10 different spots for a given sample, and then an average atom concentration and its standard deviation were calculated for each of the identified elements. The electron energy of 20 keV was used in investigations.

4.5. Assessments of Plant Physiological Activity and Growth

The effects of fertilization treatments were assessed by periodical height measurements of whole plants during the vegetative season and evaluation of their biomass yield and energy value in the autumn. Assessments of the physiological activity of plants (gas exchange, index of chlorophyll content, enzyme activity) were carried out on fully developed leaves situated under the top of plants. In each experimental variant, one leaf from each of 10 plants was taken for the evaluation of gas exchange and enzyme activity. The material was collected in the third week of July in a temperature range of 25–30 °C, in sunshine and air humidity of 50–60%.

The height of the plants was measured at monthly intervals throughout the growing season [17]. The weights of fresh green biomass, and dry (dried at 130 °C for 3 days) were assessed in November on the basis of 5 plants taken from each experimental variant. The data presented were calculated for one plant as an average for the treatment [17]. Assessments of gas exchange (net photosynthesis, transpiration, stomatal conductance, and intercellular CO_2 concentration) were performed using a TPS-2 -Portable Photosynthesis System (PP Systems, Amesbury, MA, USA) [17,20].

An index of the chlorophyll content in leaves was estimated using a SPAD-502 chlorophyll meter (Konica Minolta, Osaka, Japan) [17]. The activities of acid (pH 6) (EC 3.1.3.2) and alkaline (pH 7.5) (EC 3.1.3.1) phosphorylases (U g^{-1} (FM) min^{-1}) in leaves and RNase (EC 3.1.27.5) (U g^{-1}(FM) min^{-1}) were studied using the methods demonstrated by Knypl and Kabzinska [50]. The activity of total dehydrogenases (EC 1.1.1.-) was measured using the procedure presented by Górnik and Grzesik [20,51] with a spectrophotometer (UVmini-1240, Shimadzu, Japan) for formazan determination at a wavelength of 480 nm.

4.6. Assessments of Biomass Energy Properties and Torrefied Materials

The energy crop samples were pre-prepared before the experiments, and the biomass was separated from foreign bodies, cleaned from contamination, chopped, and ground so as to reach the appropriate geometric dimensions. The plants were cut into sections of 2 to 4 cm, ground, and then sieved on a special automatic screen and dried in an electric oven at 110 °C for 4 h. Then, the samples were tightly closed and sent for technical and elemental analysis. The content of the elements carbon, hydrogen, nitrogen, and sulphur were determined and the volatility, moisture content, ash, combustion heat, and calorific values were determined (the same analysis was performed after the torrefaction process). The weight of the samples was determined before and after the drying process and the results were used to determine the moisture content. In each Jerusalem artichoke torrefaction process experiment, the dried biomass was divided into three separate samples of 20 g and each of them was evenly distributed on three horizontal screens of a reactor metal structure with different perforations made of acid-proof steel. The nitrogen was heated up by using electrical heaters to the set temperature (for this experimental research, temperatures between 220 °C and up to 280 °C were measured and the torrefied products were analyzed according to the mass loss and caloric value increase). Only selected samples were analyzed because, as a result of previous studies, a too high weight loss of more than 35% results in a very high degree of carbonization and high energy loss (above 10% of original energy content). The reverse is achieved with a mass loss below 25%, which is too low, and the biomass, in the end, is not fully roasted (it is not fragile and still has a high degree of moisture absorption and low calorific value).

4.7. Statistical Analysis

The investigations presented were performed in the field for three years (in a series) in northern Poland in three replicates for each experimental variant. The experimental plots and replicates, with differently fertilized plants, were situated randomly. Because of the similar climate conditions and growth in subsequent years, the data obtained were presented as means from the years and 10 plants (measuring height, gas exchange, index of chlorophyll content, enzyme activity) or 5 plants (taking weights of fresh green biomass) from each replicate. These were processed applying analysis of variance (ANOVA I), by Statistica 12. The means of the chosen parameters were grouped employing the Dunett's test and the contrast between the control sample and the remaining samples was used at the $\alpha = 0.05$ significance level.

5. Conclusions

The research presented shows the prospects of increasing the energy efficiency of Jerusalem artichoke crops by ecological use of the waste from corn grain biodigestion to methane, applied separately or together with Apol-humus, a new generation soil improver, and Stymjod, a nano-organic-mineral fertilizer, as an alternative to artificial fertilizers, which pollute the environment.

The research presented shows that the temperature at which we obtained a 30% weight loss and a 10% energy loss as a result of the Jerusalem artichoke torrefaction process is 245 °C. The research shows an increase in the calorific value as a result of the Jerusalem artichoke torrefaction process from 15.82 to 22.12 MJ kg^{-1}. Research on the Jerusalem artichoke torrefaction process Figure 14 has shown that the amount of ash after the torrefaction process in Jerusalem artichoke is still at a relatively low level compared to biomass not subjected to the torrefaction process (Jerusalem artichoke unprocessed as a result of the torrefaction process has an ash content of <3%) and solid fossil fuels, such as Polish hard coal, has an ash content of <15%. An SEM-EDS analysis of the ash composition after burning at 700 °C of torrefied Jerusalem artichoke showed a very favorable composition of mineral substances that can be reused as additives to organic fertilizers, a carrier of such elements as: K (20.46%) and P (3.36%) and C (22.51%).

An SEM-EDS analysis was performed on the fly ash of Jerusalem artichoke sintered at 700 °C and leaching toxicity of heavy metals was analyzed by a horizontal vibration extraction procedure (HVEP). It was found that the structure of fly ash is strengthened with an increase of the temperature, which is conducive to the stabilization of heavy metals.

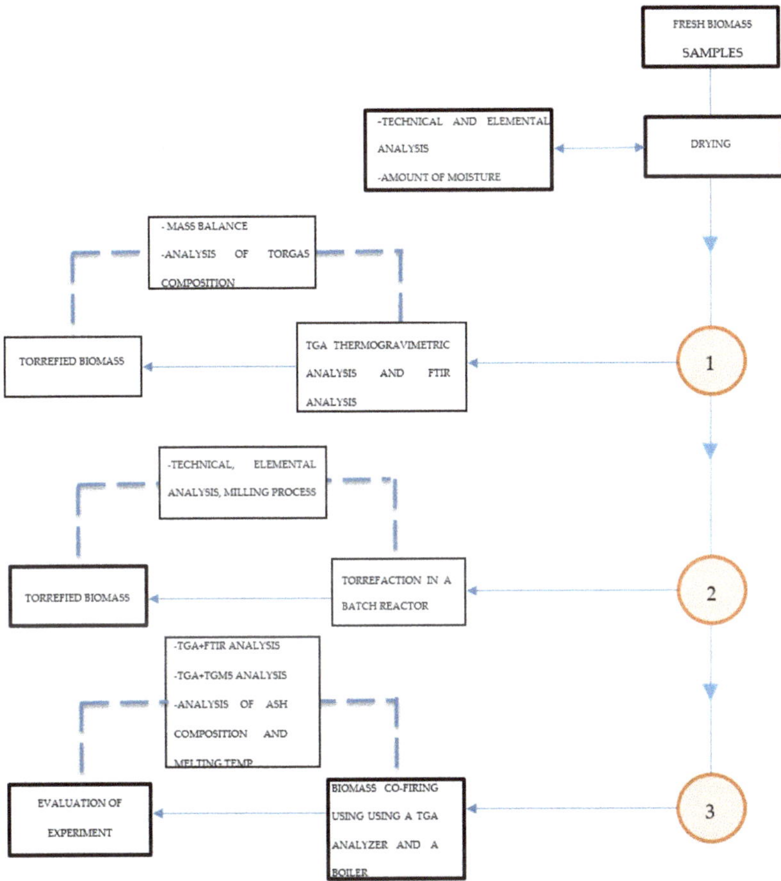

Figure 14. Scheme of the Jerusalem artichoke torrefaction process analytical method order: TGA, DTA, TG-FTiR, TG-MS torrefaction process, and biomass combustion plus SEM-EDS ash analysis.

The method of growing Jerusalem artichoke presented in the experiments carried out is ecological and improves the parameters of the torrefaction process. The Jerusalem artichoke produced as a result of ecological fertilization is characterized by low cultivation costs and the heat energy produced in this carbonized solid biofuel requires about 25% less expenditure than during fertilization with chemical fertilizers.

The research also demonstrates that biomass obtained under the influence of such fertilization can be used for energy purposes and, for example, can be converted to high energy density torrefied solid biofuel. It was demonstrated that Jerusalem artichoke could be a promising high-yielding energy crop. Its ecological fertilization with the waste from biogas plants (30–40 m^3 ha^{-1}), Apol-humus, and Stymjod positively influenced gas exchange (net photosynthesis, transpiration, stomatal conductance, and intercellular CO_2 concentration), the index of chlorophyll content, the activity of the selected enzymes (acid and alkaline phosphorylase, RNase, and dehydrogenase), and energetic parameters, which markedly affected the Jerusalem artichoke growth kinetics during the whole vegetative season as well as the yield of fresh and dry biomass. The waste studied can be used as a cost-effective and environmentally friendly biofertilizer, if it is applied in defined doses and in agreement with the national legal regulations on the safe application of these components. The research shows that a torrefaction process temperature of around 245 °C is one of the most optimal temperatures for the production of carbonized solid biofuel from Jerusalem artichoke. Compared to research results on other energy crops and straw biomass, an isothermal temperature of 245 °C during torrefaction for carbonized solid biofuel of Jerusalem artichoke biomass fertilized with biogas plant waste is relativlely low. In the near future, biomass plants will be a major source of biofuel production and may inhibit the growth of oil prices [52].

Author Contributions: S.S., Z.R.-D., M.G.; methodology, P.P., Z.R.-D., K.P., J.S.; software, S.S., Z.R.-D., M.G.; validation, S.S., P.P., J.S. and M.G.; formal analysis, S.S., Ł.A. and K.P.; investigation, S.S., Z.R.-D. and M.G.; resources, S.S., P.P.; data curation, S.S., W.L., J.S. and Z.R.-D.; writing—original draft preparation, S.S., Z.R.-D., M.G., K.P.; writing—review and editing, S.S., P.P., W.L. and Ł.A.; visualization, S.S., Z.R.-D.; supervision, S.S. and P.P.; project administration, S.S.; funding acquisition, S.S., M.G. and Z.R.-D. All authors have read and agreed to the published version of the manuscript.

Funding: The research results in this article were financed by the National Centre for Research and Development Grant from Warsaw Poland—LIDER (grant no. 0155/L-9/2017). The research was supported by the National Centre for Research and Development Grant No. BIOSTRATEG2/296369/5/NCBR/2016 and Nr: 2011/03/N/ST8/02776.

Conflicts of Interest: The authors declare no conflict of interest.

References

1. Biofuels Progress Report. Report on the progress made in the use of biofuels and other renewable fuels in the Member States of the European Union. 2007. Available online: https://www.ebb-eu.org/legis/biofuels%20progress%20report%20100107%20provisional%20version.pdf (accessed on 2 April 2020).
2. Biofuels in the, E.U. A Vision for 2030 and Beyond. 2006. Available online: https://ec.europa.eu/research/energy/pdf/draft_vision_report_en.pdf (accessed on 1 April 2020).
3. Faaij, A.P.C. Bio-energy in Europe: Changing technology choices. *Energy Policy* **2006**, *34*, 322–342. [CrossRef]
4. Fernando, A.L.; Retteenmaier, N.; Soldatos, P.; Panoutsou, C. *Sustainability of Perennial Crops Production for Bioenergy and Bioproducts*; Academic Press: London, UK, 2018; pp. 245–283.
5. Zawadzka, A.; Imbierowicz, M. *Inwestowanie w Energetykę Odnawialną*; PAN Oddział w Łodzi, Komisja Ochrony Środowiska: Łódź, Poland, 2010; pp. 169–184. (In Polish)
6. Bridgeman, T.G.; Jones, J.M.; Shield, I.; Williams, P.T. Torrefaction of reed canary grass, wheat straw and willow to enhance fuel qualities and combustion properties. *Fuel* **2008**, *87*, 844–856. [CrossRef]
7. Prins, M.J. Thermodynamic analysis of biomass gasification and torrefaction. Ph.D. Thesis, Eindhoven University of Technology, Eindhoven, The Netherlands, 2005.
8. Lin, Y.-L. Effects of Microwave—Induced torrefaction on waste straw upgrading. *Int. J. Chem. Eng. Appl.* **2015**, *6*, 401–404.
9. Gao, K.; Zhang, Z.; Zhu, T.; Tian, X.; Gao, Y.; Zhao, L.; Li, T. The influence of leaf removal on tuber yield and fuel characteristics of Helianthus tuberosus L. in a semi-arid area. *Ind. Crop. Prod.* **2019**, *131*, 8–13. [CrossRef]
10. Krzystek, L.; Wajszczuk, K.; Pazera, A.; Matyka, M.; Slezak, R.; Ledakowicz, S. The influence of plant cultivation methods on biogas production: Energy efficiency. In Proceedings of the WasteEng 2018 Conference, Prague, Czech Republic, 2–5 July 2018.
11. Alburquerque, A.J.; Fuente, C.; Ferrer-Costa, A.; Carrasco, L.; Cegarra, J.; Abad, M.; Bernal, P.M. Assessment of the fertilizer potential of digestates from farm and agroindustrial residues. *Biomass Bioen.* **2012**, *40*, 181–189. [CrossRef]
12. Jasiulewicz, M.; Janiszewska, A.D. Possible opportunities for the development of biogas plants on the example of the Zachodniopomorskie province. *Inżynieria Rol.* **2013**, *2*, 91–102.
13. Łagocka, A.; Kamiński, M.; Cholewiński, M.; Pospolita, W. Health and environmental benefits of utilization of post-fermentation pulp from agricultural biogas plants as a natural fertilizer. *Kosmos* **2016**, *65*, 601–607.
14. Pszczolkowska, A.; Pszczolkowski, W.; Romanowska-Duda, Z. Potential of *Chlorella vulgaris* culture for waste treatment from anaerobic biomass biodigestion at the Piaszczyna (Poland) integrated facility. *J. Phycol.* **2019**, *55*, 816–829. [CrossRef]
15. Romanowska-Duda, Z.; Piotrowski, K.; Jagiełło, N.; Dębowski, M.; Zieliński, M. Development of new Lemnaceae breeding technology using Apol-humus and biogas plant waste. *Inter. Agrophys.* **2019**, *33*, 297–302. [CrossRef]

16. Dębowski, M.; Rusanowska, P.; Zieliński, M.; Dudek, M.; Romanowska-Duda, Z. Biomass production and nutrient removal by *Chlorella vulgaris* from anaerobic digestion effluents. *Energies* **2018**, *11*, 1654. [CrossRef]

17. Grzesik, M.; Romanowska-Duda, Z.; Kalaji, H.M. Effectiveness of cyanobacteria and green algae in enhancing the photosynthetic performance and growth of willow (*Salix viminalis* L.) plants under limited synthetic fertilizers application. *Photosynthetica* **2017**, *55*, 510–521. [CrossRef]

18. Lalak, J.; Kasprzycka, A.; Paprota, E.M.; Tys, J.; Murat, A. Development of optimum substrate compositions in the methane fermentation process. *Int. Agrophys.* **2015**, *29*, 313–321. [CrossRef]

19. Romanowska-Duda, Z.; Grzesik, M.; Janas, R. Maximal efficiency of PSII as a marker of sorghum development fertilized with waste from a biomass biodigestion to methane. *Front. Plant Sci.* **2019**, *9*, 1920. [CrossRef] [PubMed]

20. Grzesik, M.; Górnik, K.; Janas, R.; Lewandowki, M.; Romanowska-Duda, Z.; van Duijn, B. High efficiency stratification of apple cultivar Ligol seed dormancy by phytohormones, heat shock and pulsed radio frequency. *J. Plant. Physiol.* **2017**, *21*, 81–90. [CrossRef] [PubMed]

21. Grzesik, M.; Romanowska-Duda, Z.B. Ability of Cyanobacteria and microalgae in improvement of metabolic activity and development of willow plants. *Pol. J. Environ. Stud.* **2015**, *24*, 1003–1012. [CrossRef]

22. Kalaji, H.; Oukarroum, A.; Kouzmanova, M.; Brestic, M.; Zivcak, M.; Samborska, I.A.; Cetner, M.D.; Allakhverdiev, S.I.; Goltsev, V. Identification of nutrient deficiency in maize and tomato plants by in vivo chlorophyll a fluorescence measurements. *Plant Physiol. Bioch.* **2014**, *81*, 16–25. [CrossRef]

23. Herrmann, I.; Karnieli, A.; Bonfil, D.J.; Cohen, Y.; Alchanatis, V. SWIR-based spectral indices for assessing nitrogen content in potato fields. *Int. J. Remote Sens.* **2010**, *31*, 5127–5143. [CrossRef]

24. Homolová, L.; Malenovský, Z.; Clevers, J.G.P.W.; García-Santos, G.; Schaepman, M.E. Review of optical-based remote sensing for plant trait mapping. *Ecol. Complex* **2013**, *15*, 1–16. [CrossRef]

25. Camino, C.; González-Dugo, V.; Hernández, P.; Sillero, J.C.; Zarco-Tejada, P.J. Improved nitrogen retrievals with airborne-derived fluorescence and plant traits quantified from VNIR-SWIR hyperspectral imagery in the context of precision agriculture. *Int. J. Appl. Earth Obs. Geoinf.* **2018**, *70*, 105–117. [CrossRef]

26. Hamann, F.A.; Czaja, S.; Hunsche, M.; Noga, G.; Fiebiga, A. Monitoring physiological and biochemical responses of two apple cultivars to water supply regimes with non-destructive fluorescence sensors. *Scientia Hort.* **2018**, *242*, 51–61. [CrossRef]

27. Dick, W.A.; Tabatabai, M.A. Significance and potential uses of soil enzymes. *Environ. Manag.* **1992**, 95–125.

28. Grzesik, M.; Romanowska-Duda, Z.B.; Piotrowski, K. The effect of potential climatic changes, *Cyanobacteria*, Biojodis and Asahi SL on development of the Virginia fanpetals (*Sida hermaphrodita*) plants. *Pamiętnik Puławski* **2009**, *151*, 483–491.

29. Grzesik, M.; Romanowska-Duda, Z. Improvements in germination, growth, and metabolic activity of corn seedlings by grain conditioning and root application with cyanobacteria and microalgae. *Pol. J. Environ. Stud.* **2014**, *23*, 1147–1153.

30. Romanowska-Duda, Z.; Grzesik, M.; Kalaji, H.M. Phytotoxkit test in growth assessment of corn as an energy plant fertilized with sewage sludge. Environ. *Prot. Eng.* **2010**, *36*, 73–81.

31. Romanowska-Duda, Z.; Janas, R.; Grzesik, M. Application of Phytotoxkit in the quick assessment of ashes suitability as fertilizers in sorghum crops. *Inter. Agrophys.* **2019**, *33*, 145–152. [CrossRef]

32. Kordas, L.; Giemza-Mikoda, M.; Jabłońska, M. Evaluation of energy value of sorghum varieties depending on the time, sowing density and fertilization. *Fragm. Agron.* **2012**, *29*, 114–119.

33. Fahramand, M.; Moradi, H.; Noori, M.; Sobhkhizi, A.; Adibian, M.; Abdollahi, S.; Rigi, K. Influence of humic acid on increase yield of plants and soil properties. *Inter. J. Farm. Alli. Sci.* **2014**, *3*, 339–341.

34. Górnik, K.; Grzesik, M.; Romanowska–Duda, B.Z. The effect of chitosan on rooting of grapevine cuttings and on subsequent plant growth under drought and temperature stress. *J. Fruit Ornam. Plant Res.* **2008**, *16*, 333–343.

35. Piotrowski, K.; Romanowska-Duda , Z.; Grzesik, M. How biojodis and cyanobacteria alleviate the negative influence of predicted environmental constraints on growth and physiological activity of corn plants 2016. *Pol. J. Environ. Stud.* **2016**, *25*, 741–751. [CrossRef]

36. Jeznach, A. General Description of Organic and Mineral Fertilizer "Stymjod" Intended for Foliar Fertilization of Plants. 2015. Available online: http://www.phu-jeznach.com.pl/doc/Stymjod/Opis_preparatu_STYMJOD_2.pdf (accessed on 28 July 2018).

37. Smoleń, S.; Sady, W.; Rożek, S.; Ledwożyw-Smoleń, I.; Strzetelski, P. Preliminary evaluation of the influence of iodine and nitro gen fertilization on the effectiveness of iodine biofortification and mineral composition of carrot storage roots. *J. Elementol.* **2011**, *16*, 275–285.

38. Pszczółkowska, A.; Romanowska-Duda, Z.; Pszczółkowski, W.; Grzesik, M.; Wysokińska, Z. Biomass production of selected energy plants: Economic analysis and logistic strategies. *Comp. Econ. Res.* **2012**, *15*, 77–103.

39. Romanowska-Duda, Z.; Piotrowski, K.; Wolska, B.; Dębowski, M.; Zieliński, M.; Dziugan, P.; Szufa, S. *Stimulating Effect of Ash from Sorghum on the Growth of Lemnaceae—A New Source of Energy Biomass*; Springer International Publishing: New York, NY, USA, 2020; pp. 341–349.

40. Kołecka, K.; Gajewska, M.; Obarska-Pempkowiak, H. Integrated dewatering and stabilization system as the environmentally friendly technology in sewage sludge management in Poland. *Ecol. Eng.* **2017**, *98*, 346–353. [CrossRef]

41. Dzikuć, M.; Kuryło, P.; Dudziak, R.; Szufa, S.; Dzikuć, M.; Godzisz, K. Selected aspects of combustion optimization of coal in power plants. *Energies* **2020**, *13*, 2208. [CrossRef]

42. Szufa, S.; Wielgosiński, G.; Piersa, P.; Czerwińska, J.; Dzikuć, M.; Adrian, Ł.; Lewandowska, W.; Marczak, M. Torrefaction of straw from oats and maize for use as a fuel and additive to organic fertilizers—TGA analysis, kinetics as products for agricultural purposes. *Energies* **2020**, *13*, 2064. [CrossRef]

43. Jewiarz, M.; Wróbel, M.; Mudryk, K.; Szufa, S. Impact of the drying temperature and grinding technique on biomass grindability. *Energies* **2020**, *13*, 3392. [CrossRef]

44. Szufa, S.; Dzikuć, M.; Adrian, Ł.; Piersa, P.; Romanowska-Duda, Z.; Lewandowska, W.; Marczak, M.; Błaszczuk, A.; Piwowar, A. Torrefaction of oat straw to use as solid biofuel, an additive to organic fertilizers for agriculture purposes and activated carbon—TGA analysis, kinetics. In Proceedings of the 6th International Conference—Renewable Energy Sources, Krynica, Poland, 12–14 June 2019.

45. Szufa, S.; Adrian, Ł.; Piersa, P.; Romanowska-Duda, Z.; Ratajczyk-Szufa, J. Torrefaction process of millet and cane using batch reactor. In *Renewable Energy Sources: Engineering, Technology, Innovation, Springer Proceedings in Energy*; Wróbel, M., Jewiarz, M., Szlęk, A., Eds.; Springer Nature: Cham, Switzerland, 2019; pp. 371–379.

46. Szufa, S.; Adrian, Ł.; Piersa, P.; Romanowska-Duda, Z.; Grzesik, M.; Cebula, A.; Kowalczyk, S. Experimental studies on energy crops torrefaction process using batch reactor to estimate torrefaction temperature and residence time. In *Renewable Energy Sources: Engineering, Technology, Innovation*; Springer: Cham, Switzerland, 2018; pp. 365–373.

47. Szufa, S.; Romanowska-Duda, B.Z.; Grzesik, M. Torrefaction proces of the phragmites communis growing in soil contaminated with cadmium. In Proceedings of the 20th European Biomass Conference and Exibition, Milan, Italy, 18–22 June 2014; pp. 628–634.

48. Ławińska, K.; Szufa, S.; Obraniak, A.; Olejnik, T.; Siuda, R.; Kwiatek, J.; Ogrodowczyk, D. Disc granulation process of carbonation lime mud as a method of post-production waste management. *Energies* **2020**, *13*, 3419. [CrossRef]

49. Arias, B.; Pevida, C.; Fermoso, J.; Plaza, M.G.; Rubiera, F.; Pis, J.J. Influence of torrefaction on the grindability and reactivity of woody biomass. *Fuel Process. Technol.* **2008**, *89*, 169–175. [CrossRef]

50. Knypl, J.S.; Kabzińska, E. Growth, phosphatase and ribonuclease activity in phosphate deficient Spirodela oligorrhiza cultures. *Biochem. Physiol. Pflanz.* **1977**, *171*, 279–287. [CrossRef]

51. Górnik, K.; Grzesik, M. Effect of Asahi SL on China aster 'Aleksandra' seed yield, germination and some metabolic events. *Acta Physiol. Plant.* **2002**, *24*, 379–383. [CrossRef]

52. Szufa, S.; Piersa, P.; Czerwińska, J.; Adrian, Ł.; Dzikuć, M.; Lewandowska, W.; Wróbel, M.; Jewiarz, M.; Knapczyk, A. Drying and torrefaction processes of miscanthus for use as a pelletized solid biofuel and carrier to organic fertilizers—Kinetics, TGA-analysis and VOC-analysis. *Energies* **2020**, in press.

Sample Availability: Samples of the compounds of untreated and torrefied Jerusalem Artichoke plus their ashes are available from the authors.

molecules

Article

Effective Utilisation of Halophyte Biomass from Saline Soils for Biorefinering Processes

Jolanta Batog, Krzysztof Bujnowicz, Weronika Gieparda, Aleksandra Wawro * and Szymon Rojewski

Institute of Natural Fibres and Medicinal Plants—National Research Institute, Wojska Polskiego 71B, 60-630 Poznan, Poland; jolanta.batog@iwnirz.pl (J.B.); krzysztof.bujnowicz@iwnirz.pl (K.B.); weronika.gieparda@iwnirz.pl (W.G.); szymon.rojewski@iwnirz.pl (S.R.)
* Correspondence: aleksandra.wawro@iwnirz.pl; Tel.: +48-618455814

Abstract: The salinity of European soil is increasing every year, causing severe economic damage (estimated 1–3 million hectares in the enlarged EU). This study uses the biomass of halophytes—tall fescue (grass) and hemp of the Białobrzeskie variety from saline soils—for bioenergy, second generation biofuels and designing new materials—fillers for polymer composites. In the bioethanol obtaining process, in the first stage, the grass and hemp biomass were pretreated with 1.5% NaOH. Before and after the treatment, the chemical composition was determined and the FTIR spectra and SEM pictures were taken. Then, the process of simultaneous saccharification and fermentation (SSF) was carried out. The concentration of ethanol for both the grass and hemp biomass was approx. 7 g·L^{-1} (14 g·100 g^{-1} of raw material). In addition, trials of obtaining green composites with halophyte biomass using polymers (PP) and biopolymers (PLA) as a matrix were performed. The mechanical properties of the composites (tensile and flexural tests) were determined. It was found that the addition of a compatibilizer improved the adhesion at the interface of PP composites with a hemp filler. In conclusion, the grass and hemp biomass were found to be an interesting and promising source to be used for bioethanol and biocomposites production. The use of annually renewable plant biomass from saline soils for biorefinering processes opens up opportunities for the development of a new value chains and new approaches to sustainable agriculture.

Keywords: halophyte biomass; saline soils; pretreatment; SSF; biorefinering process; bioethanol; biocomposites

Citation: Batog, J.; Bujnowicz, K.; Gieparda, W.; Wawro, A.; Rojewski, S. Effective Utilisation of Halophyte Biomass from Saline Soils for Biorefinering Processes. *Molecules* **2021**, *26*, 5393. https://doi.org/10.3390/molecules26175393

Academic Editors: Alejandro Rodriguez Pascual, Eduardo Espinosa Víctor and Carlos Martín

Received: 2 August 2021
Accepted: 31 August 2021
Published: 5 September 2021

Publisher's Note: MDPI stays neutral with regard to jurisdictional claims in published maps and institutional affiliations.

1. Introduction

The dynamic development of economic activities often causes changes in the environment and human life quality. The world population is anticipated to grow 40% within 40–50 years with unprecedented demands for energy, food, freshwater and a clean environment. At 43% of the total landmass, exploiting the Earth's arid and semi-arid lands becomes a matter of necessity [1].

Soil salinisation effects are seen as a major cause of desertification and, therefore, are a serious form of soil degradation, endangering the potential use of European soils. The contamination of the environment by the excessive salinity of soils may be induced during coastal flooding, by the intensive agriculture practices or under the pressure of climate changes (aridisation phenomena). Such soils have lower or zero suitability for traditional agriculture production, influencing the decline of regional economics [2].

The remediation of these sites or the implementation of new ways to use them are necessary to achieve the restoration of agricultural production and thus economic growth. This can be met by the use of halophytes in the development of an integrated system of soil bioremediation and biomass biorefinery [3].

Tall fescue (*Festuca arundinacea*) and hemp (*Cannabis sativa* L.) are characterised by high biomass production and wide ecological amplitude, including tolerance to soil salinity.

Tall fescue creates large, dense and deeply rooted clumps, showing considerable tolerance to unfavourable habitat conditions. It occurs on roadsides, meadows (including saline meadows) and alluvia.

Hemp is a plant with a short growing season, resistant to diseases and unfavourable environmental conditions, with a dry matter yield of 10–15 t·ha^{-1}. They improve soil quality and are useful for the restoration of brownfield sites; 1 ha of hemp binds about 11 t of CO_2 [4–6].

This study addresses the challenge of improving the knowledge and understanding of halophyte species in order to better use the biomass from these plants for the production of new environmentally friendly materials [7].

EU Member States have been obliged to achieve a certain share of biofuels in transport and to take measures to reduce greenhouse gas emissions. According to the EU RED II Directive, the contribution of advanced biofuels and biogas produced, among others, from lignocellulosic raw materials, as a share of final energy consumption in the transport sector, it is expected to amount to at least 0.2% by 2022, 1% by 2025 and 3.5% by 2030 [8]. Therefore, it is very interesting and important to try to obtain bioethanol from the halophyte biomass from saline soils [3].

The production of lignocellulosic ethanol involves the breakdown of cell walls into individual polymers and the hydrolysis of carbohydrates into simple sugars. Plant biomass contains the lignocellulose complex (cellulose, hemicellulose, lignin), which is relatively resistant to biodegradation. The process of converting halophyte biomass to bioethanol involves several steps, from the preparation of the plant material (physical and chemical treatment), through enzymatic hydrolysis, to ethanol fermentation. The purpose of the pretreatment of biomass is to crush the solid phase and loosen the compact structure of the lignocellulose. Moreover, the method combining cellulose hydrolysis with sugar fermentation in one bioreactor (SSF process), where enzymes must be adapted to the conditions of the fermentation process (30–40 °C), seems to be effective and economical [9–11].

Another important point is that around 25 million tons of plastic waste are produced annually in the EU. The market is dominated by products made from materials such as PE, PP, PET, PS and PVC, that take several hundred years to decompose. More and more emphasis is placed on the need to significantly reduce the amount of plastic waste (Single-Use Plastics Directive) [12,13].

The plastics processing industry is looking for material innovations that meet customer expectations and legal requirements; therefore, in the next few years, an increased interest in environmentally friendly biodegradable materials is expected. An example is the production of polylactic acid (PLA), which is one of the leading bioplastics on the market [14–16].

The barrier to the wider use of biodegradable plastics is the high production costs, higher than traditionally used plastics.

A method to improve the economics may be the use of natural fillers from the biomass of annual plants (flax and hemp biomass), which are cheaper than biodegradable plastics [17,18]. Materials composed of thermoplastic polymers with natural fillers have functional properties that do not differ from pure polymers and may be used in various fields of the economy, including the automotive industry, transport, construction and the furniture industry.

Due to the expected increase in the production of biocomposite materials in the EU in the coming years, an increase in the demand for biocomponents based on natural resources, including natural fibres, should be expected.

Moreover, to fully understand the impacts of biorefineries, taking a comprehensive view of sustainability, it is necessary to consider a wider range of factors including economics and social and environmental issues. The potential benefits of strengthening the halophyte based biorefineries are certainly reduced greenhouse gas emissions, less dependence on fossil resources, better land and natural resource management and improved food security and soil quality. In addition, a significant positive effect of biobased industries is the creation of employment opportunities in rural areas, as well as the possibility of creating new markets for agriculture (biofuels, green composites) in synergy with the existing ones.

The aim of the presented study is to indicate the possibility of using tall fescue and hemp from saline soils in the process of obtaining second generation biofuels and for the design of green biocomposites. Thus far, only a few reports of the use of halophyte biomass from degraded areas in biorefining processes have been published [3,19,20]. A novelty in this manuscript is an indication of the possibility of managing degraded, saline soils through the production of biomass for industrial purposes, including bioenergy, which is in line with the idea of sustainable development, that is now a priority policy of the European Union.

2. Results and Discussion

2.1. Bioethanol Production Process

2.1.1. Halophyte Biomass Pretreatment

The biomass crushed on a knife mill with a mesh of 2 mm was treated with alkaline. One of the most popular alkaline reagents used in the treatment of raw plant material is sodium hydroxide. The main purpose of the pretreatment of lignocellulosic biomass is to crush the solid phase and loosen the compact lignocellulose structure. After a chemical pretreatment test with sodium hydroxide (1.5–3%, 90 °C), sulfuric acid (1–2%, 121 °C) and hot water (135 °C), it turned out that the most effective treatment was alkaline with 1.5% NaOH. Table 1 shows the alkaline treatment of the grass and hemp biomass from saline soils using 1.5% NaOH.

Table 1. Content of reducing sugars before (BP) and after (AP) sodium hydroxide treatment.

Halophyte	Sample	Reducing Sugars (mg·g^{-1})
Grass	BP	100.20 ± 0.09
	AP	354.59 ± 0.01
Hemp	BP	62.95 ± 0.10
	AP	187.95 ± 0.13

It was found that grass from the saline soil was characterised by an almost two times higher content of reducing sugars than hemp. The results indicate that grass is more susceptible to alkaline treatment than hemp.

An effective chemical treatment should ensure the destruction of the crystalline structure of cellulose, the separation of lignin from carbohydrates and, thus, an increase in the availability of the substrate for further biofuel production processes [21].

To confirm the efficiency of the alkaline treatment, a determination of the chemical composition of the halophyte biomass after the NaOH treatment was performed and compared to the chemical composition of the biomass before the pretreatment. The results are presented in Table 2.

Table 2. Chemical composition of halophyte biomass (% of dry matter); BP: before pretreatment; AP: after pretreatment.

Halophyte	Sample	Cellulose (%)	Hemicellulose (%)	Lignin (%)
Grass	BP	33.69 ± 0.40	34.74 ± 0.39	17.08 ± 0.16
	AP	50.41 ± 0.18	25.23 ± 0.37	12.35 ± 0.07
Hemp	BP	47.34 ± 0.40	33.49 ± 0.68	13.94 ± 0.05
	AP	58.46 ± 0.29	22.12 ± 0.13	17.35 ± 0.26

The analysis of the chemical composition of the halophyte biomass before and after the treatment showed that the alkaline effect caused a visible increase in the cellulose content by over 10%, especially for grass, which saw a 17% increase. In addition, a partial degradation of hemicellulose was found (as much as 11% for hemp). In the case of the lignin content for the hemp biomass, an increase was observed after the alkaline pretreatment (over 3%), and in the case of grass—an almost 5% reduction. Similar observations regarding the hemp biomass were presented by Stevulova et al. [22], who showed that the lignin content after pretreatment with sodium hydroxide was 7% higher than before.

It was found that in the case of the grass biomass, both the higher cellulose growth and the reduction in lignin content positively influenced the higher values of the reducing sugars released after the pretreatment with sodium hydroxide compared to the hemp biomass (see Table 1). It should be emphasised that one of the main goals of the chemical treatment of lignocellulosic materials is the removal of lignin, which is a strong obstacle in the process of biomass conversion.

The effect of alkaline treatment on the grass and hemp biomass from saline soils was confirmed using Fourier Transform Infrared Spectrometer (FTIR) between 600–4000 cm^{-1} shown in the Figure 1a,b and using Scanning Electron Microscopy (SEM).

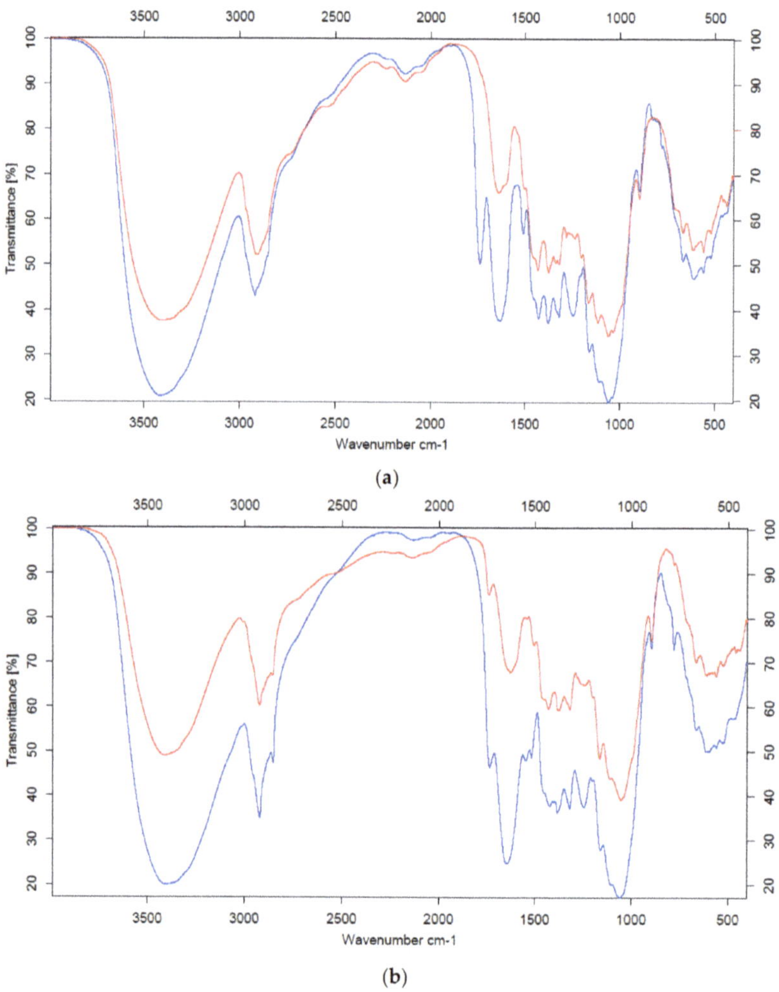

Figure 1. (**a**) FTIR spectra of hemp biomass from saline soil before (blue) and after (red) NaOH pretreatment. (**b**) FTIR spectra of grass biomass from saline soil before (blue) and after (red) NaOH pretreatment.

The broad band, ranging from about 3500 to 3000 cm^{-1}, comes from the O–H stretching vibrations in the cellulose molecule. In all the cases, the band after pretreatment is less intense and narrower. The band at 2900 cm^{-1}, resulting from the stretching vibrations of the C–H group (cellulose), is also less intense after pretreatment. The decrease in the intensity of these bands, despite the increase in the cellulose content after treatment (confirmed by results from the chemical composition), may be caused by the reduction in the crystalline structure of cellulose [23,24]. These bands, in the case of different materials, have a different shape (for grass, two characteristic peaks appear in this place, while for hemp, one broad band is visible). The vibration band visible at 1730 cm^{-1}, resulting from the C=O stretching vibrations of the acetyl group in hemicellulose [25], disappears in the case of hemp after treatment, while in the case of grass, it is significantly reduced. This is due to the degradation of hemicellulose during the alkaline pretreatment of biomass. In turn, the lower intensity of the band at 1600 cm^{-1} corresponding to the O–H stretching vibrations, reflecting the amount of water absorbed in the sample, is probably caused by the loss of water during the drying process of the samples after pretreatment [26]. The absorption band at 1510 cm^{-1}, resulting from the vibrations of the aromatic ring in lignin in the case of grass, decreases after alkaline treatment, which is reflected in the chemical composition of the biomass. Significant changes are visible in the band at 1230 cm^{-1}. This band is attributed to the vibration of the guaiacyl ring in lignin, as well as to the vibration of the C–O groups in pectin. After the alkaline treatment, this band disappears completely or significantly loses its intensity, which may indicate a reduction in both lignin and pectin. Going to the lower wavenumber values, the following three more characteristic bands for the cellulose molecule appear: 1160 cm^{-1} (asymmetric C–O–C stretching vibrations), 1110 cm^{-1} (C–OH skeletal vibrations) and 1050 cm^{-1} (C–O–C skeletal vibrations of the pyranose ring) [27]. These bands, despite the increase in cellulose content after chemical treatment, are reduced. This is characteristic of each biomass tested and can be attributed to the reduction in the crystalline structure of the cellulose after treatment.

Significant changes on the surface of the grass and hemp biomass were observed and presented in the SEM images taken before and after pretreatment (Figure 2).

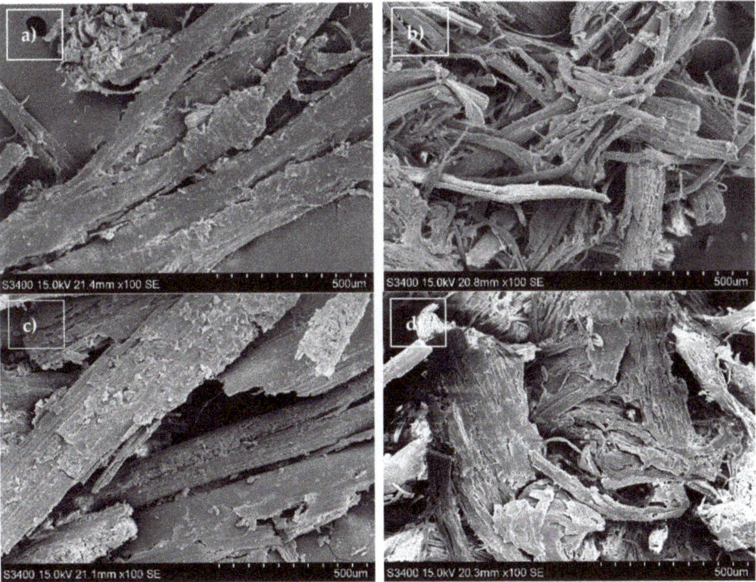

Figure 2. SEM images of halophyte biomass: (**a**) hemp biomass before NaOH pretreatment, (**b**) hemp biomass after NaOH pretreatment, (**c**) grass biomass before NaOH pretreatment, (**d**) grass biomass after NaOH pretreatment.

Figure 2a,c show that the untreated grass and hemp biomass have intact and rigid structures with well-ordered fibrous skeletons, effectively blocking access to lignocellulose [28]. After the pretreatment of the grass and hemp biomass with sodium hydroxide for, similar changes were observed on the surface of the biomass. The SEM images of the halophyte biomass after alkaline treatment (Figure 2b,d) show damage to the structure of the biomass and the appearance of hollow areas, which increases its surface and positively affects the enzymatic availability and digestibility of the biomass [29,30].

To sum up, the main goal of the chemical treatment of lignocellulosic materials for biofuel is to increase the availability of the biomass structure by the decrystallisation of cellulose and the removal of lignin.

2.1.2. Enzyme Complex

The second stage of the process of obtaining bioethanol from plant biomass is the enzymatic hydrolysis process. The breakdown of cellulose into simple sugars requires the synergistic action of cellulases—endoglucanases, cellobiohydrolases and ß-glucosidases. Two enzyme preparations were selected for the research—Flashzyme Plus 200 and Celluclast 1.5 L [31]. In order to select the enzyme complex for the SSF process, tests were performed using selected enzymes and their supplementation with glucosidase and xylanase (Table 3).

Table 3. Content of reducing sugars after the enzymatic test.

Enzyme	Reducing Sugar (mg·g^{-1})	
	Hemp	**Grass**
Flashzyme Plus 200	338 ± 0.04	846 ± 1.00
Celluclast 1.5 L	342 ± 0.05	696 ± 0.52
Flashzyme/Celluclast 1.5 L (70/30)	420 ± 0.06	892 ± 0.02
Flashzyme/Celluclast 1.5 L (50/50)	430 ± 0.05	800 ± 0.44
Flashzyme/Celluclast 1.5 L (30/70)	355 ± 0.38	810 ± 0.34
Flashzyme/Celluclast 1.5 L (50/50)/β-glucosidase	351 ± 0.14	-
Flashzyme/Celluclast 1.5 L (50/50)/xylanase	324 ± 0.65	-
Flashzyme/Celluclast 1.5 L (50/50)/β-glucosidase/xylanase	343 ± 0.16	-
Flashzyme/Celluclast 1.5 L (70/30)/β-glucosidase	-	472 ± 3.29
Flashzyme/Celluclast 1.5 L (70/30)/xylanase	-	458 ± 1.81
Flashzyme/Celluclast 1.5 L (70/30)/β-glucosidase/xylanase	-	735 ± 1.86

For the hydrolysis of the solid fraction in the SSF process, the Flashzyme Plus 200/Celluclast 1.5 L complex in the proportion of 50/50 was selected for the hemp biomass, and the Flashzyme Plus 200/Celluclast 1.5 L complex in the proportion of 70/30 for the grass biomass. On the basis of the enzymatic test, it was found that the content of the released reducing sugars for the grass biomass (892 mg·g^{-1}) was two times higher than for the hemp biomass (430 mg·g^{-1}).

2.1.3. Simultaneous Saccharification and Fermentation (SSF)

The SSF process, consisting of simultaneous hydrolysis and fermentation, takes place under conditions ensuring the optimal synergy of enzymes and distillery yeast. After carrying out the fermentation tests with selected parameters, the amount of ethanol was determined (HPLC). Figure 3 shows the ethanol concentration after the SSF process for the hemp and grass biomass from saline soils.

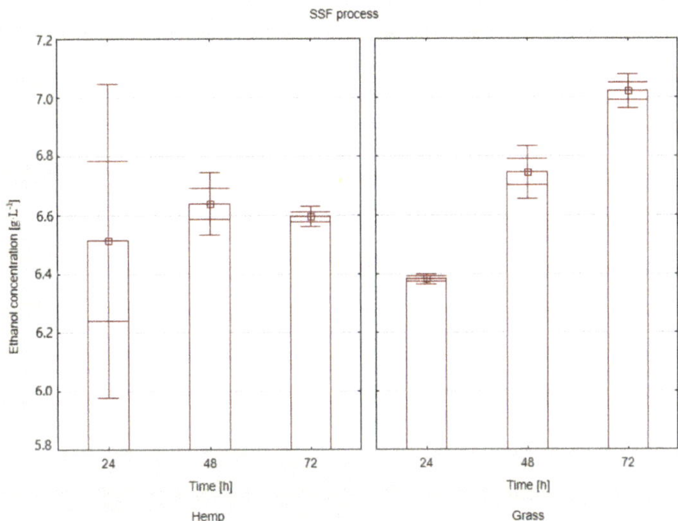

Figure 3. Ethanol concentration of hemp and grass biomass after the SSF process (Statistica 13.0).

It was noted that at 48 h, there was a slight decrease in the ethanol concentration for the hemp biomass but an increase for the grass biomass. The highest ethanol concentration for the hemp biomass was observed at 48 h and it was 6.6 g·L^{-1}, which is 13.3 g·100 g^{-1} of raw material. In turn, for the grass biomass, the highest ethanol concentration, equal to 7.0 g·L^{-1} (14.0 g·100 g^{-1} of raw material), was recorded after 72 h.

According to the literature reports, SSF is a beneficial process due to its short processing time, small reactor volume and high ethanol yield, as bioethanol is produced immediately with glucose conversion [32].

Taufikurahman and Sherly [33] conducted similar research on the production of bioethanol from the biomass of Napier grass. After 96 h of enzymatic hydrolysis and fermentation, they obtained an ethanol concentration of 1.25 g·L^{-1}. In turn, Riadi et al. [34] conducted research on lignocellulosic biomass in the form of sugarcane bagasse. They first optimised the alkaline pretreatment, followed by enzymatic hydrolysis and ethanol fermentation. After the 48-h fermentation process, they obtained an ethanol concentration of 5.84 g·L^{-1}. Obtaining bioethanol from hemp biomass was carried out by Orlygsson [35] and after the pretreatment with 0.5% NaOH and SSF process, the ethanol concentration was approx. 1 g·L^{-1}.

Summing up, it should be emphasised that both the hemp and grass biomass obtained from crops on saline soils have great potential in the process of obtaining bioethanol.

2.2. Biocomposite Production Process

2.2.1. Fillers from Halophyte Biomass

Natural fillers with particles less than 1 mm were obtained from the grinding of the halophyte biomass. It was assumed that the fraction below 1 mm would be a compromise, allowing for the effective processing of the obtained biocomposites in the injection moulding process, taking into account the diameters of the most commonly used injection nozzles, the simplification of the biomass grinding process and the lower costs of preparing natural fillers. The sieve analysis of the natural fillers was performed, and their humidity was determined. The detailed share of the individual fractions is presented in Table 4.

Table 4. Particle size distribution and humidity of fillers from halophyte biomass.

Plant Biomass	Humidity (%)	Particle Size Distribution (%)						
		1 mm	0.5 mm	0.4 mm	0.25 mm	0.2 mm	0.1 mm	Below 0.1 mm
Grass	8.73	1.2	48.3	8.3	30.5	2.5	2.7	6.5
Hemp	7.65	1.1	53.8	14.7	15.8	8.5	2.6	3.5

The bulk density of the hemp fillers (0.19 g·mL^{-1}) and the grass fillers (0.22 g·mL^{-1}) were determined.

2.2.2. Mechanical Properties of Biocomposites

The multipurpose test specimens—type A in accordance with ISO 3167 [36], used to determine the mechanical properties of composites and biocomposites with halophyte biomass—are shown in Figure 4.

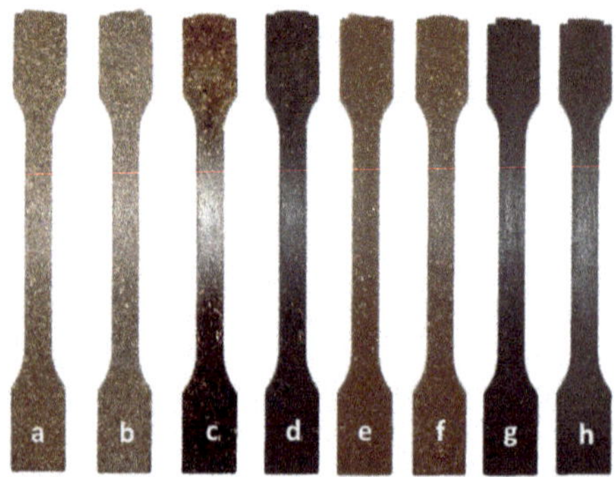

Figure 4. Test specimens of composites and biocomposites with halophyte biomass: (**a**) PP/Hemp20, (**b**) PP/Hemp30, (**c**) PP/Grass20, (**d**) PP/Grass30, (**e**) PLA/Hemp20, (**f**) PLA/Hemp30, (**g**) PLA/Grass20, (**h**) PLA/Grass30.

Tables 5 and 6 show the effect of the different fillers from halophyte biomass on both the tensile and flexural properties of the polypropylene composites.

Table 5. Tensile and flexular properties of PP/hemp composites.

Sample	Tensile Strength δ_M (MPa)	Tensile Modulus E_t (GPa)	Flexural Strength δ_{fM} (MPa)	Flexural Modulus E_f (GPa)
PP HP648T	31.0 ± 0.27	1.5 ± 0.02	41.5 ± 0.48	1.2 ± 0.13
PP-H20	23.3 ± 0.25	1.7 ± 0.04	39.5 ± 0.60	1.8 ± 0.17
PP-H20S5	27.5 ± 0.57	1.8 ± 0.05	40.7 ± 0.75	2.7 ± 0.16
PP-H30	19.3 ± 0.21	2.3 ± 0.03	37.0 ± 0.70	2.5 ± 0.22
PP-H30S5	31.4 ± 0.21	2.8 ± 0.05	51.6 ± 0.77	3.1 ± 0.14

Table 6. Tensile and flexular properties of PP/grass composites.

Sample	Tensile Strength δ_M (MPa)	Tensile Modulus E_t (GPa)	Flexural Strength δ_{fM} (MPa)	Flexural Modulus E_f (GPa)
PP HP648T	31.0 ± 0.27	1.5 ± 0.02	41.5 ± 0.48	1.2 ± 0.13
PP-G20	23.6 ± 0.19	1.7 ± 0.04	39.5 ± 0.27	1.3 ± 0.08
PP-G20S5	28.1 ± 0.25	1.8 ± 0.07	40.3 ± 0.35	1.4 ± 0.09
PP-G30	20.4 ± 0.14	2.0 ± 0.03	41.0 ± 0.40	1.9 ± 0.17
PP-G30S5	24.4 ± 0.27	2.0 ± 0.05	44.4 ± 0.27	1.9 ± 0.11

The 24–38% reduction in the tensile strength, respectively, was demonstrated for the composites based on PP with hemp fillers in the amount of 20% and 30%. The use of 5%

interfacial adhesion promoter (Scona 8112) indicates a positive effect on material adhesion, which ensures more effective mechanical properties. In the presence of a compatibilizer, the PP composites with 30% of the hemp filler showed better mechanical properties: the tensile strength is the same as that of pure PP, and the flexural strength is 24% higher.

Depending on the amount of hemp filler in the presence of the compatibilizer, the composites showed a significant increase in the tensile modulus during elongation (20–87%) and bending (17–58%).

The use of grass biomass fillers in the amount of 20 and 30% in composites based on polypropylene resulted in the reduction in tensile strength by 24 and 34%, respectively. The use of the 5% interfacial adhesion promoter (Scona 8112) improved the tensile strength of the composites by an average of 14%. However, the composites show a higher (13–33%) modulus of elasticity in relation to the pure PP. The use of grass fillers in the amount of 20 and 30% did not significantly affect the flexural strength of the composites. The modulus of elasticity determined during bending increased by almost 58% when the filler was used in the amount of 30%.

The effect of the different fillers from the halophyte biomass on both the tensile and flexural properties of the PLA composites are shown in Tables 7 and 8.

Table 7. Tensile and flexular properties of PLA/hemp composites.

Sample	Tensile Strength δ_M (MPa)	Tensile Modulus E_t (GPa)	Flexular Strength δ_{fM} (MPa)	Flexural Modulus E_f (GPa)
PLA 3260HP	64.5 ± 1.25	3.5 ± 0.07	108.6 ± 0.99	3.4 ± 0.14
PLA-H20	51.4 ± 0.99	5.5 ± 0.04	88.0 ± 2.37	4.9 ± 0.21
PLA-H30	53.0 ± 1.17	6.7 ± 0.05	94.7 ± 1.76	6.6 ± 0.18

Table 8. Tensile and flexular properties of PLA/grass composites.

Sample	Tensile Strength δ_M (MPa)	Tensile Modulus E_t (GPa)	Flexular Strength δ_{fM} (MPa)	Flexural Modulus E_f (GPa)
PLA 3260HP	64.5 ± 1.25	3.5 ± 0.07	108.6 ± 0.99	3.4 ± 0.14
PLA-G20	49.1 ± 0.30	3.9 ± 0.02	83.6 ± 0.43	3.8 ± 0.11
PLA-G30	41.2 ± 1.22	4.1 ± 0.02	78.1 ± 0.93	4.0 ± 0.20

The use of hemp fillers in the amount of 20 and 30% in composites based on PLA–Ingeo 3260HP resulted in a reduction in tensile strength by about 20 and 18%, respectively. The reduction in the flexural strength by 13–19% was demonstrated by the composites of PLA with hemp fillers. Depending on the amount of the hemp filler, the composites showed a significant increase in the modulus during elongation (57–91%) and bending (44–94%) in relation to pure PLA.

The use of grass biomass fillers in the amount of 20 and 30% in composites based on PLA–Ingeo 3260HP resulted in a reduction in tensile strength by about 24 and 36%, respectively. The composites show, respectively, an 11 and 17% increase in the modulus of elasticity in relation to pure PLA. The flexural strength of the composites decreased by about 28% for the 30% filler content. The modulus of elasticity determined during bending increased by 12 and 18% in relation to pure PLA.

It should be noted that the interaction between the matrix and the filler is an important factor influencing the mechanical properties. Despite the hydrophilic nature of the polylactide polymer matrix surface, similar to natural fillers, PLA composites with hemp and grass fillers do not show effective interface adhesion. The interfacial bonding strength of the natural fillers with the polymer matrix were lower, and at a lower value of the force acting on the samples, the fillers were detached from the matrix, and the entire load acted on the polymer matrix [37,38].

Moreover, it was found that the value of the reduction in tensile strength can be correlated with the content of natural fillers. For composites with 20% by weight of hemp filler, the reduction in tensile strength was 24.8% for PP composites and 20% for

PLA composites, respectively. For PP composites with 30% by weight of hemp filler, the reduction in tensile strength was 38%. For composites containing 20 and 30% by weight of grass filler, the reduction in tensile strength was 24% and 34–36% for PP and PLA composites, respectively.

Summing up, it should be emphasised that the addition of a compatibilizer improved the adhesion at the interface of the PP composites with the hemp filler, which is confirmed by the better mechanical properties of composites in terms of elongation and bending.

3. Materials and Methods

3.1. Halophyte Biomass

The raw material used in the study was the biomass of grass—tall fescue and hemp of the Białobrzeskie variety from a saline field in the Kuyavian-Pomeranian Voivodeship (Poland), the average salinity of which, calculated as sodium chloride, was $3.5\ g \cdot L^{-1}$.

3.2. Bioethanol Production Process

3.2.1. Halophyte Biomass Pretreatment

Grass and hemp biomass was subjected to preliminary crushing to particles of size 20–40 mm and dried at 50–55 °C for 24 h. Then, the material was disintegrated on knife mill SM-200 (Retsch, Hann, Germany) with a sieve of the mesh size of 2 mm.

The next step was the alkaline treatment of the halophyte biomass for 5 h with 1.5% sodium hydroxide at 90 °C [39]. NaOH/biomass weight ratio was 10:1. After the alkaline pretreatment was carried out, the biomass solution was filtered on a Büchner funnel, then washed with distilled water until neutralised, and dried in a laboratory dryer at 50 °C for 24 h. The alkali effect on the content of the released reducing sugar was determined using Miller's method with 3,5-dinitrosalicylic acid (DNS) [40]. The raw material was incubated at 40 °C in 0.05 M citrate buffer pH 4.8 for 2 h using the enzyme preparation Flashzyme Plus 200 (AB Enzyme) at the dose of 20 $FPU \cdot g^{-1}$. The absorbance of the supernatant was measured at 530 nm on UV–VIS Spectrophotometer V-630, (Jasco, Pfungstadt, Germany).

3.2.2. Enzyme Complex

In order to select the enzyme complex for the SSF process, tests were performed using selected enzymes—Flashzyme Plus 200 and Celluclast 1.5 L (Novozymes, Bagsværd, Denmark) and their supplementation with glucosidase 20 $CBU \cdot g^{-1}$ and xylanase 500 $XU \cdot g^{-1}$ (Sigma-Aldrich, Darmstadt, Germany).

The composition of Flashzyme Plus 200 (90 $FPU \cdot mL^{-1}$, 2430 $XU \cdot mL^{-1}$) is endoglucanase, cellobiohydrolase, cellobiase, xylanase and mannanase, and Celluclast 1.5 L (62 $FPU \cdot mL^{-1}$, 278 $XU \cdot mL^{-1}$) consists of cellulase from *Trichoderma reesei*.

Enzymatic tests were carried out for 5% of biomass with the enzyme in the amount of 10 $FPU \cdot g^{-1}$, at pH 4.8 and for 24 h at 38 °C. The selection criterion was the content of reducing sugars determined using the Miller's method.

3.2.3. Simultaneous Saccharification and Fermentation (SSF)

The SSF process was carried out in bioreactor Biostat B Plus (Sartorius, Goettingen, Germany) in 2-litre vessel equipped with pH, temperature, stirring and foaming controls. The temperature was maintained at 37 °C and stirred at 900 rpm; pH was controlled at 4.8 by adding 1 M NaOH or 1 M HCl. The mixture of Flashzyme Plus 200 and Celluclast 1.5 L enzymes in the amount of 10 $FPU \cdot g^{-1}$ was used for the hydrolysis process of biomass. The fermentation process was performed with the use of not hydrated, freeze-dried yeast *S. cerevisiae* at a dose of 1 $g \cdot L^{-1}$, which corresponded to cell concentration after inoculation of about 1×10^7 $cfu \cdot mL^{-1}$. Duration of the SSF process was 24, 48 and 72 h.

Ethanol yield from 100 g of raw material Y_s ($g \cdot 100\ g^{-1}$ of raw material) was calculated according to the following Equation (1) [41]:

$$Y_s = \frac{Et \times 100}{M} \tag{1}$$

where *Et*—amount of ethanol in 1000 mL of tested sample (g); and *M*—mass of material weighed in 1000 mL fermentation sample (g).

3.3. Biocomposite Production Process

3.3.1. Natural Fillers from Halophyte Biomass

Samples of dried halophyte biomass were grounded using a Rekord A (Jehmlich, Nossen, Germany) mill with a sieve separator with a mesh diameter of 1 mm. The sieve analysis of the natural fillers obtained in the milling process was performed using Analysette 3 Spartan (Fritsch, Idar-Oberstein, Germany) and their humidity were determined with moisture analyser MA.X2.A (Radwag, Radom, Poland).

3.3.2. Polymer Matrix

The polymer matrix consisted of traditional thermoplastic polymer—polypropylene Moplen HP648T (Basell Orlen Polyolefins, Płock, Poland): density 0.9 $g \cdot mL^{-1}$ and mass flow rate (MFR) 53 $g \cdot 10 \, min^{-1}$ (190 °C, 2.16 kg) and biodegradable polymer—poly(lactic acid) Ingeo 3260HP (NatureWorks, Blair, NE, USA): density 1.24 $g \cdot mL^{-1}$ and MFR 65 $g \cdot 10 \, min^{-1}$ (210 °C, 2.16 kg).

To improve adhesion between the natural filler and the polypropylene matrix, the interfacial adhesion promoter Scona 8112 (S) (BYK-Chemie GmbH, Germany)—PP grafted with maleic anhydride—was used in an amount of 5% by weight.

3.3.3. Preparation of Composites

Composites and biocomposites contained 20 and 30 wt% hemp (H) and grass (G) fillers were compounded with polypropylene (PP) and polylactide (PLA) in co-rotating twin screw extruder Leistritz MICRO 27 GL/GG-44D (Leistritz Extrusionstechnik, Nürnberg, Germany) with Brabender gravimetric feeding system (Brabender Technologie, Duisburg, Germany). Compounding parameters: barrel temperature profile 170–200 °C, extruder rotation speed of 150 rpm, throughput 16 $kg \cdot h^{-1}$.

Composites pellets were dried to achieve the appropriate process humidity. Polypropylene composites were dried in flow dryer EHD-25BT (Enmair Automation Machinery, Guangdong, China) at temperatures of 105 °C to a humidity level below 0.15%. Biocomposites based on biodegradable polymer were dried in molecular dehumidifier Drywell DW25/40 (Digicolor, Herford, Germany) at 70 °C (dew point −40 °C) to a humidity level below 0.02%.

Multipurpose test specimens—type A in accordance with ISO 3167 [36]—were moulded by hydraulic injection moulding machine Haitian Mars II Eco 600 kN (Haitian Plastics Machinery, Zhejiang, China). Barrel temperature profile: 180 °C (hopper), 185 °C, 190 °C and 190 °C (nozzle). Mould temperature was set at 40 °C.

3.4. Analytical and Testing Methods

The chemical composition of halophyte biomass before and after pretreatment was determined, i.e., cellulose acc. to TAPPI T17 m-55 [42], hemicellulose as the difference holocellulose acc. to TAPPI T9 m-54 [43] and cellulose, and lignin acc. to TAPPI T13 m-54 [44].

In order to provide a more complete picture of the molecular structure of grass and hemp biomass before and after the alkaline pretreatment, the analysis of FTIR spectroscopy was performed using a Fourier Transform Infrared Spectrometer ISS 66v/S (Bruker, Bremen, Germany) at wavenumbers of 400–4000 cm^{-1} [22].

The physical morphologies of halophyte biomass before and after the chemical treatment were performed by using Scanning Electron Microscope S-3400N (Hitachi, Japan) in high vacuum conditions. The samples were covered with gold dust.

The content of ethanol was determined using High Performance Liquid Chromatography on Elite LaChrom (Hitachi, Tokio, Japan) using an RI L-2490 detector, Rezex ROA

300×7.80 mm column (Phenomenex, Torrance, CA, USA), as the mobile phase used $0.005N\ H_2SO_4$ at a flow rate of 0.6 mL·min^{-1}, at 40 °C.

Composites' tensile and flexural tests were carried out at room temperature with a universal testing machine Inspekt Table 50 (Hegewald & Peschke MPT, Nossen, Germany) as recommended by ISO 527 [45,46] and ISO 178 [47], respectively. A crosshead speed was set to 5 mm·min^{-1} in both tests. Tensile tests were performed using an MFA clip-on extensometer (Mess- & Feinwerktechnik, Velbert, Germany) with a nominal length of 50 mm.

3.5. Statistical Analysis

The experiments of ethanol fermentation were carried out in triplicates. Standard deviations were calculated using the analysis of variance ANOVA, Statistica 13.0 software ($p < 0.05$).

4. Conclusions

This study took on the challenge of broadening the knowledge and understanding of the halophyte species to achieve better use of the biomass from these plants. To sum, the biomass of both the tall fescue and the hemp of the Białobrzeskie variety from saline areas turned out to be a suitable source for the production of second-generation bioethanol and a natural filler for polymer composites based on traditional polymers (polypropylene) and biodegradable polymers (polylactide).

The SSF process made it possible to obtain an ethanol concentration for grass and hemp biomass at the level of approx. 7 g·L^{-1} (14 g·100 g^{-1} of raw material). In the case of composites, studies have shown that despite lowering the mechanical strength in terms of stretching and bending, it is possible to improve interfacial adhesion by modifying the compatibility between the phases of the natural filler and the polymer matrix.

This study will certainly have a positive impact on raising public opinion about the social and economic benefits of the bioproducts obtained from biomass sources using land unsuitable for agriculture.

Author Contributions: Conceptualisation, J.B. and K.B.; methodology, J.B., K.B., W.G., A.W. and S.R.; software, A.W. and W.G.; validation, J.B. and K.B.; formal analysis, J.B. and K.B.; investigation, K.B., W.G., A.W. and S.R.; resources, J.B., K.B., W.G., A.W. and S.R.; data curation, J.B., K.B., W.G., A.W. and S.R.; writing—original draft preparation, J.B. and K.B.; writing—review and editing, J.B., K.B., W.G. and A.W.; visualisation, J.B.; supervision, J.B.; project administration, J.B.; funding acquisition, J.B. All authors have read and agreed to the published version of the manuscript.

Funding: This research was funded by National Centre for Research and Development, Poland, grant ERA-NET CO-FOUND FACCE SURPLUS 2 (2018-2021).

Institutional Review Board Statement: Not applicable.

Informed Consent Statement: Not applicable.

Data Availability Statement: Not applicable.

Acknowledgments: The study was conducted as research project ERA-NET CO-FOUND FACCE SURPLUS 2 (2018–2021): Integrated system of bioremediation—biorefinering using halophyte species, and financed by National Centre for Research and Development, Poland.

Conflicts of Interest: The authors declare no conflict of interest.

Sample Availability: Samples of the compounds are not available from the authors.

References

1. Hendricks, R.C.; Bushnell, D.M. Atmospheric and soil carbon and halophytes. In Proceedings of the 13th International Symposium on Transport Phenomena and Dynamics of Rotating Machinery (ISROMAC-13), Honolulu, HI, USA, 4–7 April 2010; p. 113.
2. Shrivastava, P.; Kumar, R. Soil salinity: A serious environmental issue and plant growth promoting bacteria as one of the tools for its alleviation. *Saudi J. Biol. Sci.* **2015**, *22*, 123–131. [CrossRef]
3. Hameed, A.; Khan, M.A. Halophytes: Biology and economic potentials. *Karachi Univ. J. Sci.* **2011**, *39*, 40–44.

4. Zadrożniak, B.; Radwańska, K.; Baranowska, A.; Mystkowska, I. Possibility of industrial hemp cultivation in areas of high nature value. *Econ. Reg. Stud.* **2017**, *10*, 114–127. [CrossRef]

5. Wawro, A.; Batog, J.; Gieparda, W. Chemical and enzymatic treatment of hemp biomass for bioethanol production. *Appl. Sci.* **2019**, *9*, 5348. [CrossRef]

6. Hu, H.; Liu, H.; Liu, F. Seed germination of hemp (*Cannabis sativa* L.) cultivars responds differently to the stress of salt type and concentration. *Ind. Crop. Prod.* **2018**, *123*, 254–261. [CrossRef]

7. Mohapatra, S.; Mishra, C.; Behera, S.S.; Thatoi, H. Application of pretreatment, fermentation and molecular techniques for enhancing bioethanol production from grass biomass—A review. *Renew. Sustain. Energy Rev.* **2017**, *78*, 1007–1032. [CrossRef]

8. Available online: https://eur-lex.europa.eu/legal-content/EN/ALL/?uri=CELEX:32009L0028 (accessed on 8 April 2021).

9. Abo, B.O.; Gao, M.; Wang, Y.; Wu, C.; Ma, H.; Wang, Q. Lignocellulosic biomass for bioethanol: An overview on pretreatment, hydrolysis and fermentation processes. *Rev. Environ. Health* **2019**, *34*, 57–68. [CrossRef]

10. Rahmati, S.; Doherty, W.; Dubal, D.; Atanda, L.; Moghaddam, L.; Sonar, P.; Hessel, V.; Ostrikov, K. Pretreatment and fermentation of lignocellulosic biomass: Reaction mechanisms and process engineering. *React. Chem. Eng.* **2020**, *5*, 2017–2047. [CrossRef]

11. Vasić, K.; Knez, Ž.; Leitgeb, M. Bioethanol production by enzymatic hydrolysis from different lignocellulosic sources. *Molecules* **2021**, *26*, 753. [CrossRef] [PubMed]

12. European Council. Directive (EU) 2019/904 of the European Parliament and of the Council of 5 June 2019 on the Reduction of the Impact of Certain Plastic Products on the Environment. Available online: https://eur-lex.europa.eu/legal-content/EN/TXT/HTML/?uri=CELEX:32019L0904&from=HU (accessed on 1 February 2021).

13. Moore, C. Plastic Pollution. Encyclopedia Britannica. Available online: https://www.britannica.com/science/plastic-pollution (accessed on 18 June 2021).

14. Getme, A.S.; Patel, B. A review: Bio-fiber's as reinforcement in composites of polylactic acid (PLA). *Mater. Today Proc.* **2019**, *26*, 2116–2122. [CrossRef]

15. Das, P.P.; Chaudhary, V. Moving towards the era of bio fibre based polymer composites. *Clean. Eng. Technol.* **2021**, *4*, 100182. [CrossRef]

16. Manral, A.; Ahmad, F.; Chaudhary, V. Static and dynamic mechanical properties of PLA bio-composite with hybrid reinforcement of flax and jute. *Mater. Today Proc.* **2020**, *25*, 577–580. [CrossRef]

17. Papadopoulou, E.; Bikiaris, D.; Chrysafis, K.; Władyka-Przybylak, M.; Wesołek, D.; Mankowski, J.; Kołodziej, J.; Baraniecki, P.; Bujnowicz, K.; Gronberg, V. Value-added industrial products from bast fiber crops. *Ind. Crop. Prod.* **2015**, *68*, 116–125. [CrossRef]

18. Gąsiorowski, R.; Rojewski, S.; Wesołek, D.; Maciejewski, H.; Bujnowicz, K. The influence of lignocellulosic filler modification with silicon compounds on flammability and adhesion with a composite polymer matrix. *Polym. Process.* **2013**, *19*, 336–339.

19. Akinshina, N.; Toderich, K.; Azizov, A.; Saito, L.; Ismail, S. Halophyte biomass—A promising source of renewable energy. *J. Arid Land Stud.* **2014**, *24*, 231–235.

20. Turcios, A.E.; Cayenne, A.; Uellendahl, H.; Papenbrock, J. Halophyte plants and their residues as feedstock for biogas production-chances and challenges. *Appl. Sci.* **2021**, *11*, 2746. [CrossRef]

21. Kumar, D.; Murthy, G.S. Impact of pretreatment and downstream processing technologies on economics and energy in cellulosic ethanol production. *Biotechnol. Biofuels* **2011**, *4*, 27. [CrossRef]

22. Stevulova, N.; Cigasova, J.; Estokova, A.; Terpakova, E.; Geert, A.; Kacik, F.; Singovszka, E.; Holub, M. Properties characterization of chemically modified hemp hurds. *Materials* **2014**, *7*, 8131–8150. [CrossRef] [PubMed]

23. Araújo, D.; Vilarinho, M.; Machado, A. Effect of combined dilute-alkaline and green pretreatments on corncob fractionation: Pretreated biomass characterization and regenerated cellulose film production. *Ind. Crop. Prod.* **2019**, *141*, 111785. [CrossRef]

24. Ciolacu, D.; Ciolacu, F.; Popa, V.I. Amorphous cellulose—Structure and characterization. *Cellul. Chem. Technol.* **2011**, *45*, 13–21.

25. Sun, X.F.; Xu, F.; Sun, R.C.; Fowler, P.; Baird, M.S. Characteristics of degraded cellulose obtained from steam-exploded wheat straw. *Carbohydr. Res.* **2005**, *340*, 97–106. [CrossRef] [PubMed]

26. Gupta, A.D.; Pandey, S.; Kumar Jaiswal, V.; Bhadauria, V.; Singh, H. Simultaneous oxidation and esterification of cellulose for use in treatment of water containing Cu (II) ions. *Carbohydr. Polym.* **2019**, *222*, 114964. [CrossRef] [PubMed]

27. Zhao, J.; Xu, Y.; Wang, W.; Griffin, J.; Wang, D. High Ethanol concentration (77 g/L) of industrial hemp biomass achieved through optimizing the relationship between ethanol yield/concentration and solid loading. *ACS Omega* **2020**, *5*, 21913–21921. [CrossRef] [PubMed]

28. Zhao, J.; Xu, Y.; Zhang, M.; Wang, D. Integrating bran starch hydrolysates with alkaline pretreated soft wheat bran to boost sugar concentration. *Bioresour. Technol.* **2020**, *302*, 122826. [CrossRef] [PubMed]

29. Abraham, R.E.; Barrow, C.J.; Puri, M. Relationship to reducing sugar production and scanning electron microscope structure to pretreated hemp hurd biomass (*Cannabis sativa*). *Biomass Bioenerg.* **2013**, *58*, 180–187. [CrossRef]

30. Kumar, D.; Murthy, G.S. Pretreatments and enzymatic hydrolysis of grass straws for ethanol production in the Pacific Northwest U.S. *Biol. Eng.* **2011**, *3*, 97–110. [CrossRef]

31. Wawro, A.; Batog, J.; Gieparda, W. Efektywność enzymatycznej konwersji biomasy sorgo i konopi do glukozy. *Przemysł Chem.* **2020**, *99*, 1731–1734.

32. Robak, K.; Balcerek, M. Current state-of-the-art in ethanol production from lignocellulosic feedstocks. *Microbiol. Res.* **2020**, *240*, 126534. [CrossRef]

33. Taufikurahman, T.; Xie, S. Production of Bioethanol and Crude Cellulase Enzyme Extract from Napier Grass (*Pennisetum purpureum* S.) through Simultaneous Saccharification and Fermentation. *Bio J. Biol. Sci. Technol. Man.* **2020**, *2*, 18–29. [CrossRef]
34. Riadi, L.; Hansen, Y.; Pratiwi, J.; Goretti, M.; Purwanto, M. Mild Alkaline Pretreatment on Sugarcane Bagasse: Effects of Pretreatment Time and Lime to Dry Bagasse Ratio. *B Life Environ. Sci.* **2020**, *57*, 43–50.
35. Orlygsson, J. Ethanol production from biomass by a moderate thermophile, Clostridium AK1. *Icel. Agric. Sci.* **2012**, *25*, 25–35.
36. International Standards Organization. *PN-EN ISO 3167:2014, Plastics—Multipurpose Test Specimens*; Polish Committee for Standardization: Geneva, Switzerland, 2014.
37. Yan, Z.L.; Wang, H.; Lau, K.T.; Pather, S.; Zhang, J.C.; Lin, G.; Ding, Y. Reinforcement of polypropylene with hemp fibres. *Compos. Part B Eng.* **2013**, *46*, 221–226. [CrossRef]
38. Yu, T.; Jiang, N.; Li, Y. Study on short ramie fiber/poly(lactic acid) composites compatibilized by maleic anhydride. *Compos. Part A Appl. Sci. Manuf.* **2014**, *64*, 139–146. [CrossRef]
39. Batog, J.; Wawro, A. Chemical and biological deconstruction in the conversion process of sorghum biomass for bioethanol. *J. Nat. Fibers* **2021**, 1–12. [CrossRef]
40. Miller, G.L. Use of dinitrosalicylic acid reagent for determination of reducing sugars. *Anal. Chem.* **1959**, *31*, 426–428. [CrossRef]
41. Kawa-Rygielska, J.; Pietrzak, W. Zagospodarowanie odpadowe pieczywa do produkcji bioetanolu. *Żywność Nauka Technol. Jakość* **2011**, *79*, 105–118.
42. TAPPI 17 m-55. In *Cellulose in Wood*; TAPPI Press: Atlanta, GA, USA, 1955.
43. TAPPI T9 m-54. In *Holocellulose in Wood*; TAPPI Press: Atlanta, GA, USA, 1998.
44. Bagby, M.O.; Nelson, G.H.; Helman, E.G.; Clark, T.F. Determination of lignin non-wood plant fiber sources. *TAPPI J.* **1971**, *54*, 11.
45. International Standards Organization. *PN-EN ISO 527-1:2020, Plastics—Determination of Tensile Properties—Part 1: General Principles*; Polish Committee for Standardization: Geneva, Switzerland, 2020.
46. International Standards Organization. *PN-EN ISO 527-2:2012, Plastics—Determination of Tensile Properties—Part 2: Test Conditions for Moulding and Extrusion Plastics*; Polish Committee for Standardization: Geneva, Switzerland, 2012.
47. International Standards Organization. *PN-EN ISO 178:2019, Plastics—Determination of Flexural Properties*; Polish Committee for Standardization: Geneva, Switzerland, 2019.

Article

Copper Adsorption on Lignin for the Removal of Hydrogen Sulfide

Miroslav Nikolic [1], **Marleny Cáceres Najarro** [1,*], **Ib Johannsen** [1], **Joseph Iruthayaraj** [1], **Marcel Ceccato** [1,2] **and Anders Feilberg** [1,*]

1 Department of Engineering, Aarhus University, Hangøvej 2, 8200 Aarhus, Denmark; nmiroslav@ymail.com (M.N.); ibj@eng.au.dk (I.J.); josedeva76@gmail.com (J.I.); mceccato@inano.au.dk (M.C.)
2 Interdisciplinary Nanoscience Center (iNANO), Aarhus University, Gustav Wieds Vej 14, 8000 Aarhus, Denmark
* Correspondence: mcaceres@eng.au.dk (M.C.N.); af@eng.au.dk (A.F.)

Academic Editor: Alejandro Rodríguez
Received: 28 September 2020; Accepted: 23 November 2020; Published: 27 November 2020

Abstract: Lignin is currently an underutilized part of biomass; thus, further research into lignin could benefit both scientific and commercial endeavors. The present study investigated the potential of kraft lignin as a support material for the removal of hydrogen sulfide (H_2S) from gaseous streams, such as biogas. The removal of H_2S was enabled by copper ions that were previously adsorbed on kraft lignin. Copper adsorption was based on two different strategies: either directly on lignin particles or by precipitating lignin from a solution in the presence of copper. The H_2S concentration after the adsorption column was studied using proton-transfer-reaction mass spectrometry, while the mechanisms involved in the H_2S adsorption were studied with X-ray photoelectron spectroscopy. It was determined that elemental sulfur was obtained during the H_2S adsorption in the presence of kraft lignin and the differences relative to the adsorption on porous silica as a control are discussed. For kraft lignin, only a relatively low removal capacity of 2 mg of H_2S per gram was identified, but certain possibilities to increase the removal capacity are discussed.

Keywords: kraft lignin; adsorbent material; biobased materials; copper adsorption; H_2S adsorption; H_2S removal

1. Introduction

In recent times, significant efforts are being placed into the substitution of fossil-based natural gas, also known as synthetic natural gas (SNG), with renewable biomethane (CH_4), which is mainly obtained from upgrading biogas via the removal of CO_2, H_2S, and other trace gases [1,2]. Before upgrading, the biogas composition can contain between 80 and 4000 ppm of H_2S, depending on the feedstock material [3]. The corrosive and harmful nature of H_2S [3,4] requires it to be efficiently removed for further uses of biogas/biomethane. Numerous technologies already exist for the removal of H_2S from gaseous streams, including solid reagents (oxides and salts), activated carbon, biological removal, solvent absorption, and chelating solutions; however, all of them have various deficiencies, such as incomplete absorption, methane absorption, operational difficulties, or high costs involved [5]. Therefore, novel approaches are being explored in order to uncover more effective solutions.

Lignin, the second most abundant biopolymer after cellulose, is currently an underutilized component of lignocellulosic biomass due to its heterogeneity, recalcitrant nature, and difficulties in processing, amongst others [6–8]. Lignin is often treated as waste material and the largest quantities are currently being burnt for energy as a low-value fuel [9]. With the emergence of biorefineries, the availability of lignin is only expected to increase in the future, making the valorization of lignin to various chemicals and materials of high importance [10].

In material applications, lignin has been studied for many years as a precursor for activated carbon adsorbents [11,12]. Activated carbons with structured pores and large internal surface areas have high adsorption capacities toward various substances [13], including H_2S [14]. The high cost of activated carbons [15], together with increased environmental considerations [16], has influenced research to move toward lower-cost natural adsorbent materials. Lignin has also been studied for its natural adsorbent properties, without the utilization of carbonization processes, but mainly for adsorbing metal ions for use in, e.g., water purification [17]. The adsorption of metal ions on lignin is mostly driven by a large fraction of functional groups that are available due to various processes of lignin isolation from nature [17]. For example, milled wood lignin, which is a reasonable representation of native lignin, contains mostly aliphatic alcohol groups and only around 1 mmol/g of phenolic OH and around 0.2 mmol/g of carboxyl groups [18]. On the other hand, the most abundant technical lignins, such as kraft lignin and lignosulfonates, differ from that. Kraft process leads to around 3 mmol/g of phenolic OH and around 0.5 mmol/g of carboxyl groups [18,19], while lignosulfonates are typically characterized by 5–8 wt% of sulfur all in the form of a low pKa sulfonic groups, which can represent 13% of lignosulfonates' weight [20]. Since transition metals are also effective in the removal of H_2S [3,5], we have chosen to premiere a study of utilizing non-carbonized lignin as a support for copper ions in the removal of H_2S.

Ferric oxide and chelated ferric solutions are the most often used among the H_2S removal technologies centered on transition metals [5,14] due to the high removal capacity and the more benign environmental effects of iron compared to most of the other transition metals. However, we have opted for copper in our study for the following reasons. First, a recent study has shown the good removal capacity of H_2S by a copper (II) solution due to the precipitation of copper (II) sulfide [21]. When utilized heterogeneously (as adsorbed on other solid supports), copper has shown better results compared to several metal ions, including iron [22]. In addition, if the use of lignin in water purification is commercialized in the future, large amounts of lignin saturated with heavy metal ions, such as copper, can easily become available. To the best of our knowledge, the application of lignin as a support material for copper in gas purification has not been reported previously. The present study focused on establishing copper adsorption mechanisms on a lignin structure and determined their removal capacity toward H_2S. The results offer an initial insight into potentially extending the value chain of lignin into biogas cleaning, going beyond water purification. The system was tested for the treatment of a model system consisting of H_2S in N_2 in order to characterize the process under controlled conditions. H_2S is the dominant sulfur compound in many applications, including biogas [23–25], and its removal will not be significantly influenced by the presence of small amounts of other sulfur compounds. A lignin-Cu adsorbent would typically be applied as a final purification step to protect, e.g., catalysts from sulfur poisoning [24,25]. It is therefore relevant to test a model system containing only H_2S.

2. Materials and Methods

Softwood kraft lignin BioPiva 100 was purchased from UPM (Helsinki, Finland). Copper (II) bromide and tetrahydrofurane (THF) were purchased from Sigma Aldrich (Vandtårnsvej, Denmark) and used as obtained. Davison silica gel grade 12 60/80 mesh was sourced from Supelco Analytical (Bellefonte, PA, USA). A hydrogen sulfide (H_2S) gas cylinder (720 ppm) with nitrogen as a carrier gas was procured from AGA (Fredericia, Denmark).

2.1. Copper Adsorption

Before further treatment, the kraft lignin was washed three times with Milli-Q water (Millipore, Denmark) and dried at 45 °C. This process allowed for removing soluble compounds, e.g., sulfate ions [26]. In one set of experiments, 2 g of washed kraft lignin was dispersed in a 0.15 M solution of $CuBr_2$ and stirred overnight. The following day, lignin was separated via centrifugation, washed several times, and dried at 45 °C before further use.

In the second set of experiments, washed kraft lignin was dissolved in THF, mixed with a $CuBr_2$ water solution, and stirred overnight. The copper concentration (0.15 M) was kept the same as in the previous experiment, while a particular THF:water ratio of 5:2 by weight was selected based on trial and error to avoid precipitation. The following day, the THF was removed using a rotary evaporator and the lignin particles were washed with water several times and dried at 45 °C before further use.

Thermogravimetric analysis (TGA) was used to establish the amount of copper adsorbed on solid lignin. Kraft lignin powder before and after adsorption was heated from room temperature to 800 °C at a rate of 10 °C/min under nitrogen.

2.2. H_2S Removal Experiments

For the H_2S removal experiments, lignin was mixed with Davison porous silica in a 1:3 weight ratio and the adsorption experiments were repeated in the same manner. The H_2S removal experiments were performed in a laboratory-developed test at room temperature to determine the capacity of lignin (with adsorbed copper) for H_2S removal. Controls utilizing only silica with adsorbed copper, silica without adsorbed copper, and a silica/lignin mixture without adsorbed copper were tested to enable normalization and to calculate the adsorption capacity of lignin.

The material was packed in a Teflon column (length 90 mm, inner diameter 3.2 mm, 0.5 g of adsorbent material) by placing glass wool on both ends of the column to prevent material loss during the experiments. The H_2S was diluted with compressed air and the flow rates were controlled with gas mass flow controllers (Bronkhorst EL FLOW, Ruurlo, The Netherlands). Air containing 5 ppm of H_2S (the H_2S concentration was chosen after initial screening tests) was passed through a Teflon column at 300 mL/min. To determine the influence of moisture on the H_2S removal, the experiments were performed utilizing non-conditioned compressed air (14% relative humidity (RH)) and air passed through a water impinger to obtain 90–100% relative humidity.

A proton-transfer-reaction mass spectrometer (PTR-MS, Ionicon Analytik GmbH, Innsbruck, Austria), which allowed for online data collection, was used for monitoring the H_2S (mass-to-charge ratio (m/z) of 35) removal by measuring after the adsorption column. The instrument used was a high-sensitivity quadrupole PTR-MS. The PTR-MS was operated under standard drift tube conditions with a voltage at 600 V and pressure close to 2.2 mbar, while the inlet and chamber temperature was set to 75 °C.

The experiments were stopped when a 10% (0.5 ppm of H_2S) breakthrough was reached. An empty column was run before the actual experiment in order to establish the maximum measured level of H_2S with PTR-MS under the set conditions. Although silica gel is hydrophilic, the humidity did not affect any of the packed columns over the course of the experiments to any significant degree. The H_2S signal was therefore not corrected for humidity dependency in order to determine percentage removal efficiency [27]. The removal capacity was calculated based on the inlet and outlet H_2S concentrations, flow rate, breakthrough time, and the volume of adsorbent material.

2.3. X-ray Photoelectron Spectroscopy (XPS)

The XPS analyses were performed with a Kratos Axis UltraDLD spectrometer (Kratos Analytical Ltd., Manchester, UK) using a monochromated Al X-ray source (1486.7 eV) operated at 150 W and pressure in the main chamber in the 10^{-9} torr range. The information collected was related to approximately the top 10 nm of the sample surface. Survey scans were performed with a 1.0 eV step size and a 160 eV analyzer pass energy, while the high-resolution scans were recorded at a 20 eV pass energy and a 0.1 eV step size. The spectra were processed using CasaXPS software (Version 2.3.20, Casa Software Ltd., Terrace Teignmouth, UK) and calibrated by setting the main C1s peak to 284.8 eV. The atomic surface concentration (atom%) was determined from the survey spectra and represents the average of the measurements taken at two different spots on each sample.

3. Results and Discussion

3.1. Copper Adsorption

The formation of stable surface complexes between the Cu(II) cations and the lignin substrate has been documented [28]. The surface of lignin particles is rich in functional groups, such as carboxyl and alcohol groups, that have a high affinity toward metal ions. The softwood kraft lignin used in this study was previously characterized to have 0.5 mmol/g of carboxyl groups, 2.0 mmol/g of aliphatic OH, and 4.8 mmol/g of various phenolic OH species [29]. Past research has shown higher adsorption of copper ions on lignin at higher pH levels in general [30,31], suggesting the importance of deprotonating surface carboxyl groups for obtaining a high adsorption capacity. At the same time, for certain metal ions, such as Fe(III), coordination with phenolic oxygen was established [32] and should not be neglected. Figure 1 shows the thermogravimetric analysis for kraft lignin and the samples with adsorbed copper. TGA is a technique that is broadly used to study the adsorption of materials [33–36], and in the present study, it allowed for establishing the amount of copper adsorbed on powder lignin. Kraft lignin had a residual weight of around 40% at 800 °C, which is in line with other research results for softwood kraft lignin with the same applied heating rate [37]. As expected, samples with adsorbed copper had a higher residual weight at 800 °C compared to pure kraft lignin. The copper adsorption capacity of lignin was calculated based on the TGA curves as the difference between the residual weight at 800 °C for lignin before and after adsorption. When Cu(II) was adsorbed on lignin particles, a capacity of 14 ± 5 mg/g of lignin was determined, in agreement with previous results for various lignin particles [30,31,38–40].

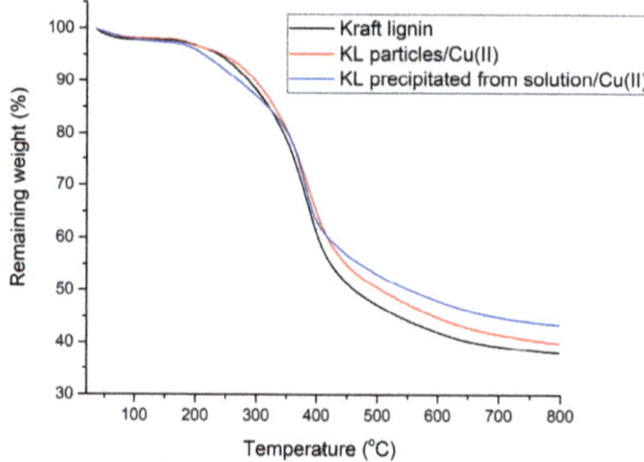

Figure 1. Thermogravimetric analysis for kraft lignin (KL) and the samples with adsorbed copper.

On the other hand, by dissolving lignin in the THF/H_2O mixture and later precipitating it, the capacity for copper adsorption increased several times to 72 ± 13 mg/g of lignin. By solubilizing lignin, thus disentangling the polymer structure and solvating all the functional groups [41], the number of functional groups accessible to copper ions increased compared to the particle form, thereby increasing the total number of adsorption sites for copper ions. This effect can be considered as a general concept regarding the accessibility of functional groups. For example, Sun et al. [42] have shown with the use of qNMR combined with wet chemistry that the accessible amine content is lower on modified silica particles compared to the amine content of the same dissolved silica. It is well known that in solution, Cu(II) can form complexes of various stabilities with amines, acids, and hydroxy-acids [43]. For hydroxy-acids, such as dihydroxybenzoic acid, it has been shown that Cu complexes can remain stable without being dissolved through a combination of electrostatic interactions and an extended network of hydrogen bonds due to retention of water molecules [44]. Similar mechanisms are likely involved for macromolecular lignin-Cu complexes; thus, once copper is complexed in solution, it can remain stable upon precipitation. Considering all these, it is not surprising that the adsorption capacity increased for the same lignin. Metal ion adsorption on lignin is most often exploited for water purification, in which case, adsorption on already-formed particles is the most logical route [17]. However, this result shows that for other applications, alternative approaches should also be considered.

3.2. H_2S Removal Experiments

The performances of the materials for H_2S adsorption were characterized in terms of the breakthrough capacity using the method described in Section 2.2. The utilized PTR-MS instrument allowed for continuous and direct measurement of the H_2S levels with a time resolution of 100 ms. Figure 2 shows an example of breakthrough curves during screening tests for a system with and without adsorbed copper. It was clear from the tested controls that without copper, breakthroughs were occurring practically immediately and that copper was responsible for the H_2S removal from the air stream. Initially, tests were done by filling the column with only lignin. However, due to the kraft lignin powder packing density and low porosity of the particles, it was not possible to obtain stable airflow through the column (because of the large pressure build-up and likely occurrence of air channeling). The typical surface area of kraft lignin powder is in the range of 0.1–30 m^2/g [45–47], which is low for powder adsorbents, and for this reason, Davison porous silica gel with a surface area of 750 m^2/g (data obtained from the supplier) was introduced. The selected 3:1 weight ratio of silica powder to lignin powder provided a distribution of particles in the column such that there was no airflow fluctuation through the column during the experiment. Tests were also made with only porous silica as the adsorbent material (with and without copper) to enable the normalization of the results for lignin/silica mixtures.

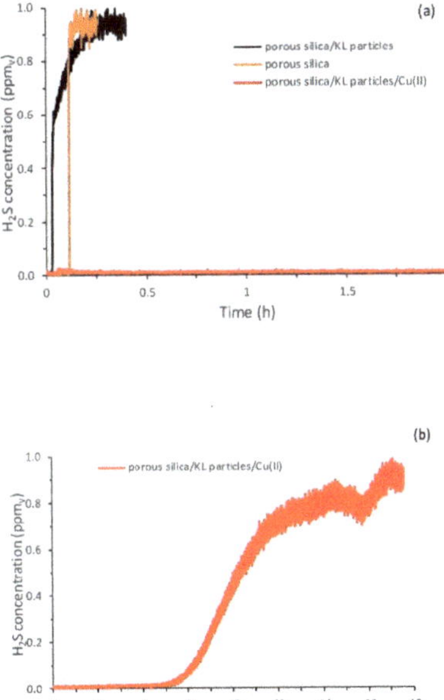

Figure 2. (**a**) Adsorption curves of H_2S for the silica particles (orange line) and a blend of silica and kraft lignin particles with (red line) and without (black line) adsorbed copper. (**b**) Continuation of the experiment for the blend of particles with adsorbed copper (red line).

Table 1 shows the determined breakthrough capacities for H_2S removal. It can be seen that the higher removal capacity was obtained for the system that could adsorb more copper since the native materials (silica and kraft lignin) did not show tendencies toward adsorbing H_2S at significant levels (data not shown). Past research has shown that the levels of copper adsorption by non-modified amorphous silica are around 5 mg/g [48–50], which is lower than the determined value for the two lignin systems herein and is likely the reason for the lower H_2S removal capacity of silica.

Table 1. H_2S adsorption capacities in milligrams per gram of adsorbent material for copper-adsorbed silica, copper-adsorbed kraft lignin particles, and copper-adsorbed kraft lignin obtained via precipitation from a solution, which was tested at low and high relative humidities.

Sample	H_2S Adsorption Capacity (mg/g)	
	Low Relative Humidity (14% RH)	**High Relative Humidity (90–100% RH)**
Porous silica/Cu(II)	0.62 ± 0.02	0.38 ± 0.05
KL particles/Cu(II)	0.90 ± 0.39	0.52 ± 0.11
KL precipitated from solution/Cu(II)	2.05 ± 0.40	1.80 ± 0.60

The highest removal capacity was determined for lignin precipitated from solution with adsorbed copper at the level of 2 mg H_2S/g of lignin, which is multiple times lower compared to, e.g., activated carbon [51]. In the current work, our main goal was to study the reactivity and no effort has been made to increase the porosity and surface area of the material. Therefore, it was not expected that the non-modified kraft lignin could reach the performance of specifically activated carbons with very high surface areas. However, certain H_2S removal technologies for gas purification, such as amine absorption and biological purification, struggle to completely remove H_2S [5], while implementation in national grids requires H_2S concentrations under 5 ppm and applications in fuel cells can require levels below 0.1 ppm [3,52]. This suggests that if the processing conditions and the price levels are convenient, one can potentially utilize the lignin-Cu adsorbents as an additional post-treatment purification system for H_2S removal for, e.g., fuel cell applications or biogas upgrading that require particularly low levels of H_2S. This would mean that lignin-Cu would be used for the treatment of pretreated relatively clean gas (with respect to co-contaminants). Regarding processing, solid adsorbents are easy to handle, but in order to use lignin as a sole powder adsorbent (and without carbonization), the porosity and surface area of the particles would need to be increased without substantially reducing the concentration of functional groups. Examples of such methods could be related to the works of Li et al. [53] or Cantu et al. [54]. Specific modification of lignin can further increase the metal ion capacity, where values as high as 400 mg of Cu(II) per gram of lignin-based adsorbent have already been achieved via the carboxymethylation of lignin [55], where the H_2S removal capacity could be enhanced in that manner. Increasing the capacity would also make studying the copper recovery relevant, which regarding lignin-based adsorbents, currently relies on desorption at low pH levels [56,57].

Biogas is typically saturated with water that is removed during the upgrading process [58]. H_2S removal technologies, such as iron oxides and activated carbons, require a certain level of humidity in order to achieve high removal capacities, but too high a water content can negatively influence the process [58,59]. The humidity is believed to enable H_2S dissociation on the surface of the carbon, which promotes redox reactions with dissociatively adsorbed oxygen [51]. For these reasons and to obtain a better understanding of our system, it was important to study whether the humidity level influenced the adsorption capacity and the H_2S removal mechanisms. In our experiments, by saturating the air with humidity, the adsorption capacity decreased by 10–40% (Table 1) for all the samples; this issue is further discussed in the following section.

3.3. X-ray Photoelectron Spectroscopy

After the H_2S adsorption, the silica samples and silica/kraft lignin samples with a higher capacity were collected and studied using XPS in order to understand the mechanisms and determine whether copper (II) sulfide was formed or a reaction was occurring on the adsorbent material. The determined peak positions are summarized in Table 2, while Figure 3 shows the copper and sulfur 2p XPS spectra for the silica sample.

Table 2. X-ray photoelectron spectroscopy data for adsorbent materials after H_2S removal.

Sample	Cu $2p_{3/2}$	S $2p_{3/2}$	Cu (LMM *)
Porous silica/Cu(II)	932.4, 933.5	162.7	568.4, 571.7
Porous silica/Cu(II) (adsorption at 90–100% RH)	932.4, 933.5	162.9	568.9, 572.3
Porous silica/KL precipitated from solution/Cu(II)	933.1	163.8	571.6
Porous silica/KL precipitated from solution/Cu(II) (adsorption at 90–100% RH)	933.6	164.0	571.6

* LMM = Auger transition.

The two Cu $2p_{3/2}$ peaks of 932.4 eV and 933.5 eV were both assigned to monovalent Cu(I) with a formal oxidation state +1 ($Cu_{2-x}S$) [60]. The binding energy of 933.5 eV is also characteristic of Cu(II) [61], but the absence of a shake-up satellite peak (at approximately 944 eV) [62] excluded this possibility. The first peak at 932.4 eV was 0.2–0.3 eV lower than the value expected for CuS and indicated a S-rich nonstoichiometric environment. This was confirmed by the S $2p_{3/2}$ peak at 162.7–162.9 eV, which is higher than the typical S $2p_{3/2}$ peaks for Cu_2S that are expected at 161.7–161.9 eV [63]. The presence of the two peaks in Figure 3b arose from a spin–orbit splitting between the S $2p_{1/2}$ and S$2p_{3/2}$ states. Finally, the Cu (LMM) peak at 568.7–568.9 eV confirmed that the formal oxidation state of copper was Cu(I), since for Cu (0), the peak is expected at 568.1 eV [64]. To summarize, on the porous silica sample after H_2S adsorption, the formal oxidative states could be confirmed as +1 for copper and −1 for sulfur.

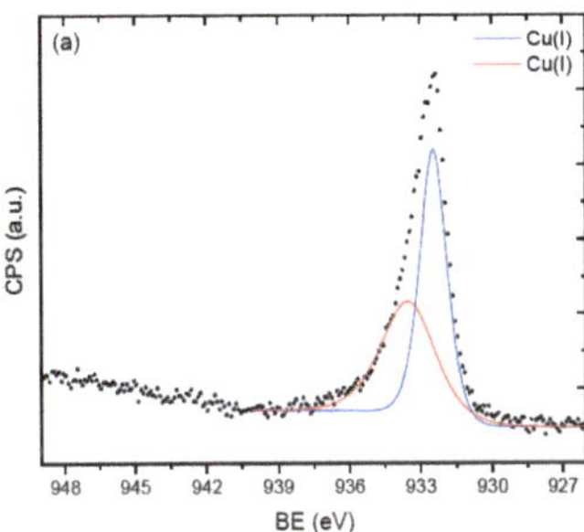

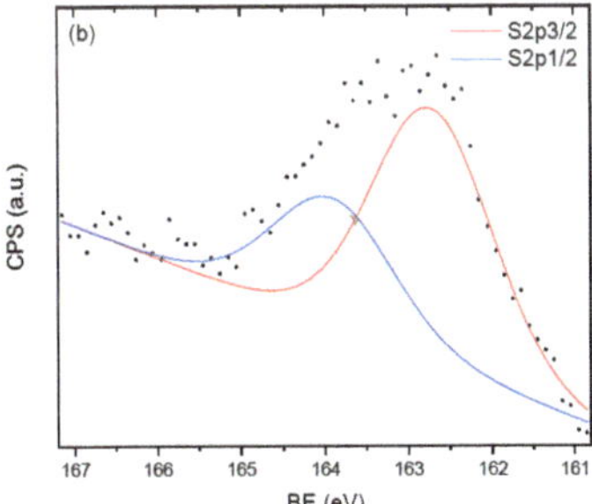

Figure 3. The (**a**) copper $2p_{3/2}$ and (**b**) sulfur $2p_{3/2,1/2}$ X-ray photoelectron spectra of the silica sample after H_2S adsorption. CPS: Counts per second, BE: Binding Energy.

For the sample that was a silica/lignin mixture, the situation was found to be somewhat different (Figure 4). The position of the S2p$_{3/2}$ 163.8–164.0 peak could be ascribed to elemental S^0 [65]. Another possibility does exist, as unpassivated sulfur atoms in a Cu$_{2-x}$S compound were found at a similar position [60]. Regarding copper, similar observations were made as that of the previous sample. Cu(II) and Cu^0 oxidation states were excluded in the same manner. The Cu 2p$_{3/2}$ peaks at 933.1–933.6 eV were assigned to Cu(I) in a Cu-rich environment, which could explain the high binding energy compared to Cu$_2$S, while the Cu (LMM) at 571.6 eV was higher than expected for Cu(I), as has been observed before [64]. It was interesting to see that while the situation with copper did not change substantially, in the presence of lignin, the dominant form of sulfur was its elemental form. It seems that in the presence of lignin, the electrochemical potential of the reaction changed, allowing for a higher oxidative state of sulfur to emerge. It was previously found that the adsorption of transition metal ions, such as iron, on polymeric lignin model compounds can be responsible for the oxidation of phenolic groups to species like semiquinonic radicals [66]. The occurrence of these and other types of radicals, for example, hydroxyl radicals, can explain the formation of elemental sulfur.

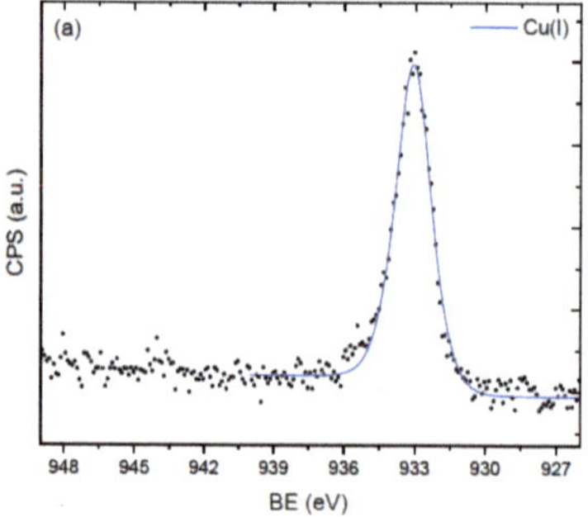

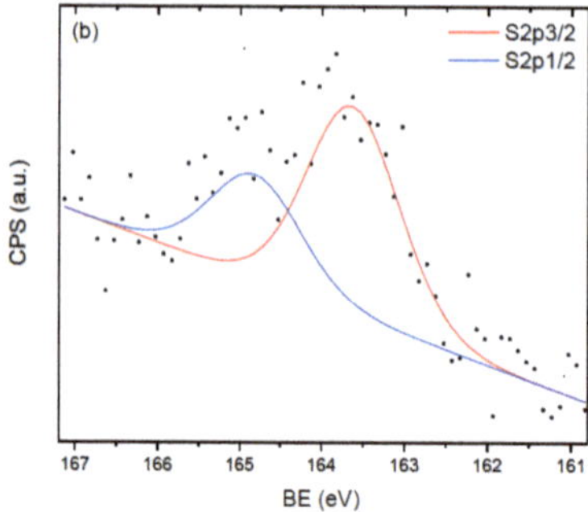

Figure 4. The (**a**) copper 2p$_{3/2}$ and (**b**) sulfur 2p$_{3/2,1/2}$ X-ray photoelectron spectra of porous silica/KL precipitated from the solution/Cu(II) sample after H$_2$S adsorption.

When we observe the peak positions for all the samples in Table 2, it is clear that the mechanisms involved were the same, irrespective of the relative air humidity. It appears that low relative humidity is sufficient for effective changes in oxidative states of copper and sulfur, likely by enabling the dissociation of H$_2$S on the surface of the adsorbent material. As the mechanisms are the same, a lower H$_2$S adsorption capacity with 100% relative humidity of air (Table 1) is possibly explained by water molecules competing for the same sites on the adsorbent material. Another possibility could be the development of a similar situation observed with iron oxides, where water is necessary for the redox process, but a high water content leads to a decrease in available adsorption sites since the oxide surface area decreases due to the adsorbent material sticking together [58].

4. Conclusions

This study has shown that during H_2S adsorption in the presence of kraft lignin, elemental sulfur was formed, while the oxidation state of copper changed to +1. By first solubilizing lignin, it was possible to increase the copper adsorption by several times compared to the adsorption on already-formed lignin particles, and thus, improve the H_2S removal capacity. The observed H_2S removal capacity of 2 mg/g of kraft lignin was low and it is clear that to consider the utilization of lignin from this perspective, further studies should be done. For instance, introducing porosity into lignin particles, along with increasing the adsorption of transition metal ions by specific chemical modifications, would positively influence the removal capacity.

Author Contributions: Conceived and designed the experiments, all authors; performed the experiments, M.N., M.C.N., M.C., and A.F.; analyzed and interpreted the data, all authors; writing—original draft preparation, M.N.; writing—review and editing, M.N., M.C.N., M.C., I.J., J.I., I.B., & A.F. All authors have read and agreed to the published version of the manuscript.

Funding: The research was funded by the NordForsk organization through project number 82214 within the Nordic Green Growth Research and Innovation Programme.

Acknowledgments: The authors would like to thank project partners within the Nordic Green Growth Research and Innovation Programme for helpful discussions during this work.

References

1. Jensen, M.B.; Strubing, D.; de Jonge, N.; Nielsen, J.L.; Ottosen, L.D.M.; Koch, K.; Kofoed, M.V.W. Stick or leave-Pushing methanogens to biofilm formation for ex situ biomethanation. *Bioresour. Technol.* **2019**, *291*, 121784. [CrossRef]

2. Wall, D.M.; Dumont, M.; Murphy, J.D. *Green Gas: Facilitating a Future Green Gas Grid through the Production of Renewable Gas*; Murphy, J.D., Ed.; IEA Bioenergy: Paris, France, 2018; ISBN 978-1-910154-37-3.

3. Bailon, L.; Hinge, J. *Biogas Upgrading. Evaluation of Methods for H2S Removal*; Danish Technological Institute, 2014. Available online: www.teknologisk.dk/_/media/60599_Biogas%20upgrading.%20Evaluation%20of% 20methods%20for%20H2S%20removal.pdf (accessed on 1 September 2020).

4. European Chemicals Agency: Homepage. Available online: https://echa.europa.eu/substance-information/-/ substanceinfo/100.029.070. (accessed on 8 June 2020).

5. Huertas, J.I.; Giraldo, N.; Izquierdo, S. Removal of H2S and CO2 from Biogas by Amine Absorption. In *Mass Transfer in Chemical Engineering Processes*; Markos, J., Ed.; InTech: Rijeka, Croatia, 2011; pp. 133–151, ISBN 978-953-307-619-5.

6. Gosselink, R.J.A.; de Jong, E.; Guran, B.; Abächerli, A. Co-ordination network for lignin—standardisation, production and applications adapted to market requirements (EUROLIGNIN). *Ind. Crops Prod.* **2004**, *20*, 121–129. [CrossRef]

7. Himmel, M.E.; Ding, S.Y.; Johnson, D.K.; Adney, W.S.; Nimlos, M.R.; Brady, J.W.; Foust, T.D. Biomass recalcitrance: Engineering plants and enzymes for biofuels production. *Science* **2007**, *315*, 804–807. [CrossRef] [PubMed]

8. Vardon, D.R.; Franden, M.A.; Johnson, C.W.; Karp, E.M.; Guarnieri, M.T.; Linger, J.G.; Salm, M.J.; Strathmann, T.J.; Beckham, G.T. Adipic acid production from lignin. *Energy Environ. Sci.* **2015**, *8*, 617–628. [CrossRef]

9. De Jong, E.; Gosselink, R.J.A. *Lignocellulose-Based Chemical Products*; Elsevier Inc. Academic Press: Cambridge, MA, USA, 2014; pp. 277–313.

10. Zakzeski, J.; Bruijnincx, P.C.; Jongerius, A.L.; Weckhuysen, B.M. The catalytic valorization of lignin for the production of renewable chemicals. *Chem. Rev.* **2010**, *110*, 3552–3599. [CrossRef] [PubMed]

11. Hayashi, J.; Kazehaya, A.; Muroyama, K.; Watkinson, A.P. Preparation of activated carbon from lignin by chemical activation. *Carbon N. Y.* **2000**, *38*, 1873–1878. [CrossRef]

12. Khezami, L.; Chetouani, A.; Taouk, B.; Capart, R. Production and characterisation of activated carbon from wood components in powder: Cellulose, lignin, xylan. *Powder Technol.* **2005**, *157*, 48–56. [CrossRef]

13. Suhas; Carrott, P.J.M.; Ribeiro Carrott, M.M.L. Lignin-from natural adsorbent to activated carbon: A review. *Bioresour. Technol.* **2007**, *98*, 2301–2312. [CrossRef]

14. Abatzoglou, N.; Boivin, S. A review of biogas purification processes. *Biofuels Bioprod. Biorefining* **2009**, *3*, 42–71. [CrossRef]

15. Rafatullah, M.; Sulaiman, O.; Hashim, R.; Ahmad, A. Adsorption of methylene blue on low-cost adsorbents: A review. *J. Hazard. Mater.* **2010**, *177*, 70–80. [CrossRef]

16. O'Connell, D.W.; Birkinshaw, C.; O'Dwyer, T.F. Heavy metal adsorbents prepared from the modification of cellulose: A review. *Bioresour. Technol.* **2008**, *99*, 6709–6724. [CrossRef] [PubMed]

17. Ge, Y.; Li, Z. Application of Lignin and Its Derivatives in Adsorption of Heavy Metal Ions in Water: A Review. *ACS Sustain. Chem. Eng.* **2018**, *6*, 7181–7192. [CrossRef]

18. Balakshin, M.; Capanema, E. On the Quantification of Lignin Hydroxyl Groups With 31 P and 13 C NMR Spectroscopy. *J. Wood Chem. Technol.* **2015**, *35*, 220–237. [CrossRef]

19. Gosselink, R.J.A.; van Dam, J.E.G.; de Jong, E.; Scott, E.L.; Sanders, J.P.M.; Li, J.; Gellerstedt, G. Fractionation, analysis, and PCA modeling of properties of four technical lignins for prediction of their application potential in binders. *Holzforschung* **2010**, *64*, 193–200. [CrossRef]

20. Sumerskii, I.; Korntner, P.; Zinovyev, G.; Rosenau, T.; Potthast, A. Fast track for quantitative isolation of lignosulfonates from spent sulfite liquors. *RSC Adv.* **2015**, *5*, 92732–92742. [CrossRef]

21. Yan, G.; Weng, H.; Yang, J.; Bao, W.; Gao, Y.; Yin, Y. Hydrogen sulfide removal by copper sulfate circulation method. *Korean J. Chem. Eng.* **2016**, *33*, 2359–2365. [CrossRef]

22. Nguyen-Thanh, D.; Bandosz, T.J. Activated carbons with metal containing bentonite binders as adsorbents of hydrogen sulfide. *Carbon N. Y.* **2005**, *43*, 359–367. [CrossRef]

23. Montebello, A.M.; Fernández, M.; Almenglo, F.; Ramírez, M.; Cantero, D.; Baeza, M.; Gabriel, D. Simultaneous methylmercaptan and hydrogen sulfide removal in the desulfurization of biogas in aerobic and anoxic biotrickling filters. *Chem. Eng. J.* **2012**, *200*, 237–246. [CrossRef]

24. Calbry-Muzyka, A.S.; Gantenbein, A.; Schneebeli, J.; Frei, A.; Knorpp, A.J.; Schildhauer, T.J.; Biollaz, S.M.A. Deep removal of sulfur and trace organic compounds from biogas to protect a catalytic methanation reactor. *Chem. Eng. J.* **2019**, *360*, 577–590. [CrossRef]

25. Dannesboe, C.; Hansen, J.B.; Johannsen, I. Removal of sulfur contaminants from biogas to enable direct catalytic methanation. *Biomass Conv. Bioref.* **2019**. [CrossRef]

26. Svensson, S. Minimizing the Sulphur Content in Kraft Lignin. Master's Thesis, Mälardalen University, Västerås, Sweden, 2008.

27. Feilberg, A.; Liu, D.; Adamsen, A.P.S.; Hansen, M.J.; Jonassen, K.E.N. Odorant Emissions from Intensive Pig Production Measured by Online Proton-Transfer-Reaction Mass Spectrometry. *Environ. Sci. Technol.* **2010**, *44*, 5894–5900. [CrossRef] [PubMed]

28. Merdy, P.; Guillon, E.; Aplincourt, M.; Dumonceau, J.; Vezin, H. Copper sorption on a straw lignin: Experiments and EPR characterization. *J. Colloid Interface Sci.* **2002**, *245*, 24–31. [CrossRef] [PubMed]

29. Abbadessa, A.; Oinonen, P.; Henriksson, G. Characterization of Two Novel Bio-based Materials from Pulping Process Side Streams: Ecohelix and CleanFlow Black Lignin. *BioResources* **2018**, *13*, 7606–7627. [CrossRef]

30. Harmita, H.; Karthikeyan, K.G.; Pan, X. Copper and cadmium sorption onto kraft and organosolv lignins. *Bioresour. Technol.* **2009**, *100*, 6183–6191. [CrossRef] [PubMed]

31. Mohan, D.; Pittman, C.U., Jr.; Steele, P.H. Single, binary and multi-component adsorption of copper and cadmium from aqueous solutions on Kraft lignin—A biosorbent. *J. Colloid Interface Sci.* **2006**, *297*, 489–504. [CrossRef]

32. Merdy, P.; Guillon, E.; Aplincourt, M.; Duomonceau, J. Interaction of metallic cations with lignins. Part 1: Stability of iron (III), manganese (II) and copper (II) complexes with phenolic lignin model compounds: Coumaric, ferulic and sinapic acids and coniferyl alcohol. *J. Chem. Res.* **2000**, *2000*, 76–77. [CrossRef]

33. Badruddoza, A.Z.M.; Tay, A.S.H.; Tan, P.Y.; Hidajat, K.; Uddin, M.S. Carboxymethyl-β-cyclodextrin conjugated magnetic nanoparticles as nano-adsorbents for removal of copper ions: Synthesis and adsorption studies. *J. Hazard. Mater.* **2011**, *185*, 1177–1186. [CrossRef]

34. Pajchel, L.; Kolodziejski, W. Solid-state MAS NMR, TEM, and TGA studies of structural hydroxyl groups and water in nanocrystalline apatites prepared by dry milling. *J. Nanoparticle Res.* **2013**, *15*, 1868. [CrossRef]

35. Nikolic, M.; Nguyen, H.D.; Daugaard, A.E.; Löf, D.; Mortensen, K.; Barsberg, S.; Sanadi, A.R. Influence of surface modified nano silica on alkyd binder before and after accelerated weathering. *Polym. Degrad. Stab.* **2016**, *126*, 134–143. [CrossRef]

36. Roonasi, P.; Holmgren, A. A Fourier transform infrared (FTIR) and thermogravimetric analysis (TGA) study of oleate adsorbed on magnetite nano-particle surface. *Appl. Surf. Sci.* **2009**, *255*, 5891–5895. [CrossRef]

37. Ház, A.; Jablonský, M.; Šurina, I.; Kačík, F.; Bubeníková, T.; Ďurkovič, J. Chemical Composition and Thermal Behavior of Kraft Lignins. *Forests* **2019**, *10*, 483. [CrossRef]

38. Todorciuc, T.; Bulgariu, L.; Popa, V.I. Adsorption of cu(ii) from aqueous solution on wheat straw lignin: Equilibrium and kinetic studies. *Cell. Chem. Technol.* **2015**, *49*, 439–447.

39. Sciban, M.; Klasnja, M. Study of the Adsorption of Copper(II) Ions from Water onto Wood Sawdust, Pulp and Lignin. *Adsorpt. Sci. Technol.* **2016**, *22*, 195–206. [CrossRef]

40. Šćiban, M.B.; Klašnja, M.T.; Antov, M.G. Study of the biosorption of different heavy metal ions onto Kraft lignin. *Ecol. Eng.* **2011**, *37*, 2092–2095. [CrossRef]

41. Miller-Chou, B.A.; Koenig, J.L. A review of polymer dissolution. *Prog. Polym. Sci.* **2003**, *28*, 1223–1270. [CrossRef]

42. Sun, Y.; Kunc, F.; Balhara, V.; Coleman, B.; Kodra, O.; Raza, M.; Chen, M.; Brinkmann, A.; Lopinski, G.P.; Johnston, L.J. Quantification of amine functional groups on silica nanoparticles: A multi-method approach. *Nanoscale Adv.* **2019**, *1*, 1598–1607. [CrossRef]

43. Motschi, H. Correlation of EPR-parameters with thermodynamic stability constants for copper(II) complexes Cu(II)-EPR as a probe for the surface complexation at the water/oxide interface. *Colloids Surf.* **1984**, *9*, 333–347. [CrossRef]

44. Cariati, F.; Erre, L.; Micera, G.; Panzanelli, A.; Ciani, G.; Sironi, A. Interaction of metal ions with humic-like models. Part. I. Synthesis, spectroscopic and structural properties of diaquabis(2,6-dihydroxybenzoato) copper(II) and hexaaquaM(II) bis(2,6-dihydroxybenzoate) dihydrate (M = Mn, Fe, Co, Ni, Cu and Zn). *Inorganica Chim. Acta* **1983**, *80*, 57–65. [CrossRef]

45. Abdelaziz, O.Y.; Ravi, K.; Mittermeier, F.; Meier, S.; Riisager, A.; Lidén, G.; Hulteberg, C.P. Oxidative Depolymerization of Kraft Lignin for Microbial Conversion. *ACS Sustain. Chem. Eng.* **2019**, *7*, 11640–11652. [CrossRef]

46. Ghavidel, N.; Fatehi, P. Synergistic effect of lignin incorporation into polystyrene for producing sustainable superadsorbent. *RSC Adv.* **2019**, *9*, 17639–17652. [CrossRef]

47. Klapiszewski, Ł.; Wysokowski, M.; Majchrzak, I.; Szatkowski, T.; Nowacka, M.; Siwińska-Stefańska, K.; Szwarc-Rzepka, K.; Bartczak, P.; Ehrlich, H.; Jesionowski, T. Preparation and Characterization of Multifunctional Chitin/Lignin Materials. *J. Nanomater.* **2013**, *2013*, 1–13. [CrossRef]

48. Ibrahim, H.S.; Ammar, N.S.; Abdel Ghafar, H.H.; Farahat, M. Adsorption of Cd(II), Cu(II) and Pb(II) using recycled waste glass: Equilibrium and kinetic studies. *Desalin. Water Treat.* **2012**, *48*, 320–328. [CrossRef]

49. Chiron, N.; Guilet, R.; Deydier, E. Adsorption of Cu(II) and Pb(II) onto a grafted silica: Isotherms and kinetic models. *Water Res.* **2003**, *37*, 3079–3086. [CrossRef]

50. Lee, J.Y.; Chen, C.H.; Cheng, S.; Li, H.Y. Adsorption of Pb(II) and Cu(II) metal ions on functionalized large-pore mesoporous silica. *Int. J. Environ. Sci. Technol.* **2015**, *13*, 65–76. [CrossRef]

51. Bandosz, T.J. On the adsorption/oxidation of hydrogen sulfide on activated carbons at ambient temperatures. *J Colloid Interface Sci* **2002**, *246*, 1–20. [CrossRef]

52. Barelli, L.; Bidini, G.; de Arespacochaga, N.; Pérez, L.; Sisani, E. Biogas use in high temperature fuel cells: Enhancement of KOH-KI activated carbon performance toward H_2S removal. *Int. J. Hydrogen Energy* **2017**, *42*, 10341–10353. [CrossRef]

53. Li, Z.; Ge, Y.; Wan, L. Fabrication of a green porous lignin-based sphere for the removal of lead ions from aqueous media. *J. Hazard. Mater.* **2015**, *285*, 77–83. [CrossRef]

54. Perez-Cantu, L.; Liebner, F.; Smirnova, I. Preparation of aerogels from wheat straw lignin by cross-linking with oligo(alkylene glycol)-α,ω-diglycidyl ethers. *Microporous Mesoporous Mater.* **2014**, *195*, 303–310. [CrossRef]

55. Tian, J.; Ren, S.; Fang, G.; Ma, Y.; Ai, Q. Preparation and Performance of Dimethyl-Acetoxy-(2-Carboxymethyl Ether)-Lignin Ammonium Chloride Amphoteric Surfactant. *BioResources* **2014**, *9*, 14. [CrossRef]

56. Ge, Y.; Qin, L.; Li, Z. Lignin microspheres: An effective and recyclable natural polymer-based adsorbent for lead ion removal. *Mater. Des.* **2016**, *95*, 141–147. [CrossRef]

57. Qin, L.; Ge, Y.; Deng, B.; Li, Z. Poly (ethylene imine) anchored lignin composite for heavy metals capturing in water. *J. Taiwan Inst. Chem. Eng.* **2017**, *71*, 84–90. [CrossRef]

58. Wellinger, A.; Lindberg, A. Biogas Upgrading and Utilization. IEA Bionergy. Task 24. 2000. Available online: http://www.iea-biogas.net/files/daten-redaktion/download/publi-task37/Biogas%20upgrading.pdf (accessed on 1 September 2020).

59. Sitthikhankaew, R.; Chadwick, D.; Assabumrungrat, S.; Laosiripojana, N. Effects of humidity, O_2, and CO_2 on H2S adsorption onto upgraded and KOH impregnated activated carbons. *Fuel Process. Technol.* **2014**, *124*, 249–257. [CrossRef]

60. Majeski, M.W.; Bolotin, I.L.; Hanley, L. Cluster beam deposition of Cu_2-XS nanoparticles into organic thin films. *ACS Appl. Mater. Interfaces* **2014**, *6*, 12901–12908. [CrossRef]

61. Silvester, E.J.; Grieser, F.; Sexton, B.A.; Healy, T.W. Spectroscopic studies on copper sulfide sols. *Langmuir* **1991**, *7*, 2917–2922. [CrossRef]

62. Biesinger, M.C.; Hart, B.R.; Polack, R.; Kobe, B.A.; Smart, R.S.C. Analysis of mineral surface chemistry in flotation separation using imaging XPS. *Miner. Eng.* **2007**, *20*, 152–162. [CrossRef]

63. Kundu, M.; Hasegawa, T.; Terabe, K.; Yamamoto, K.; Aono, M. Structural studies of copper sulfide films: Effect of ambient atmosphere. *Sci. Technol. Adv. Mater.* **2008**, *9*, 35011. [CrossRef]

64. Mott, D.; Yin, J.; Engelhard, M.; Loukrakpam, R.; Chang, P.; Miller, G.; Bae, I.-T.; Chandra Das, N.; Wang, C.; Luo, J.; et al. From Ultrafine Thiolate-Capped Copper Nanoclusters toward Copper Sulfide Nanodiscs: A Thermally Activated Evolution Route. *Chem. Mater.* **2010**, *22*, 261–271. [CrossRef]

65. Moulder, J.; Stickle, W.; Sobol, P. *Handbook of X-ray Photoelectron Spectroscopy*; Chastain, K.B.J., Ed.; Perkin-Elmer Corporation: Waltham, MA, USA, 1992.

66. Guillon, E.; Merdy, P.; Aplincourt, M.; Dumonceau, J.; Vezin, H. Structural Characterization and Iron(III) Binding Ability of Dimeric and Polymeric Lignin Models. *J. Colloid Interface Sci.* **2001**, *239*, 39–48. [CrossRef]

Sample Availability: Samples of the compounds are not available from the authors.

Publisher's Note: MDPI stays neutral with regard to jurisdictional claims in published maps and institutional affiliations.

 *molecules*

Article

Production and Mechanical Characterisation of TEMPO-Oxidised Cellulose Nanofibrils/β-Cyclodextrin Films and Cryogels

Bastien Michel [1], Julien Bras [1], Alain Dufresne [1], Ellinor B. Heggset [2] and Kristin Syverud [2,3,*]

1   Univeristy Grenoble Alpes, CNRS, Grenoble INP*, LGP2, 38000 Grenoble, France;
    bastien.michel@lgp2.grenoble-inp.fr (B.M.); julien.bras@grenoble-inp.fr (J.B.);
    alain.dufresne@grenoble-inp.fr (A.D.)
2   RISE PFI, NO-7491 Trondheim, Norway; ellinor.heggset@rise-pfi.no
3   Departments of Chemical Engineering, NTNU, 7491 Trondheim, Norway
*   Correspondence: kristin.syverud@rise-pfi.no

Received: 20 April 2020; Accepted: 18 May 2020; Published: 20 May 2020

Abstract: Wood-based TEMPO-oxidised cellulose nanofibrils (toCNF) are promising materials for biomedical applications. Cyclodextrins have ability to form inclusion complexes with hydrophobic molecules and are considered as a method to bring new functionalities to these materials. Water sorption and mechanical properties are also key properties for biomedical applications such as drug delivery and tissue engineering. In this work, we report the modification with β-cyclodextrin (βCD) of toCNF samples with different carboxyl contents viz. 756 ± 4 μmol/g and 1048 ± 32 μmol/g. The modification was carried out at neutral and acidic pH (2.5) to study the effect of dissociation of the carboxylic acid group. Films processed by casting/evaporation at 40 °C and cryogels processed by freeze-drying were prepared from βCD modified toCNF suspensions and compared with reference samples of unmodified toCNF. The impact of modification on water sorption and mechanical properties was assessed. It was shown that the water sorption behaviour for films is driven by adsorption, with a clear impact of the chemical makeup of the fibres (charge content, pH, and adsorption of cyclodextrin). Modified toCNF cryogels (acidic pH and addition of cyclodextrins) displayed lower mechanical properties linked to the modification of the cell wall porosity structure. Esterification between βCD and toCNF under acidic conditions was performed by freeze-drying, and such cryogels exhibited a lower decrease in mechanical properties in the swollen state. These results are promising for the development of scaffold and films with controlled mechanical properties and added value due to the ability of cyclodextrin to form an inclusion complex with active principle ingredient (API) or growth factor (GF) for biomedical applications.

Keywords: nanocellulose; β-cyclodextrin; cryogels; films

1. Introduction

Cellulose nanofibrils (CNFs) are high-aspect ratio nanoparticles formed by bundles of cellulose chains that are a succession of glucose subunits linked by β-1-4 glycosidic bonds. CNFs are produced from a cellulosic raw material, usually wood, the most abundant and renewable polymer available on earth. CNFs are produced by a combination of chemical/enzymatic pretreatments and mechanical treatment, usually using a homogeniser [1], a microfluidiser [2], or a grinder [3]. The variety of existing pretreatments [4] allows for a variety of surface chemistries, making CNF materials suitable for many applications [5]. CNFs pretreated in the presence of (2,2,6,6-tetramethyl-piperidin-1-yl)oxyl, also known as TEMPO, proposed by Saito et al., 2006, which consists of the regioselective oxidation of C6 primary hydroxyls of cellulose to C6 carboxylate groups, have been considered in a wide variety of applications due to its carboxyl content and reduced size [6,7]. CNFs are generally used in two different forms: either as films/nanopapers or as gels and can be used as rheology modifiers or emulsion stabilisers and additives in many applications. Films are obtained by solvent casting [8–10], and nanopapers are obtained by filtration [11,12]. Three types of gels can be identified: hydrogels, cryogels obtained by freeze-drying, and aerogels obtained by supercritical drying [13,14]. For many applications, water sorption properties are important. The TEMPO-oxidised cellulose nanofibrils (toCNF), like cellulose, are hygroscopic materials, which means that they can attract and retain water molecules from their environment by absorption or adsorption [10]. The impact of process parameters on cryogels mechanical properties have been previously studied [14–19], highlighting the importance of density and preparation method on the mechanical properties.

As a natural, biodegradable, and abundant polymer with reactive surface chemistry and good biocompatibility, nanocellulose is a promising material within the medical field. In recent years, applications in wound healing [20,21], drug delivery [9], and tissue engineering [22] have been investigated. In tissue engineering, the scaffold should stimulate cells to differentiate, proliferate, and form tissue. The interplay between the matrix and cells should be driven by the action of signals, which can be a mechanical stimulation, chemical compounds, or growth factors (usually proteins) [23]. To be suited for tissue engineering applications, scaffolds needs to exhibit stiffness similar to the natural extra-cellular matrix (ECM) of the tissue to be repaired and measured by the elastic modulus E. Typical values of stiffness of ECM are 0.1–1 KPa for brain tissue, 8–17 KPa for muscle tissue, and 25–40 KPa for the cross-linked collagen matrix [24,25]. Another crucial aspect for tissue engineering application is the scaffold architecture. A high porosity is needed to promote the cellular penetration and an adequate diffusion of nutrients to the cells [26].

The utilisation of wood-based CNFs for tissue-engineering applications is encouraged by recent studies that confirmed the safety of CNFs [27–30], the construction of cell-friendly porous structures [14], and the control of mechanical properties [31,32]. In addition, CNFs, in the form of a highly entangled network, have shown the ability to retain the active principle ingredient up to several months [9]. However, major challenges for biomedical applications are yet to be overcome, such as increasing the bioavailability of drugs, as most new drugs are described as poorly soluble [33], and controlling the delivery kinetics of active principle ingredient (API). To address these issues, this research study proposes the use of cyclodextrins (CD).

Cyclodextrins are cyclic oligosaccharides consisting of glucose subunits linked by α-1-4 glycosidic bonds. Due to their conformation, with a hydrophobic interior and a hydrophilic exterior, these macromolecules exhibit cage-like properties and can form an inclusion complex with hydrophobic compounds [34,35]. These properties have led to their use in various fields, such as cosmetics, food, environment, and medicine [36–38]. Regarded as safe, they are widely used as an excipient in the pharmaceutical field [39,40]. For such applications, β-cyclodextrin (βCD), a cyclodextrin with seven glucose subunits, and its derivatives are the most commonly used [33,41]. βCD are also of a great interest for tissue engineering applications, with their properties to encapsulate lipophilic compounds proven to improve the performance of scaffolds [42–44]. This property could also lead to the immobilisation of the growth factor [45] or drug delivery [39] during the cell growth to optimise the effect of the scaffold. Previous studies reported the association of cyclodextrin with various cellulose derivatives [46,47]. The association between CDs and CNFs or cellulose nanocrystals (CNCs) has been attempted in a very few and recent studies, summarised in Table 1. To the best of our knowledge, no study presents the impact of βCD on both the sorption and the mechanical properties of toCNF substrates (films or cryogels).

The aim of the present study is to modify toCNF with βCD (preferably with covalent linkage) and to see what kind of effect this surface functionalisation has on the sorption and mechanical properties. Thus, a comparison with the same structures using unmodified toCNF is necessary. For that purpose, two suspensions of toCNF with different charge contents were prepared. Fibre modification with cyclodextrins was carried out at neutral and acidic (pH 2.5) to study the effect of the dissociation of the carboxylic acid group. Films, processed by casting/evaporation at 40 °C and cryogels, processed by freeze-drying were prepared from βCD-modified toCNF and compared with reference samples of unmodified toCNF. Water sorption was evaluated gravimetrically for both films and cryogels. The impact of density on the mechanical properties of the cryogels was assessed for cryogels obtained from unmodified toCNF and prepared by freeze-drying from suspensions at different dry matter contents for both charge contents. Compression tests in the dry and swollen state were performed on cryogels from all suspensions, and microscopic observation (SEM) was carried out to link the mechanical behaviour to the macroscopic structure of the materials.

Table 1. Previous works on the association nanocellulose-cyclodextrin. CNFs: cellulose nanofibrils, toCNFs: TEMPO-oxidised cellulose nanofibrils, HP-CNFs: hydroxypropyl cellulose nanofibrils, CNCs: cellulose nanocrystals, βCD: β-cyclodextrin, CMβCD: carboxymethyl- β-cyclodextrin, HPβCD: hydroxypropyl-β-cyclodextrin.

Nanocellulose	CD	Functionalisation Strategy	Application	Source
toCNFs	βCD	Direct grafting	Release of essential oil	[48]
CNFs	βCD	Cross-linking with citric acid	Depollution	[49]
toCNFs	CMβCD	Amidation via EDC/NHS	Depollution	[50]
HP-CNFs	HPβCD	Electrospinning	Drug release	[51]
CNFs	βCD	Coating/Adsorption	Drug release	[52]
CNFs	βCD	Cross-linking with citric acid	Antibacterial packaging	[53]
toCNF	βCD	Noncovalent interaction	Drug Delivery/ Tissue Engineering	[29]
toCNCs	βCD/HPβCD	Direct grafting	Release of essential oil	[54]
CNCs	βCD	Grafting with epichlorohydrin	Tissue engineering	[55]
CNCs	βCD	Crosslinking with fumaric and succinic acid	Release of essential oil	[56]
CNCs	βCD	Ionic interaction	Drug delivery	[57]
CNCs	βCD	Grafting with epichlorohydrin	Supramolecular hydrogels	[58]

2. Results and Discussion

TEMPO-oxidised cellulose nanofibril suspensions were successfully produced. The amounts of carboxylic groups were determined to be 756 ± 4 μmol/g and 1048 ± 32 μmol/g. Films and cryogels were processed from the two different toCNF suspensions in four different conditions presented in Table 2.

Table 2. Samples codes.

Sample Code	βCD	pH
1	0	Neutral
2		Acidic
3	10 wt%	Neutral
4		Acidic

The suspension with 756 μmol/g carboxyl content will be referred as L-toCNF in the text; hence, samples from L-toCNF will be labelled L1, L2, L3, and L4. Similarly, the suspension with 1048-μmol/g carboxyl content will be referred as H-toCNF; i.e., samples from H-toCNF will be labelled H1, H2, H3, and H4. Figure 1 shows atomic force microscopy (AFM) pictures of the nanofibers obtained for both charge contents. Similar and slightly thinner fibrils were obtained for the most oxidised cellulose, as expected.

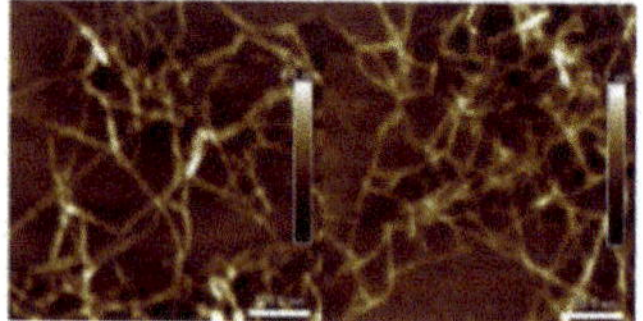

Figure 1. Atomic force microscopy (AFM) images of L-toCNF (left) and H-toCNF (right). toCNFs: TEMPO-oxidised cellulose nanofibrils. L and H are low and high contents, respectively.

2.1. Water Sorption Analysis

2.1.1. Films

Water sorption of films was assessed gravimetrically in a Percival climatic chamber at 25 °C and 90% relative humidity (RH) for 48 h. Figure 2 displays the time dependence of water sorption for films with various carboxyl contents and casting conditions (casting pH and amount of cyclodextrin). Sorption equilibriums and % of sorption equilibriums after 30 min are reported in Table 3.

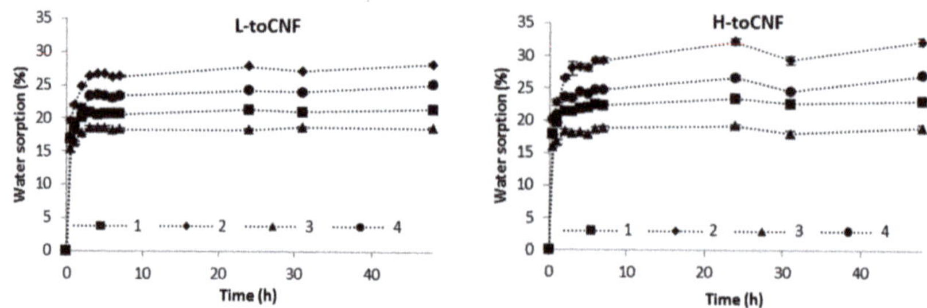

Figure 2. Water sorption as a function of time of conditioning at 25 °C 90% relative humidity (RH) for L-toCNF (left) and H-toCNF (right).

Table 3. Sorption equilibrium after 48 h and % of sorption after 30 min for low-charge content cellulose nanofibrils (L-toCNF) films and high-charge content cellulose nanofibrils (H-toCNF) films.

Sample	Water Sorption Equilibrium after 48 h (wt%)		% of Sorption after 30 min	
	L-toCNF	**H-toCNF**	**L-toCNF**	**H-toCNF**
1	21.5 ± 0.2	23.2 ± 0.2	77.6 ± 0.2	76.7 ± 1.2
2	28.3 ± 0.3	32.3 ± 0.6	69.3 ± 2.0	62.9 ± 2.1
3	18.6 ± 0.1	19.0 ± 0.4	82.0 ± 1.5	83.9 ± 0.9
4	25.2 ± 0.5	27.1 ± 0.5	67.9 ± 2.0	74.1 ± 0.5

For each sample of both L-toCNF and H-toCNF, the sorption equilibrium was reached after approximately 4 h, which indicates that this property is not dependent of any of the variable parameters in this study. Table 3 reports the sorption equilibrium obtained after 48 h. It is slightly higher for H-toCNF samples compared to L-toCNF: + 1.9% for raw suspension samples (condition 1), + 4% for pH 2.5 samples (condition 2), + 0.4% for 10 wt% CD samples (condition 3), and + 1.9% for 10 wt% CD/pH 2.5 samples (condition 4). For both charge contents, the same dynamic between the different casting conditions was observed: sample 3 < 1 < 4 < 2. The carboxylic content at acidic pH is in its carboxylic acid form, hence increasing the water sorption by forming more H-bound with water molecules than in neutral conditions. Cyclodextrins, by adsorbing onto the toCNF fibres, decreases the water sorption ability. For hygroscopic cellulosic materials, the two mechanisms of sorption, namely adsorption and absorption, need to be considered. In addition, depending on the surface chemistry of the fibres, sorption can be either slow or fast [59]. In the case of toCNF, due to their hydroxyl and carboxylic surface groups, sorption occurs quite fast. We can distinguish also two mechanism of sorption: direct sorption, which corresponds to the water molecules that form hydrogen bonds directly with toCNF, and indirect sorption, which corresponds to water molecules that bind with already bound water molecules [10]. Interfibril interactions, on the other hand, may inhibit swelling [60], because the increase in volume of hydrated nanofibrils can be slowed down by other nanofibrils and water binding to the nanofibrils reduces the interfibril binding. While adsorption is a fast process, absorption occurs more slowly, with less water molecules penetrating the inner surface and amorphous regions [61]. SEM images of the cross-section of films (Figure 3) show the same laminar structure with a similar density, regardless the carboxyl content and casting conditions.

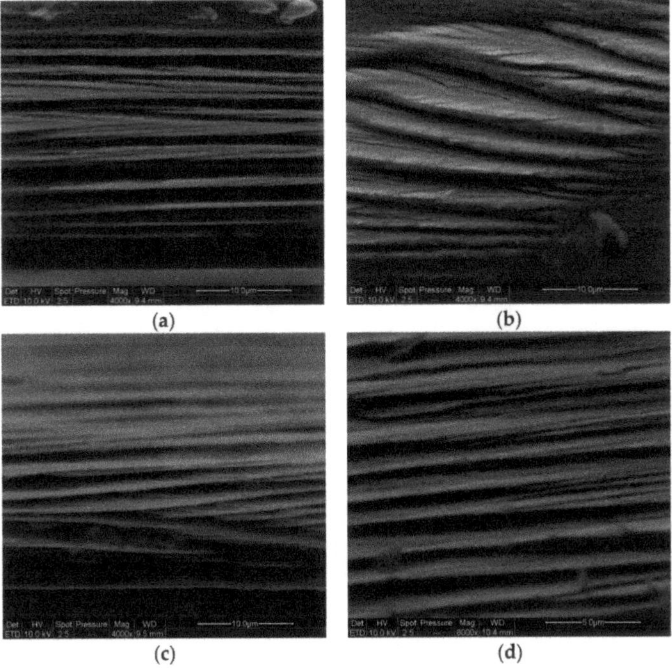

Figure 3. SEM images of the cross-section of films casted from suspensions H1 (**a**), H2 (**b**), H3 (**c**), and H4 (**d**).

In addition, the sorption equilibrium was achieved within a relatively short period of time, with about 65% of the sorption equilibrium reached after 30 min for films cast in acidic conditions, up to around 80% for films containing cyclodextrins. The difference in sorption equilibrium observed between L-toCNF and H-toCNF can be linked to the number of carboxylic functions prone to form H-bonds with water molecules, which is higher for H-toCNF than for L-toCNF. The increase in sorption for films prepared under acidic conditions is explained by the acid form of the carboxylic groups, which is more prone to form H-bonds with water molecules than the carboxylate form at neutral pH. Finally, the decrease in sorption with the addition of cyclodextrin is thought to be due to the adsorption of β-CD on the surface of the fibres, which could decrease the amount of water bounded to the fibres. Water sorption is mainly driven by adsorption and chemical makeup-dependent. For a given drying temperature, the water sorption can be slightly tuned by varying the process parameters, which is an interesting property for drug delivery applications, where swelling and sorption properties are important for the delivery kinetics.

2.1.2. Cryogels

Water sorption tests for cryogels were conducted gravimetrically. The cryogels were weighted and immersed in distilled water and then removed at different time intervals. Tissue paper was used to remove excess water prior to weighing. For both charge contents, the swelling equilibrium was reached after the first measurement at 15 min, as shown in Figure 4. The swelling equilibrium after 3 h was 4871 ± 471% and 4890 ± 141% for L-toCNF and H-toCNF samples, respectively. According to these results, cryogels were immersed 1 h before compression to study the mechanical properties of swollen cryogels so that they would be at swelling equilibrium. It also appeared that the charge content does not have a significant impact on the sorption equilibrium, suggesting that the absorption mechanism in immersion is mainly driven by the porosity and the pore morphology, which are not strongly affected by the charge content.

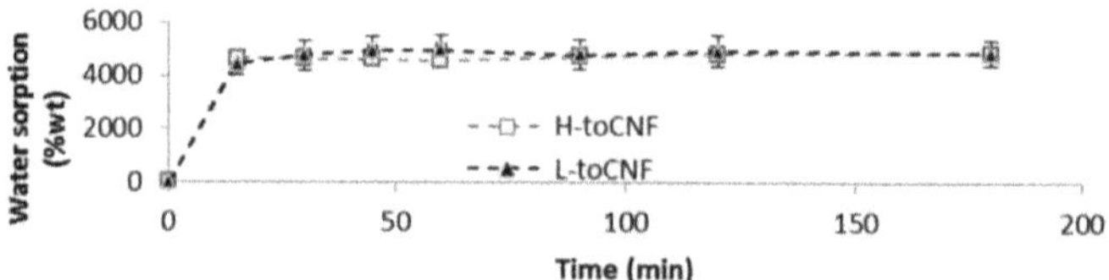

Figure 4. Water sorption for toCNF cryogels in immersion.

2.2. Mechanical Characterisation

Compression tests were carried out on cryogels of cylindrical shape. The density of each sample was determined by dividing the mass of each cryogel by its volume. The volume of the cryogels was measured from height and diameter measurements using a calliper. In each case, the compression curve can be divided into three different regions: For low-strain values, the compression stress increases linearly with the strain in the elastic domain up to the yield point. The compression modulus was calculated at a strain corresponding to half the yield stress in order to be reproducible between all samples. For strains higher than the yield point, the plastic region was reached. In this region the stress increases with the strain with significant residual deformation after unloading. For high compressive strains, the curves exhibited a sharp increase in the compressive stress, typical of a densification regime.

This behaviour has previously been reported for cellulose-based foam materials [15,16,18,19]. The cryogels obtained are closed-cell wall foams, as observed in Figure 5. In such materials, elasticity is caused by the stretching of the cell walls, plastic deformation occurs when the cells are starting to lose their integrity, and densification occurs when cells collapse on themselves, reducing porosity and causing the cryogels to behave like the solid itself [62,63].

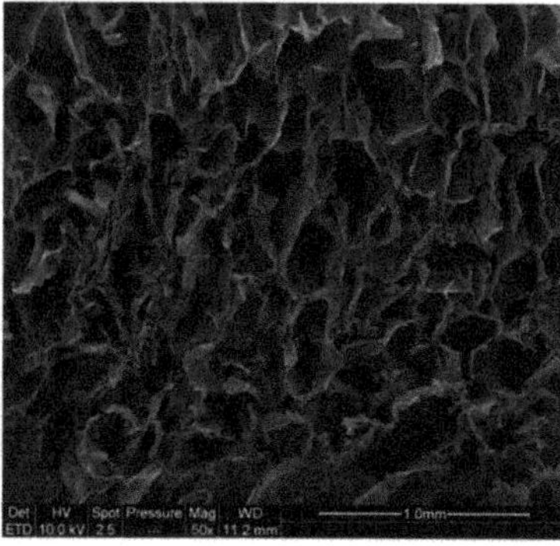

Figure 5. SEM images of the cross-section of toCNF cryogel.

2.2.1. Impact of Density

Compression tests were carried out on cryogels with four different densities prepared from toCNF with both charge densities. Variations of the different properties (compression modulus, normalised compression modulus, and maximum stress at 70% deformation) with the relative density are presented in Figure 6, and numerical values are summarised in Table 4. The results for L-toCNF and H-toCNF are similar for all densities, indicating that the charge content has no major impact on the mechanical properties under the process conditions tested. All the properties increased with the density, quite linearly for the compression modulus, and the maximum stress at 70% deformation, while the normalised compression modulus seems to stabilise for densities higher than 15 mg/cm^3 (which corresponds to a relative density of 0.01). It is worth noting that the normalised compression modulus withstands a huge decrease for relative densities lower than 0.008. Density is of major importance for the mechanical behaviour of cryogels, and by controlling the density, it is possible to tailor the mechanical properties of the cryogel.

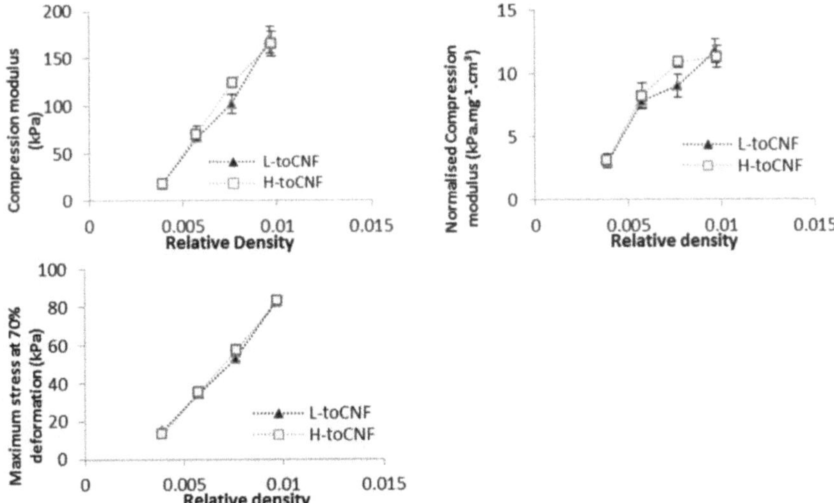

Figure 6. Relative density dependence of the compression modulus, normalised compression modulus, and maximum stress at 70% deformation for L-toCNF and H-toCNF cryogels.

Table 4. Mechanical properties for toCNF at various densities.

Initial Dry Content	Porosity (%)		Compression Modulus (kPa)		Normalised Compression Modulus (kPa·mg^{-1}·cm^3)		Maximum Stress at 70% Deformation (kPa)	
	L-toCNF	H-toCNF	L-toCNF	H-toCNF	L-toCNF	H-toCNF	L-toCNF	H-toCNF
0.4 wt%	99.6	99.6	17 ± 2	18 ± 3	3.0 ± 0.3	3.1 ± 0.6	15 ± 1	13 ± 1
0.6 wt%	99.4	99.4	68 ± 5	71 ± 9	7.8 ± 0.6	8.2 ± 1.0	35 ± 1	35 ± 1
0.8 wt%	99.2	99.2	102 ± 10	125 ± 4	9.0 ± 0.9	10.9 ± 0.4	53 ± 1	57 ± 2
1 wt%	99.0	99.0	170 ± 14	165 ± 13	11.8 ± 0.9	11.3 ± 0.9	83 ± 1	84 ± 1

2.2.2. Impact of pH and Cyclodextrins on Dry Cryogels

Cryogels were prepared from the four compositions by freeze-drying the nanofibril suspensions. SEM images of the cross-section of the cryogels are presented in Figure 7. For all cryogels, closed-cell wall structures are observed, organised as an alveolar structure.

This specific orientation of porosity is due to the freeze-drying process. Indeed, freezing occurred from the bottom part of the freeze-dryer, resulting in the growth of ice in a specific direction, leading to anisotropy in the pore orientation.

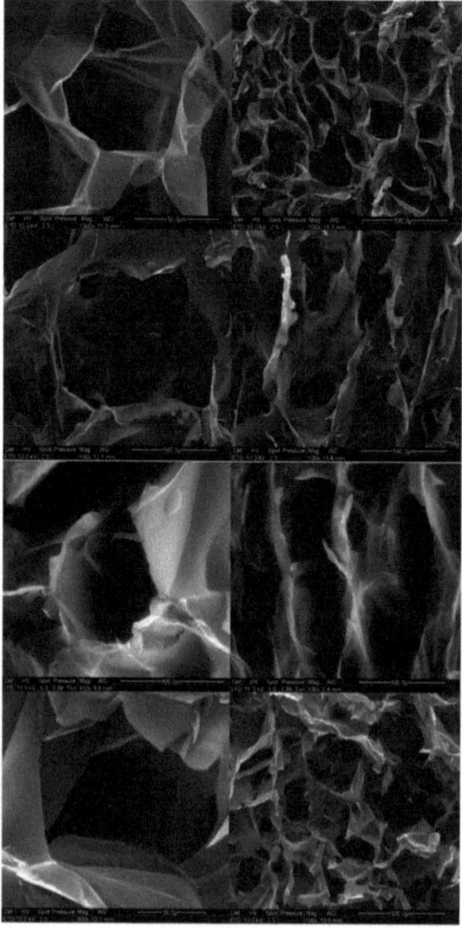

Figure 7. SEM images for toCNF cryogels. From Top to bottom: cryogels from suspensions 1/2/3/4.

In addition, we can observe that the structure of the cell walls for modified toCNF cryogels presents more structural deflects (holes and folds) than the unmodified toCNF one. Compression tests were carried out on cryogels prepared from the four compositions for both charge contents. Typical stress-strain curves for each composition are presented in Figure 8.

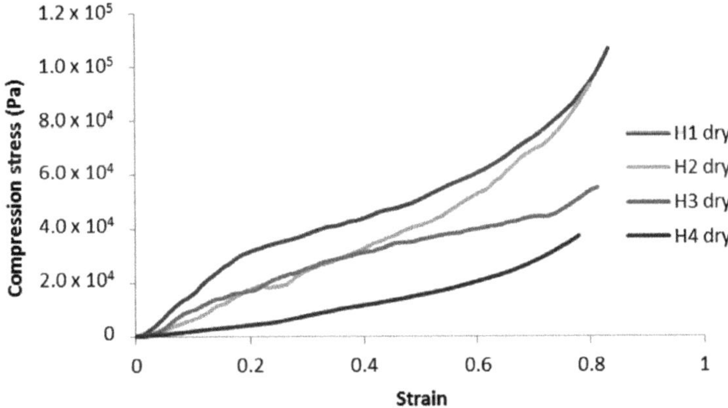

Figure 8. Representative compression curves for H-toCNF cryogels.

Unmodified toCNF cryogels (H1 and H2) exhibit clear elastomeric behaviours, with well-defined linear elastic zones up to a strain of about 0.2, followed by a compression plateau and a densification for higher strains. For modified toCNF cryogels (H3 and H4), the linear elastic zone is restrained to lower strain values, and the yield point is less marked. The deflects observed in Figure 7 in the cell walls for modified toCNF cryogels oppose the elastic buckling of the cells, explaining the small elastic region for modified cryogels. The density, compression modulus, normalised compression modulus, and maximum stress at 70% deformation for each type of cryogel and for both charge contents are summarised in Table 5.

Table 5. Mechanical properties of to-CNF cryogels of different compositions.

	Cryogel Density (mg/cm³)		Compression Modulus (kPa)		Normalised Compression Modulus (kPa·mg⁻¹·cm³)		Maximum Stress at 70% Deformation (kPa)	
	L-toCNF	H-toCNF	L-toCNF	H-toCNF	L-toCNF	H-toCNF	L-toCNF	H-toCNF
1	14.89 ± 0.47	16.07 ± 0.57	104 ± 13	150 ± 10	7.0 ± 0.8	9.4 ± 0.8	64 ± 3	65 ± 3
2	25.32 ± 0.84	20.82 ± 1.08	34 ± 6	76 ± 6	1.4 ± 0.2	3.7 ± 0.3	58 ± 7	63 ± 3
3	17.91 ± 0.60	18.45 ± 1.66	59 ± 10	84 ± 13	3.3 ± 0.5	4.7 ± 0.3	42 ± 2	43 ± 2
4	20.85 ± 0.96	26.41 ± 2.19	37 ± 3	17 ± 3	1.8 ± 0.2	0.6 ± 0.1	59 ± 5	24 ± 4

An increase in density can be observed for cryogels cast under acidic pH and with cyclodextrins. The carboxyl content increases the interactions between the fibres in its carboxylic form, as it forms a more densely packed structure. The adsorption of cyclodextrins on the surface of the fibres also increases the interaction between fibres or creates local deflects in the fibres' arrangement. Both mechanisms impact the mechanical properties. The mechanical properties decrease when the density increases, and the elastic region in the stress-strain curve also decreases for cryogels cast under acidic pH or when containing cyclodextrins. It is also worth noting that the yield point is lower for modified cryogels (pH and cyclodextrins) than for the unmodified one and that the plateau is less pronounced. Since elasticity of foams is linked to the stretching of the cell walls and the plastic behaviour prior to densification is linked to the compression of cells, the modification of fibre-fibre interactions under acidic pH and/or with adsorption of cyclodextrin on the fibre surface is responsible for the modification of the cell wall, thus the mechanical properties. Both charge contents exhibit a similar behaviour for toCNF, toCNF pH 2.5, and toCNF 10 wt% CD, but a noticeable difference is observed for toCNF 10 wt% CD pH 2.5. For L-toCNF, the density decreases between pH 2.5 and 10 wt% CD pH 2.5, while the normalised compression modulus and the maximum stress at deformation are quite similar. For H-toCNF, the density increases between pH 2.5 and 10 wt% CD pH 2.5, while the normalised compression modulus and the maximum stress at 70% deformation decrease. Under these conditions (10 wt% CD and pH 2.5), esterification occurred between the carboxylic acid of the toCNF and hydroxyl groups of cyclodextrins, as evidenced by the attenuated total reflectance-Fourier transform infrared (ATR-FTIR) spectra shown in Figure 9.

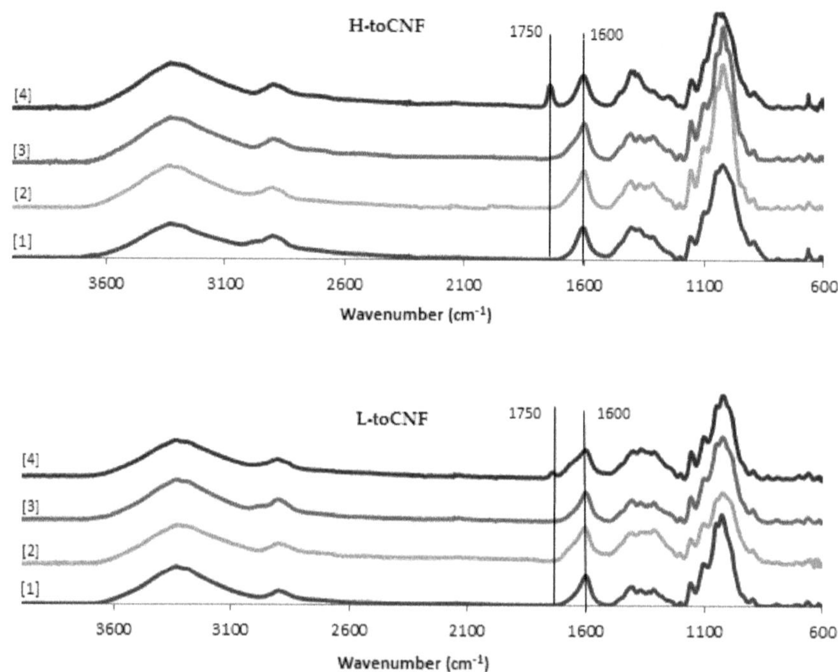

Figure 9. Attenuated total reflectance-Fourier transform infrared (ATR-FTIR) spectra for toCNF cryogels.

The peak observed at 1600 cm^{-1} corresponds to the carboxylate ions present on the surface of the fibres introduced during TEMPO-mediated oxidation. For toCNF 10wt% CD pH 2.5, a peak at 1750 cm^{-1} can be observed which corresponds to the ester groups. The presence of this esterification peak only for toCNF 10wt% CD pH 2.5 indicates that the esterification reaction occurred between the hydroxyl groups of the cyclodextrins and the carboxylic acid groups of toCNF under acidic pH, suggesting that the lyophilisation process allows the reaction by the removal of water. Considering the respective charge contents for L-toCNF and H-toCNF, the efficiency of esterification might be higher for H-toCNF, which can explain the decrease in mechanical properties observed for H4.

2.2.3. Difference Between Dry and Swollen Cryogels

Cryogels were immersed in water for 1 h prior to the experiment, and tissue paper was used to remove excess water before compression. The changes in mechanical properties between the dry and swollen states are summarised in Table 6. Comparison of the typical stress-strain curves for dry and swollen cryogels are presented in Figure 10. For swollen cryogels, no clear elastic zone is observed. The water molecules, by binding to the fibrils, inhibit interfibril interactions and, thus, largely reduce the elastic behaviour. As a result, a significant decrease in mechanical properties is observed for each sample tested, as illustrated in Table 6.

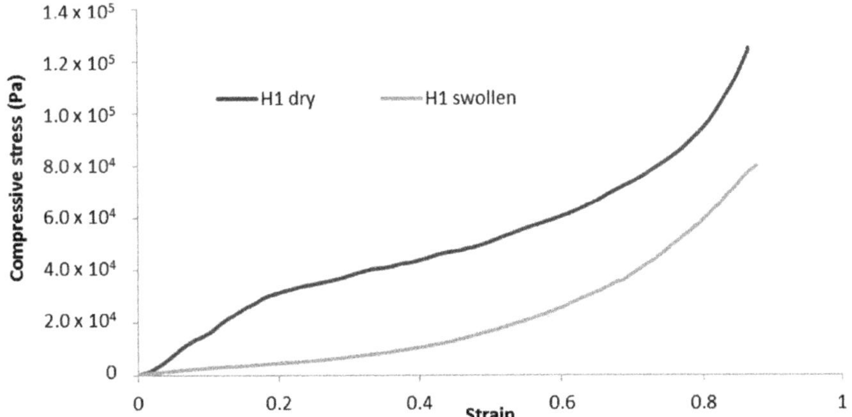

Figure 10. Representative compression curves for dry and swollen H-toCNF cryogels.

Table 6. Diminution of mechanical properties between the dry and swollen states.

Casting Conditions	Compression Modulus (kPa)		Diminution of Compression Modulus (%)		Diminution of Maximum Stress at 70% Deformation (%)	
	L-toCNF	H-toCNF	L-toCNF	H-toCNF	L-toCNF	H-toCNF
1	45 ± 4	33 ± 4	−57 %	−78%	−4%	−54%
2	9 ± 1	8 ± 1	−75 %	−90%	−82%	−90%
3	6 ± 1	7 ± 1	−89%	−92%	−90%	−89%
4	9 ± 1	9 ± 4	−77%	−49%	−82%	−73%

Since no clear elastic deformation zone can be observed for swollen cryogels, the compression modulus was calculated at a low strain (< 0.1) for comparison with dry cryogels. The decrease in mechanical properties is more significant for H-toCNF (H1, H2, and H3 cryogels), which can be explained by the higher charge content, making the fibres more sensitive to the adsorption of water molecules. It is also worth noting that the decrease in mechanical properties is reduced for H4 in comparison with L4, with respectively −49% and −77% decreases in the compression modulus compared to their dry state. This behaviour could be explained by the esterification between cyclodextrin and toCNF, cross-linking fibres, and making the structure less sensitive to water ad/absorption.

Esterification of βCD with toCNF under acidic conditions by freeze-drying was proven. Nevertheless, some questions remain about the yield of grafting and the adsorption mechanism between unbound cyclodextrins and toCNF. Given the chemical similarity between toCNF and βCD, direct characterisation and quantification of grafting seems impossible, as a cyclodextrin with several hydroxyl functions is likely to bind with the carboxylic acid of toCNF. Nevertheless, the presence of multiple hydroxyl functions on both toCNF and βCD indicates a strong adsorption between these two components. However, some cyclodextrins might be trapped in the ice phase during the freezing process and, therefore, are not available for an esterification reaction during the freeze-drying phase. In order to better quantify this adsorption and to be able to confirm that no material will be released from the materials produced, further studies, in particular with the means of Quartz Crystal Microbalance with Dissipation monitoring (QCM-D) and Isothermal Titration Calorimetry (ITC), will be conducted. Additionally, a method adapted from [56] will be implemented, with the use of phenolphthalein (PhP). The interactions between PhP and βCD, described by the authors of [64], could lead to an indirect estimation of βCD available in the materials, and released measurements of βCD could lead to an estimation of the portion of cyclodextrins linked (adsorbed/grafted) to toCNF.

The compression modulus for swollen modified toCNF cryogels ranges between 6 and 9 kPa, i.e., in the range of mechanical stiffness of muscle tissue, and the high porosities obtained (> 99% for all conditions), which are mandatory to promote a good vascularisation, the diffusion of nutrients to the cells, and tissue growth, made this structurally suited for tissue engineering applications. Further studies will be focused on the mechanical properties of such materials under successive stress. Nevertheless, to confirm the potential for tissue-engineering applications, cytotoxicity and degradation studies need to be done in further work. However, βCD toCNF materials could be of a great interest for other applications, such as filtration or depollution, using cyclodextrins to capture molecules of interest rather than release them.

3. Materials and Methods

3.1. Materials

A mixture of bleached and never-dried spruce (picea abies ca. 75%) and pine (pinus sylvestries, ca. 25%) cellulose pulp from Södra (Växjö, Sweden) was used as raw material. All chemicals used in this study were of laboratory-grade quality purchased from Sigma-Aldrich, St. Louis, MO, USA.

3.2. Preparation of toCNF With Two Different Charge Contents

TEMPO-oxidised cellulose nanofibrils were produced according to a protocol adapted from [6]. Never-dried cellulose (110 g of cellulose content) was suspended in water (3 L) and stored overnight at 4 °C. The suspension was dispersed with a blender and mixed with a solution (400 mL) containing TEMPO (1.375 g) and sodium bromide (13.75 g). Water was added to obtain a total volume of 8250 mL (75-mL/g cellulose). TEMPO-mediated oxidation of cellulose was started by adding different amounts of 13% NaClO: 2.5 mmol/g cellulose for a charge content of 750 µmol/g and 3.3 mmol/g cellulose for a charge content of 1100 µmol/g. NaClO were added gradually, and the pH was maintained at 10.5 by adding 0.5M NaOH. The slurry was stirred for 15 min after the complete addition of NaClO, and the pH was then dropped to 7 with 0.1-M HCl. Methanol (100 mL) was then added to the slurry. The product was thoroughly washed with water by filtration until the conductivity of the filtrate was below 5 µS/cm. Homogenisation was conducted using a Rannie 15 type 12.56 × homogeniser (APV, SPX Flow Technology, Silkeborg, Denmark). The suspension was diluted to 1.2 wt% and dispersed with an electric mixer. The fibres underwent two passes in the homogeniser at 600 bar and 1000 bar, respectively. The final suspensions were stored at 4 °C.

3.3. Determination of the Charge Content

The carboxyl group content was determined by conductometric titration as described in previous studies (e.g., [7,32,65]). NaCl (5 mL 0.1M) was added to a toCNF dispersion with 0.2-g solid content in 450 mL. The pH was adjusted to approximately 2.5 by addition of 0.1-M HCl and further diluted with water to a total volume of 500 mL. The dilution was titrated with 0.05-M NaOH solution added at a rate of 0.15 mL/min under stirring up to a pH of 11. An automatic titrator (902 Titrando, Methrom AG, Herisau, Switzerland) was used, and the conductivity of the sample was automatically measured (856 Conductivity Module, Methrom AG, Herisau, Switzerland) for increments of 0.02 mL. Data were recorded by Tiamo Titration software. The carboxyl content was calculated from the titration curve using the Gran plot. Duplicates were made for both suspensions and NaOH titration (control).

3.4. Material Processing

3.4.1. Film Processing

Dry toCNF (0.25 g) was weighed and diluted with water to a total volume of 50 mL. The suspension was dispersed for 2 min at 7000 rpm with an UltraTurrax (IKA-Werke, Staufen, Germany), at room temperature. βCD (0.025 g) and 0.1-M HCl (2 mL) were added to the relevant samples. The suspensions were magnetically stirred for 1 h and placed in an ultra-sonic bath for 3 min. The suspension was then cast in petri dishes (9-cm diameter) and stored in an oven at 40 °C for 18 h. The resulting films were stored in closed petri dishes at room temperature.

3.4.2. Cryogel Processing

Impact of pH, cyclodextrin, and comparison dry/swollen: Fifty millilitres of 0.8 wt% toCNF suspensions were prepared and dispersed 2 min at 7000 rpm with an UltraTurrax. βCD (0.04 g) and 0.1-M HCl (3 mL) were added if required. The suspensions were magnetically stirred for 1 h and placed in an ultra-sonic bath for 3 min. The suspensions were poured into a 24-well plate (3 mL per well) and freeze-dried for 24 h at −20 °C and 0.3 mbar (BK FD12S, Biobase Biodustry, Jinan, China). The resulting cryogels were stored in closed well plates.

Impact of density: Fifty millilitres of L-toCNF and H-toCNF suspensions at 1 wt%, 0.8 wt%, 0.6 w%, and 0.4 wt% were prepared and dispersed for 2 min at 7000 rpm with an UltraTurrax. The suspensions were magnetically stirred for 1 h and placed in an ultra-sonic bath for 3 min. The suspensions were then poured into a 24-well plate (3mL per well) and put in a freezer at −20 °C for 24 h before freeze-drying (ALPHA 2-4 LDplus, Christ ®, Osterode am Harz, Germany).

3.5. Water Sorption Analysis

Water sorption tests on films were carried out gravimetrically in a Percival climatic chamber at 25 °C and 90% RH (relative humidity). The samples were weighted every hour at the beginning of the experiment and at selected times thereafter. Water sorption experiments were conducted after 48 h, with at least 3 replicates for each sample. The samples were put in a desiccator for 16 h prior to the experiment. The water sorption was characterised by the weight change between the initial sample weight (m_0) and the weight after a certain time t (m_t), according to Equation (1):

$$Water\ sorption\ [\%] = \frac{m_t - m_0}{m_0} \times 100 \tag{1}$$

Water sorption test on cryogels were conducted gravimetrically. The cryogels were weighted and immersed in distilled water and then removed at different times. Excess water was removed before weighting. Water sorption was calculated using Equation (1).

3.6. Microscopy

Atomic force microscopy images were recorded on a Dimension icon® (Bruker, Billerica, MA, USA). The concentration of the suspension was adjusted to 10^{-3} wt% by diluting the CNF dispersion using the high shear mixer Ultra-Turrax. A drop of this suspension was deposited on a freshly cleaved mica plate before drying overnight under a fume hood at room temperature. The acquisition was performed in tapping mode using a silica-coated cantilever (OTESPA® 300 kHz-42 N/m, Bruker, Billerica, MA, USA). Zones of 1.1*1.1 μm^2 were analysed.

Scanning electron microscopy images were performed with ESEM (Quanta 200, FEI, Japan). Film and cryogel cross-sections were cut with a razor blade. SEM observation was carried out on cross-sections after carbon sputter coating of 5 nm, with a tension of 10 kV and a spot size of 3.5. The working distance was set between 9.5 mm and 11.5 mm depending on the sample.

For both microscopy techniques, at least 5 different images were performed to check the consistency in various zones of the sample, and the most representatives were selected for the discussion.

3.7. Mechanical Characterisation

Impact of density: Compression tests were performed using a TA Instruments RSA 3 (New Castle, DE, USA) dynamic mechanical analyser fitted with a 100-N load cell. Samples prepared as cylinders were individually measured and compressed with a crosshead speed of 0.1 mm/s at room temperature. At the least, triplicates were performed, and the average is presented.

Impact of pH, cyclodextrin, and comparison dry/swollen: Compression tests were performed with a Stable Micro Systems TA-XT2 texture-analyser (Stable Micro Systems, Godalming, UK), equipped with a P/35 probe and with a crosshead speed of 0.1 mm/s, as previously described by Heggset et al., 2018 [66].

A minimum of 6 cryogels were tested for each sample. The compression modulus was calculated in the elastic region at half the strain of the beginning of the plateau region, and the stress at 70% strain was directly read from the data. The normalised compression modulus was calculated by dividing the compression modulus by the cryogel density. The cryogel density ρ was determined by dividing the mass of each cryogel by its volume. The volume of produced cryogels was measured from height and diameter measurements using a calliper. For each sample, the two extreme values were removed. The relative density of the cryogels was calculated from the ratio ρ/ρ_c, where ρ_c is the density of cellulose, 1.5 g/cm^3 [67]. The porosity was calculated from Equation (2):

$$Porosity\ [\%] = \left(1 - \frac{\rho}{\rho_C}\right) * 100 \tag{2}$$

3.8. Fourier Transform Infrared Spectroscopy

Infrared spectra were recorded in attenuated total reflectance (ATR) mode, using a Perkin Elmer Spectrum 65 (Perkin Elmer, Wellesley, MS, USA). Spectra were recorded between 4000 and 600 cm^{-1}, with 16 scans and a resolution of 4 cm^{-1}. Since this technique is used to determine the possible esterification between the cyclodextrins and the toCNF, and given the proximity between the carboxylic peak and the ester peak (respectively, $\approx$1720 cm^{-1} and $\approx$1750 cm^{-1}), each cryogel was dipped in 0.05 M NaOH for 10 s to convert carboxylic acid groups to carboxylate groups (1600 cm^{-1}) and dried in the oven for 30 min prior to analysis. As the control sample, the neat samples and neat samples after 30 min drying in the oven were also analysed to ensure that esterification was only due to the freeze-drying process. At least 5 different zones of the sample were analysed, and the most representative spectra were used for discussion.

4. Conclusions

In this work, films and cryogels of β-cyclodextrin-modified TEMPO-oxidised cellulose nanofibrils were produced. Water sorption analysis and mechanical characterisation were conducted on both modified and unmodified materials under dry and wet conditions. Two unmodified nanofibrils suspensions were prepared with different charge contents (750 μmol/g and 1050 μmol/g), and modification was carried out under neutral and acidic conditions. The sorption equilibrium was reached after 4 h for all films tested, but the charge content and acidic casting pH were shown to increase the water sorption, while cyclodextrins decreased it. Density, process pH, and the addition of cyclodextrins had major impacts on the mechanical properties, related to the modification of the cell wall structure. Finally, covalent esterification binding between β-cyclodextrin and toCNF under acidic pH by freeze-drying was achieved and had an interesting impact on the mechanical properties in the swollen state. This study is a step towards the production of mechanically tailored cryogels containing cyclodextrin, making them promising materials for the sustained delivery of active principle ingredients.

Author Contributions: Conceptualisation, B.M., J.B., E.B.H., and K.S.; methodology, B.M., J.B., A.D., E.B.H., and K.S.; validation, B.M.; formal analysis, B.M.; investigation, B.M.; resources, J.B., A.D., E.B.H., and K.S.; data curation, B.M.; writing—original draft preparation, B.M.; writing—review and editing, B.M., J.B., A.D., E.B.H., and K.S.; visualisation, B.M.; supervision, J.B., A.D., E.B.H., and K.S.; project administration, J.B., A.D., E.B.H., and K.S.; and funding acquisition, J.B., E.B.H., and K.S. All authors have read and agreed to the published version of the manuscript.

Funding: This work is supported by the French National Research Agency in the framework of the "Investissements d'avenir" program Glyco@Alps (ANR-15-IDEX-02) and NTNU through its Department of Chemical Engineering. LGP2 is part of the LabEx Tec 21 (Investissements d'Avenir—Grant Agreement No. ANR-11-LABX-0030) and of the PolyNat Carnot Institute (Investissements d'Avenir—Grant Agreement No. ANR-16-CARN-0025-01). This research was made possible thanks to the facilities of the TekLiCell platform funded by the Région Rhône-Alpes (ERDF: European regional development fund). The authors thank the IDEX Université Grenoble Alpes for funding B.M.'s travel grant.

Acknowledgments: The authors acknowledge Cecile Sillard (LGP2) for AFM images; Berthine Khelifi (LGP2) for SEM images; and Ingebjørg Leirset (RISE PFI), Johnny Kvakland Melbø (RISE PFI), and Anne Marie Reitan (RISE PFI) for lab support.

Conflicts of Interest: The authors declare no conflicts of interest.

References

1. Turbak, A.F.; Snyder, F.W. *Microfibrillated Cellulose*; Patent and Trademark Office: Washington, DC, USA, 1983.

2. Taipale, T.; Österberg, A.; Nykänen, M.; Ruokolainen, J.; Laine, J. Effect of microfibrillated cellulose and fines on the drainage of kraft pulp suspension and paper strength. *Cellulose* **2010**, *17*, 1005–1020. [CrossRef]

3. Josset, S.; Orsolini, P.; Siqueira, G.; Tejado, A.; Tingaut, P.; Zimmermann, T. Energy consumption of the nanofibrillation of bleached pulp, wheat straw and recycled newspaper through a grinding process. *Nord. Pulp Pap. Res. J.* **2014**, *29*, 167–175. [CrossRef]

4. Rol, F.; Belgacem, M.N.; Gandini, A.; Bras, J. Recent advances in surface-modified cellulose nanofibrils. *Prog. Polym. Sci.* **2018**. [CrossRef]

5. Abitbol, T.; Rivkin, A.; Cao, Y.; Nevo, Y.; Abraham, E.; Ben-Shalom, T.; Lapidot, S.; Shoseyov, O. Nanocellulose, a tiny fiber with huge applications. *Curr. Opin. Biotechnol.* **2016**, *39*, 76–88. [CrossRef] [PubMed]

6. Saito, T.; Nishiyama, Y.; Putaux, J.-L.; Vignon, M.; Isogai, A. Homogeneous suspensions of individualized microfibrils from TEMPO-catalyzed oxidation of native cellulose. *Biomacromolecules* **2006**, *7*, 1687–1691. [CrossRef] [PubMed]

7. Isogai, A.; Saito, T.; Fukuzumi, H. TEMPO-oxidized cellulose nanofibers. *Nanoscale* **2011**, *3*, 71–85. [CrossRef] [PubMed]

8. Hubbe, M.A.; Ferrer, A.; Tyagi, P.; Yin, Y.; Salas, C.; Pal, L.; Rojas, O.J. Nanocellulose in thin films, coatings, and plies for packaging applications: A Review. *BioResources* **2017**, *12*, 2143–2233. [CrossRef]

9. Kolakovic, R.; Peltonen, L.; Laukkanen, A.; Hirvonen, J.; Laaksonen, T. Nanofibrillar cellulose films for controlled drug delivery. *Eur. J. Pharm. Biopharm.* **2012**, *82*, 308–315. [CrossRef]

10. Torstensen, J.Ø.; Liu, M.; Jin, S.-A.; Deng, L.; Hawari, A.I.; Syverud, K.; Spontak, R.J.; Gregersen, Ø.W. Swelling and free-volume characteristics of TEMPO-oxidized cellulose nanofibril films. *Biomacromolecules* **2018**, *19*, 1016–1025. [CrossRef]

11. Kontturi, K.S.; Biegaj, K.; Mautner, A.; Woodward, R.T.; Wilson, B.P.; Johansson, L.-S.; Lee, K.-Y.; Heng-Orcid, J.Y.Y.; Bismarck, A.; Kontturi, E. Noncovalent surface modification of cellulose nanopapers by adsorption of polymers from aprotic solvents. *Langmuir* **2017**, *33*, 5707–5712. [CrossRef]

12. Orsolini, P.; Michen, B.; Huch, A.; Tingaut, P.; Caseri, W.R.; Zimmermann, T. Characterization of pores in dense nanopapers and nanofibrillated cellulose membranes: A critical assessment of established methods. *ACS Appl. Mater. Interfaces* **2015**, *7*, 25884–25897. [CrossRef] [PubMed]

13. Darpentigny, C.; Nonglaton, G.; Bras, J.; Jean, B. Highly absorbent cellulose nanofibrils aerogels prepared by supercritical drying. *Carbohydr. Polym.* **2020**, *229*, 115560. [CrossRef]

14. De France, K.J.; Hoare, T.; Cranston, E.D. Review of hydrogels and aerogels containing nanocellulose. *Chem. Mater.* **2017**, *29*, 4609–4631. [CrossRef]

15. Buchtová, N.; Pradille, C.; Budtova, J.-L. Mechanical properties of cellulose aerogels and cryogels. *Soft Matter* **2019**, *15*, 7901–7908. [CrossRef] [PubMed]

16. Darpentigny, C.; Molina-Boisseau, S.; Nonglaton, G.; Bras, J.; Jean, B. Ice-templated freeze-dried cryogels from tunicate cellulose nanocrystals with high specific surface area and anisotropic morphological and mechanical properties. *Cellulose* **2020**, *27*, 233–247. [CrossRef]

17. Lavoine, N.; Bergström, L. Nanocellulose-based foams and aerogels: Processing, properties, and applications. *J. Mater. Chem. A* **2017**, *5*, 16105–16117. [CrossRef]

18. Gupta, S.; Martoïa, F.; Orgéas, L.; Dumont, P. Ice-templated porous nanocellulose-based materials: Current progress and opportunities for materials engineering. *Appl. Sci.* **2018**, *8*, 2463. [CrossRef]

19. Martoïa, F.; Cochereau, T.; Dumont, P.J.J.; Orgéas, L.; Terrien, M.; Belgacem, M.N. Cellulose nanofibril foams: Links between ice-templating conditions, microstructures and mechanical properties. *Mater. Des.* **2016**, *104*, 376–391. [CrossRef]

20. Rees, A.; Powell, L.C.; Chinga-Carrasco, G.; Gethin, D.T.; Syverud, K.; Hill, K.E.; Thomas, D.W. 3D Bioprinting of carboxymethylated-periodate oxidized nanocellulose constructs for wound dressing applications. *Biomed Res. Int.* **2015**, *2015*, 1–7. [CrossRef]

21. Hakkarainen, T.; Koivuniemi, R.; Kosonen, M.; Escobedo-Lucea, C.; Sanz-Garcia, A.; Vuola, J.; Valtonen, J.; Tammela, P.; Mäkitie, A.; Luukko, K.; et al. Nanofibrillar cellulose wound dressing in skin graft donor site treatment. *J. Control. Release* **2016**, *244*, 292–301. [CrossRef]

22. Campodoni, E.; Heggset, E.B.; Rashad, A.; Ramírez-Rodríguez, G.B.; Mustafa, K.; Syverud, K.; Tampieri, A.; Sandri, M. Polymeric 3D scaffolds for tissue regeneration: Evaluation of biopolymer nanocomposite reinforced with cellulose nanofibrils. *Mater. Sci. Eng. C* **2019**, *94*, 867–878. [CrossRef] [PubMed]

23. Howard, D.; Buttery, L.D.; Shakesheff, K.M.; Roberts, S.J. Tissue engineering: Strategies, stem cells and scaffolds. *J. Anat.* **2008**, *213*, 66–72. [CrossRef] [PubMed]

24. Discher, D.E. Tissue cells feel and respond to the stiffness of their substrate. *Science* **2005**, *310*, 1139–1143. [CrossRef] [PubMed]

25. Engler, A.J.; Sen, S.; Sweeney, H.L.; Discher, D.E. Matrix elasticity directs stem cell lineage specification. *Cell* **2006**, *126*, 677–689. [CrossRef]

26. O'Brien, F.J. Biomaterials & scaffolds for tissue engineering. *Mater. Today* **2011**, *14*, 88–95. [CrossRef]

27. Alexandrescu, L.; Syverud, K.; Gatti, A.; Chinga-Carrasco, G. Cytotoxity tests of cellulose nanofibril-based structures. *Cellulose* **2013**, *20*, 1765–1775. [CrossRef]

28. Basu, A.; Celma, G.; Strømme, M.; Ferraz, N. In vitro and in vivo evaluation of the wound healing properties of nanofibrillated cellulose hydrogels. *ACS Appl. Bio Mater.* **2018**, *1*, 1853–1863. [CrossRef]

29. Fiorati, A.; Contessi-Negrini, N.; Baschenis, E.; Altomare, L.; Farè, S.; Giacometti-Schieroni, A.; Piovani, D.; Mendichi, R.; Ferro, M.; Castiglione, F.; et al. TEMPO-nanocellulose/Ca2+ hydrogels: Ibuprofen drug diffusion and in vitro cytocompatibility. *Materials* **2020**, *13*, 183. [CrossRef]

30. Rashad, A.; Mustafa, K.; Heggset, E.B.; Syverud, K. Cytocompatibility of wood-derived cellulose nanofibril hydrogels with different surface chemistry. *Biomacromolecules* **2017**, *18*, 1238–1248. [CrossRef]

31. Syverud, K.; Kirsebom, H.; Hajizadeh, S.; Chinga-Carrasco, G. Cross-linking cellulose nanofibrils for potential elastic cryo-structured gels. *Nanoscale Res. Lett.* **2011**, *6*, 626. [CrossRef]

32. Syverud, K.; Pettersen, S.R.; Draget, K.; Chinga-Carrasco, G. Controlling the elastic modulus of cellulose nanofibril hydrogels—Scaffolds with potential in tissue engineering. *Cellulose* **2015**, *22*, 473–481. [CrossRef]

33. Loftsson, T.; Brewster, M.E. Pharmaceutical applications of cyclodextrins: Basic science and product development: Pharmaceutical applications of cyclodextrins. *J. Pharm. Pharmacol.* **2010**, *62*, 1607–1621. [CrossRef] [PubMed]

34. Kurkov, S.V.; Loftsson, T. Cyclodextrins. *Int. J. Pharm.* **2013**, *453*, 167–180. [CrossRef] [PubMed]

35. Szejtli, J. Introduction and general overview of cyclodextrin chemistry. *Chem. Rev.* **1998**, *98*, 1743–1754. [CrossRef]

36. Del Valle, E.M.M. Cyclodextrins and their uses: A review. *Process. Biochem.* **2004**, *39*, 1033–1046. [CrossRef]

37. Crini, G.; Fourmentin, S.; Fourmentin, M.; Morin-Crini, N. *Principales Applications des Complexes d'Inclusion Cyclodextrine/Substrat*; French Society of Chemistry: Paris, France, 2019; p. 21.

38. Morin-Crini, N.; Winterton, P.; Fourmentin, S.; Wilson, L.D.; Fenyvesi, É.; Crini, G. Water-insoluble β-cyclodextrin–epichlorohydrin polymers for removal of pollutants from aqueous solutions by sorption processes using batch studies: A review of inclusion mechanisms. *Prog. Polym. Sci.* **2018**, *78*, 1–23. [CrossRef]

39. Otero-Espinar, F.J.; Torres-Labandeira, J.J.; Alvarez-Lorenzo, C.; Blanco-Méndez, J. Cyclodextrins in drug delivery systems. *J. Drug Deliv. Sci. Technol.* **2010**, *20*, 289–301. [CrossRef]

40. Jambhekar, S.S.; Breen, P. Cyclodextrins in pharmaceutical formulations I: Structure and physicochemical properties, formation of complexes, and types of complex. *Drug Discov. Today* **2016**, *21*, 356–362. [CrossRef]

41. Cusola, O.; Tabary, N.; Belgacem, M.N.; Bras, J. Cyclodextrin functionalization of several cellulosic substrates for prolonged release of antibacterial agents. *J. Appl. Polym. Sci.* **2013**, *129*, 604–613. [CrossRef]

42. Lukášek, J.; Hauzerová, Š.; Havlíčková, K.; Strnadová, K.; Mašek, K.; Stuchlík, M.; Stibor, I.; Jenčová, V.; Řezanka, M. Cyclodextrin-polypyrrole coatings of scaffolds for tissue engineering. *Polymers* **2019**, *11*, 459. [CrossRef]

43. Venuti, V.; Venuti, V.; Rossi, B.; Mele, A.; Melone, L.; Punta, C.; Majolino, D.; Masciovecchio, C.; Caldera, F.; Trotta, F. Tuning structural parameters for the optimization of drug delivery performance of cyclodextrin-based nanosponges. *Expert Opin. Drug Deliv.* **2017**, *14*, 331–340. [CrossRef] [PubMed]

44. Alvarez-Lorenzo, C.; García-González, C.A.; Concheiro, A. Cyclodextrins as versatile building blocks for regenerative medicine. *J. Control. Release* **2017**, *268*, 269–281. [CrossRef]

45. Grier, W.K.; Tiffany, A.S.; Ramsey, M.D.; Harley, B.A.C. Incorporating β-cyclodextrin into collagen scaffolds to sequester growth factors and modulate mesenchymal stem cell activity. *Acta Biomaterialia* **2018**, *76*, 116–125. [CrossRef] [PubMed]

46. Nada, A.A.; Abdellatif, F.H.H.; Ali, E.A.; Abdelazeem, R.A.; Soliman, A.A.S.; Abou-Zeid, N.Y. Cellulose-based click-scaffolds: Synthesis, characterization and biofabrications. *Carbohydr. Polym.* **2018**, *199*, 610–618. [CrossRef]

47. Kim, K.O.; Kim, G.J.; Kim, J.H. A cellulose/β-cyclodextrin nanofiber patch as a wearable epidermal glucose sensor. *RSC Adv.* **2019**, *9*, 22790–22794. [CrossRef]

48. Saini, S.; Quinot, D.; Lavoine, N.; Belgacem, M.N.; Bras, J. β-Cyclodextrin-grafted TEMPO-oxidized cellulose nanofibers for sustained release of essential oil. *J. Mater. Sci.* **2017**, *52*, 3849–3861. [CrossRef]

49. Yuan, G.; Prabakaranb, M.; Sunc, Q.; Jung, L.; Chung, S.; Mayakrishnan, I.-M.; Song, G.; Kim, K.-H.; Soo, I. Cyclodextrin functionalized cellulose nanofiber composites for the faster adsorption of toluene from aqueous solution. *J. Taiwan Inst. Chem. Eng.* **2017**, *70*, 352–358. [CrossRef]

50. Ruiz-Palomero, C.; Soriano, M.L.; Valcárcel, M. β-Cyclodextrin decorated nanocellulose: A smart approach towards the selective fluorimetric determination of danofloxacin in milk samples. *Analyst* **2015**, *140*, 3431–3438. [CrossRef]

51. Aytac, Z.; Sen, H.S.; Durgun, E.; Uyar, T. Sulfisoxazole/cyclodextrin inclusion complex incorporated in electrospun hydroxypropyl cellulose nanofibers as drug delivery system. *Colloids Surf. B Biointerfaces* **2015**, *128*, 331–338. [CrossRef]

52. Lavoine, N.; Tabary, N.; Desloges, I.; Martel, B.; Bras, J. Controlled release of chlorhexidine digluconate using β-cyclodextrin and microfibrillated cellulose. *Colloids Surf. B Biointerfaces* **2014**, *121*, 196–205. [CrossRef]

53. Lavoine, N.; Givord, C.; Tabary, N.; Desloges, I.; Martel, B.; Bras, J. Elaboration of a new antibacterial bio-nano-material for food-packaging by synergistic action of cyclodextrin and microfibrillated cellulose. *Innov. Food Sci. Emerg. Technol.* **2014**, *26*, 330–340. [CrossRef]

54. De Castro, D.O.; Tabary, N.; Martel, B.; Gandini, A.; Belgacem, N.; Bras, J. Controlled release of carvacrol and curcumin: Bio-based food packaging by synergism action of TEMPO-oxidized cellulose nanocrystals and cyclodextrin. *Cellulose* **2018**, *25*, 1249–1263. [CrossRef]

55. Jimenez, A.; Jaramillo, F.; Hemraz, U.; Boluk, Y.; Ckless, K.; Sunasee, R. Effect of surface organic coatings of cellulose nanocrystals on the viability of mammalian cell line. *Nanotechnol. Sci. Appl.* **2017**, *10*, 123–136. [CrossRef] [PubMed]

56. Castro, D.O.; Tabary, N.; Martel, B.; Gandini, A.; Belgacem, N.; Bras, J. Effect of different carboxylic acids in cyclodextrin functionalization of cellulose nanocrystals for prolonged release of carvacrol. *Mater. Sci. Eng. C* **2016**, *69*, 1018–1025. [CrossRef] [PubMed]

57. Ndong-Ntoutoume, G.M.A.; Graneta, R.; Pierre, J.; Frédérique, M.; Légera, D.Y.; Fidanzi-Dugasa, C.; Lequartb, V.; Jolyb, N.; Liagrea, V.; Chaleixa, B.; et al. Development of curcumin–cyclodextrin/cellulose nanocrystals complexes: New anticancer drug delivery systems. *Bioorganic Med. Chem. Lett.* **2016**, *26*, 941–945. [CrossRef]

58. Lin, N.; Dufresne, A. Supramolecular hydrogels from in situ host–guest inclusion between chemically modified cellulose nanocrystals and cyclodextrin. *Biomacromolecules* **2013**, *14*, 871–880. [CrossRef]

59. Okubayashi, S.; Griesser, U.J.; Bechtold, T. A kinetic study of moisture sorption and desorption on lyocell fibers. *Carbohydr. Polym.* **2004**, *58*, 293–299. [CrossRef]

60. Dufresne, A. *Nanocellulose: From Nature to High Performance Tailored Materials*; De Gruyter: Berlin, Germany, 2012.

61. Belbekhouche, S.; Bras, J.; Siqueira, G.; Chappey, C.; Lebrun, L.; Khelifi, B.; Marais, S.; Dufresne, A. Water sorption behavior and gas barrier properties of cellulose whiskers and microfibrils films. *Carbohydr. Polym.* **2011**, *83*, 1740–1748. [CrossRef]

62. Gibson, L.A.; Ashby, M. *Cellular Solids, Structure and Properties*; Cambridge University Press: Cambridge, UK, 1999.
63. Mills, N. *Polymer Foams Handbook: Engineering and Biomechanics Applications and Design Guide*; Elsevier: Amsterdam, The Netherlands, 2007.
64. Goel, A.; Nene, S.N. Modifications in the phenolphthalein method for spectrophotometric estimation of beta cyclodextrin. *Starch Stärke* **1995**, *47*, 399–400. [CrossRef]
65. Orelma, H.; Filpponen, I.; Johansson, L.-S.; Österberg, M.; Rojas, O.J.; Laine, J. Surface functionalized nanofibrillar cellulose (NFC) film as a platform for immunoassays and diagnostics. *Biointerphases* **2012**, *7*, 61. [CrossRef]
66. Heggset, E.B.; Strand, B.L.; Sundby, K.W.; Simon, S.; Chinga-Carrasco, G.; Syverud, K. Viscoelastic properties of nanocellulose based inks for 3D printing and mechanical properties of CNF/alginate biocomposite gels. *Cellulose* **2018**. [CrossRef]
67. Weishaupt, R.; Siqueira, G.; Schubert, M.; Tingaut, P.; Maniura-Weber, K.; Zimmermann, T.; Thöny-Meyer, L.; Faccio, G.; Ihssen, J. TEMPO-oxidized nanofibrillated cellulose as a high density carrier for bioactive molecules. *Biomacromolecules* **2015**, *16*, 3640–3650. [CrossRef] [PubMed]

 *molecules*

Article

Optimization with Response Surface Methodology of Microwave-Assisted Conversion of Xylose to Furfural

Carmen Padilla-Rascón [1,2], Juan Miguel Romero-García [1,2] , Encarnación Ruiz [1,2,*] and Eulogio Castro [1,2]

[1] Department of Chemical, Environmental and Materials Engineering, Universidad de Jaén, Campus Las Lagunillas, 23071 Jaén, Spain; cpadilla@ujaen.es (C.P.-R.); jrgarcia@ujaen.es (J.M.R.-G.); ecastro@ujaen.es (E.C.)

[2] Centre for Advanced Studies in Earth Sciences, Energy and Environment (CEACTEMA), Universidad de Jaén, Campus Las Lagunillas, 23071 Jaén, Spain

* Correspondence: eruiz@ujaen.es; Tel.: +34-953212779

Academic Editors: Alejandro Rodríguez, Eduardo Espinosa and Sylvain Caillol

Received: 9 June 2020; Accepted: 30 July 2020; Published: 6 August 2020

Abstract: The production of furfural from renewable sources, such as lignocellulosic biomass, has gained great interest within the concept of biorefineries. In lignocellulosic materials, xylose is the most abundant pentose, which forms the hemicellulosic part. One of the key steps in the production of furfural from biomass is the dehydration reaction of the pentoses. The objective of this work was to assess the conditions under which the concentration of furfural is maximized from a synthetic, monophasic, and homogeneous xylose medium. The experiments were carried out in a microwave reactor. $FeCl_3$ in different proportions and sulfuric acid were used as catalysts. A two-level, three-factor experimental design was developed for this purpose. The results were further analyzed through a second experimental design and optimization was performed by response surface methodology. The best operational conditions for the highest furfural yield (57%) turned out to be 210 °C, 0.5 min, and 0.05 M $FeCl_3$.

Keywords: lignocellulosic material; xylose; furfural; iron chloride; microwave reactor; biorefinery

1. Introduction

Owing to the depletion of fossil resources, it is necessary to look for renewable sources for the production of fuels and chemicals. Lignocellulosic biomass is an alternative natural resource that presents multiple advantages, because of its abundance, renewability, low cost, and the fact that it is a waste product with no competing uses. Lignocellulosic materials are mainly composed of hemicellulose, cellulose, and lignin. Other fractions also present in a lower proportion are extractives, which are low molecular weight organic compounds, and the ashes, which are inorganic compounds. Although the chemical composition of the biomass is variable, hemicellulose is usually the second most abundant compound with multiple uses, after cellulose. Xylose is the most abundant pentose in the hemicelluloses of hardwoods and agricultural plants [1].

In recent years, the production of furfural from renewable sources has gained great attention within the concept of sustainable biorefineries, giving added value to lignocellulosic materials and advancing the sustainable bioeconomy. The U.S. Department of Energy identified furfural as one of the top 30 platform chemicals [2]. Its composition, formed by unsaturated bonds and aldehyde groups, makes it a versatile platform molecule, from which a great variety of chemical compounds can be obtained, with applications in oil refining, production of plastics, food, and pharmaceutical and agricultural industries [3–5].

Currently, the most commonly used lignocellulosic biomasses for production of furfural are corn cob, sugar cane bagasse, rice hull, and wheat bran, considering that they are rich in hemicelluloses and are available in large quantities [1]. Olive subproducts, specifically olive stones or pits, have also been used to obtain furfural [6]. Among the advantages of using olive pits, the high availability in Mediterranean areas, their composition with a high proportion of hemicelluloses, and their easy acid hydrolysis are frequently cited.

Furfural is obtained from lignocellulosic materials in two steps, as shown briefly in Figure 1. The first step reaction, hydrolysis, consists of the transformation of polysaccharides into monosaccharides, obtaining a hydrolysate rich in sugars; in this case, the predominant sugar is xylose. The second step is the dehydration reaction of pentoses, mainly xylose, to furfural. This process can be done within the same reactor, one-stage, or separately in two steps. When the process is carried out in two steps, the hydrolysate can be separated from the insolubilized solid after the first step, allowing greater utilization of the lignocellulosic material. Furfural is obtained by dehydration of the pentoses (mainly xylose) and the unhydrolyzed solid can be employed for other uses, such as enzymatic hydrolysis and fermentation [7].

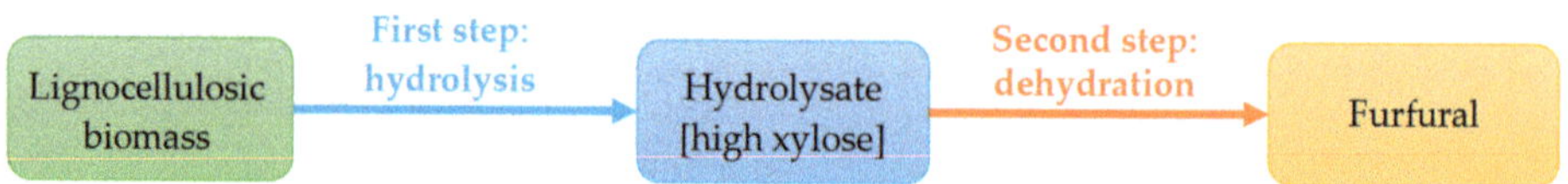

Figure 1. Schematic diagram of furfural production from lignocellulosic biomass.

The pentose dehydration reaction involves the protonation of three carbon atoms in the sugar ring, removing three water molecules and obtaining the furanic molecules. Normally, these reactions take place in an aqueous medium, owing to its polarity, availability, sustainability, and low price. In the dehydration reaction, undesired secondary reactions also occur, by condensation between the furfural and the intermediate products of dehydration and by the degradation of the pentoses to low molecular weight products, generating soluble and insoluble secondary products (called humins) [8]. The use of catalysts facilitates the dewatering reaction. Most of the catalysts studied have Brønsted sites, which directly dehydrate the initial carbohydrate. Furthermore, the use of Lewis acid sites favors the isomerization of the carbohydrate and the subsequent dehydration through the Brønsted sites. These catalysts can be homogeneous, the disadvantages of which are high pressure and the production of a corrosive acid stream that needs to be neutralized. Alternatively, heterogeneous catalysts, limited by their cost and synthesis, are also used and may be subjected to deactivation owing to the deposit of subproducts on their surface [9,10].

Hemicellulose-rich raw materials can be hydrolyzed by various processes, among which the dilute acid hydrolysis stands out as one of the most efficient to selectively recover hemicellulose sugars [7,11,12]. These acids, like sulfuric acid, also enhance the dehydration reaction, in which it acts as Lewis acid [13]. Furthermore, the potential of salts as a catalyst for these reactions has been corroborated in multiple studies. Metal salts, and especially metal chlorides, such as $FeCl_3$, have been used in the pretreatment step of lignocellulosic materials to improve the generation of sugars that can in turn be converted into valuable chemicals, acting also as a catalyst for the dehydration of sugars into furfural, as a prominent example. According to the proposed mechanism by which this general process takes place, $FeCl_3$ acts as Brønsted acid in several reactions whose equilibrium is influenced by the pH and the initial concentration of the metal salt, improving the performance in obtaining furfural [4,14,15]. A complete description of the way metal salts can act for pretreating lignocellulosic materials along with relevant examples is available elsewhere [16].

Microwave-assisted reactions are booming, as they have many advantages, for example, they are versatile, need shorter reaction times, present uniform heating, and can be more efficient reactions. In multiple studies, microwave-assisted heating resulted in better yields compared with traditional heating methods [10,17,18]. Microwaves are widely used with lignocellulosic materials for the production of value-added chemicals such as furfural [5,19]. Another application is the fractionation of lignocellulosic materials. For example, Zhang et al. [20] reported on a microwave-assisted organoslv pretreatment using raw poplar. In another study, rye and wheat stillage were submitted to a first microwave treatment to produce a highly concentrated sugar solution, which was further fermented to ethanol [21]. Microwaves were also used to assist hemicellulose extraction from corn fiber [22], or for extracting phenolics from cocoa pod husk [23]. Recently, microwave-assisted treatment by deep eutectic solvents was reported for the delignification of garlic skin [24].

Response surface methodology is a tool that allows obtaining a mathematical model to establish a relationship between the parameters of interest studied (temperature, time, and so on) and the results obtained (yields, conversions, and so on). The interrelationships between the factors studied and their influence on the responses obtained are analyzed statistically to determine their significance. It also allows finding the best experimental conditions based on the results. The use of this methodology is widespread when optimizing the process conditions for furfural production [25–27].

As a previous step to get useful information on the best conditions to produce furfural from the pentose fraction of lignocellulosic materials, this paper addresses the use of sulfuric acid and ferric chloride to obtain furfural from a solution with a high concentration of xylose (30 g/L) in a monophasic and homogeneous medium. The microwave heating technique is used because of its advantages. The main objective is to optimize the conditions for the highest furfural production. For this purpose, two experimental designs were made, with temperature, time, and concentration of ferric chloride as factors.

2. Results

2.1. Two-Level Factorial Design

The results obtained from the factorial design of 11 experiments (2^3 + 3 central points) with the three factors studied (concentration of $FeCl_3$, time, and temperature) are shown in Table 1. The initial concentration of xylose was 30 g/L, while the consumption values ranged between 7.52 and 29.44 g/L, which is equivalent to conversion between 25.06 and 98.13% (above 90% in most cases). As for the furfural obtained, the values were between 3.02 and 10.35 g/L, which represents a yield of 15.63 and 53.59%, respectively. Finally, the selectivity ranged between 25.83 and 62.38%, with the average above 45%.

Table 1. Two-level, three-factor experimental design. Experimental conditions and response results.

Run	Factors			Responses				
	A: $FeCl_3$ (M)	B: Time (min)	C: Temp. (°C)	Xylose Consumed (g/L)	Furfural (g/L)	Yield(%)	Conversion (%)	Selectivity (%)
2	0.3	1	170	16.68	4.79	24.87	55.61	44.73
4	0.1	5	200	29.33	7.84	38.99	97.77	39.88
6	0.1	5	170	18.11	6.52	33.87	60.38	56.10
7	0.3	1	200	29.44	7.96	41.13	98.13	41.91
8	0.3	5	170	26.75	7.72	39.93	89.15	44.79
9	0.1	1	200	29.07	10.35	53.59	96.91	55.30
10	0.1	1	170	7.52	3.02	15.63	25.06	62.38
11	0.3	5	200	28.99	4.82	24.96	96.63	25.83
1	0.2	3	185	29.04	8.91	45.61	96.80	47.12
3	0.2	3	185	29.13	9.05	45.70	97.12	47.06
5	0.2	3	185	28.55	9.16	45.71	95.18	48.02

An overview of the results shows that increasing temperature also produced an increase in the xylose consumed and the furfural produced, indicating that the temperature favours the dehydration of xylose and the production of furfural, but it can also be seen that this does not only depend on the temperature factor. For determining the effect of the factors studied (concentration of $FeCl_3$, time, and temperature) on yield (furfural yield), conversion (xylose conversion), and selectivity (furfural selectivity), the results were statistically analyzed and modelled according to a linear model with interaction between the factors. The models obtained in coded terms together with the *p*-value, R^2, coefficient of variation (CV), and standard deviation (SD) are shown in Table 2. As can be seen, for all models, the R^2 obtained is in the order of 0.99 (99% of the change is produced by the factors studied) and CV in all cases is lower than 3.2%.

Table 2. Models in coded terms obtained for yield, conversion, and selectivity in a factorial design. A = concentration of $FeCl_3$, B = time, and C = temperature. CV, coefficient of variation.

Response	Model (Coded Terms)	*p*-Value	R^2	CV (%)	SD (%)
Yield (%)	Yield = 34.12 − 1.40·A + 0.32·B + 5.55·C − 5.22·AC − 8.01·BC	<0.0001	0.997	2.38	0.89
Conversion (%)	Conversion = 77.45 + 7.43·A + 8.53·B + 19.91·C − 7.41·AC − 8.69·BC	<0.0001	0.999	1.26	1.04
Selectivity (%)	Selectivity = 46.36 − 7.05·A − 4.71·B − 5.63·C − 3.16·BC	<0.0001	0.988	3.18	1.48

The temperature (C) is the most influential factor on the yield, as can be deduced from the highest value of its coefficient (contribution >20%); moreover, the total contribution of the temperature, if the interaction concentration $FeCl_3$-temperature (AC) and the interaction time-temperature (BC) are taken into consideration, represents more than 75%. At a low temperature, the yield increases with the concentration of $FeCl_3$ (A), while at a high temperature, the yield decreases with the increasing concentration of $FeCl_3$. Similar yield results are obtained at high concentrations of $FeCl_3$, independently of the temperature (Figure 2a). As far as the interaction between temperature and time is concerned, yields increase with time (B) in the low temperature region, while the behavior is the opposite at a high temperature (Figure 2b).

Regarding the conversion, a similar pattern to that in the yield was observed, that is, the temperature is the most influential factor. In this case, the contribution of the linear term is more than 52% and, together with AC and BC interactions, it represents 70%. In the case of conversion, the other two factors have a greater influence, contributing 17%. At a high temperature, the conversion is high regardless of the $FeCl_3$ concentration and, at a low temperature, the conversion increases with the $FeCl_3$ concentration, not reaching as high as at a high temperature (Figure 2c). The interaction between temperature and time follows the same trend described above for conversion (Figure 2d).

Finally, in the case of selectivity, the influence of the factors studied is different from the two previous cases, because the most influential factor is the $FeCl_3$ concentration, with a contribution of more than 43%, with that of temperature and time being 27% and 19%, respectively. These last two factors interact in such a way that, at a low temperature, the selectivity decreases slightly over time and, at a high temperature, the selectivity decreases more drastically over time (Figure 2e).

When looking for the conditions that could maximize the variables studied only in the case of selectivity, we can say, without the need to make calculations, that they would be the lowest of all the factors studied ([$FeCl_3$] = 0.1 M, time = 1 min, and temperature = 170 °C), because all the coefficients of the model obtained are negative (Table 2). In the case of yield and conversion, calculations would be necessary because, among the coefficients of their models, there are positive and negative values, but, if Figure 2 is observed, they can also be deduced. For the conversion, it can be easily deduced from Figure 2c,d that the best conditions are the high temperature (200 °C) and, in that case, the concentration of $FeCl_3$ and time has little influence, so one would opt for low values ([$FeCl_3$] = 0.1 M, time = 1 min). In the case of the yield, observing Figure 2a,b, it is also easy to deduce that the best conditions would be 200 °C, [$FeCl_3$] = 0.1 M, and 1 min.

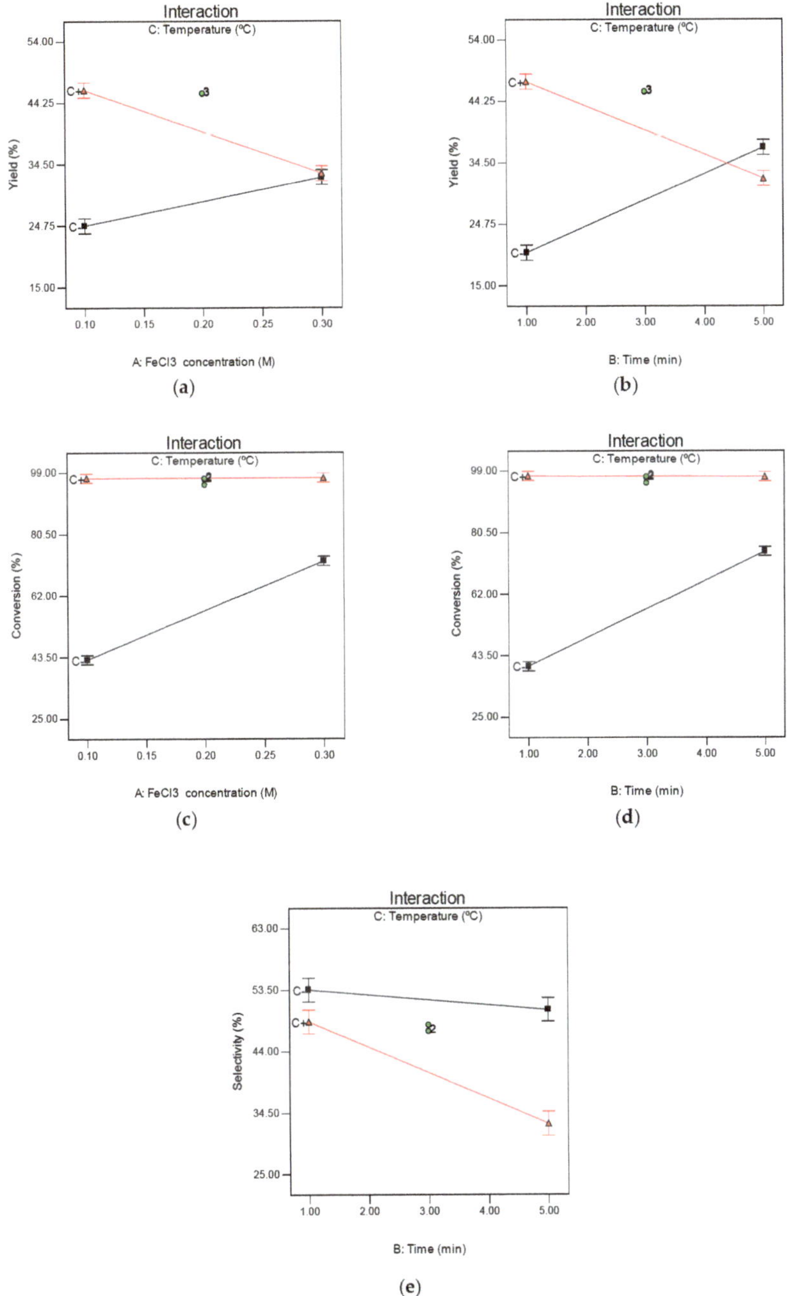

Figure 2. Interaction of factors studied in the factorial design: (**a**) interaction of FeCl₃ concentration and temperature in yield; (**b**) interaction of time and temperature in yield; (**c**) interaction of FeCl₃ concentration and temperature in conversion; (**d**) interaction of time and temperature in conversion; (**e**) interaction of time and temperature in selectivity. Red line, 200 °C; black line, 170 °C.

2.2. Central Composite Design

On the basis of the results obtained in the two-level factorial design, a rotatable composite central design was designed to further assess the influence of the selected factors and to optimize the conditions for obtaining furfural. The experimental range was changed to shorter times (0.5–1 min) and lower $FeCl_3$ concentrations (0.05–1 M), and the temperature range was reduced to a shorter one (190–200 °C). The results (xylose consumption, furfural produced, yield, conversion, and selectivity) of the 25 experiments carried out according to the design conditions are shown in Table 3.

Table 3. Results (consumed xylose, furfural produced, yield, conversion, and selectivity) of the rotatable composite central design (duplicates of the star points +5 central points).

Run	A: $FeCl_3$ (M)	B: Time (min)	C: Temp. (°C)	Consumed Xylose (g/L)	Furfural (g/L)	Yield (%)	Conversion (%)	Selectivity (%)
1	0.040	0.75	195	21.24	8.73	45.39	70.68	64.22
2	0.1	1	190	23.60	10.01	51.77	78.16	66.23
4	0.05	0.5	200	25.43	10.04	52.07	84.40	61.69
5	0.05	1	190	20.31	8.11	41.87	67.14	62.36
6	0.1	1	200	28.88	10.32	53.35	95.54	55.84
7	0.075	1.10	195	26.66	9.93	51.38	88.28	58.20
8	0.075	0.75	202.07	28.45	10.76	55.78	94.37	59.11
9	0.075	0.75	187.93	19.07	7.64	39.55	63.15	62.63
11	0.1	0.5	200	27.97	10.30	53.35	92.68	57.56
12	0.040	0.75	195	22.05	8.99	46.64	73.18	63.73
14	0.110	0.75	195	26.64	9.61	49.69	88.22	56.33
15	0.075	0.40	195	21.56	8.72	45.10	71.38	63.19
17	0.05	0.5	190	15.57	6.37	33.16	51.85	63.96
18	0.110	0.75	195	25.98	9.67	50.30	86.51	58.15
19	0.075	0.75	187.93	18.32	7.29	37.95	61.00	62.22
20	0.075	0.40	195	22.31	8.97	46.68	74.31	62.82
21	0.1	0.5	190	20.73	8.06	41.71	68.63	60.78
22	0.05	1	200	26.09	10.18	52.67	86.40	60.95
23	0.075	0.75	202.07	27.28	10.50	54.41	90.50	60.12
25	0.075	1.10	195	26.06	9.86	51.05	86.35	59.12
3	0.075	0.75	195	25.08	9.73	50.27	82.91	60.63
10	0.075	0.75	195	25.05	9.49	49.08	82.94	59.18
13	0.075	0.75	195	24.73	9.53	49.54	82.30	60.19
16	0.075	0.75	195	24.50	9.91	51.33	81.20	63.22
24	0.075	0.75	195	23.54	9.18	47.56	78.06	60.93

The initial concentration of xylose was maintained at around 30 g/L, resulting in this design displaying higher consumption values than those in the factorial design, ranging from 15.57 to 28.88 g/L or conversion of between 51.85 and 95.54%. As for the furfural obtained, values between 6.37 and 10.76 g/L were obtained, which represents a yield of 33.16 and 55.78%, respectively, a shorter range than in the factorial design and higher on average. Finally, the selectivity showed values between 55.84 and 66.23%, being on average around 61%, compared with the factorial design where it was around 46%.

For the analysis of the influence on the responses (yield, conversion and selectivity) of the factors studied (concentration of $FeCl_3$, time, and temperature), the results were evaluated using the response surface methodology and modelled according to a quadratic model. The models obtained in coded and real terms together with the p-value, R^2, coefficient of variation (CV), and standard deviation (SD) are shown in Table 4. The three models show R^2 surpassing 0.91, yield and conversion above 0.97 (97% of the change is produced by the factors studied), and a CV in all cases below 2.5% (below 1.3 in the case of selectivity), that is, good values that allow us to use the models obtained for analysis.

Table 4. Models in coded and real terms obtained for performance, conversion, and selectivity in the rotatable composite central design. A = concentration of $FeCl_3$, B = time, and C = temperature.

Response	Model (Coded and Real Terms)	p-Value	R^2	CV (%)	SD (%)
Yield (%)	(Coded) = 49.10 + 1.98·A + 2.15·B + 5.57·C − 2.06·AC − 2.27·BC − 0.54·A^2 − 1.09·C^2 (Real) = −2345.52 + 3423.72·A + 362.70·B + 20.67·C − 16.48·AC − 1.82·BC − 871.20·A^2 − 0.04·C^2	<0.0001	0.970	2.38	1.14
Conversion (%)	(Coded) = 80.34 + 5.55·A + 4.41·B + 11.20·C − 1.30·AC − 2.50·BC − 1.78·C^2 (Real) = 3533.74 + 2246.99·A + 406.90·B + 32.25·C − 10.38·AC − 2.00·BC − 0.07·C^2	<0.0001	0.977	2.46	1.95
Selectivity (%)	(Coded) = 60.49 − 2.23·A − 1.19·B − 1.07·C (Real) = 112.51 − 89.22·A − 4.74·B − 0.21·C	<0.0001	0.911	1.26	0.77

The most influential factor in yield is temperature, given its higher coded coefficient of 5.57, followed by time and $FeCl_3$ concentration with similar coefficients of 2.15 and 1.98, respectively. As for the interactions between factors, it can be seen that the yield at a high temperature varies little with the concentration of $FeCl_3$, being slightly higher at an intermediate concentration owing to the slight curvature it presents (quadratic term, A^2), while at a low temperature, the yield increases with the concentration of $FeCl_3$ (Figure 3a). In the interaction with time, the behaviour is similar to that already described; the performance decreases in a negligible way with the increase of time at a high temperature, while at a low temperature, it increases with time (Figure 3b). The yield is higher at a high temperature for both interactions. The conversion and the yield have similar behaviour in front of the studied factors, as can be seen in the obtained models (Table 4) and the response surfaces (Figure 3), sharing the temperature as the most influential factor. Analyzing more in detail, the conversion increases with the increase of $FeCl_3$ concentration, as well as with the increase of time, with this increase being more accentuated at low temperatures in both cases (Figure 3c,d). At a high temperature, the highest conversion values occur, and it is at this point that the increase in time has little influence on the increase in conversion (Figure 3d).

Selectivity shows different behaviour than that found in yield and conversion (Figure 3). Selectivity has, as its most influential factor, the concentration of $FeCl_3$, followed by time and temperature at a considerable distance, according to the values of their coefficients of 2.23, 1.19, and 1.07, respectively (Table 4). Furthermore, in the selectivity model, only the linear terms were significant and all of them have a negative sign, which indicates that the lowest conditions of the factors are going to be the most favourable ones.

Table 5 shows the results of the best conditions and the values obtained from the different responses (performance, conversion, and selectivity) for different optimization proposals, giving the same weight to the responses in the case of multiple optimizations. The highest value of selectivity would be 64.98%, and would be obtained under the lowest conditions of the design, that is, 190 °C, $[FeCl_3]$ = 0.05 M, and 0.5 min. On the contrary, the highest conversion value would be obtained under the highest design conditions, that is, 200 °C, $[FeCl_3]$ = 0.1 M, and 1 min, with a value of 95.93%. The highest yield value, 53.71%, is obtained at 200 °C, $[FeCl_3]$ = 0.07 M, and 0.5 min, a value close to that obtained under the conditions of maximizing conversion, 52.84%; on the contrary, the conversion and selectivity values are more distant.

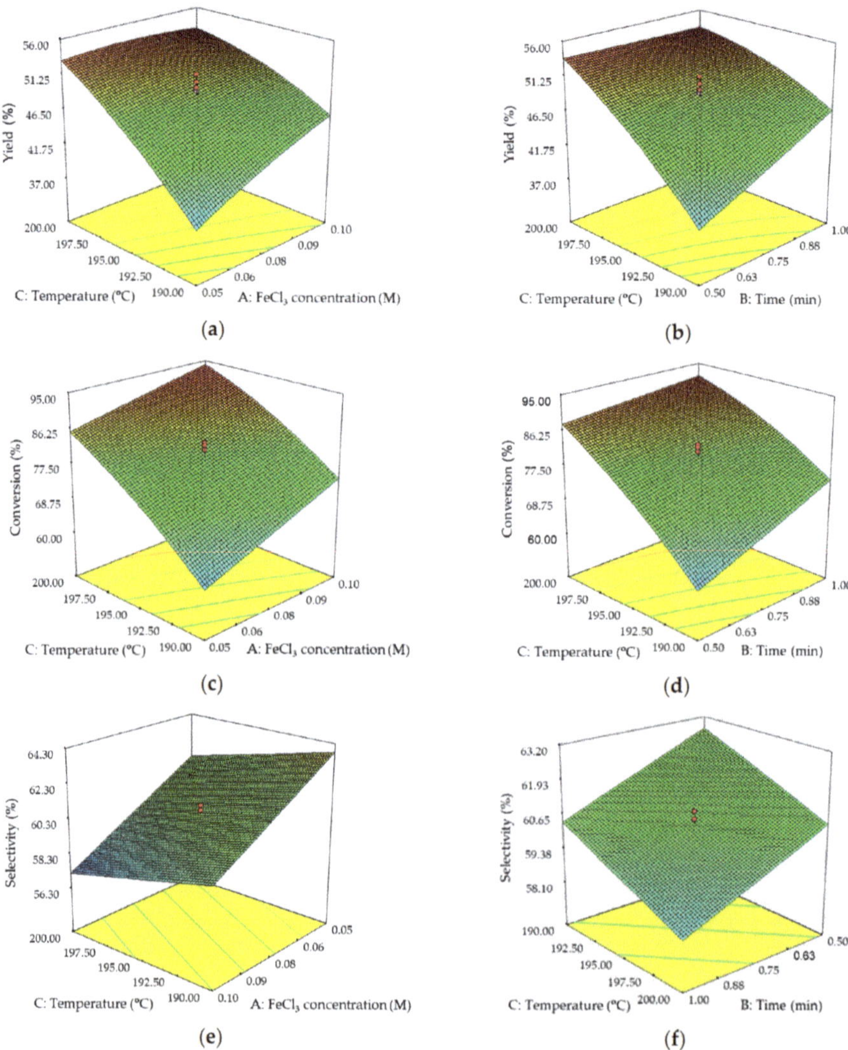

Figure 3. Response surfaces obtained for the composite central design: (**a**) influence of $FeCl_3$ concentration and temperature on yield; (**b**) influence of time and temperature on yield; (**c**) influence of $FeCl_3$ concentration and temperature on conversion; (**d**) influence of time and temperature on conversion; (**e**) influence of $FeCl_3$ concentration and temperature on selectivity; (**f**) influence of time and temperature on selectivity.

Table 5. Results of the optimization of the rotatable composite central design.

Maximise	A: FeCl$_3$ (M)	B: Time (min)	C: Temp. (°C)	Yield (%)	Conversion (%)	Selectivity (%)
Yield	0.07	0.5	200	53.71	87.53	60.77
Conversion	0.1	1	200	52.84	95.93	56.01
Selectivity	0.05	0.5	190	33.44	53.61	64.98
Yield and selectivity	0.05	0.5	200	53.24	83.59	62.84
Yield and conversion	0.1	1	200	52.84	95.93	56.01
Yield, conversion, and selectivity	0.05	0.5	200	53.24	83.59	62.84
Star point experimental average	0.075	0.75	202.07	55.10	92.43	59.62
Predict star point	0.075	0.75	202.07	54.81	92.62	58.98
Out of range	0.05	0.5	210	64.35	99.34	60.70
Experimental out of range	0.05	0.5	210	57.12	98.51	57.98

Different optimizations have been made with more than one response, with yield as the common response in all cases. In the first of these, yield and selectivity have been jointly optimized, obtaining conditions of 200 °C, [FeCl$_3$] = 0.05 M, and 0.5 min and values of 53.24% and 62.84%, respectively, these values being very close to the individual optimums. In the second one, performance and conversion have been jointly optimized, obtaining conditions of 200 °C, [FeCl$_3$] = 0.1 M, and 1 min, resulting in these conditions having a greater difference compared with the individual performance optimum. Finally, in the case of the triple optimization of the three responses, the same conditions result as in the joint optimization of yield and selectivity, that is, 200 °C, [FeCl$_3$] = 0.05 M, and 0.5 min, with a final yield of 53.24%. Reviewing the experimental results of the composite central design, it has been found that, under the conditions of one of the star points (202.07 °C, [FeCl$_3$] = 0.075 M, and 0.75 min), the yield value obtained is on average 55.10% higher than that found under the optimum design conditions. The values of yield, conversion, and selectivity (54.81%, 92.62%, and 58.98%) according to the conditions of this star point were obtained with models resulting in errors of less than 1% (Table 5).

As can be seen in Figure 4, the efficiency and conversion show a good linear adjustment with R^2 = 0.92, resulting in a 58% efficiency for the 100% conversion, higher than the value obtained in the star point conditions. Given the above, using the model obtained for the conversion, the temperature for which 100% would be obtained was sought. The concentration of FeCl$_3$ was set at 0.05 M and 0.5 min (triple optimum conditions) and a temperature of 210 °C was obtained (at 211 °C, the conversion is higher than 100%). The conditions of 210 °C, [FeCl$_3$] = 0.05 M, and 0.5 min were tested experimentally, obtaining a conversion of 98.51%, very close to that predicted by the model of 99.34%. Under these conditions, a yield value of 57.12% was obtained, confirming the proposed hypothesis. This value is very close to 57.30%, which would result from substituting the conversion value of 98.51% in the regression of Figure 4. If the experimental values and those predicted by the models for this temperature condition outside the range are compared, the experimental values of yield and selectivity are lower, by 11.24% and 4.47%, respectively, and it can be said that the yield model cannot be used with temperatures so far from those in the initial range.

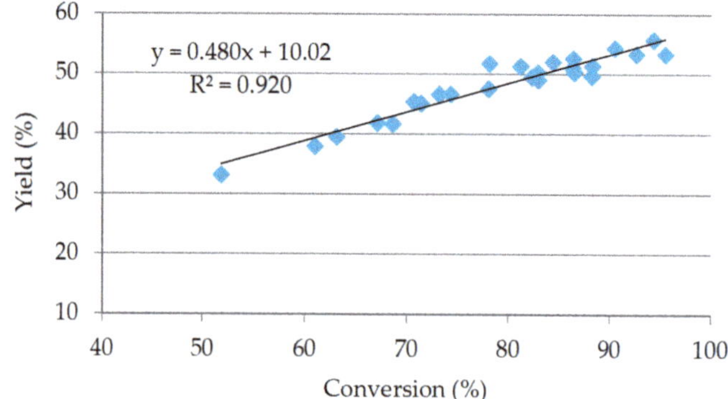

Figure 4. Linear performance regression versus conversion of the values of the composite central design.

3. Discussion

Different authors have made designs of experiments to study the influence of different factors and to determine the best conditions to obtain the maximum performance of furfural. In the cases that are going to be considered next, the response surface methodology and quadratic models were used as in our work.

Lamminpää et al. [28] studied four variables by a central composite circumscribed design using a preheating oven (360–420 °C) and a fluidized sand bath. The variables were time (20–40 min), temperature (140–200 °C), initial xylose concentration (0.067–0.20 M), and formic acid concentration (10–30% *w/w*). The design contained 29 experiments, including five central points. In this design, the authors obtained conversions ranging from 6.2% to 98.2% and selectivities from 42% to 73%, values similar to those obtained in this work. Conversion and yield were mainly influenced positively by temperature and, as in our case, followed by acid concentration. The maximum yield value obtained was 65% at 200 °C, 20 min, 30 wt% of formic acid, and with an initial xylose concentration of 10 g/L. This yield in the same conditions, but increasing the initial xylose concentration to 30 g/L (the same as in our work), drops to 56.8%, very similar to the value obtained in our work of 57.1%, although with a much shorter time of only 0.5 min in our case. The industrial process adopted by China with corncobs achieves around 50% yield with 3–4 wt% sulfuric acid at 153 °C for 5 h [9].

Yang et al. [27] also performed a central composite design in this case in a stainless steel autoclave, with an electric jacket, a constant initial xylose concentration of 80 g/L, and varying the proportion of water-o-nitrotoluene. The reaction temperature (170–210 °C), formic acid concentration (5–25 g/L, $pH_{25\ °C}$ 2.32–1.96), o-nitrotoluene volume percentage (20–80 vol%), and residence time (40–200 min) were analyzed in the experimental design. The design consisted of 27 experiments, including three central points. The results included selectivities between 51.1 and 99.7% and yields between 20.3 and 71.2%, which are higher than those obtained in the previous design and those presented in this work. Among the factors studied in this design, temperature and o-nitrotoluene percentage are the most influential, with time and formic acid concentration having a low influence on yield and selectivity. Both yield and selectivity present a maximum of around 190 °C. On the contrary, the increase of o-nitrotoluene percentage (0 vol% to 80 vol%) was positive in both yield and selectivity, with increases from about 30% to 70% and 70% to 99%, respectively. The maximum furfural yield (74%) and selectivity (86%) were obtained in 75 min at 190 °C for 20 g/L formic acid concentration and 75 vol% o-nitrotoluene. This yield is higher than that obtained in our work, but the time is much higher, that is, 75 min compared with 0.5 min, and it also uses an organic solvent from petroleum, which makes the process less sustainable. To reduce the reaction time, the authors added different halides (NaCl, KCl, KBr, and KI) with a concentration of 0.5 M, achieving an optimum time in 60 min, or in other words, a reduction of 15 min. This time is still far from that obtained in our work, even though it achieved a slight improvement in performance and selectivity, for example, with NaCl, of 78% and 90%, respectively.

Another type of design was used to study the effect of lignin on the production of furfural from xylose (30 g/L), at D-optimal design [29]. In the design, four variables were included: three quantitative variables, time (20–80 min), temperature (160–180 °C), and initial lignin concentration (0–20 g/L); and one qualitative variable, the acid catalyst (10% *w/w*) formic acid or 0.2% *w/w* sulfuric acid (initial pH similar). The design contained 26 experiments, including 3 central points, plus 4 extra points, for a total of 30 experiments. The xylose conversion varied from 12% to 90% with the formic acid and from 6% to 85% with the sulfuric acid, values somewhat lower than those obtained in our work. The furfural yield varied from 9% to 58% in HCOOH and from 4% to 53% in H_2SO_4, in the order of our work. The selectivities were 65–80% and 63–76% in HCOOH and H_2SO_4, respectively, higher than those obtained in our work, with the least variation between large and small values compared with the other responses. The most influential factors in yield are temperature and time in both HCOOH and H_2SO_4 cases. The difference in the influence of lignin is found to be more significant in the case of H_2SO_4 than in HCOOH; for example, at 180 °C in sulfuric acid, the yield is higher than 50% without lignin, but in the presence of 20 g/L lignin, the yield is lower than 40%, whereas in formic acid, the yield stays about the same, at around 62%. The maximum yields are given at 180 °C, 80 min, and without lignin, with values around 64% and 56% with HCOOH and H_2SO_4, respectively, values in the order of those obtained in this work, but with a much longer time. The authors concluded that lignin affected furfural formation in two ways: (1) the lignin has an acid neutralization capacity (the maximum rise in pH was 0.13 units in formic acid and 0.23 units in sulfuric acid; time and temperature are insignificant parameters for the pH change response model), and (2) the lignin inhibits xylose dehydration into furfural (the neutralizing effect does not explain all of the decreases in conversion and yield, so some additional mechanism must be involved) [29]. Rasmussen et al. [30] concluded that there are at least three possible routes for furfural formation from xylose, and the reaction conditions determine which mechanism dominates.

Overview of the Production of Furfural from Xylose

Table 6 shows a varied representation of the processes used to produce furfural from xylose (maximize furfural yield) in monophase systems with green solvents, as in our case, to have processes without petroleum derivatives.

Among those green solvents are water, seawater, gamma-valerolactone (GVL), GVL-water, dimethyl sulfoxide (DMSO), N,N-dimethylacetamide (DMA), and some ionic liquids ([emin]Br, [emim]HSO_4, [bmim]PF_6) that are not chloride based as they are considered toxic and corrosive [31]. In most cases, some kind of catalyst was used to increase the reaction speed and improve the yield in furfural.

These catalysts can be both homogeneous and heterogeneous. Among the homogeneous catalysts are inorganic acids (HCl, H_2SO_4); organic acids (formic acid, maleic acid); salts ($FeCl_3$, $SnCl_4$); heteropolyacid salts (MP34CsPW, $H_3PW_{12}O_{40}$ (PW)); a combination of mineral acid + salt (HCl-NaCl, H_2SO_4-$FeCl_3$), as in our case; or combinations of different salts ($CrCl_2$-LiBr). Among the heterogeneous catalysts, we find zeolites (H-ZSM-5, H-Mordenite, H-Beta) or mesoporous acid-catalysts (MSHS-SO_3H, Nafion 117).

Table 6. Yield results of furfural production from xylose in green solvents in the monophase system.

Xylose	Solvent	Catalyst	Temperature/Time	Heating	Furfural Yield (%)	Ref.
30 g/L= 200 mM	H_2O	MSHS-SO_3H (3.3 g/L)	190 °C/1 h	Autoclave	43.5	[32]
10 wt%	H_2O	H-ZSM-5 (catalyst-xylose ratio, 0.3 *w/w*)	200 °C/20 min	Autoclave	46	[33]
35 mM	H_2O	HCl (50 mM)-NaCl (850 mM)	200 °C/5 min	Oil bath	81.3	[14]
35 mM	H_2O	HCl (50 mM)-NaCl (3.5 wt% = 599 mM)	200 °C/440 s (7.3 min)	Microwave	76	[34]
740 mM	H_2O	HCl (100 mM)	170 °C/30 min	Microwave	40	[17]
667 mM	H_2O	HCl (100 mM)	180 °C/30 min	Microwave	39	[19]
67 mM = 10 g/L	H_2O	Maleic acid (250 mM)	200 °C/28 min	Microwave	67	[35]
57 mM	H_2O	Not used	200 °C/60 min	Microwave	49	[36]
30 g/L	H_2O	H_2SO_4 (2% *w/v*)-$FeCl_3$ (0.05 M)	210 °C/0.5 min	Microwave	57.1	Present study
67 mM	H_2O	Formic acid (30 wt%)	200 °C/20 min	Oven preheating (360–420 °C) and a fluidized sand bath	65	[28]
200 mM	H_2O	Formic acid (30 wt%)	200 °C/20 min	Oven preheating (360–420 °C) and a fluidized sand bath	56.8	[28]
30 g/L	H_2O	Formic acid (30 wt%)	180 °C/80 min	Oven preheating (360–420 °C) and a fluidized sand bath	~63.8	[29]
30 g/L	H_2O	H_2SO_4 (0.2 wt%)	180 °C/80 min	Oven preheating (360–420 °C) and a fluidized sand bath	~56.2	[29]
18 wt%	H_2O	H_2SO_4 (20 mM)	250 °C/1 min	Supercritical flow reactor system	64	[37]
50 mM	Seawater (salts (26,46 g/kg)	HCl (50 mM)	200 °C/10 min	Oil bath	71.7	[38]
150 mM	GVL	H_2SO_4	175 °C	Not specified	75	[39]
2.4 wt%	GVL	$FeCl_3 \cdot 6H_2O$ (0.6 wt%)	180 °C/9 min	Oil bath	83.6	[13]
2 wt%	GVL-H_2O (10 wt% H_2O)	H_2SO_4 (0.05 M)	170 °C/15 min	Oil bath	87	[40]
2 wt%	GVL-H_2O (10 wt% H_2O)	H-Mordenite	175 °C/2 h	Oil bath	81	[41]
2 wt%	GVL-H_2O (10 wt% H_2O)	H-Beta (3.75 wt%)	160 °C/1 h	Oil bath	71	[40]
200 mM	DMSO	H-Mordenite (100 g/L)	140 °C/4 h	Autoclave	39	[42]
200 mM	DMSO	MP34CsPW (30 g/L)	140 °C/4 h	Oil bath	45	[3]
200 mM	DMSO	$H_3PW_{12}O_{40}$ (PW) (20 g/L)	140°C/4 h	Oil bath	67	[43]
9.1 wt%	DMSO	Nafion 117 (20 wt% of initial xylose)	150 °C/2 h	Oil bath	60	[44]
10 wt%	DMSO	HCl (0.1 M)-Sn-beta	110 °C/3 h	Not specified	14.3	[45]
10 wt%	DMA	$CrCl_2$ (6 mol% of xylose)-LiBr (10 wt%)	100 °C/4 h	Oil bath	56	[46]
20 wt%	[emin]Br	$SnCl_4$ (10 mol% of xylose)	130 °C/1 h	Oil bath	71.1	[15]
100 g/L	[emim]HSO_4	Not used	100 °C/30 min	Oil bath	62	[47]
37.5 g/L	[bmim]PF_6	PEG-OSO_3H (50 mM)-$MnCl_2$ (75 mM)	120 °C/18 min	Not specified	75	[48]

GVL, gamma-valerolactone; DMSO, dimethyl sulfoxide; DMA, N,N-dimethylacetamide; PEG, polyethylene glycol.

There is also a combination of homogeneous and heterogeneous catalysts, such as inorganic acid + zeolite (HCl-Sn-beta) or a polymer bound an acid + salt (PEG-OSO$_3$H-MnCl$_2$). As for the forms of heating used, they could be grouped into two large groups, conventional ones like autoclave, oil bath or oven preheating (360–420 °C), and a fluidized sand bath; and non-conventional ones, such as microwaves (our case) and supercritical systems.

The initial xylose concentration used (Table 6) shows a wide range from just 5 g/L to 200 g/L, with the average being around 50 g/L and the median of 30 g/L, which coincides with the value used in this work. The temperatures used range from 100 °C to 250 °C, with the average being around 170 °C and the median being 180 °C. If only microwave heating is taken into account, the temperature range is shorter, 170–210 °C, where the average is just over 193 °C and the median is 200 °C, showing that optimum temperatures with microwaves are higher when using other forms of heating. The times used also show a wide range from 0.5 min (this work) to 240 min, with an average of 70 min, while the average in the case of using microwaves around a third is about 25 min, showing itself as a faster process and one that would also save energy [17,34]. The yields obtained range from 14% to 87%, with the average being around 61%; meanwhile, taking into account only microwave heating, this average would be around 55%; the value obtained in this work of 57% is in the order of these averages. If all the values shown in Table 6 are analyzed to see how they affect the furfural yield, the following is found: the highest values are given to short times (less than 20 min), high temperatures (180–200 °C) (same trend found in this work), and the increase of the initial xylose concentration having a slight negative effect.

With the idea of being able to increase the yield of furfural, one of the proposals is the biphasic systems in such a way that furfural is produced in one phase and is selectively transferred to the insoluble organic phase, avoiding its degradation and displacing the reaction balance. Several authors have reported that they have doubled the performance of furfural when using a biphase system (with methyl isobutyl ketone (MIBK) or cyclopentylmethyl ether (CPME)) versus a monophase system with water only [10,17,49]. Table 7 shows some biphase processes with green solvents for the production of furfural.

Table 7. Yield results of furfural production from xylose in green solvents in biphase systems.

Xylose	Solvent	Catalyst	Temperature/Time	Heating	Furfural Yield (%)	Ref.
1.85 wt%	[bmim]HSO$_4$-MIBK (1:4.4, *w/w*)	Not used	140 °C/4 h	Oil bath	80.3	[50]
400 mmol/L H$_2$O	H$_2$O-2MTHF (1:1, *v/v*)	FeCl$_3$ (80 mM)-NaCl (20 wt%)	140 °C/4 h	Oil bath	71	[51]
4 wt% H$_2$O phase	H$_2$O-CPME (1:2.33, *v/v*)	H$_2$SO$_4$ (1 wt% H$_2$O phase)-NaCl (40 wt% H$_2$O phase)	170 °C/100 min	Oil bath	100	[52]
1.25 mol/L H$_2$O	H$_2$O-CPME (1:3, *v/v*)	FeCl$_3$ (5.08 g/L)-NaCl (18.13 g/L)	170 °C/20 min	Microwave	74	[10]
1 mol/L H$_2$O	H$_2$O-CPME (1:3, *v/v*)	NaCl (23.75 g/L)-Nafion NR50 (23.75 g/L)	170 °C/40 min	Microwave	80	[49]
200 g/L H$_2$O	H$_2$O-DMSO (1:1, *v/v*)	SnCl$_4$ (catalyst/xylose molar ratio 0.5)	130 °C/6 h	Oil bath	63	[53]
10 wt% H$_2$O phase	H$_2$O-DMSO-SBP (5:1:5, *v/v/v*)	Sn-MMT (xylose-catalyst, 5:1 *w/w*)-NaCl (satured solution, aprox. 36 g/100 g of H$_2$O)	180 °C/30 min	Autoclave	76.8	[54]
1 g/L, 5 mL H$_2$O	H$_2$O-MIBK (1.5:8, *v/v*)	[Sbmim]HSO$_4$ (0.5g/1.5 mL H$_2$O)	150 °C/25 min	Autoclave	91.4	[55]
740 mmol/L H$_2$O	H$_2$O-MIBK (1:1, *w/w*)	HCl (0.1 mol/L H$_2$O)	170 °C/30 min	Microwave	80	[17]
10 wt% of H$_2$O	H$_2$O-THF (1:2, *w/w*)	[SbPy]BF$_4$ (100 wt% of initial xylose)	180 °C/1 h	Microwave	85	[56]
250 mmol/L H$_2$O	H$_2$O-THF (1:3, *v/v*)	AlCl$_3$-6H$_2$O (25 mM)-NaCl (1.5 M)	140 °C/45 min	Microwave	75	[57]

MIBK, methyl isobutyl ketone; 2MTHF, 2 methyltetrahydrofuran; CPME, cyclopentylmethyl ether; DMSO, dimethyl sulfoxide; SBP, 2-s-butilfenol; THF, tetrahydrofuran.

In these processes, mainly water, but also ionic liquids ([bmim]HSO$_4$) were used as the reaction phase solvents and organic cosolvents as follows: methyl isobutyl ketone (MIBK), 2-methyltetrahydrofuran (2MTHF), cyclopentylmethyl ether (CPME), dimethyl sulfoxide (DMSO), 2-s-butilfenol (SBP), and tetrahydrofuran (THF). In most cases, some kind of catalyst was used, either a homogeneous catalyst such as inorganic acids (HCl), salts (SnCl$_4$), a combination of mineral acid + salt (H$_2$SO$_4$-NaCl), combinations of different salts (FeCl$_3$-NaCl, AlCl$_3$-NaCl) or ionic liquids ([Sbmim]HSO$_4$, [SbPy]BF$_4$); or a combination of heterogeneous and homogeneous (Sn-MMT-NaCl, Nafion NR50-NaCl). As for the forms of heating used, the processes could be grouped into conventional ones, including autoclave and oil bath, and non-conventional ones such as microwaves. The initial xylose concentration assayed shows a very wide range from 18.5 g/L to more than 600 g/L, with the average being around 150 g/L, or around 120 g/L in the case of microwave heating. The range of temperature was from 130 °C to 180 °C, with the median being 170 °C, the same median value as in the case of microwave heating. Concerning the values of the process time, a wide range from 20 min to 360 min was used, with an average close to 110 min, with the average in the case of using microwaves around a third being about 39 min.

Microwave-assisted processes were faster than conventional heated processes, as in the case of monophasic, and also more selective [34]. The yields obtained ranged from 63% to 100%, with the average being around 80%, with the average value being very similar with microwave heating, at 79%. The average value obtained in biphasic processes is 30% higher than the average value obtained in monophasic processes (80% versus 60%). If all the values shown in Table 7 were analyzed to see how they affected the furfural yield, it is found that the highest values were given at high temperatures (170–180 °C) and at short times (less than 60 min), while the initial xylose concentration had no apparent influence.

4. Materials and Methods

4.1. Chemical Reagents

Xylose was purchased from Fagron (Terrassa, Spain); $FeCl_3$ hexahydrate (99%) was purchased from Emsure (Darmstadt, Germany); H_2SO_4 (98%) was purchased from Honeywell Fluka (Seelze, Germany). All materials were used without further purification. Aqueous solutions were prepared with deionized H_2O.

4.2. General Procedure for Dehydration Treatment

All experiments were conducted using a microwave reactor (Anton Paar Monowave 400, Graz, Austria). The xylose and the catalysts in different concentrations were charged in the 10 mL glass vessel. The reaction volume was 4 mL; the xylose concentration was 30 g/L; and the catalysts used were H_2SO_4 at 2% *w/v* acting as Brønsted acid and $FeCl_3$ as Lewis acid, with a variable concentration [10,13].

The heating dynamic followed was to heat the sample to the set temperature in 2 min, maintaining the temperature for the experiment time, and cooling down to 40 °C with compressed air. The magnetic agitation during the heating and maintenance was 600 rpm, and it was 800 rpm during the cool down period. The temperature was measured by an IR sensor (Anton Paar Monowave 400, Graz, Austria). The pressure inside the glass vessel was also monitored throughout the experiments, through the septum that covers it. The experimental conditions were set based on previous experiments not included.

4.3. Methodology Based on the Design of Experiments

A two-level factorial design with three central points was carried out. The $FeCl_3$ concentration in the range of 0.1 and 0.3 M, the process time from 1 to 5 min, and the temperature between 170 and 200 °C were selected as factors, with a total of 11 experiments (Table 1).

The samples obtained in the microwave reactor were analyzed in high performance liquid chromatography (HPLC), as detailed in Section 4.4, and the results in terms of xylose conversion, furfural yield, and selectivity, as defined below, were taken as responses and statistically analyzed with the commercial software Design Expert 7.0.0, Stat-Ease Inc (Minneapolis, MN, USA).

On the basis of the data obtained from this first factorial design, a second rotatable composite central design was performed, where the range of variation for the factors was reduced as follows: the concentration of $FeCl_3$, 0.05–0.1 M; time, 0.5–1 min; and temperature, 190–200 °C. Moreover, five center and star points (values above and below the experimental range for each of the factors) were added (Table 3).

The liquors obtained were measured in HPLC and analyzed statistically as in the previous design. These results were optimized to obtain the maximum yield of furfural and selectivity.

4.4. Analysis of the Liquid Fractions and Quantification of the Yield, Selectivity, and Conversion

The liquors obtained in the microwave reactor were analyzed by high performance liquid chromatography (HPLC). The compounds were determined using an Agilent Technologies 1260 model (Santa Clara, CA, USA) with ICSep ICE-COREGEL 87H3 column operating at 65 °C with 5 mM sulfuric acid as the mobile phase (0.6 mL/min). Samples were previously filtered through 0.45 µm nylon membranes.

Conversion of xylose, furfural selectivity, and yield were defined as follows:

$$\text{Xylose conversion } (\%) = \frac{\text{Consumed xylose}}{\text{Initial xylose}} \times 100 \tag{1}$$

$$\text{Furfural yield } (\%) = \frac{\text{Furfural produced}}{\text{Furfural stoichiometric potential}(*)} \times 100 \tag{2}$$

$$(*)\ \text{Furfural stoichiometric potential} = \text{Initial xylose} \times 0.64 \qquad (3)$$

$$\text{Xylose conversion (\%)} = \frac{\text{Consumed xylose}}{\text{Initial xylose}} \times 100 \qquad (4)$$

5. Conclusions

The analysis and comparison of results demonstrated the following:

Furfural can be obtained by dehydration of xylose, the main pentose found in lignocellulosic materials, thus making use of these renewable sources of energy and biobased chemicals. Temperature, time, and the concentration of iron chloride were identified as relevant factors of the conversion process. Following a two experimental design approach and optimization by response surface methodology, a highest yield of 57.1% along with 98.5% conversion was obtained. The experimental conditions leading to the best result, including the highest temperature assayed (210 °C) and lowest iron chloride concentration (0.05 M) and process time (0.5 min), can be used as a starting point for the study of the conversion process using the whole lignocellulosic material, after being subjected to the required operations.

The processes used to obtain xylose-rich liquors from lignocellulosic materials should be selective on the hemicelluloses to solubilize the minimum amount of lignin, as this negatively affects the production of furfural.

The use of microwave heating in both monophase and biphase systems allows shorter reaction times, around one-third compared with conventional heating methods, which saves time and probably also energy.

The highest furfural yield values are obtained at high temperatures and at short times in both monophase and biphase systems.

Biphase systems increase the furfural yield on average by about 30% compared with monophase systems, in each case, using green solvents.

Future investigations will be focused on the use of microwaves for heating and a biphase system with green solvents on a xylose-rich liquor obtained from lignocellulosic materials. The methodology found here to obtain the optimal conditions of temperature, time, and concentration of catalyst will be applied to obtain furfural with a high yield. According to the current results, we would propose that high temperatures, above 200 °C, should be tested.

Author Contributions: Conceptualization, E.R. and E.C.; methodology, C.P.-R.; software, J.M.R.-G.; validation, E.R. and E.C.; formal analysis, E.R.; investigation, C.P.-R.; resources, E.R. and E.C.; data curation, C.P.-R. and J.M.R.-G.; writing—original draft preparation, C.P.-R. and J.M.R.-G.; writing—review and editing, E.C.; visualization, C.P.-R. and J.M.R.-G.; supervision, E.R.; project administration, E.R.; funding acquisition, E.R. and E.C. All authors have read and agreed to the published version of the manuscript.

Funding: The authors want to acknowledge the financial support from Agencia Estatal de Investigación (MICINN, Spain) and Fondo Europeo de Desarrollo Regional, reference project ENE2017-85819-C2-1-R.

Acknowledgments: C.P.-R. was supported by Universidad de Jaén (research grant R5/04/2017).

Conflicts of Interest: The authors declare no conflict of interest.

References

1. Machado, G.; Leon, S.; Santos, F.; Lourega, R.; Dullius, J.; Mollmann, M.E.; Eichler, P. Literature Review on Furfural Production from Lignocellulosic Biomass. *Nat. Resour.* **2016**, *07*, 115–129. [CrossRef]

2. Cai, C.M.; Zhang, T.; Kumar, R.; Wyman, C.E. Integrated furfural production as a renewable fuel and chemical platform from lignocellulosic biomass: Furfural production from lignocellulosic biomass. *J. Chem. Technol. Biotechnol.* **2014**, *89*, 2–10. [CrossRef]

3. Dias, A.S.; Lima, S.; Pillinger, M.; Valente, A.A. Acidic cesium salts of 12-tungstophosphoric acid as catalysts for the dehydration of xylose into furfural. *Carbohydr. Res.* **2006**, *341*, 2946–2953. [CrossRef] [PubMed]

4. Rong, C.; Ding, X.; Zhu, Y.; Li, Y.; Wang, L.; Qu, Y.; Ma, X.; Wang, Z. Production of furfural from xylose at atmospheric pressure by dilute sulfuric acid and inorganic salts. *Carbohydr. Res.* **2012**, *350*, 77–80. [CrossRef]

5. Sánchez, C.; Serrano, L.; Andres, M.A.; Labidi, J. Furfural production from corn cobs autohydrolysis liquors by microwave technology. *Ind. Crops Prod.* **2013**, *42*, 513–519. [CrossRef]

6. Montané, D.; Salvadó, J.; Torras, C.; Farriol, X. High-temperature dilute-acid hydrolysis of olive stones for furfural production. *Biomass Bioenerg.* **2002**, *22*, 295–304. [CrossRef]

7. Mesa, L.; Morales, M.; González, E.; Cara, C.; Romero, I.; Castro, E.; Mussatto, S.I. Restructuring the processes for furfural and xylose production from sugarcane bagasse in a biorefinery concept for ethanol production. *Chem. Eng. Process.* **2014**, *85*, 196–202. [CrossRef]

8. Sádaba, I.; Ojeda, M.; Mariscal, R.; Granados, M.L. Silica-poly(styrenesulphonic acid) nanocomposites for the catalytic dehydration of xylose to furfural. *Appl. Catal. B Environ.* **2014**, *150–151*, 421–431. [CrossRef]

9. Agirrezabal-Telleria, I.; Gandarias, I.; Arias, P.L. Heterogeneous acid-catalysts for the production of furan-derived compounds (furfural and hydroxymethylfurfural) from renewable carbohydrates: A review. *Catal. Today* **2014**, *234*, 42–58. [CrossRef]

10. Le Guenic, S.; Delbecq, F.; Ceballos, C.; Len, C. Microwave-assisted dehydration of D-xylose into furfural by diluted inexpensive inorganic salts solution in a biphasic system. *J. Mol. Catal. A Chem.* **2015**, *410*, 1–7. [CrossRef]

11. Padilla-Rascón, C.; Ruiz, E.; Romero, I.; Castro, E.; Oliva, J.M.; Ballesteros, I.; Manzanares, P. Valorisation of olive stone by-product for sugar production using a sequential acid/steam explosion pretreatment. *Ind. Crops Prod.* **2020**, *148*, 112279. [CrossRef]

12. Kim, S.B.; Lee, J.H.; Yang, X.; Lee, J.; Kim, S.W. Furfural production from hydrolysate of barley straw after dilute sulfuric acid pretreatment. *Korean J. Chem. Eng.* **2015**, *32*, 2280–2284. [CrossRef]

13. Zhang, L.; Yu, H.; Wang, P.; Li, Y. Production of furfural from xylose, xylan and corncob in gamma-valerolactone using FeCl$_3$·6H$_2$O as catalyst. *Bioresour. Technol.* **2014**, *151*, 355–360. [CrossRef] [PubMed]

14. Marcotullio, G.; De Jong, W. Chloride ions enhance furfural formation from d-xylose in dilute aqueous acidic solutions. *Green Chem.* **2010**, *12*, 1739. [CrossRef]

15. Nie, Y.; Hou, Q.; Li, W.; Bai, C.; Bai, X.; Ju, M. Efficient Synthesis of Furfural from Biomass Using SnCl$_4$ as Catalyst in Ionic Liquid. *Molecules* **2019**, *24*, 594. [CrossRef] [PubMed]

16. Romero, I.; Ruiz, E.; Castro, E. Pretreatment with Metal Salts. In *Biomass Fractionation Technologies for a Lignocellulosic Feedstock Based Biorefinery*; Elsevier: Amsterdam, The Netherlands, 2016; pp. 209–227. ISBN 978-0-12-802323-5.

17. Weingarten, R.; Cho, J.; Conner, W.C., Jr.; Huber, G.W. Kinetics of furfural production by dehydration of xylose in a biphasic reactor with microwave heating. *Green Chem.* **2010**, *12*, 1423. [CrossRef]

18. Hricovíniová, Z. Xylans are a valuable alternative resource: Production of d-xylose, d-lyxose and furfural under microwave irradiation. *Carbohydr. Polym.* **2013**, *98*, 1416–1421. [CrossRef]

19. Yemiş, O.; Mazza, G. Acid-catalyzed conversion of xylose, xylan and straw into furfural by microwave-assisted reaction. *Bioresour. Technol.* **2011**, *102*, 7371–7378. [CrossRef]

20. Zhang, Y.; Wang, Z.; Feng, J.; Pan, H. Maximizing utilization of poplar wood by microwave-assisted pretreatment with methanol/dioxane binary solvent. *Bioresour. Technol.* **2020**, *300*, 122657. [CrossRef]

21. Mikulski, D.; Kłosowski, G. Microwave-assisted dilute acid pretreatment in bioethanol production from wheat and rye stillages. *Biomass Bioenerg.* **2020**, *136*, 105528. [CrossRef]

22. Benko, Z.; Andersson, A.; Gáspár, M.; Réczey, K.; Stålbrand, H. Heat extraction of corn fiber hemicellulose. *Appl. Biochem. Biotechnol.* **2007**, *136*, 14.

23. Mashuni; Hamid, F.H.; Muzuni; Kadidae, L.O.; Jahiding, M.; Ahmad, L.O.; Saputra, D. The determination of total phenolic content of cocoa pod husk based on microwave-assisted extraction method. *AIP Conf. Proc.* **2020**, *2243*, 030013. [CrossRef]

24. Ji, Q.; Yu, X.; Yagoub, A.E.-G.A.; Chen, L.; Zhou, C. Efficient removal of lignin from vegetable wastes by ultrasonic and microwave-assisted treatment with ternary deep eutectic solvent. *Ind. Crops Prod.* **2020**, *149*, 112357. [CrossRef]

25. Riansa-ngawong, W.; Prasertsan, P. Optimization of furfural production from hemicellulose extracted from delignified palm pressed fiber using a two-stage process. *Carbohydr. Res.* **2011**, *346*, 103–110. [CrossRef] [PubMed]

26. Yang, W.; Li, P.; Bo, D.; Chang, H. The optimization of formic acid hydrolysis of xylose in furfural production. *Carbohydr. Res.* **2012**, *357*, 53–61. [CrossRef]

27. Yang, W.; Li, P.; Bo, D.; Chang, H.; Wang, X.; Zhu, T. Optimization of furfural production from d-xylose with formic acid as catalyst in a reactive extraction system. *Bioresour. Technol.* **2013**, *133*, 361–369. [CrossRef]

28. Lamminpää, K.; Tanskanen, J. Study of furfural formation using formic acid. In Proceedings of the 8th World Congress of Chemical Engineering, Montréal, QC, Canada, 23 August 2009.

29. Lamminpää, K.; Ahola, J.; Tanskanen, J. Acid-catalysed xylose dehydration into furfural in the presence of kraft lignin. *Bioresour. Technol.* **2015**, *177*, 94–101. [CrossRef]

30. Rasmussen, H.; Sørensen, H.R.; Meyer, A.S. Formation of degradation compounds from lignocellulosic biomass in the biorefinery: Sugar reaction mechanisms. *Carbohydr. Res.* **2014**, *385*, 45–57. [CrossRef]

31. Peleteiro, S.; Rivas, S.; Alonso, J.L.; Santos, V.; Parajó, J.C. Furfural production using ionic liquids: A review. *Bioresour. Technol.* **2016**, *202*, 181–191. [CrossRef]

32. Jeong, G.H.; Kim, E.G.; Kim, S.B.; Park, E.D.; Kim, S.W. Fabrication of sulfonic acid modified mesoporous silica shells and their catalytic performance with dehydration reaction of d-xylose into furfural. *Micropor. Mesopor. Mat.* **2011**, *144*, 134–139. [CrossRef]

33. O'Neill, R.; Ahmad, M.N.; Vanoye, L.; Aiouache, F. Kinetics of Aqueous Phase Dehydration of Xylose into Furfural Catalyzed by ZSM-5 Zeolite. *Ind. Eng. Chem. Res.* **2009**, *48*, 4300–4306. [CrossRef]

34. Xiouras, C.; Radacsi, N.; Sturm, G.; Stefanidis, G.D. Furfural Synthesis from d -Xylose in the Presence of Sodium Chloride: Microwave versus Conventional Heating. *Chem. Sus. Chem.* **2016**, *9*, 2159–2166. [CrossRef] [PubMed]

35. Kim, E.S.; Liu, S.; Abu-Omar, M.M.; Mosier, N.S. Selective Conversion of Biomass Hemicellulose to Furfural Using Maleic Acid with Microwave Heating. *Energ. Fuel* **2012**, *26*, 1298–1304. [CrossRef]

36. Möller, M.; Schröder, U. Hydrothermal production of furfural from xylose and xylan as model compounds for hemicelluloses. *RSC Adv.* **2013**, *3*, 22253. [CrossRef]

37. Antal, M.J.; Leesomboon, T.; Mok, W.S.; Richards, G.N. Mechanism of formation of 2-furaldehyde from d-xylose. *Carbohydr. Res.* **1991**, *217*, 71–85. [CrossRef]

38. Hongsiri, W.; Danon, B.; Jong, W. de Kinetic Study on the Dilute Acidic Dehydration of Pentoses toward Furfural in Seawater. *Ind. Eng. Chem. Res.* **2014**, *53*, 5455–5463. [CrossRef]

39. Mellmer, M.A.; Sener, C.; Gallo, J.M.R.; Luterbacher, J.S.; Alonso, D.M.; Dumesic, J.A. Solvent Effects in Acid-Catalyzed Biomass Conversion Reactions. *Angew. Chem. Int. Ed.* **2014**, *53*, 11872–11875. [CrossRef]

40. Gallo, J.M.R.; Alonso, D.M.; Mellmer, M.A.; Yeap, J.H.; Wong, H.C.; Dumesic, J.A. Production of Furfural from Lignocellulosic Biomass Using Beta Zeolite and Biomass-Derived Solvent. *Top. Catal.* **2013**, *56*, 1775–1781. [CrossRef]

41. Gürbüz, E.I.; Gallo, J.M.R.; Alonso, D.M.; Wettstein, S.G.; Lim, W.Y.; Dumesic, J.A. Conversion of Hemicellulose into Furfural Using Solid Acid Catalysts in γ-Valerolactone. *Angew. Chem. Int. Ed.* **2013**, *52*, 1270–1274. [CrossRef]

42. Kim, S.B.; You, S.J.; Kim, Y.T.; Lee, S.; Lee, H.; Park, K.; Park, E.D. Dehydration of D-xylose into furfural over H-zeolites. *Korean J. Chem. Eng.* **2011**, *28*, 710–716. [CrossRef]

43. Dias, A.S.; Pillinger, M.; Valente, A.A. Liquid phase dehydration of d-xylose in the presence of Keggin-type heteropolyacids. *Appl. Catal. A Gen.* **2005**, *285*, 126–131. [CrossRef]

44. Lam, E.; Majid, E.; Leung, A.C.W.; Chong, J.H.; Mahmoud, K.A.; Luong, J.H.T. Synthesis of Furfural from Xylose by Heterogeneous and Reusable Nafion Catalysts. *Chem. Sus. Chem.* **2011**, *4*, 535–541. [CrossRef] [PubMed]

45. Choudhary, V.; Pinar, A.B.; Sandler, S.I.; Vlachos, D.G.; Lobo, R.F. Xylose Isomerization to Xylulose and its Dehydration to Furfural in Aqueous Media. *ACS Catal.* **2011**, *1*, 1724–1728. [CrossRef]

46. Binder, J.B.; Blank, J.J.; Cefali, A.V.; Raines, R.T. Synthesis of Furfural from Xylose and Xylan. *Chem. Sus. Chem.* **2010**, *3*, 1268–1272. [CrossRef] [PubMed]

47. Lima, S.; Neves, P.; Antunes, M.M.; Pillinger, M.; Ignatyev, N.; Valente, A.A. Conversion of mono/di/polysaccharides into furan compounds using 1-alkyl-3-methylimidazolium ionic liquids. *Appl. Catal. A Gen.* **2009**, *363*, 93–99. [CrossRef]

48. Zhang, Z.; Du, B.; Quan, Z.-J.; Da, Y.-X.; Wang, X.-C. Dehydration of biomass to furfural catalyzed by reusable polymer bound sulfonic acid (PEG-OSO3H) in ionic liquid. *Catal. Sci. Technol.* **2014**, *4*, 633. [CrossRef]

49. Le Guenic, S.; Gergela, D.; Ceballos, C.; Delbecq, F.; Len, C. Furfural Production from d-Xylose and Xylan by Using Stable Nafion NR50 and NaCl in a Microwave-Assisted Biphasic Reaction. *Molecules* **2016**, *21*, 1102. [CrossRef]

50. Peleteiro, S.; da Costa Lopes, A.M.; Garrote, G.; Parajó, J.C.; Bogel-Łukasik, R. Simple and Efficient Furfural Production from Xylose in Media Containing 1-Butyl-3-Methylimidazolium Hydrogen Sulfate. *Ind. Eng. Chem. Res.* **2015**, *54*, 8368–8373. [CrossRef]

51. vom Stein, T.; Grande, P.M.; Leitner, W.; Domínguez de María, P. Iron-Catalyzed Furfural Production in Biobased Biphasic Systems: From Pure Sugars to Direct Use of Crude Xylose Effluents as Feedstock. *Chem. Sus. Chem.* **2011**, *4*, 1592–1594. [CrossRef]

52. Campos Molina, M.J.; Mariscal, R.; Ojeda, M.; López Granados, M. Cyclopentyl methyl ether: A green co-solvent for the selective dehydration of lignocellulosic pentoses to furfural. *Bioresour. Technol.* **2012**, *126*, 321–327. [CrossRef]

53. Wang, W.; Li, H.; Ren, J.; Sun, R.; Zheng, J.; Sun, G.; Liu, S. An efficient process for dehydration of xylose to furfural catalyzed by inorganic salts in water/dimethyl sulfoxide system. *Chinese J. Catal.* **2014**, *35*, 741–747. [CrossRef]

54. Li, H.; Ren, J.; Zhong, L.; Sun, R.; Liang, L. Production of furfural from xylose, water-insoluble hemicelluloses and water-soluble fraction of corncob via a tin-loaded montmorillonite solid acid catalyst. *Bioresour. Technol.* **2015**, *176*, 242–248. [CrossRef] [PubMed]

55. Tao, F.; Song, H.; Chou, L. Efficient process for the conversion of xylose to furfural with acidic ionic liquid. *Can. J. Chem.* **2011**, *89*, 83–87. [CrossRef]

56. Serrano-Ruiz, J.C.; Campelo, J.M.; Francavilla, M.; Romero, A.A.; Luque, R.; Menéndez-Vázquez, C.; García, A.B.; García-Suárez, E.J. Efficient microwave-assisted production of furfural from C5 sugars in aqueous media catalysed by Brönsted acidic ionic liquids. *Catal. Sci. Technol.* **2012**, *2*, 1828. [CrossRef]

57. Yang, Y.; Hu, C.-W.; Abu-Omar, M.M. Synthesis of Furfural from Xylose, Xylan, and Biomass Using AlCl3·6 H2O in Biphasic Media via Xylose Isomerization to Xylulose. *Chem. Sus. Chem.* **2012**, *5*, 405–410. [CrossRef] [PubMed]

Sample Availability: Samples of the compounds are not available from the authors.

 *molecules*

Article

Effects of Biosurfactants on Enzymatic Saccharification and Fermentation of Pretreated Softwood

Alfredo Oliva-Taravilla [1], Cristhian Carrasco [2], Leif J. Jönsson [1] and **Carlos Martín [1,*]**

[1] Department of Chemistry, Umeå University, SE-901 87 Umeå, Sweden; alfredoolivat@gmail.com (A.O.-T.); leif.jonsson@umu.se (L.J.J.)

[2] Instituto de Investigación y Desarrollo de Procesos Químicos (IIDEPROQ), Chemical Engineering, Faculty of Engineering, Universidad Mayor de San Andrés, La Paz 12958, Bolivia; cristhian.carrasco@gmail.com

* Correspondence: carlos.martin@umu.se; Tel.: +46-90-786 6099

Academic Editors: Alejandro Rodríguez and Eduardo Espinosa

Received: 30 June 2020; Accepted: 2 August 2020; Published: 5 August 2020

Abstract: The enzymatic hydrolysis of cellulose is inhibited by non-productive adsorption of cellulases to lignin, and that is particularly problematic with lignin-rich materials such as softwood. Although conventional surfactants alleviate non-productive adsorption, using biosurfactants in softwood hydrolysis has not been reported. In this study, the effects of four biosurfactants, namely horse-chestnut escin, *Pseudomonas aeruginosa* rhamnolipid, and saponins from red and white quinoa varieties, on the enzymatic saccharification of steam-pretreated spruce were investigated. The used biosurfactants improved hydrolysis, and the best-performing one was escin, which led to cellulose conversions above 90%, decreased by around two-thirds lignin inhibition of Avicel hydrolysis, and improved hydrolysis of pretreated spruce by 24%. Red quinoa saponins (RQS) addition resulted in cellulose conversions above 80%, which was around 16% higher than without biosurfactants, and it was more effective than adding rhamnolipid or white quinoa saponins. Cellulose conversion improved with the increase in RQS addition up to 6 g/100 g biomass, but no significant changes were observed above that dosage. Although saponins are known to inhibit yeast growth, no inhibition of *Saccharomyces cerevisiae* fermentation of hydrolysates produced with RQS addition was detected. This study shows the potential of biosurfactants for enhancing the enzymatic hydrolysis of steam-pretreated softwood.

Keywords: biosurfactants; cellulose; enzymatic saccharification; fermentation; quinoa saponins; steam-pretreated spruce

---

1. Introduction

Biochemical conversion of lignocellulosic feedstocks by process steps including pretreatment, enzymatic saccharification, and microbial fermentation is a major route to advanced biofuels and bio-based chemicals and polymers. Softwood is a major potential feedstock for biorefining processes. Softwood species include, for example, Norway spruce (*Picea abies*), which is among the predominant tree species in Nordic forestry, and different varieties of pine, which is common both in boreal forests and in plantation forestry in the Southern Hemisphere. Softwood typically contains more lignin than hardwood or agricultural residues. The high lignin content greatly contributes to making softwood recalcitrant to bioconversion [1]. Enzymatic saccharification of softwood typically results in lower yields than what is obtained with other lignocellulosic materials treated with similar enzyme dosages [2]. Although the exact mechanisms behind inhibition by lignin of enzymatic saccharification of cellulose are not yet fully elucidated [3], a major issue behind that phenomenon is non-specific, catalytically non-productive, and sometimes irreversible, binding of cellulases to lignin during enzymatic hydrolysis [4]. Lignin hydrophobicity is an important factor behind this phenomenon [5].

The non-specific adsorption of an enzyme to a substrate in a heterologous biocatalytic system, such as the enzymatic hydrolysis of cellulose, can be alleviated by introducing a surfactant into the medium [6,7]. A surfactant is an amphiphilic molecule composed of a hydrophobic portion covalently linked to a hydrophilic moiety [8]. It is known that the addition of non-ionic surfactants, i.e., those having a non-ionic hydrophilic moiety, improves the enzymatic hydrolysis of lignocellulosic materials [9]. However, most of the surfactants that have so far been used as additives for enhancing the enzymatic hydrolysis of cellulose are synthetic substances that are not biodegradable, and therefore they are toxic to the environment [7]. Using biosurfactants, which are amphiphilic substances based on renewable resources, often of microbial origin, has less impact on the environment, and it is, therefore, a more sustainable approach than using conventional synthetic surfactants [10]. Due to their specificity, biodegradability, and biocompatibility, biosurfactants have become very useful in different sectors, such as bioremediation [11], and in the pharmaceutical, cosmetic [12], food, and petroleum industries [13].

There are some examples of using biosurfactants in cellulose hydrolysis [14,15], but so far that application has been relatively poorly investigated. Two groups of biosurfactants that have attracted interest for lignocellulose bioconversion are rhamnolipids and saponins. Rhamnolipids, which contain hydrophilic rhamnose moieties linked to hydrophobic β-hydroxylated fatty-acid chains [16], have been used to prevent non-productive binding of enzymes to lignin during hydrolysis of herbaceous lignocellulosic feedstocks and hardwood [7,17,18]. Wang et al. [17] observed that the addition of rhamnolipids improved the activity and stability of cellulases, which resulted in an increase of the release of reducing sugars, and they proposed that there is more than one mechanism involved in that positive effect. Saponins, which are composed of a hydrophobic unit, a triterpenoid aglycone known as sapogenin, and at least two hydrophilic glycoside moieties [19], have recently been used to improve the enzymatic hydrolysis of residues from production of furfural from corn cobs [15,20,21].

Although the use of biosurfactants, such as rhamnolipids or saponins, has shown interesting results in the hydrolysis of herbaceous materials, their application to more recalcitrant lignocellulosic feedstocks, for instance, softwood, has, to our knowledge, not yet been reported. It is especially interesting to investigate the effects of saponins that are discarded in residual streams of agro-processing, for example, quinoa saponins. The coat covering quinoa seeds contains saponins that are related with bitterness, and, therefore, the seed coats are removed when processing quinoa grain for food use. In that operation, large amounts of saponin-rich residues are generated [22,23]. Although saponins have potential utility because of their detergent activity, toxicity against viral diseases, and cholesterol-lowering effects among other properties [24], the quinoa seed coat residue is currently underutilized. A way of giving value to that waste stream could be by extracting the saponins, and using them as biosurfactants, for instance, as enhancers of the enzymatic hydrolysis of cellulose. However, this would be a new way to utilize quinoa saponins, and because of the antimicrobial activity of saponins [25,26] including their toxic effect on yeast [27], it is not clear if it would be a viable approach.

The objective of the current work was to investigate the potential of biosurfactants for improving the enzymatic hydrolysis of cellulose contained in steam-pretreated spruce (SPS). Rhamnolipid from *Pseudomonas aeruginosa* and three sorts of saponins, namely a commercial product of high purity and two crude saponin extracts, were included in the study. Furthermore, the effects of saponin addition in the enzymatic hydrolysis step on the ethanolic fermentation by *Saccharomyces cerevisiae* were evaluated.

2. Results

2.1. Effects of Rhamnolipid and Escin on Saccharification of Avicel in the Presence of Lignin

Enzymatic saccharification of 50:50 mixtures of Avicel and lignin, with or without the addition of biosurfactants, was compared with saccharification of pure Avicel. Two commercially available biosurfactants, namely rhamnolipid from *P. aeruginosa* [16] and escin, a saponin preparation from horse-chestnut seeds [28], were investigated in experiments in which they were added at a ratio of 2 g surfactant per 100 g Avicel. The cellulose conversion during enzymatic saccharification of a mixture of Avicel and lignin corresponded to around 77.0% of the conversion achieved in the reference reaction (Avicel without lignin) (Section A of Table 1). This effect can be attributed to a large extent to the non-productive adsorption of cellulases to lignin. When biosurfactants were added, the detrimental effect of lignin decreased, and the enzymatic hydrolysis was improved, slightly (around 3%) for rhamnolipid and remarkably (more than 15%) for escin. In the presence of escin, a relative cellulose conversion of 92.4% was achieved, which corresponds to a reduction of around two thirds of the inhibition caused by lignin.

Table 1. Cellulose conversion in enzymatic hydrolysis of Avicel/lignin mixtures (A) and steam-pretreated spruce (SPS) (B) in the presence of biosurfactants.

Substrate	Surfactant	Cellulose Conversion, % (*w/w*) [1]		
		D^2: 0	D^2: 2	D^2: 4
A				
Avicel + lignin	None	76.9 (0.4)	-	-
Avicel + lignin	Rhamnolipid	-	80.2 (3.3)	-
Avicel + lignin	Escin	-	92.4 (3.1)	-
B				
Pretreated solids	None	70.7 (5.2)	-	-
Pretreated solids	Rhamnolipid	-	75.7 (1.1)	79.6 (1.2)
Pretreated solids	Escin	-	87.2 (0.4)	87.4 (0.8)
Pretreated solids	RQ saponin[3]	-	75.3 (2.0)	81.9 (0.5)
Pretreated solids	WQ saponin[4]	-	73.0 (0.2)	76.8 (0.7)

[1] The values shown in section A are percentages of the conversion in the lignin-free experiment; [2] biosurfactant dosage (in g/100 g substrate); [3] crude saponin from red quinoa; [4] crude saponin from white quinoa. Mean of three replicates. The standard deviations are shown in parentheses.

2.2. Effects of Rhamnolipid and Saponins on Saccharification of Steam-Pretreated Spruce

As a further investigation of the positive effects of the addition of biosurfactants observed in the enzymatic saccharification of Avicel/lignin mixtures, experiments were made with a real lignocellulosic substrate, viz. steam-pretreated spruce. The range of biosurfactants was extended with two crude saponin extracts from seed coats of red and white quinoa varieties. Crude quinoa saponins are products of potential industrial interest and they have functional similarities with escin, which was the best-performing biosurfactant in the saccharification of Avicel. The biosurfactant dosage of 2 g/100 g was used for facilitating the comparison with the Avicel experiment, but a larger dosage, 4 g/100 g, was also evaluated considering that crude saponin extracts might have a weaker effect compared to that of pure rhamnolipid and escin.

As in the experiment with Avicel/lignin mixtures (Table 1, Section A), rhamnolipid (RL) and escin improved the hydrolytic conversion of the cellulose in steam-pretreated spruce (SPS) (Table 1, Section B). The trend observed in the hydrolysis of SPS was comparable to that of Avicel hydrolysis. Escin was similarly effective at both tested dosages (improvements of 23–24%), whereas the improvement caused by RL was about twice as high (13%) at the higher dosage compared to the lower dosage (7% improvement). Both quinoa saponin extracts also improved the enzymatic saccharification, and red quinoa saponins (RQS) more than white quinoa saponins (WQS) (Table 1). As with RL, the improvements caused by the quinoa saponins were highly dose-dependent (improvements of 16% for RQS and 9% for WQS at high dosages, compared to 7% for RQS and 3% for WQS at low dosages). The highest dosage of biosurfactant addition (4 g/100 g) resulted in cellulose conversion levels of 87.4,

79.6, 81.9, and 76.8% for escin, RL, RQS, and WQS, respectively. The corresponding increases compared to the reference reaction without a biosurfactant (24, 16, and 13% for escin, RQS, and RL, respectively) were statistically significant (p-values ≤ 0.04).

2.3. Evaluation of Red Quinoa Saponins (RQS) Dosage during Enzymatic Saccharification

Although escin gave the best results, RQS, which was the second most effective among the tested biosurfactants, also deserves attention as an additive to the enzymatic hydrolysis of steam-pretreated spruce. Quinoa saponins, which can be easily extracted from quinoa seed coats, are a low-cost alternative to more expensive biosurfactants. Furthermore, using quinoa saponins in lignocellulose bioconversion might contribute to upgrading seed coats, which are an abundant and underutilized byproduct in quinoa-producing countries [29]. Therefore, the dosages of RQS under different saccharification conditions were further evaluated and optimized.

In order to elucidate how the effect of RQS changes when enzymatic saccharification is performed at enzyme and substrate loadings higher than those assayed in the first experiment, three different levels for the RQS dosage and for the loadings of enzymes and pretreated spruce were set according to a Box–Behnken factorial design. The highest RQS dosage used for the experiments with different biosurfactants (4 g/100 g biomass) was chosen as the center in the experimental design (Table 2).

Table 2. Experimental conditions used for the evaluation of the dosage of red quinoa saponins (RQS) in the enzymatic hydrolysis of steam-pretreated spruce.

Experimental Reaction	RQS Dosage, g/100 g	Biomass Loading, % (w/w)	Enzyme Loading, FPU and CBU/g dry Weight Substrate
1	4	5	5
2	2	5	7.5
3	6	5	7.5
4	4	5	10
5	2	10	5
6	6	10	5
7	4	10	7.5
8	2	10	10
9	6	10	10
10	4	15	5
11	2	15	7.5
12	6	15	7.5
13	4	15	10
Control	0	10	7.5

The Pareto chart of standardized effects shows that the dosages of both enzyme and RQS exerted directly proportional effects on cellulose conversion, whereas substrate loading affected it inversely, and none of the factor interactions exerted any important influence on the response variable (Figure 1a). The response surface plots showed that the cellulose conversion decreased with the increase in the substrate load (Figure 1b) and increased with the increase in the enzyme load (Figure 1c).

Independently of the load of enzyme and substrate, increasing the RQS dosage always led to a clear increase in cellulose conversion. For instance, at 5% solids and 7.5 FPU/g, increasing the RQS dosage from 2 g/100 g (experimental reaction 2) to 6 g/100 g (reaction 3) enhanced the conversion by around 10% (from 66 to 73%, Figure 2a). The same applies for the experiments with 10% solids load, at either 5 or 10 FPU/g, where the conversion increased, respectively, from 51% (reaction 5) to 57% (reaction 6) and from 65% (reaction 8) to 71% (reaction 9) (Figure 2b) when the RQS dosage increased from 2 to 6 g/100 g. The effect was also evident at the highest load of substrate (15%), where increasing the saponins dosage from 2 g/100 g (reaction 11) to 6 g/100 g (reaction 12) at an enzyme loading of 7.5 FPU/g resulted in an 8% improvement (from 55.6% to 60.1%) of the cellulose conversion (Figure 2b).

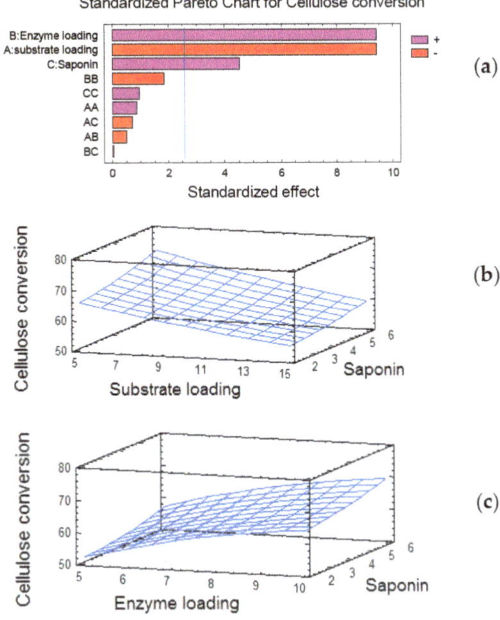

Figure 1. Cellulose conversion during the enzymatic hydrolysis of SPS with different RQS dosages under different loadings of enzymes and substrate. Pareto chart of standardized effects on cellulose conversion (positive, +; negative −) (**a**), and response surface plots at constant loading of enzyme (**b**) and substrate (**c**).

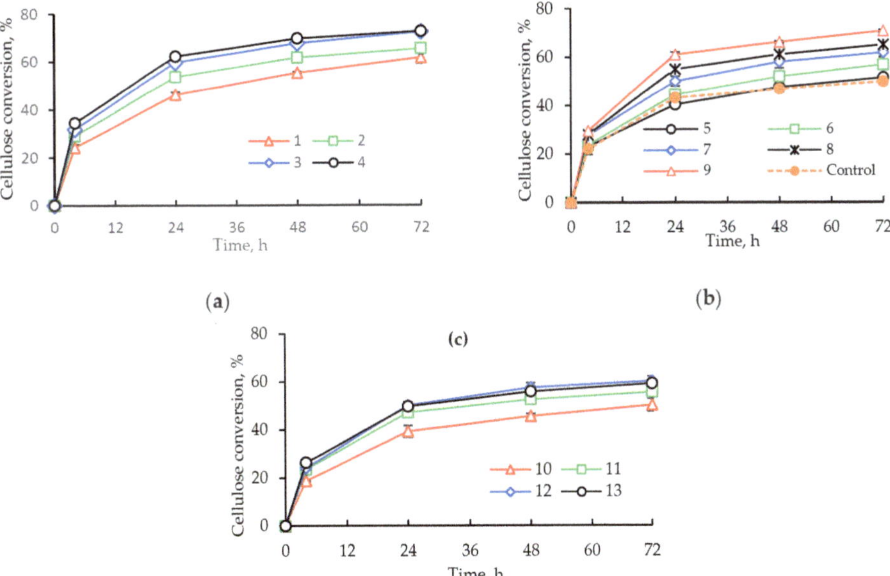

Figure 2. Time course of the enzymatic hydrolysis of steam-pretreated spruce at substrate loadings of 5% (**a**), 10% (**b**), and 15% (**c**). The numbers in the legend correspond to the experimental reactions as indicated in Table 2. Mean of three replicates. The error bars correspond to standard deviations.

Another positive effect of addition of saponins was the reduction in the enzyme dosage required for achieving similar cellulose conversion. For instance, in the experiments with 5% solids load, a 6 g/100 g RQS dosage combined with 7.5 FPU/g cellulase loading (run 3) gave a conversion comparable with that achieved by using 4 g/100 g saponins and 10 FPU/g (run 7) (Figure 2a). Similarly, at 10% solids, both 2 and 6 g/100 g RQS addition combined with the lowest cellulase loading (5 FPU/g) (reactions 5 and 6) resulted in better conversion than that achieved with a 50% higher cellulase dosage and no saponin (Control) (Figure 2b). This is in good agreement with previous reports showing that in the enzymatic hydrolysis of furfural production residues, the addition of *Gleditsia* saponin alleviated the negative effect on cellulose conversion caused by increased substrate loading [21].

Further details were revealed by examining the concentration of free protein in the reaction mixtures at the end of hydrolysis in trials with similar enzyme loadings but different saponin dosages. Independently of the amount of added enzyme, the final protein concentration was always higher for reaction mixtures with higher saponin dosages (Table 3). For instance, for 7.5 FPU/g loading and no saponin, the protein concentration at the end of the reaction was 34.6% of the initial, while in the presence of RQS, the fraction of free protein was 38.1%. Similarly, in experiments with cellulase loadings of 5 and 10 FPU/g, the fraction of protein remaining in the supernatant at the end of the saccharification reaction increased by 10% by increasing the saponin loading from 2 to 6 g/100 g.

Table 3. Total soluble protein before and after enzymatic hydrolysis of steam-pretreated spruce.

Enzyme Loading, FPU/g	Saponins Dosage, g/100 g	Protein Concentration, mg/mL		Free Protein, %
		Initial	Final	
5	2	1.24 (0.01)	0.48 (0.01)	38.7
5	6	1.24 (0.01)	0.58 (0.01)	47.7
7.5	0	1.39 (0.10)	0.48 (0.01)	34.6
7.5	4	1.39 (0.10)	0.53 (0.01)	38.1
10	2	1.69 (0.10)	0.56 (0.01)	33.3
10	6	1.69 (0.10)	0.76(0.04)	43.3

Neither the surface response plots (Figure 1b,c) nor a complementary experiment with a wider RQS dosage range (between 2 and 12 g/100 g) revealed any optimal saponins loadings for the hydrolysis of steam-pretreated spruce. Higher dosages than those included in the experimental design only slightly affected the saccharification. For instance, the cellulose conversion achieved by an RQS addition of 8 g/100 g showed no statistical difference compared to the value obtained with 6 g/100 g, and higher dosages resulted in only marginal improvements (data not shown).

In order to estimate the fraction of RQS that adsorbed onto the lignin surface, the concentrations of RQS remaining in the medium after the saccharification reaction were assessed (Table 4). The result shows that RQS recovery after saccharification was higher when the initial dosage was low. For instance, when the initial concentration was 2 g/L, half of the RQS remained free in the reaction mixture, which points towards adsorption of the other half onto lignin. The RQS remaining in the medium after hydrolysis increased with increasing initial concentrations, indicating that after a certain dosage, no more saponin was adsorbed onto lignin.

Table 4. Change of the concentration of red quinoa saponins (RQS) after enzymatic hydrolysis.

RQS$_{initial}$, g/L	RQS$_{final}$, g/L	RQS Recovery, %
2.0	1.0 (±0.1)	50.0
6.0	5.0 (±0.0)	83.3
8.0	7.5 (±0.1)	93.8
10.0	9.5 (±0.1)	95.0
12.0	11.6 (±0.1)	96.6

The enzyme loading was 7.5 FPU/g 100 g.

2.4. Effect of Red Quinoa Saponins (RQS) on Ethanol Production by Saccharomyces cerevisiae

Although our experiments demonstrated the positive effect of biosurfactants on the enzymatic saccharification of steam-pretreated spruce, it was not clear how red quinoa saponins can affect the ethanolic fermentation of hydrolysates. Saponins can be toxic to different microorganisms. For example, *Quillaja saponaria* saponins inhibit the growth of some yeast strains [27], and saponins from quinoa seeds have been shown to exert antimicrobial, antifungal, vermicide, and anti-yeast activity [26,27]. Therefore, an evaluation of the possible inhibitory effect of RQS on the fermentation of hydrolysates was performed.

To evaluate saponin toxicity, *S. cerevisiae* Ethanol Red, a robust yeast strain especially developed for the ethanol industry [30], was used as the fermenting microorganism, and two parallel sets of experiments were run. In the first set of fermentations, aimed at evaluating the effect of the remaining saponins from that added in the enzymatic hydrolysis, spruce hydrolysates were used as the fermentation medium. In the second experimental series, focusing on the effect of the saponins added directly at the start of the fermentation, synthetic media supplemented with RQS were used. The RQS additions to the synthetic medium were equivalent to the amounts added before the enzymatic reaction for preparing the hydrolysates.

The glucose consumption and ethanol production during the course of the fermentation of the hydrolysates showed that no inhibition occurred independently of the RQS dosage used in the hydrolysis (Figure 3a). There was a short lag phase at the beginning of the fermentation, but it was not caused by saponins since it was observed also in the RQS-free hydrolysate. After that, glucose was consumed at a high rate in all the hydrolysates, and it was depleted after 24 h. Ethanol concentrations of around 12 g/L were reached in the fermentation of the RQS-free hydrolysate, and between 16 and 18 g/L in the RQS-containing hydrolysates. That divergence was due to different initial glucose concentrations, which were 24 g/L in the hydrolysate produced without saponins supplementation, and 32–36 g/L in those produced with RQS addition (Figure 3a).

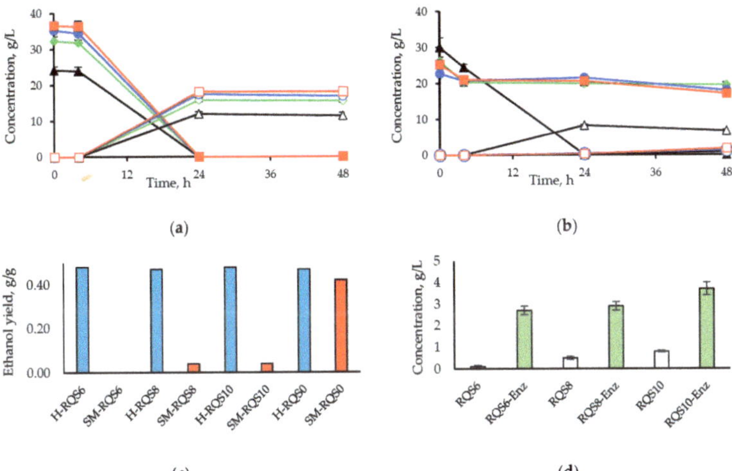

Figure 3. Concentration of glucose (filled symbols) and ethanol (open symbols) during fermentation of hydrolysates produced by enzymatic hydrolysis of SPS in presence of RQS (**a**) and synthetic media with directly-added RQS (**b**). RQS dosages: 6 g/100 g biomass (rhombs), 8 g/100 g biomass (circles), 10 g/100 g biomass (squares), RQS-free control (triangles). Ethanol yield in the fermentations of hydrolysates (blue bars) and synthetic media (red bars) (**c**). Final glucose concentration after incubation of RQS with (green bars) and without (white bars) cellulases (**d**). RQS concentrations in the synthetic media are equivalent to the dosages used in hydrolysis. Mean of three replicates. The error bars correspond to standard deviations.

A different picture was observed when RQS was added to the synthetic medium right before yeast inoculation. A strong inhibition was noticeable in all the saponin-containing assays, whereas glucose was readily consumed in the saponin-free control (Figure 3b). In the presence of freshly added saponins, no signs of metabolic activity were detected during the first 24 h, and some minor sugar consumption and ethanol formation were evident only after 48 h. The ethanol yields per initial glucose in the synthetic medium containing freshly-added RQS were negligible compared with that of the saponin-free control, while in the hydrolysates, the yields were comparable independently of the saponin concentrations used in the hydrolysis (Figure 3c). The statistical significance of the differences between ethanol concentrations achieved in the fermentations of hydrolysates produced with RQS addition and those achieved in fermentations of synthetic media with freshly added RQS was confirmed by one-way analysis of variance (Table 5).

Table 5. One-way ANOVA for the ethanol concentration in the fermentation of hydrolysates produced by enzymatic hydrolysis of SPS in the presence of RQS and in synthetic media with directly added RQS.

Source of Variation	Sum of Squares	Degree of Freedom	Mean Squares	*F*-Ratio	*P*-Value	F Critical Value
Between groups	1136.787	3	378.929	587.721	1.16×10^{-19}	3.090
Within groups	12.895	20	0.645			
Total	1149.681	23				

In an additional experiment for helping to clarify the above-stated difference, incubation of RQS with cellulases led to a dose-dependent glucose release (Figure 3d). This can be interpreted as an evidence of saponin degradation as a result of the splitting of the glycoside moieties from sapogenins during the enzymatic saccharification reaction.

3. Discussion

3.1. Effect of Biosurfactants on Enzymatic Hydrolysis of Avicel-Lignin Mixtures and Steam-Pretreated Spruce

The observed improvement in enzymatic hydrolysis can be attributed to the amphiphilic nature of the added biosurfactants, whose hydrophilic head groups would protrude into the aqueous liquid phase, while the hydrophobic tails would adsorb to exposed lignin on the surface of the pretreated solids. As a consequence, lignin–cellulase interactions are obstructed, and the protein-binding activity of lignin is reduced [15,31]. This results in an increase of the availability of effective cellulases to catalyze cellulose hydrolysis, and, therefore, the hydrolytic conversion is improved.

The differences between the effects of rhamnolipid and escin might be connected to the different chemical nature of their aglycones. Rhamnolipids are glycolipid biosurfactants with hydroxylated fatty acid tails [32], while escin, being a saponin, contains a triterpenoid sapogenin as aglycone [33]. It seems that rhamnolipid aglycone, probably due to the presence of a free carboxyl group and relatively short carbon chains compared with the more hydrophobic escin sapogenin (Figure 4), is less effective in adsorbing to lignin and thereby preventing the interference of lignin with the enzymatic saccharification of cellulose. The pH of the media used in the experiments, pH 5.0, is around or above the normal pK_a values of carboxylic acid groups, which might further explain why short fatty acid tails of rhamnolipids would be somewhat less effective in adsorbing to hydrophobic surfaces. The effect of *P. aeruginosa* rhamnolipid in enzymatic hydrolyses of hardwood cellulose was recently reported for eucalyptus wood chips [18], but no studies about using escin in lignocellulose hydrolysis are available to date.

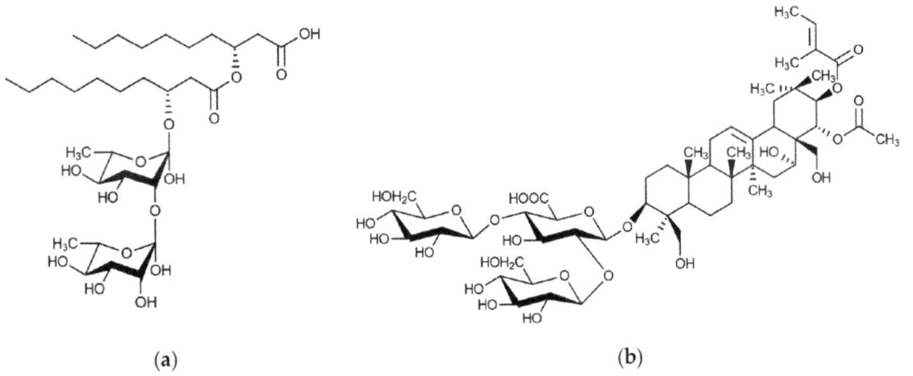

(a) (b)

Figure 4. Molecular structures of rhamnolipid from *P. aeruginosa* (**a**) and horse-chestnut escin (**b**).

The cellulose conversions achieved in the enzymatic saccharification of steam-pretreated spruce in the presence of escin, RL, and quinoa saponins (~77–87%) compare favorably with that of 74.9% reported by Xing et al. [21] for *Gleditsia* saponin-supplemented enzymatic saccharification of residues from production of furfural using corn cobs. Although the results by Xing et al. were achieved in experiments with a high solids loading (20%), our results merit attention considering that the enzyme load we used (15.4 FPU/g cellulose) was lower than theirs (30 FPU/g cellulose), our hydrolysis was shorter (72 h instead of 120 h), and the lignin content of the SPS (48.8%) was higher than that of the corn cob residues (45.3%).

Although all the tested biosurfactants exerted at least some positive effect on enzymatic hydrolysis, the magnitude of the effect was different. The differences between the effects of escin and rhamnolipid might be attributed to the different chemical functionalities of their aglycones (Figure 4). The different effect between quinoa saponins and escin saponin, which have the same chemical nature, are probably linked to their purity, i.e., the used escin was a chemically pure compound, while quinoa saponins were used in the form of crude extracts that were not subjected to any major purification. It is more difficult to explain why the effects of saponins from different quinoa varieties were so different, but one reason could be that the chemical properties of saponins differ slightly from each other.

As it occurs with other biosurfactants, when saponins are added to the enzymatic hydrolysis mixture, the hydrophobic aglycone would adsorb to lignin and block unspecific adsorption of cellulases. With increasing hydrophobicity of the aglycone, the propensity of the saponin to adsorb to lignin would increase, and a further reduction in the unspecific binding of cellulases to lignin would occur. Quinoa saponins are rather heterogeneous, and they can differ in either the sapogenin moiety, the glycoside component, or both of them [24]. It can be hypothesized that saponins with highly hydrophobic aglycones would better prevent non-productive adsorption than those whose aglycones are less hydrophobic. The presence of hederagenin, a triterpenoid with certain polarity due to a free hydroxyl group (Figure 5a), as a major aglycone in WQS [34] might make them slightly less hydrophobic than those from red quinoa varieties, where sapogenins without polar functionalities, such as oleanolic acid (Figure 5b), are predominant. Furthermore, the hydrophobicity of the red quinoa pigments, which are apparently retained in the saponin extract, might also have played a role in increasing the RQS affinity for lignin. The hydrophobic binding properties of amaranthine, one of the main pigments in red quinoa, have been previously shown [35]. Amaranthine and iso-amaranthine are abundant in saponins from red quinoa but are absent in saponins from white quinoa [34].

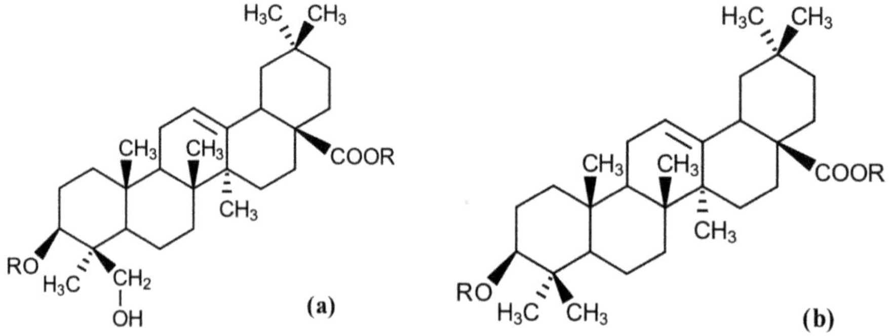

Figure 5. Molecular structures of sapogenins found in different varieties of quinoa: hederagenin (**a**) and oleanolic acid (**b**).

Overall, the results confirm our hypothesis of a positive effect of biosurfactants on the enzymatic hydrolysis of pretreated softwood. As for less recalcitrant materials [14,15,17,18,20,21], RL and saponins enhanced the enzymatic hydrolysis of cellulose contained in steam-pretreated spruce. The potential of chemically synthesized surfactants for improving the enzymatic hydrolysis of pretreated softwood has been shown before [6,36], but the current study is, to the best of our knowledge, the first time that the use of biosurfactants is reported for that kind of feedstock.

3.2. Evaluation of Red Quinoa Saponins (RQS) Dosage during Enzymatic Saccharification

It has been shown before that the problem of catalytically non-productive binding of cellulases to lignin is more pronounced when low enzyme loadings, as those required in the industry, are used [37]. On the other hand, lignin from different biomass sources can bind cellulases to a different extent [38], and there are indications that the binding capacity of softwood lignin is higher than that of lignins from other materials [39]. Some of our experiments with pretreated softwood were conducted with cellulase loadings as low as 5 FPU/g biomass, and even at those low loadings, the positive effect of RQS was evident (Figure 1c). This, together with the low cost of quinoa saponins, provides a solid argument favoring their use as additives for the large-scale bioconversion of softwood.

The higher concentrations of free protein at the end of the saccharification reactions in experiments with higher saponin dosages (Table 3) can be linked to the observed saponin-induced reduction in the enzyme requirement for achieving comparable cellulose conversion (Figure 2). This phenomenon indicates that the surfactant effect of saponin weakens the cellulase-binding capacity of lignin. This would result in more cellulase becoming available for catalyzing the saccharification of cellulose [6].

The fact that no optimal RQS dosage was found in enzymatic hydrolysis might be related to micellar phenomena and lignin saturation resulting from high saponin concentrations in water. The experiment revealed that when the saccharification reaction is carried out using increasing RQS dosages, a threshold is reached, after which no further positive effects of the added surfactant are discernible. This might be because at a high RQS dosage the critical micelle concentration is reached, and the formed micelles hinder the diffusion of reactants and restrict the enzymatic reaction [15,40]. Since the physical and physico-chemical properties of saponin micelles depend on their biological origin [41], a further study in this area for saponins from different quinoa varieties would be of interest. It is also possible that lignin gets saturated, and no more saponins are adsorbed onto its surface independently of the amount added (Table 4). Consequently, RQS dosages above the saturation threshold would not affect the enzymatic hydrolysis of cellulose. This explains why no significant improvement of the enzymatic conversion was observed with saponin dosages above 6 g/100 g. Based on this, one can consider 6 g/100 g as an optimal RQS dosage within the experimental conditions investigated in this study. Our results are in line with a recent report showing that the glucose yield in enzymatic hydrolysis of organosolv-pretreated poplar sawdust increased proportionally with the saponin dosage only until a certain threshold, and after that it levelled off [42].

3.3. Effect of Red Quinoa Saponins (RQS) on Ethanol Production by Saccharomyces cerevisiae

The observed inhibition of the fermentation of synthetic medium with freshly-added RQS agrees with previous studies of the effects of saponins on the yeast microflora in winemaking [27]. That toxicity is related to the increase of yeast cell membrane permeability induced by saponins when their aglycones interact with sterols and hydrophobic tails of phospholipids, damaging the integrity of the plasma membrane and promoting the leakage of proteins and other intracellular constituents [25]. With that as background, the result that there was no inhibition of the fermentation of hydrolysates that were produced using RQS as a supplement was surprising and interesting. The different response of yeast to the saponin added to the preceding saccharification compared to that directly added to the fermentation reaction can be attributed to transformations happening during enzymatic hydrolysis.

We hypothesize that the absence of inhibition of the fermentation of the hydrolysates might be explained by either lowered concentrations of saponins in the medium or by their decreased toxicity as a result of the chemical reactions. One explanation could be that the saponin molecules, after adsorption to the lignin on the surface of pretreated solids, remain attached to it, and are removed from the liquid phase when it is separated from the solid phase after the reaction. In that case, the actual saponins concentration in the hydrolysates subjected to fermentation would be lower than the concentration added before enzymatic hydrolysis and lower than the RQS concentration in the synthetic medium. Another explanation could be that enzymes split off RQS sugar moieties from aglycones during the saccharification reaction, which decreases the saponins concentration and their detrimental effect on plasma membrane integrity. It has been shown that removal of sugars from saponins reduces their membrane-permeabilizing ability and antimicrobial activity. For instance, the heterologous expression of tomatinase, an enzyme that cleaves sugar moieties, in *S. cerevisiae* resulted in increased resistance to saponins [43]. In the current study, splitting of saponin sugar moieties was confirmed in an experiment, where incubation of RQS with cellulases for 72 h resulted in a clear increase in the glucose concentration compared with that observed when RQS was incubated without the enzyme preparation (Figure 3d). It is evident that cellulases, or perhaps other enzymes in the enzyme preparations used, can hydrolyze the β-glucosidic bonds between the D-glucopyranosyl moieties of saponin and the triterpenoid sapogenin.

Another hypothesis that could be considered is related to the potential fermentation-enhancing capacity of the released sapogenins. It could be so those sapogenins resulting from partial splitting of saponin sugar moieties during enzymatic hydrolysis, and still remaining in the hydrolysate, could be favorable for yeast in a similar way as for structurally related substances, such as certain sterols, that provide protection against ethanol toxicity during winemaking [44].

4. Materials and Methods

4.1. Pretreatment of Material

Norway spruce chips were pretreated by SEKAB E-Technology using a 30-L stainless steel reactor operating in continuous mode in the Biorefinery Demonstration Plant (Örnsköldsvik, Sweden). Unbarked wood chips were treated at 204 °C with addition of 1.2–1.3 kg SO_2/h, corresponding to a 2% (*w/w*) load of SO_2 per dry wood chips. The residence time in the reactor was 7 min, and the resulting pH was 1.5. The slurry obtained after the pretreatment was vacuum-filtered in order to separate the pretreatment liquid and the pretreated solids, which were washed with deionized water, and stored at 4 °C. The dry weight of the pretreated solids was determined by drying triplicate aliquots in a moisture analyzer (Mettler Toledo, Greifensee, Switzerland). The pretreated solids used as substrate in the enzymatic saccharification experiments contained (mass fraction, % dry weight): cellulose, 43.9; mannan, 2.2; xylan, 1.1; lignin, 48.8.

4.2. Enzymes and Chemicals

The enzyme preparations used were Celluclast 1.5L, with 50 FPU (filter paper units) per mL, and Novozyme 188, with 510 IU of β-glucosidase activity per mL [procured from Sigma-Aldrich (St. Louis, MO, USA) and produced by Novozymes A/S (Bagsværd, Denmark)]. Cellulase and β-glucosidase activities were determined by the filter paper method and the *p*-nitrophenyl-β-ᴅ-glucopyranoside (*p*NPG) assay, respectively [45]. Microcrystalline cellulose (Avicel PH-101, Sigma Aldrich), hereafter referred to as Avicel, was used in some enzymatic hydrolysis experiments. Rhamnolipid (RL) from *Pseudomonas aeruginosa* was acquired from Sigma-Aldrich. The saponins used in the saccharification experiments were escin from horse-chestnut (*Aesculus hippocastanum*), which is a high-purity saponin commercialized by Sigma-Aldrich, and raw saponins from red (RS) and white (WS) varieties of royal quinoa (*Chenopodium quinoa* Willd.). Raw saponins were kindly provided by Irupana Andean Organic Food S.A. (El Alto, Bolivia) and further sieved (1–1.7 mm) using a portable sieve shaker at the Instituto de Investigación y Desarrollo de Procesos Químicos, Universidad Mayor de San Andrés (La Paz, Bolivia). Saponin from *Quillaja saponaria* bark, supplied by Sigma-Aldrich, was used for analytical purposes. The used lignin was produced at Thünen Institute of Wood Research (Hamburg, Germany) by sulfuric acid-assisted organosolv pulping of beech wood [46].

4.3. Microorganism and Media

S. cerevisiae Ethanol Red (Fermentis Ltd., Marcq-en-Baroeul, France) was used as the fermentative organism. Yeast precultures in Erlenmeyer flasks (100 mL) were obtained by cultivation in 25 mL of growth medium containing (in g/L) glucose (20), yeast extract (5), and Bacto Peptone (5). The cultivation was pursued for 16 h at 150 rpm and 28 °C in an Ecotron orbital incubator (INFORS HT, Bottmingen, Switzerland). The cells were separated by centrifugation at 9000 rpm and 20 °C for 10 min, washed with sterile deionized water, and then suspended in sodium chloride 0.9% (*w/v*) right before inoculation.

4.4. Enzymatic Saccharification of Avicel in Presence of Lignin and Biosurfactants

Around 50 mg (DW, dry weight) Avicel and 50 mg lignin were suspended in 50 mM citrate buffer (pH 5.0) in 2-mL Eppendorf tubes giving a final solids load of 5% DW (*w/w*). The enzyme preparations Celluclast 1.5 L and Novozyme 188 were added at dosages corresponding to 15.4 FPU and 15.6 CBU per g cellulose. Rhamnolipid and escin were added at a ratio of 2 g per 100 g of Avicel (Table 1). A reference reaction with no lignin and no surfactants was also included. The reaction mixtures were incubated at 50 °C and 180 rpm for 72 h in the Ecotron orbital incubator, and each assay was performed in triplicate. Representative samples, withdrawn at the start and the end of the reaction, were centrifuged, and the glucose concentration in the supernatants was quantitated by HPLC, and used for calculating the enzymatic conversion of cellulose.

4.5. Enzymatic Saccharification of Steam-Pretreated Solids in Presence of Biosurfactants

Five hundred mg (DW) pretreated solids were mixed with 50 mM citrate buffer (pH 5.0) in 15-mL Falcon tubes giving a final solids load of 5% DW (*w/w*). Saccharification reactions were performed using Celluclast 1.5 L and Novozyme 188 added in the same loadings and under the same conditions as described above (Section 4.4). RL and the three different sorts of saponins (escin and crude saponins from either red or white quinoa) were added simultaneously with the enzymes at loadings of 2 and 4 g/100 g pretreated solids. A reaction without any added biosurfactants was performed as control. The enzymatic conversion of cellulose was calculated according to the following expression:

$$\mathrm{EC} = \frac{Glu/1.111}{CellPS * MassPS/100} \tag{1}$$

where: EC, enzymatic conversion, % (*w/w*); *Glu*, glucose mass in the hydrolysate, g; 1.111, coefficient considering water addition to anhydroglucose units during hydrolysis; *CellPS*, cellulose content in the pretreated solids, % (*w/w*); *MassPS*, mass of pretreated solids sample in the enzymatic hydrolysis assay, g.

4.6. Assessing the Red Quinoa Saponins (RQS) Dosage on Enzymatic Hydrolysis

In a first step, the effects of the RQS dosage (2, 4, and 6 g/100 g DW pretreated solids) at different enzyme loadings (5, 7.5, and 10 FPU and CBU/g substrate) and loadings of pretreated solids (5, 10, and 15% (*w/w*)) on cellulose conversion were evaluated using a Box–Behnken factorial design (Table 2). Standardized Pareto charts and response surface plots were used for analyzing the effects of the independent factors on the enzymatic conversion of cellulose. An experiment without saponins, and at the conditions corresponding to the central point of the experiment, was carried out as reference. As some sugar can be released from RQS during enzymatic hydrolysis, a substrate blank, i.e., a reaction with RQS and enzyme but without substrate, was used for correcting the glucose values in the rest of the experiment. An enzyme blank was also included. The hydrolysis was run following the same protocol as described above (Section 4.4), with the exception that sampling was performed at 0, 4, 24, 48, and 72 h. The software Statgraphics Plus 5.0 for Windows (Manugistics Inc., Rockville, MD, USA) was used for designing the experiment and analyzing the results.

In a second step, a wider range of saponin dosages (0, 6, 8, 10, and 12 g/100 g pretreated solids) was evaluated in a study, where the other parameters were kept constant. The loads of solids and enzymes corresponded to the values of the central point of the above-described experiment.

4.7. Fermentation of Synthetic Medium and Hydrolysate in Presence of Red Quinoa Saponins (RQS)

Fermentations of either the hydrolysate containing RQS added before enzymatic hydrolysis or synthetic medium with freshly added RQS were performed. The hydrolysate was produced using substrate loading 10% (*w/w*), enzyme loading 7.5 FPU, and 7.5 CBU/g substrate, and RQS dosages of either 6, 8, or 10 g/100 g DW substrate. The hydrolysate was supplemented with the following nutrients (in g/L): yeast extract (5), NH_4Cl (2), KH_2PO_4 (1), and $MgSO_4 \cdot 7H_2O$ (0.3). The synthetic medium contained (in g/L): glucose (20), yeast extract (5), and Bacto Peptone (5). The synthetic medium was supplemented with RQS in concentrations that were equivalent to the dosages used in the preparation of the hydrolysate medium. The pH of both media was adjusted to 5.0 with 10 M NaOH. The hydrolysate was filter-sterilized using Acrodisc syringe filters (Pall, Ann Arbor, MI, USA), while the synthetic medium was sterilized by autoclaving at 110 °C for 5 min. Eight mL of hydrolysate or 20 mL of synthetic medium in Erlenmeyer flasks were inoculated with 1 g/L of yeast cells, and the fermentations were carried out at 30 °C and 150 rpm for 72 h. Samples were taken after 4, 24, 48, and 72 h of fermentation. All fermentations were performed in triplicates.

4.8. Analytical Methods

The composition of the pretreated solids was analyzed by using two-step treatment with sulfuric acid according to the NREL protocol [47]. The sugars were quantified by high-performance anion-exchange chromatography (HPAEC) with pulsed amperometric detection (PAD). A Dionex ICS-3000 system (Sunnyvale, CA, USA) equipped with a 3 × 30 mm guard column and a 3 × 150 mm separation column (CarboPac PA20, Dionex) was used. Acid-insoluble lignin (Klason lignin) was determined gravimetrically, whereas acid-soluble lignin (ASL) was determined spectrophotometrically at 240 nm using a UV-1800 spectrophotometer (Shimadzu, Kyoto, Japan). Glucose and ethanol concentrations from enzymatic hydrolysis and fermentations were quantified by using a high-performance liquid chromatography (HPLC) instrument equipped with a refractive index detector. An Aminex HPX-87H column (Bio-Rad Labs, Hercules, CA), operated at 55 °C with 5 mM H_2SO_4 as mobile-phase (0.6 mL/min), was employed for separation. Saponin concentration was determined by a modification of the afrosimetric method [48] using *Q. saponaria* bark saponins as calibration standard. The method is based on measuring the height of the foam column formed after successive shakings of a saponin suspension in 50 mM citrate buffer (pH 5.0) within two 15-min intervals. Total soluble protein was determined with the Bradford method using bovine serum albumin (BSA) as standard [49]. Non-adsorbed protein (%) in supernatants was calculated as the relation between the final (after 72 h) and initial concentrations of protein according to the following equation:

$$\text{Non} - \text{adsorbed protein} = \left(\frac{\text{Final protein}}{\text{Initial protein}} \right) \times 100 \qquad (2)$$

One-way analysis of variance (ANOVA) was applied for evaluating the statistical difference between the results of experiments with different biosurfactant dosages. After that, post hoc t-tests were performed to determine if the difference between specific dosages was statistically significant.

4.9. Formatting of Chemical Structures

The molecular editing software ChemDoodle 10.3.0 (https://www.ichemlabs.com/) was used for drawing chemical structures.

5. Conclusions

The results of the current study showed that adding biosurfactants to enzymatic saccharification reactions reduces non-productive binding of cellulases to lignin and effectively enhances the hydrolytic conversion not only in artificial cellulose/lignin mixtures but also in steam-pretreated softwood. Saponins, both commercial escin and crude extracts from quinoa, were highly effective as enhancers of the enzymatic hydrolysis of cellulose contained in steam-pretreated spruce. Adding red quinoa saponins to the enzymatic hydrolysis of steam-pretreated spruce did not cause any inhibitory effects on the fermentation of the resulting hydrolysates using *S. cerevisiae*. Red quinoa saponins deserve further attention as a saccharification enhancer due to their effectiveness, innocuity, low cost, and relative abundance as an agricultural by-product in quinoa-producing countries.

Author Contributions: A.O.-T., experimental design, performing the experiments, interpreting the results, and writing the original draft of the manuscript; C.C., producing saponin extracts, interpreting the results, and review of the draft; L.J.J., conceptualization and experimental design, interpreting the results, review and editing of the manuscript, and providing most of the funding; C.M., conceptualization and experimental design, interpreting the results, review and editing the manuscript, and providing part of the funding. All authors have agreed to publish the current version of the manuscript.

Funding: This research was funded by the Swedish Energy Agency (P41285-1), the Kempe Foundations (SMK-1653), the Bio4Energy research environment (www.bio4energy.se) and the Swedish Research Council (2016-05822).

Acknowledgments: The technical support provided by Stefan Stagge is gratefully acknowledged.

Conflicts of Interest: The authors declare no conflicts of interest. The funding agencies had no role in the design of the study; in the interpretation of data; in the writing of the manuscript, or in the decision to publish the results.

References

1. Schwanninger, M.; Rodrigues, J.C.; Gierlinger, N.; Hinterstoisser, B. Determination of lignin content in Norway spruce wood by Fourier transformed near infrared spectroscopy and partial least squares regression analysis. Part 2: Development and evaluation of the final model. *J. Near. Infrared Spectrosc.* **2011**, *19*, 331–341. [CrossRef]

2. Rahikainen, J.; Mikander, S.; Marjamaa, K.; Tamminen, T.; Lappas, A.; Viikari, L.; Kruus, K. Inhibition of enzymatic hydrolysis by residual lignins from softwood—Study of enzyme binding and inactivation on lignin-rich surface. *Biotechnol. Bioeng.* **2011**, *108*, 2823–2834. [CrossRef] [PubMed]

3. Vermaas, J.V.; Petridis, L.; Qi, X.; Schulz, R.; Lindner, B.; Smith, J.C. Mechanism of lignin inhibition of enzymatic biomass deconstruction. *Biotechnol. Biofuels* **2015**, *8*, 217. [CrossRef] [PubMed]

4. Palonen, H.; Tjerneld, F.; Zacchi, G.; Tenkanen, M. Adsorption of *Trichoderma reesei* CBH I and EG II and their catalytic domains on steam pretreated softwood and isolated lignin. *J. Biotechnol.* **2004**, *107*, 65–72. [CrossRef]

5. Xu, F.; Ding, H.; Osborn, D.; Tejirian, A.; Brown, K.; Albano, W.; Sheehy, N.; Langston, J. Partition of enzymes between the solvent and insoluble substrate during the hydrolysis of lignocellulose by cellulases. *J. Mol. Catal. B-Enzym.* **2008**, *51*, 42–48. [CrossRef]

6. Eriksson, T.; Börjesson, J.; Tjerneld, F. Mechanism of surfactant effect in enzymatic hydrolysis of lignocellulose. *Enzyme Microb. Technol.* **2002**, *31*, 353–364. [CrossRef]

7. Zhang, Q.; He, G.; Wang, J.; Cai, W.; Xu, Y. Mechanisms of the stimulatory effects of rhamnolipid biosurfactant on rice straw hydrolysis. *Appl. Energy* **2009**, *86*, 233–237. [CrossRef]

8. Polarz, S.; Kunkel, M.; Donner, A.; Schlötter, M. Added-value surfactants. *Chem. Eur. J.* **2018**, *24*, 18842–18856. [CrossRef]

9. Jiang, F.; Qian, C.; Esker, A.R.; Roman, M. Effect of nonionic surfactants on dispersion and polar interactions in the adsorption of cellulases onto lignin. *J. Phys. Chem. B* **2017**, *121*, 9607–9620. [CrossRef] [PubMed]

10. Henkel, M.; Geissler, M.; Weggenmann, F.; Hausmann, R. Production of microbial biosurfactants: Status quo of rhamnolipid and surfactin towards large-scale production. *Biotechnol. J.* **2017**, *12*, 1600561. [CrossRef]

11. Vatsa, P.; Sanchez, L.; Clement, C.; Baillieul, F.; Dorey, S. Rhamnolipid biosurfactants as new players in animal and plant defense against microbes. *Int. J. Mol Sci.* **2010**, *11*, 5095–5108. [CrossRef] [PubMed]

12. Bezerra, K.G.O.; Rufino, R.D.; Luna, J.M.; Sarubbo, L.A. Saponins and microbial biosurfactants: Potential raw materials for the formulation of cosmetics. *Biotechnol. Prog.* **2018**, *34*, 1482–1493. [CrossRef] [PubMed]

13. Geetha, S.J.; Banat, I.M.; Joshi, S.J. Biosurfactants: Production and potential applications in microbial enhanced oil recovery (MEOR). *Biocatal. Agric. Biotechnol.* **2018**, *14*, 23–32.

14. Parthasarathi, R.; Sivakumaar, P.K. Enhancement effects of biosurfactant produced by *Pseudomonas aeruginosa* MTCC 2297 and *Pseudomonas fluorescens* on sugar cane bagasse composting. *J. Ecobiotechnol.* **2010**, *3*, 33–39.

15. Feng, Y.; Jiang, J.; Zhu, L.; Yue, L.; Zhang, J.; Han, S. Effects of tea saponin on glucan conversion and bonding behavior of cellulolytic enzymes during enzymatic hydrolysis of corncob residue with high lignin content. *Biotechnol. Biofuels* **2013**, *6*, 161. [CrossRef]

16. Chong, H.; Li, Q. Microbial production of rhamnolipids: Opportunities, challenges and strategies. *Microb. Cell. Fact.* **2017**, *16*, 137. [CrossRef]

17. Wang, H.Y.; Fan, B.Q.; Li, C.H.; Liu, S.; Li, M. Effects of rhamnolipid on the cellulase and xylanase in hydrolysis of wheat straw. *Bioresour. Technol.* **2011**, *102*, 6515–6521. [CrossRef]

18. Bezerra, K.G.O.; Gomes, U.V.R.; Silva, R.O.; Sarubbo, L.A.; Ribeiro, E. The potential application of biosurfactant produced by *Pseudomonas aeruginosa* TGC01 using crude glycerol on the enzymatic hydrolysis of lignocellulosic material. *Biodegradation* **2019**, *30*, 351–361. [CrossRef]

19. Fiallos-Jurado, J.; Pollier, J.; Moses, T.; Arendt, P.; Barriga-Medina, N.; Morillo, E.; Arahana, V.; Torres, M.L.; Goossens, A.; Leon-Reyes, A. Saponin determination, expression analysis and functional characterization of saponin biosynthetic genes in *Chenopodium quinoa* leaves. *Plant. Sci.* **2016**, *250*, 188–197. [CrossRef]

20. Xing, Y.; Bu, L.; Sun, D.; Liu, Z.; Liu, S.; Jiang, J. Enhancement of high-solids enzymatic hydrolysis and fermentation of furfural residues by addition of *Gleditsia* saponin. *Fuel* **2016**, *177*, 142–147. [CrossRef]

21. Xing, Y.; Jia, L.; Liu, Z.; Zhang, W.; Jiang, J. Effects of *Gleditsia* saponin on high-solids enzymatic hydrolysis of furfural residues. *Ind. Crop. Prod.* **2015**, *64*, 209–214. [CrossRef]

22. Ahumada, A.; Ortega, A.; Chito, D.; Benitez, R. Saponinas de quinua (*Chenopodium quinoa* Willd.): Un subproducto con alto potencial biológico. *Rev. Colomb. Cienc. Quim. Farm.* **2016**, *45*, 438–469. [CrossRef]

23. Flores, Y.; Díaz, C.; Garay, F.; Colque, O.; Sterner, O.; Almanza, G.R. Oleanane-type triterpenes and derivatives from seed coat of Bolivian *Chenopodium quinoa* genotype "salar". *Rev. Boliv. Quím.* **2005**, *22*, 71–77.

24. Kuljanabhagavad, T.; Thongphasuk, P.; Chamulitrat, W.; Wink, M. Triterpene saponins from *Chenopodium quinoa* Willd. *Phytochemistry* **2008**, *69*, 1919–1926. [CrossRef]

25. Berlowska, J.; Dudkiewicz-Kolodziejska, M.; Pawlikowska, E.; Pielech-Przybylska, K.; Balcerek, M.; Czysowska, A.; Kregiel, D. Utilization of post-fermentation yeasts for yeast extract production by autolysis: The effect of yeast strain and saponin from *Quillaja saponaria*. *J. Inst. Brew.* **2017**, *123*, 396–401.

26. Park, J.H.; Lee, Y.J.; Kim, Y.H.; Yoon, K.S. Antioxidant and antimicrobial activities of quinoa (*Chenopodium quinoa* Willd.) seeds cultivated in Korea. *Prev. Nutr. Food. Sci.* **2017**, *22*, 195–202.

27. Fischer, M.J.C.; Pensec, F.; Demangeat, G.; Farine, S.; Chong, J.; Ramirez-Suero, M.; Mazet, F.; Bertsch, C. Impact of *Quillaja saponaria* saponins on grapevine ecosystem organisms. *Anton. Leeuw.* **2011**, *100*, 197–206. [CrossRef]

28. Gruza, M.M.; Jatczak, K.; Zagrodzki, B.; Łaszcz, M.; Koziak, K.; Malińska, M.; Cmoch, P.; Giller, T.; Zegrocka-Stendel, O.; Woźniak, K.; et al. Preparation, purification and regioselective functionalization of protoescigenin-the main aglycone of escin complex. *Molecules* **2013**, *18*, 4389–4402. [CrossRef]

29. Ruiz, K.B.; Khakimov, B.; Engelsen, S.B.; Bak, S.; Biondi, S.; Jacobsen, S.E. Quinoa seed coats as an expanding and sustainable resource of bioactive compounds: An investigation of genotypic diversity in saponin profiles. *Ind. Crops. Prod.* **2017**, *104*, 156–163. [CrossRef]

30. Wallace-Salinas, V.; Gorwa-Grauslund, M.F. Adaptative evolution of an industrial strain of *Saccharomyces cerevisiae* for combined tolerance to inhibitors and temperature. *Biotechnol. Biofuels* **2013**, *6*, 151. [CrossRef]

31. Wojciechowski, K.; Piotrowski, M.; Popielarz, W.; Sosnowski, T.R. Short- and mid-term adsorption behavior of *Quillaja* bark saponin and its mixtures with lysozyme. *Food Hydrocoll.* **2011**, *25*, 687. [CrossRef]

32. Kiefer, J.; Radzuan, M.N.; Winterburn, J. Infrared spectroscopy for studying structure and aging effects in rhamnolipid biosurfactants. *Appl. Sci.* **2017**, *7*, 533. [CrossRef]

33. Sreij, R.; Dargel, C.; Geisler, P.; Hertle, Y.; Radulescu, A.; Pasini, S.; Pérez, J.; Moleiroa, L.H.; Hellweg, T. DMPC vesicle structure and dynamics in the presence of low amounts of the saponin aescin. *Phys. Chem. Chem. Phys.* **2018**, *20*, 9070–9083. [CrossRef] [PubMed]

34. Escribano, J.; Cabanes, J.; Jiménez-Atiénzar, M.; Ibañez-Tremolada, M.; Gómez-Pando, L.R.; García-Carmona, F.; Gandía-Herrero, F. Characterization of betalains, saponins and antioxidant power in differently colored quinoa (*Chenopodium quinoa*) varieties. *Food Chem.* **2017**, *234*, 285–294. [CrossRef]

35. Rinderle, S.J.; Goldstein, I.J.; Remsen, E.E. Physicochemical properties of amaranthin, the lectin from *Amaranthus caudatus* seeds. *Biochemistry* **1990**, *29*, 10555–10561. [CrossRef]

36. Tu, M.B.; Saddler, J.N. Potential enzyme cost reduction with the addition of surfactant during the hydrolysis of pretreated softwood. *Appl. Biochem. Biotechnol.* **2010**, *161*, 274–287. [CrossRef]

37. Kumar, L.; Arantes, V.; Chandra, R.; Saddler, J.N. The lignin present in steam pretreated softwood binds enzymes and limits cellulose accessibility. *Bioresour. Technol.* **2012**, *103*, 201–208. [CrossRef]

38. Nakagame, S.; Chandra, R.P.; Saddler, J.N. The effect of isolated lignins, obtained from a range of pretreated lignocellulosic substrates, on enzymatic hydrolysis. *Biotechnol. Bioeng.* **2010**, *105*, 871–879. [CrossRef]

39. Siqueira, G.; Arantes, V.; Saddler, J.N.; Ferraz, A.; Milagres, A.M.F. Limitation of cellulose accessibility and unproductive binding of cellulases by pretreated sugarcane bagasse lignin. *Biotechnol. Biofuels* **2017**, *10*, 176. [CrossRef]

40. Pekdemir, T.; Opur, M.C.; Urum, K. Emulsification of crude oil–water systems using biosurfactants. *Process Saf. Environ. Prot.* **2005**, *83*, 38–46. [CrossRef]

41. Stanimirova, R.; Marinova, K.; Tcholakova, S.; Denkov, N.D.; Stoyanov, S.; Pelan, E. Surface rheology of saponin adsorption layers. *Langmuir* **2011**, *27*, 12486–12498. [CrossRef] [PubMed]

42. Lai, C.; Yang, C.; Zhao, Y.; Jia, Y.; Chen, L.W.; Zhou, C.F.; Yong, Q. Promoting enzymatic saccharification of organosolv-pretreated poplar sawdust by saponin-rich tea seed waste. *Bioprocess Biosyst. Eng.* **2020**. [CrossRef] [PubMed]

43. Cira, L.A.; González, G.A.; Torres, J.C.; Pelayo, C.; Gutiérrez, M.; Ramírez, J. Heterologous expression of *Fusarium oxysporum* tomatinase in *Saccharomyces cerevisiae* increases its resistance to saponins and improves ethanol production during the fermentation of *Agave tequilana* Weber var. azul and *Agave salmiana* must. *A. van Leeuw. J. Microb.* **2008**, *93*, 259–266. [CrossRef] [PubMed]

44. Vanegas, J.M.; Contreras, M.F.; Faller, R.; Longo, M.L. Role of unsaturated lipid and ergosterol in ethanol tolerance of model yeast biomembranes. *Biophys. J.* **2012**, *102*, 507–516. [CrossRef]

45. Ghose, T.K. Measurement of cellulose activities (Recommendations of Commission on Biotechnology IUPAC). *Pure Appl. Chem.* **1987**, *59*, 257–268. [CrossRef]

46. Puls, J.; Schreiber, A.; Saake, B. Conversion of beech wood into platform chemicals after organosolv treatment. In Proceedings of the 15th International Symposium on Wood, Fiber and Pulping Chemistry (ISWFPC 2009), Oslo, Norway, 15–18 June 2009; p. O-036.

47. Sluiter, A.; Hames, B.; Ruiz, R.; Scarlata, C.; Sluiter, J.; Templeton, D.; Crocker, D. Determination of structural carbohydrates and lignin in biomass. *Lab. Anal. Proced.* **2008**, *1617*, 1–16.

48. Koziol, M.J. Afrosimetric estimation of threshold saponin concentration for bitterness in quinoa (*Chenopodium quinoa* Willd). *J. Sci. Food. Agric.* **1991**, *54*, 211–219. [CrossRef]

49. Bradford, M.M. A rapid and sensitive method for the quantitation of microgram quantities of protein utilizing the principle of protein-dye binding. *Anal. Biochem.* **1976**, *72*, 248–254. [CrossRef]

Sample Availability: Samples of the compounds are not available from the authors.

 *molecules*

Article

Optimized Bioproduction of Itaconic and Fumaric Acids Based on Solid-State Fermentation of Lignocellulosic Biomass

Amparo Jiménez-Quero [1,2], Eric Pollet [1], Luc Avérous [1,*] and Vincent Phalip [1,3]

[1]   BioTeam/ICPEES-ECPM, UMR CNRS 7515, Université de Strasbourg, 25 rue Becquerel, 67087 Strasbourg, CEDEX 2, France; amparojq@kth.se (A.J.-Q.); eric.pollet@unistra.fr (E.P.)

[2]   Division of Glycoscience, Department of Chemistry, School of Engineering Sciences in Chemistry, Biotechnology and Health, KTH Royal Institute of Technology, 10044 Stockholm, Sweden

[3]   Univ. Lille, INRA, ISA, Univ. Artois, Univ. Littoral Côte d'Opale, EA 7394—ICV—Institut Charles Viollette, F-59000 Lille, France; vincent.phalip@polytech-lille.fr

*   Correspondence: luc.averous@unistra.fr

Received: 9 January 2020; Accepted: 25 February 2020; Published: 27 February 2020

Abstract: The bioproduction of high-value chemicals such as itaconic and fumaric acids (IA and FA, respectively) from renewable resources via solid-state fermentation (SSF) represents an alternative to the current bioprocesses of submerged fermentation using refined sugars. Both acids are excellent platform chemicals with a wide range of applications in different market, such as plastics, coating, or cosmetics. The use of lignocellulosic biomass instead of food resources (starch or grains) in the frame of a sustainable development for IA and FA bioproduction is of prime importance. Filamentous fungi, especially belonging to the *Aspergillus* genus, have shown a great capacity to produce these organic dicarboxylic acids. This study attempts to develop and optimize the SSF conditions with lignocellulosic biomasses using *A. terreus* and *A. oryzae* to produce IA and FA. First, a kinetic study of SSF was performed with non-food resources (wheat bran and corn cobs) and a panel of pH and moisture conditions was studied during fermentation. Next, a new process using an enzymatic cocktail simultaneously with SSF was investigated in order to facilitate the use of the biomass as microbial substrate. Finally, a large-scale fermentation process was developed for SSF using corn cobs with *A. oryzae*; this specific condition showed the best yield in acid production. The yields achieved were 0.05 mg of IA and 0.16 mg of FA per gram of biomass after 48 h. These values currently represent the highest reported productions for SSF from raw lignocellulosic biomass.

Keywords: lignocellulosic biomass; solid-state fermentation; enzymatic hydrolysis; aerated bioreactor; *Aspergillus oryzae*

---

1. Introduction

Solid-state fermentation (SSF) has emerged in the last decades as a promising industrial process for several products, especially using agricultural byproducts as the substrates [1,2]. SSF involves the growth of a microorganism on solid particles in the quasi absence of free water, and the majority of processes are performed by filamentous fungi under aerobic conditions [3]. The substrates used in SSF are often the source of nutrients for the microorganisms, and the inter-particle spaces allow gas and nutrients exchange between fungal hyphae and the medium. Fungi also behave as biocatalysts for the bioconversion of the substrates into specific target products such as bio-based fuels, commodity chemicals, enzymes, bioactive compounds, or food products [4].

SSF offers several advantages compared to submerged fermentation (SmF) such as high volumetric productivity, product concentration, simpler and smaller bioreactors because of the minimal free water, a lower sterilization cost, less generation of effluents (reduced cost of effluent treatment), and easier aeration due to lower density of the corresponding medium with high porosity [5,6]. Finally, the conditions of SSF mimic the natural environments of the filamentous fungi. However, the SSF process is slower compared to SmF, and all fermentation conditions cannot be controlled precisely. The main factors affecting fungal growth and metabolism in SSF are the selection of a suitable microorganism and substrate for the targeted generation of products, the pre-treatment of the substrate, the moisture, the temperature, and the removal of metabolic heat and gas transfers [7].

One of the most interesting biotechnological applications of SSF is the production of commodity chemicals [8,9]. The biosynthesis of chemicals from biomass creates a sustainable alternative to the conventional chemical synthesis based on fossil resources [10,11]. In the last two decades, many molecules produced from biomass with a large range of applications have been described [12,13]. Many of these building blocks are organic acids because of their capacities to generate high-value products for widespread industries such as food, pharmaceuticals or polymers [14–16]. The biosynthesis of by filamentous fungi has been studied extensively, and *Aspergilli* are often used for industrial production [17].

Among the organic acids, fumaric and itaconic acids (FA and IA, respectively) are included on the DOE's (Unites State Department of Energy) list as part of the top twelve biomass-derived platform chemicals [12]. Both acids are polyfunctional building blocks that can be polymerized for instance, to give homo- or co-polymers for applications in textile, chemical, and pharmaceutical industries. IA can be used to replace acrylic acid, an important and rather costly chemical that is non-renewable so far, while FA can be used as food additive and in psoriasis treatment [14,16]. These acids are part of the tricarboxylic acid (TCA) cycle, FA being a direct intermediate in the cycle and IA a derivate of cis-aconitate acid, and both are produced under aerobic conditions. Currently, the industrial production of FA is via catalytic isomerization of fossil-based maleic acid. However, FA could also be produced biologically as an intermediate of the TCA cycle that is present in most aerobic organisms. Laboratory-scale fermentations with *Rhizopus oryzae* have shown interesting productivities in SmF of lignocellulosic biomass, around 0.35 g/g corn straw [18,19]. IA is produced industrially by *Aspergillus terreus* in SmF with glucose as the principal carbon and energy source to a yield of 100 g/L [20–22]. The biosynthesis involves the action of the cis-aconitate decarboxylase (CAD) enzyme to transform the cis-aconitate into itaconate. The presence of CAD in *A. terreus* has been demonstrated in different studies, but this enzyme is also present in another Aspergillus species; *A. oryzae* [23,24]. This aerobic, filamentous fungus is frequently used in SSF processes due to its capacity to hydrolyze the lignocellulosic substrates by enzymatic degradation [25]. Nevertheless, the production of IA and FA by SSF with lignocellulosic biomass has not been studied extensively in the literature. A method of SSF using sugarcane pressmud as a support for IA production, which yielded 0.0003 g kg^{-1} h^{-1}, was patented in 2001 [26]. In this case, the main carbon source for the acid production has been added as liquid medium and the remaining sucrose for the sugarcane was a supplementary source. A maximum productivity of 0.021 g kg^{-1} h^{-1} of FA was reported via SSF of corn distiller grains by *R. oryzae* [27]. In a previous study, we have shown the capacity of *A. terreus* and *A. oryzae* to produce both acids by SSF process [28,29]. The yields obtained by *A. oryzae* from corn cobs were the most interesting, with 0.05 and 0.18 mg acid/g biomass of IA and FA, respectively. As expected, both productivities were lower than values reported from SmF processes utilizing soluble sugars in liquid media.

The aim of this work was to optimize the bioproduction of IA and FA in SSF by two *Aspergillus* species (*A. terreus* and *A. oryzae*) using lignocellulosic biomass as a non-food carbon source (wheat bran and corn cobs). As it is a complex and recalcitrant structure, lignocellulosic biomass is often predigested to be used as a sugar source for fermentation [30]. The bioconversion is carried out by enzymes produced under specific conditions by many microorganisms [31]. This saccharification process requires several hydrolytic enzymes such as cellulases, hemicellulases, xylanases, etc. [32]. The biomass pretreatment is one of the most expensive part of the lignocellulosic material conversion in an industrial scale. Therefore, the process could be coupled with the fermentation in a simultaneous saccharification–fermentation step, in order to improve the process yields. The study of this simultaneous process can open the possibility to decrease cost and time for an industrial activity in future. Several factors were studied to enhance organic acid production yields (pH, moisture content, enzyme hydrolysis) and large-scale fermentations were tested using these optimized factors.

2. Results

2.1. Solid-State Fermentation Kinetics

Fermentations were performed for both fungal species using both biomasses, wheat bran, and corn cobs. Organic acid production and fungal growth were studied. The determination of protein secretion level showed that *A. terreus* and *A. oryzae* present different development trends on wheat bran and corn cobs, Figure 1. The growth of *A. oryzae* on wheat bran reached a plateau after 120 h whereas *A. terreus* grew more slowly and regularly for more than 200 h on both substrates. Both species grew better on wheat bran than on corn cobs, as reported before in our previous studies, where both fungi species were capable of producing higher amounts of hydrolytic enzymes on wheat bran biomass [28,29]. These results were also confirmed by visual observations. More FA was produced from wheat bran with 0.8 and 0.6 mg/g biomass for *A. terreus* and *A. oryzae*, respectively, at the end of the fermentation with a regular increase in the yields. On corn cobs, FA production displayed a completely different profile, with a maximum yield after 48 h (0.14 and 0.12 mg/g for *A. terreus* and *A. oryzae*, respectively) and then a regular decrease.

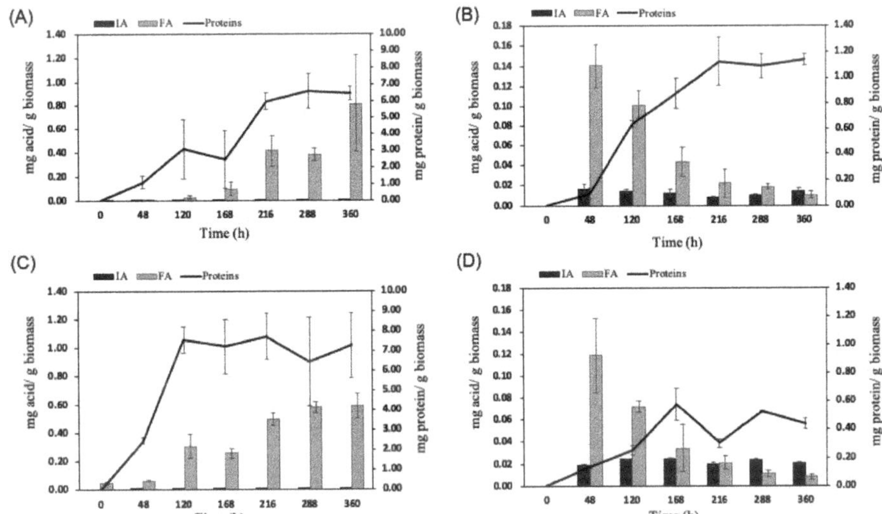

Figure 1. Fermentation kinetics on lignocellulosic biomasses: itaconic acid (IA) and fumaric acid (FA) yields and protein production (fungal growth) from wheat bran (**A,C**) and corn cobs (**B,D**) by *A. terreus* (**A** and **B**, respectively) and by *A. oryzae* (**C** and **D**, respectively).

Although the fungal growth was significantly higher on wheat bran, IA was produced only on corn cobs for both fungi, Figure 1. The different composition of both biomasses [28] may explain this behavior. A maximum IA yield of 0.025 mg/g corn cobs was produced by *A. oryzae* at 168 h of fermentation. At the same fermentation time, *A. terreus* produced half of this amount (0.012 mg IA/g biomass). The fungal biomass of *A. oryzae* was 15 times lower on corn cobs than on wheat bran and almost 2.5 times lower than the one of *A. terreus* on corn cobs.

2.2. Optimization of the SSF Steps

After testing the fermentations with both biomasses and fungi, the optimization steps were performed in two different ways. Firstly, a study with varying pH and humidity levels was carried out with corn cobs, to further improve the IA production. Secondly, an optimization of *A. oryzae* fermentation (displaying the highest IA production yield) was performed both on wheat bran and corn cobs by adding an enzyme cocktail to better hydrolyze the lignocellulosic biomasses. That could allow a more efficient conversion of accessible fermentable sugars in order to increase the yield of the fermentation products.

2.2.1. Effect of pH and Moisture Level

Optimum pH and moisture level are crucial factors in SSF processes to obtain maximum yields of the products of interest [6,33]. The initial and previously tested conditions for corn cob fermentation (Section 2.1.) were pH 5 for the inoculation and 90% humidity. To optimize the pH and moisture conditions, five different pH values and five different moisture levels were evaluated for the inoculation step of the biomass culture, Figure 2 and Figure S1 in Supplementary Materials.

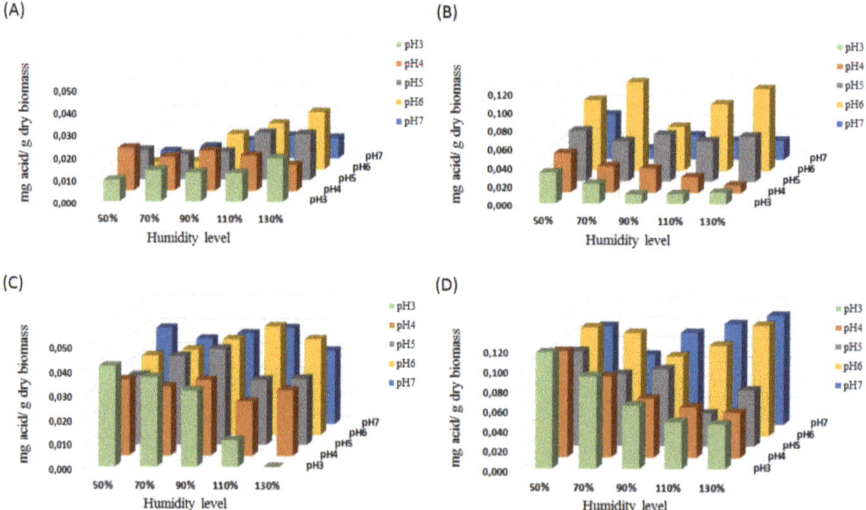

Figure 2. Solid-state fermentation (SSF) on corn cobs at different pH and moisture levels by *A. terreus* (IA and FA yields: **A** and **B**, respectively) and *A. oryzae* (**C** and **D**, respectively).

For *A. terreus*, both acids were produced with higher yields at pH 6, as reported before [28,34]. The moisture content influenced the IA and FA production differently. The best IA production, 0.025 mg IA/g corn cobs, was observed at pH 6 and 130% humidity (Figure 2A) i.e., a doubling of the production compared to the initial conditions (pH 5 and 90% humidity). FA was produced at a yield of 0.095 mg/g biomass (pH 6 and 70% humidity), which is also almost twice the production at initial conditions (Figure 2B). For both acids, a clear trend is that a neutral pH (pH = 7) seems too high (Figure 2A,B). This observation is in good agreement with previous results obtained for SmF [31].

A. oryzae also showed a preference for pH 6, for the production of both acids (Figure 2C,D). In the case of IA, the highest yield was 0.045 mg/g biomass at 110% humidity (Figure 2C), which is slightly higher than the yield under the initial conditions (0.039 mg IA/g biomass) and almost twice the yield obtained with *A. terreus*. Under the same conditions (pH 6, 110% humidity), 0.091 mg FA/g was produced (Figure 2D). It is not the highest yield since at 130% humidity, the production was even higher (0.111 mg/g biomass).

2.2.2. Enzymatic Hydrolysis

The enzymatic cocktail was obtained by SSF of wheat bran by *A. oryzae*, as shown in previous studies [29], the solid fermentation of the biomass allows to produce higher protein content with specific lignocellulolytic enzymes for biomass degradation. The cocktail can be store at −20 °C and used for simultaneous SSF.

The enzyme cocktail produced by *A. oryzae* showed the best enzymatic activity for endoxylanases (Table 1), which are responsible for hemicellulose hydrolysis. Hemicellulose is the most abundant part of corn cobs [35,36]. Moreover, cellulase and xyloglucanase activities were also found, suggesting an efficient biomass digestion. When the enzymatic cocktail was used for simultaneous saccharification-fermentation of corn cobs, *A. oryzae* rapidly secreted proteins (i.e., it grew) in the first 20 h, and a plateau was subsequently reached (Figure 3B) approximately at the same level as with the raw biomass (Figure 1D). Surprisingly, with the treated wheat bran (Figure 3A), the fungi secreted half of the proteins compared to the untreated biomass (Figure 1C). Moreover, the treatment had a dramatic negative effect on the production of FA from wheat bran (Figure 3A) with a yield (0.08 mg/g biomass) almost 8 times lower than without pretreatment (0.6 mg/g of biomass) (Figure 1C). In contrast, for corn cobs, the FA yield was feebly increased to 0.15 mg/g biomass. The profile of FA production from corn cobs (Figure 3B) was similar to the one without the enzyme cocktail (Figure 1D) with a yield reaching a maximum (after ca. 50 h) followed by a decay.

Table 1. Enzymatic activities (in ΔOD/g*min*) of the enzymatic cocktail obtained from SSF of wheat bran by *A. oryzae*.

α-Amylase activity	18.10
Cellulase activity (cellulose)	4.69
Endoxylanase activity	70.30
Cellulase activity (xyloglucan)	10.11

Enzyme activities were expressed in arbitrary units corresponding to optical density variations (ΔOD) per minute and per gram of biomass, due to the unknown extinction coefficient of AZCL substrate (Megazymes, Ireland).

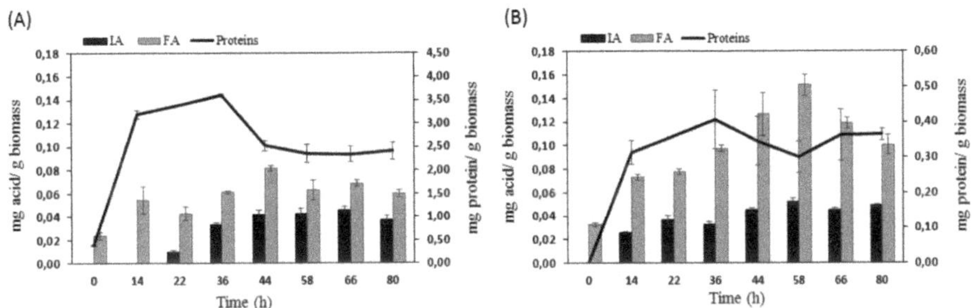

Figure 3. Kinetics of simultaneous saccharification and fermentation of wheat bran (**A**) and corn cobs (**B**) by *A. oryzae*.

The best contribution of the enzyme cocktail was observed for the production of IA. According to our knowledge, the use of such enzymatic cocktail allowing IA production from wheat bran was reported for the first time in this study. IA production was detectable after 22 h, and a maximum yield of 0.046 mg/g biomass was obtained after 66 h (Figure 3A). With corn cobs, IA production is clearly detected earlier (14 h), and a yield of 0.052 mg/g biomass was achieved (Figure 3B) that was twice the maximum yield produced without enzymatic treatment for optimized pH and moisture level, Figure 2C.

2.3. Kinetics of SSF with Optimized Conditions

According to previous results, the optimum fermentative process was 80 h at pH 6 and 110% humidity. Figure 4 presents the glass flasks and shows the development of *A. oryzae* through the fermentation (with a green color indicating a high spore concentration). The fungus grew progressively during the fermentation until 0.22 mg of protein/g of biomass was obtained, as shown earlier (at pH 5 and 90% moisture level) at the same time of fermentation (Section 2.2.1). In relatively good agreement, FA production was only slightly enhanced (+ 10%) with a maximum yield of 0.16 mg/g biomass within 48 h (Figure 4). However, for IA, the enhancement was higher because the production was more than doubled (0.051 mg/g biomass) after 48 h and higher (0.061 mg/g) after 80 h (factor 2.4). As generally described for fungi, *A. oryzae* metabolism is greatly influenced by pH and the humidity level at the start of the fermentation step [37,38].

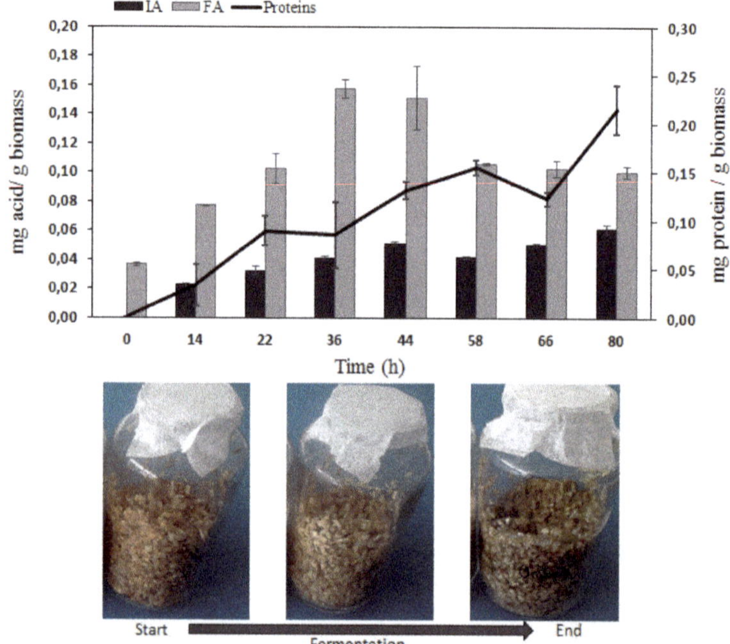

Figure 4. Kinetics of SSF on corn cobs by *A. oryzae* under optimized conditions (pH 6 and 110% moisture).

2.4. Larger Scale Fermentation

To develop and analyze the scaling up, the fermentation was performed with 200 g of corn cobs i.e., 20 times more than for the previous glass flasks experiments. The optimized conditions of pH and moisture (i.e., 110% moisture and pH = 6) were applied for the scaled-up fermentation. Aeration plays an important role in SSF for the transfer of oxygen and the evacuation of the carbon dioxide produced. Aeration is also used (Figure 5) to dissipate the metabolic heat generated by fermentation [39]. The mixture of substrates (solid lignocellulose particles and fungal mycelium) also helps to equilibrate the gas exchange, temperature, and moisture level [3], avoiding the disruption of the mycelial-substrate contact, which is particularly important for *A. oryzae*, for instance, to produce the degrading enzymes to hydrolyze the biomass.

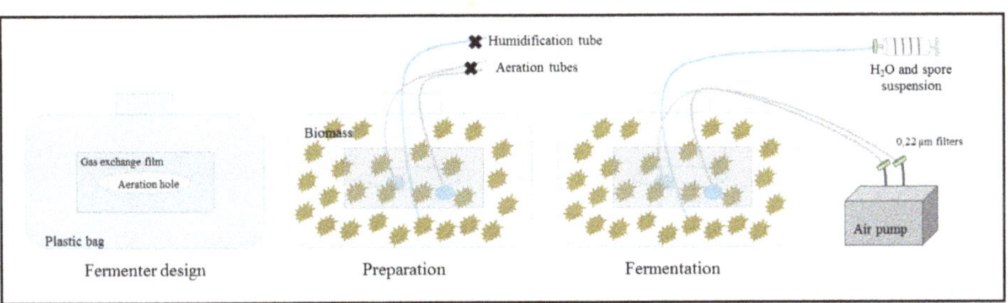

Figure 5. Illustration of aerated plastic bag fermenter (at left), made from autoclavable biohazard bags in polypropylene with an aeration hole covered by a gas exchange Miracloth film (Millipore, USA). Prior to autoclaving (middle), corn cobs were introduced as well as the air and humidification tubes (autoclavable tubes in PVC used for the liquid bioreactor). Right: Operative fermenter with aeration and the inoculum to be injected.

Two different processes were used to test the influence of both aeration and mixing. A monolayer reactor presenting the same conditions as the glass flasks except for size was compared with an aerated plastic bag fermenter. The plastic bag fermenter was gently mixed on a rocker shaker and distilled water was added in a timely manner to equilibrate the moisture level. The fungal growth was clearly different between the two fermenters (Figure 6). The aerated plastic bag yielded nearly twice the proteins concentration (0.26 mg/g biomass) as the monolayer fermenter (0.15 mg/g biomass) at the end of the fermentation. Compared with the glass flask fermentation (0.22 mg protein/g biomass, Figure 4) the monolayer fermenter produced less protein whereas the aerated one displayed an amount of protein similar to the small-scale fermentation. This difference in fungal growth was also obvious in observing *A. oryzae* sporulation, which occurred earlier for the monolayer fermenter (until the second fermentation day) than for the aerated fermenter. This premature sporulation indicates that mycelial development was interrupted by inadequate conditions. The aeration and the loss of humidity correction increased the fungal development and also delayed the sporulation.

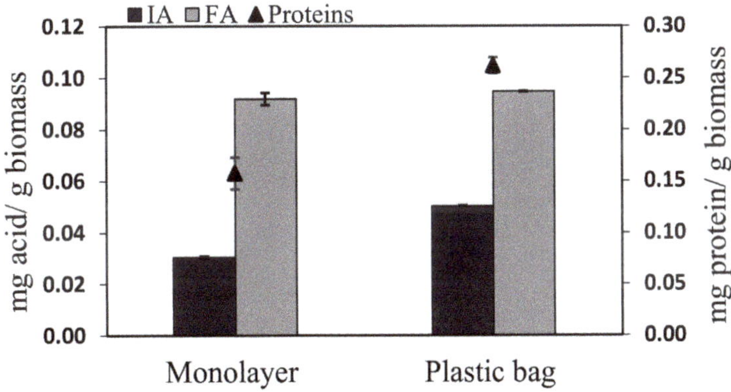

Figure 6. SSF in larger scale fermenters of *A. oryzae* from corn cobs: Organic acid productions and protein secretion (growth).

The FA production was not affected by the different conditions (0.09 mg/g biomass in both reactors), probably because 96 h of fermentation was not an optimized time to recover FA as shown in the glass flask (Figure 4) where the maximum FA yield was produced at 36 h. Conversely, IA production was 60% higher in the aerated fermenter (Figure 6) than in the monolayer reactor.

3. Discussion

From the SSF kinetics experiments, we can conclude that IA production was inversely linked to growth. Both acid yields were lower than the yields from the current SmF [31–33], then the optimization of the fermentative conditions is necessary to enhance the acid production. During the optimization steps, we could observe how low moisture content causes slower enzyme secretion from the fungus due to the lower solubility of the nutrients and the low level of growth [37,38] (Figure S2 in Supplementary Materials). However, acidic pH (3–5) and low moisture often allowed better production of the acids (Figure 2B,D). This behavior is not observed at higher pHs. These observations are in agreement with the fact that in SSF, pH variation significantly impacts the production and the stability of the enzymes [5,39], with several enzymes responsible for biomass hydrolysis during growth. The pH effect on the organic acid production was in accordance with the simultaneous fermentation and enzymatic hydrolysis study performed previously [29]. Even if the enzymatic cocktail did not improve the growth of the fungus nor FA production, the cocktail created better conditions to produce IA. FA is an intermediate metabolite of fungal fermentation, and its production is directly linked with the development of the microorganism unlike IA, a secondary metabolite [34]. The time lag between growth and IA production is perfectly consistent with a secondary metabolite behavior. These results showed the importance of a separated optimization of FA and IA production, as well as the time of recovery of the carboxylic acids during the fermentation, especially from a continuous production perspective.

The acid yields for the simultaneous saccharification and fermentation process were higher in comparison with the best results obtained for optimized pH and moisture (pH = 6 and 110% humidity) with corn cobs and *A. oryzae,* (Figure 2C,D). However, the production of the enzyme cocktail required four additional days for the entire process. Therefore, the most interesting strategy for the acid production was the kinetic fermentation of corn cob biomass by *A. oryzae* with the optimized conditions of pH and humidity. The kinetic curves showed the differential production over time for both acids, leaving the possibility to improve and optimize the time production of the acids if focusing exclusively on one of them. In this case, the pH and the moisture should be adapted from the previous results (Figure 2 and Figure S1 in Supplementary Materials).

The literature is deficient concerning SSF of FA and IA without biomass pretreatment. For IA, a mutant of *A. terreus* displays a productivity of 0.0003 mg/g h with sugarcane pressmud supplemented with sugars and nutrients [26]. The result obtained in this study was more than two times higher (0.00076 mg/g h) by *A. oryzae* (novel IA producer) with a lignocellulosic substrate without any nutrient addition [29]. Our study can contribute to further optimization on the fermentation conditions (pH, humidity, and aeration level), that combined with new studies using metabolic engineered microorganisms [40,41] providing better yields in IA production from lignocellulosic materials.

Regarding the larger scale fermentation in the plastic bag prototype reactor, the improvement in acid production could be explained by the better supply of oxygen and moisture in the fermenter. Indeed, IA fermentation is strictly aerobic, and previous studies showed that a gain in dissolved oxygen and agitation induced higher yields [42,43]. The moisture level could affect the IA production. Below 70% humidity, the nutrient transfers are limited, and the metabolism is affected [44]. In our experiment, water addition along with aeration may allow better fermentation conditions of the air-solids-water to enhance IA production. Most of these factors were studied for *A. terreus*, long known as an IA producer. However, for *A. oryzae*, conditions still need further optimization. Even if the IA yield obtained in the aerated fermenter (0.05 mg/g biomass) was similar to the small glass flask, the final production was multiplied by 20. Of course, the yields obtained in this work are still far from the industrial scale target, and further work in optimizing the fermentation conditions and down-stream processing have to be carried out. Anyway, *A. oryzae*, which showed the most interesting enzymes production for biomass degradation, seems to be an excellent candidate for further studies.

4. Materials and Methods

4.1. Feedstock and Microorganisms

Two agricultural waste biomasses used as non-food carbon sources were wheat bran and corn cobs obtained from Comptoir Agricole (Lauterbourg, France). The lignocellulosic material was milled (SX 100, Retsch) to obtain particles that were 0.5–1 mm in size. The water activity (Aw) was measured on 1 g of dry substrate by an Aw meter Fast-lab (GBX, France).

A. terreus (DSM 826) was provided by the Deutsche Sammlung von Mikroorganismen und Zellkulturen (DSMZ, Braunschweig, Germany). *A. oryzae* (UMIP 1042.72) was provided by the Fungal Culture Collection of the Pasteur Institute (France). The strains were revived on potato dextrose broth medium (PDB) for 5–6 days at 25 °C. The microorganisms were then grown and sporulated on potato dextrose agar (PDA). The spore suspensions were harvested from 5–6-day-old PDA plates with 0.2% (*v/v*) Tween-80. The spores were counted using a Malassez counting chamber and stored at −20 °C.

4.2. Initial SSF Step

In glass flasks, 5 g of solid substrate were autoclaved at 121 °C and 3 bars, for 20 min. The substrates were inoculated with spore suspensions to have an initial concentration of 10^6 spores/g of substrate. Initial moisture was adjusted to 120% for wheat bran and 90% for corn cobs (both corresponding to an Aw near 1). After thorough mixing, the flasks were covered with a porous adhesive film (VWR, Radnor, PA, USA) and incubated at 30 °C for 6 days. Unless specified otherwise, these fermentation conditions were maintained throughout the study. All the experiments were conducted in duplicate. After fermentation, the samples were recovered by mixing the fermented substrate with sterilized distilled water (7 mL/g of initial dry substrate). The preparations were centrifuged (8000× *g* for 15 min) to eliminate residual solids. Then, another centrifugation step was performed on the supernatants to remove mycelia and spores (13,600× *g* for 30 min). The resulting solution was finally filtered through a 0.2 μm membrane. The samples were analyzed by high-performance liquid chromatography (HPLC, Waters, Milford, MA, USA) and were stored at −20 °C for additional analysis.

4.2.1. pH and Humidity Level Optimization

The initial pH and humidity levels were varied to screen the best conditions for organic acid production. Citrate-phosphate buffer solutions, pH 3 to 7, were prepared. The moisture content was set at 50, 70, 90, 110, and 130 *v/w*. Five pH conditions were crossed with five relative humidity level, generating 25 different conditions performed with biological duplicates in SSF.

4.2.2. Enzyme Production for Biomass Hydrolysis

The enzymatic cocktail preparation was performed in a plate with 250 g of wheat bran by *A. oryzae*. The inocula were prepared with 300 mL of Tris buffer at pH 10 to achieve a final spore concentration of 10^6 per g of biomass, and the incubation temperature was 25 °C during the 4 days. The enzymatic cocktail was recovered with 1500 mL of sodium phosphate buffer at pH 6 and filtered with a Vivaflow 200 system (Sartorius, Göttingen, Germany). The cocktail was stored at 4 °C, and enzyme activities were determined.

4.3. Scale-Up Steps

Fermentation at a higher scale, with 200 g of biomass, was performed in two different types of reactor. A monolayer reactor consists of a glass plate covered with a gas exchange Miracloth film (Millipore, Temecula, CA, USA). The second reactor is a prototype of an aerated reactor (Figure 5). This fermenter was made from an autoclaved polypropylene laboratory bag with a central opening covered with a gas exchange film of Miracloth. Two PVC tubes were introduced and connected to stone ceramic air diffusers (3 mm in diameter). Another PVC tube was added for inoculation and double distilled water addition. The entire reactor was autoclaved with the biomass inside. During fermentation, double distilled water was added at the rate of 11 mL per day to keep the moisture constant (considering 10% evaporation/day). The aeration was provided by an air pump AC-9906 (Resun, Shenzhen, China) at a flow rate of 840 L/hour. A rocker mixer (ThermoFisher, Waltham, MA, USA) was used to shake the fermenter in order to uniformly add the water to the biomass. The influence of these controlled conditions of air and moisture content on the acid production could be studied for the plastic bag reactor but not for the monolayer reactor.

4.4. Analytical Procedure

4.4.1. Mycelial Growth (Protein Assays)

To determine the fungal proteins produced during the fermentation, Bradford method was used [45]. All the samples were centrifuged and filtered (0.22 μm) before analysis to eliminate the spores. The protein assay was calibrated using BSA (bovine serum albumin) as the standard.

4.4.2. Organic Acid Assays

A chromatographic system based on a 616 pump, a 2996 photodiode array detector operating in a range of 200 to 450 nm, and a 717 Plus autosampler (Waters, Milford, MA, USA) controlled by Empower 2 software (Waters, Milford, MA, USA) was used to analyze the samples, as previously described [29]. The columns were calibrated using commercial IA and FA samples with a 99.9% purity (Sigma-Aldrich, San Luis, MO, USA) and a UV measurement at 205 nm. Each sample was supplemented with 10 ppm IA or FA as the internal standard to confirm the acid production.

4.4.3. Enzyme Activity Assay

Chromogenic substrates, azurine-crosslinked (AZCL) polysaccharides such as AZCL-HE-cellulose, AZCL-xylan, AZCL-xyloglucan, or AZCL-amylose (Megazyme, Bray, Ireland), were used to measure the enzyme cocktail activity. The samples were collected and analyzed as previously described [28], by spectrophotometry determining absorbance of the supernatant at 595 nm that corresponds with the solubilization of dyed compounds (AZCL) by the enzymes present in the cocktail.

5. Conclusions

The IA production process appears to be ideally amenable to SSF conditions, as demonstrated in this work. However, the fermentation conditions still need further optimization to provide yields similar to the yields obtained by submerged fermentation, considering the use of lignocellulosic substrates. Additionally, the use of a novel species, *A. oryzae* (which is used industrially for enzyme production) opens up the possibility of creating a biorefinery process for the production of both organic acids and enzymes. The use of agricultural wastes and cheap and non-food substrates in the bioprocess could lower IA production costs and could therefore promote the use of bio-based IA in the polymerization process to replace petroleum-derived polymers. Furthermore, the simultaneous production of another organic acid as FA by *Aspergillus* species can open the possibility to adapt the use of different lignocellulosic biomass for particular building blocks production. Moreover, in the case of downstream purification processes for both IA and FA, the opportunity of co-polymerization could be an interesting case of study. Anyway, the results showed that time production as well as the fungi needs can differ between IA or FA production, and specific individual optimization for SSF should also be done. Moreover, further studies need to be performed in the use of commercial enzymatic cocktails to improve the production yields, but also to understand the need for better biomass hydrolysis in SSF. In this sense, *A. oryzae* seem to be a great candidate for commercial enzymes production. One possibility could be the use of metabolic engineering to guaranty the strong lignocellulolytic enzyme release combined with specific metabolic route for IA or FA, exclusively. In this work, the organic acids yields were slightly lower than those obtained in previous studies. However, the global time-lapse of the process was greatly decreased considering that no previous pretreatment steps were preformed, or enzymatic cocktails were collected.

Supplementary Materials: The following are available online. Figure S1. SSF on corn cobs at different pH and moisture levels by *A. terreus* and *A. oryzae*; Figure S2. Fungal growth (proteins productions) under different pH and moisture conditions.

Author Contributions: Conceived and designed the experiments, all authors; performed the experiments, A.J.-Q.; analyzed and interpreted the data, all authors; wrote the paper. All authors have read and agreed to the published version of the manuscript.

Funding: This work has received funding from the European Union's 7th Framework Program under grant agreement n°311815 (SYNPOL Project).

Conflicts of Interest: The authors declare no conflict of interest.

References

1. Pandey, A. *Solid-state Fermentation in Biotechnology: Fundamentals and Applications*; Asiatech Publishers: New Delhi, India, 2001.
2. Hölker, U.; Lenz, J. Solid-state fermentation—Are there any biotechnological advantages? *Curr. Opin. Microbiol.* **2005**, *8*, 301–306. [CrossRef] [PubMed]
3. Hölker, U.; Höfer, M.; Lenz, J. Biotechnological advantages of laboratory-scale solid-state fermentation with fungi. *Appl. Microbiol. Biotechnol.* **2004**, *64*, 175–186. [CrossRef] [PubMed]
4. Dashtban, M.; Schraft, H.; Qin, W. Fungal bioconversion of lignocellulosic residues; opportunities & perspectives. *Int. J. Biol. Sci.* **2009**, *5*, 578–595. [CrossRef] [PubMed]
5. Krishna, C. Solid-state fermentation systems-an overview. *Crit. Rev. Biotechnol.* **2005**, *25*, 1–30. [CrossRef]
6. Rodriguez-Leon, J.A.; Soccol, C.R.; Pandey, A.; Rodriguez, D.E. Factors Affecting Solid-state Fermentation. In *Current Developments in Solid-State Fermentation*; Pandey, A., Soccol, C.R., Larroche, C., Eds.; Springer: New York, NY, USA, 2008; pp. 26–47. [CrossRef]
7. Singh nee' Nigam, P.; Pandey, A. (Eds.) Solid-State Fermentation Technology for Bioconversion of Biomass and Agricultural Residues. In *Biotechnology for Agro-Industrial Residues Utilisation*; Springer: Dordrecht, Netherlands, 2009; pp. 197–221. [CrossRef]
8. Christensen, C.H.; Rass-Hansen, J.; Marsden, C.C.; Taarning, E.; Egeblad, K. The Renewable Chemicals Industry. *Chemsuschem.* **2008**, *1*, 283–289. [CrossRef]
9. Bozell, J.J. Feedstocks for the Future—Biorefinery Production of Chemicals from Renewable Carbon. *CLEAN Soil Air Water* **2008**, *36*, 641–647. [CrossRef]
10. Willke, T.; Vorlop, K.D. Industrial bioconversion of renewable resources as an alternative to conventional chemistry. *Appl. Microbiol. Biotechnol.* **2004**, *66*, 131–142. [CrossRef]
11. Gallezot, P. Conversion of biomass to selected chemical products. *Chem. Soc. Rev.* **2012**, *41*, 1538–1558. [CrossRef]
12. Werpy, T.; Holladay, J.; White, J. *Top Value Added Chemicals From Biomass: I. Results of Screening for Potential Candidates from Sugars and Synthesis Gas. DOE Scientific and Technical Information.*; Pacific Northwest National Lab: Richland, WA, USA, 2004; 419907. [CrossRef]
13. Bozell, J.J.; Petersen, G.R. Technology development for the production of biobased products from biorefinery carbohydrates—the US Department of Energy's 'Top 10' revisited. *Green Chem.* **2010**, *12*, 539. [CrossRef]
14. Magnuson, J.K.; Lasure, L.L. Organic Acid Production by Filamentous Fungi. In *Advances in Fungal Biotechnology for Industry, Agriculture, and Medicine*; Tkacz, J.S., Lange, L., Eds.; Springer: New York, NY, USA, 2004; pp. 307–340. [CrossRef]
15. Goldberg, I.; Rokem, J.S.; Pines, O. Organic acids: Old metabolites, new themes. *J. Chem. Technol. Biotechnol.* **2006**, *81*, 1601–1611. [CrossRef]
16. Tsao, G.T.; Cao, N.J.; Du, J.; Gong, C.S. Production of multifunctional organic acids from renewable resources. *Adv. Biochem. Eng. Biotechnol.* **1999**, *65*, 243–280. [CrossRef] [PubMed]
17. Liaud, N.; Giniés, C.; Navarro, D.; Fabre, N.; Crapart, S.; Gimbert, I.H.; Levasseur, A.; Raouche, S.; Sigoillot, J.C. Exploring fungal biodiversity: organic acid production by 66 strains of filamentous fungi. *Fungal Biol. Biotechnol.* **2014**, *1*, 1. [CrossRef]

18. Roa Engel, C.A.; Straathof, A.J.J.; Zijlmans, T.W.; Van Gulik, W.M.; Van Der Wielen, L.A.M. Fumaric acid production by fermentation. *Appl. Microbiol. Biotechnol.* **2008**, *78*, 379–389. [CrossRef] [PubMed]

19. Xu, Q.; Li, S.; Huang, H.; Wen, J. Key technologies for the industrial production of fumaric acid by fermentation. *Biotechnol. Adv.* **2012**, *30*, 1685–1696. [CrossRef]

20. Yahiro, K.; Shibata, S.; Jia, S.R.; Park, W.; Okabe, M. Efficient itaconic acid production from raw corn starch. *J. Ferment. Bioeng.* **1997**, *84*, 375–377. [CrossRef]

21. Hevekerl, A.; Kuenz, A.; Vorlop, K.D. Filamentous fungi in microtiter plates—An easy way to optimize itaconic acid production with *Aspergillus terreus*. *Appl. Microbiol. Biotechnol.* **2014**, *98*, 6983–6989. [CrossRef]

22. Willke, T.; Vorlop, K.D. Biotechnological production of itaconic acid. *Appl. Microbiol. Biotechnol.* **2001**, *56*, 289–295. [CrossRef]

23. Kanamasa, S.; Dwiarti, L.; Okabe, M.; Park, E.Y. Cloning and functional characterization of the cis-aconitic acid decarboxylase (CAD) gene from *Aspergillus terreus*. *Appl. Microbiol. Biotechnol.* **2008**, *80*, 223–229. [CrossRef]

24. Okabe, M.; Lies, D.; Kanamasa, S.; Park, E.Y. Biotechnological production of itaconic acid and its biosynthesis in *Aspergillus terreus*. *Appl. Microbiol. Biotechnol.* **2009**, *84*, 597–606. [CrossRef]

25. Begum, M.F.; Alimon, A.R. Bioconversion and saccharification of some lignocellulosic wastes by *Aspergillus oryzae* ITCC-4857.01 for fermentable sugar production. *Electron. J. Biotechnol.* **2011**, *14*, 5. [CrossRef]

26. Tsai, Y.C.; Huang, M.C.; Lin, S.F.; Su, Y.C. Method for the production of itaconic acid using Aspergillus terreus solid state fermentation. U.S. Patent 6171831, 09 January 2001.

27. West, T.P. Fumaric acid production by *Rhizopus oryzae* on corn distillers' grains with solubles. *Res. J. Microbiol.* **2008**, *3*, 35–40. [CrossRef]

28. Jiménez-Quero, A.; Pollet, E.; Zhao, M.; Marchioni, E.; Averous, L.; Phalip, V. Itaconic and fumaric acid production from biomass hydrolysates by *Aspergillus* strains. *J. Microbiol. Biotechnol.* **2016**, *26*, 1557–1565. [CrossRef] [PubMed]

29. Jiménez-Quero, A.; Pollet, E.; Zhao, M.; Marchioni, E.; Averous, L.; Phalip, V. Fungal fermentation of lignocellulosic biomass for itaconic and fumaric acid production. *J. Microbiol. Biotechnol.* **2017**, *27*, 1–8. [CrossRef] [PubMed]

30. Te Biesebeke, R.; Ruijter, G.; Rahardjo, Y.S.P.; Hoogschagen, M.J.; Heerikhuisen, M.; Levin, A.; van Driel, K.G.A.; Schutyser, M.A.I.; Dijksterhuis, J.; Zhu, Y.; et al. *Aspergillus oryzae* in solid-state and submerged fermentations. Progress report on a multi-disciplinary project. *Fems Yeast Res.* **2002**, *2*, 245–248. [CrossRef] [PubMed]

31. Ummalyma, S.B.; Supriya, R.D.; Sindhu, R.; Binod, P.; Nair, R.B.; Pandey, A.; Gnansounou, E. Biological pretreatment of lignocellulosic biomass—Current trends and future perspectives. In *Second and Third Generation of Feedstocks*; Elsevier: Alpharetta, GA, USA, 2019; pp. 197–212. [CrossRef]

32. Kumar, R.; Singh, S.; Singh, O.V. Bioconversion of lignocellulosic biomass: biochemical and molecular perspectives. *J. Ind. Microbiol. Biotechnol.* **2008**, *35*, 377–391. [CrossRef]

33. Gervais, P.; Molin, P. The role of water in solid-state fermentation. *Biochem. Eng. J.* **2003**, *13*, 85–101. [CrossRef]

34. Mondala, A.H. Direct fungal fermentation of lignocellulosic biomass into itaconic, fumaric, and malic acids: current and future prospects. *J. Ind. Microbiol. Biotechnol.* **2015**, *42*, 487–506. [CrossRef]

35. Wang, J.; Chen, X.; Chio, C.; Yang, C.; Su, E.; Jin, Y.; Qin, W. Delignification overmatches hemicellulose removal for improving hydrolysis of wheat straw using the enzyme cocktail from *Aspergillus niger*. *Bioresour. Technol.* **2019**, *274*, 459–467. [CrossRef]

36. Sandhya, C.; Sumantha, J.; Szakacs, G.; Pandey, A. Comparative evaluation of neutral protease production by *Aspergillus oryzae* in submerged and solid-state fermentation. *Process Biochem.* **2005**, *40*, 2689–2694. [CrossRef]

37. Koser, S.; Anwar, Z.; Iqbal, Z.; Anjum, A.; Aqil, T.; Mehmood, S.; Irshad, M. Utilization of *Aspergillus oryzae* to produce pectin lyase from various agro-industrial residues. *J. Radiat. Res. Appl. Sci.* **2014**, *7*, 327–332. [CrossRef]

38. Viniegra-González, G.; Favela-Torres, E.; Aguilar, C.N.; de Rómero-Gomez, S.J.; Díaz-Godínez, G.; Augur, C. Advantages of fungal enzyme production in solid state over liquid fermentation systems. *Biochem. Eng. J.* **2003**, *13*, 157–167. [CrossRef]

39. Ayyachamy, M.; Gupta, V.K.; Cliffe, F.E.; Tuohy, M.G. Enzymatic Saccharification of Lignocellulosic Biomass. In *Laboratory Protocols in Fungal Biology*; Gupta, V.K., Tuohy, M.G., Ayyachamy, M., Turner, K.M., O'Donovan, A., Eds.; Springer: New York, NY, USA, 2013; pp. 475–481. [CrossRef]

40. Zhao, C.; Chen, S.; Fang, H. Consolidated bioprocessing of lignocellulosic biomass to itaconic acid by metabolically engineering *Neurospora crassa*. *Appl. Microbiol. Biotechnol.* **2018**, *102*, 9577–9584. [CrossRef] [PubMed]

41. Tehrani, H.H.; Tharmasothirajan, A.; Track, E.; Blank, L.M.; Wierckx, N. Engineering the morphology and metabolism of pH tolerant *Ustilago cynodontis* for efficient itaconic acid production. *Metab. Eng.* **2019**, *54*, 293–300. [CrossRef] [PubMed]

42. Nemestóthy, N.; Bakonyi, P.; Komáromy, P.; Bélafi-Bakó, K. Evaluating aeration and stirring effects to improve itaconic acid production from glucose using *Aspergillus terreus*. *Biotechnol. Lett.* **2019**, *41*, 1383–1389. [CrossRef]

43. Molnar, A.P.; Németh, Z.; Kollath, I.S.; Fekete, E.; Flipphi, M.; Ag, N.; Karaffa, L. High oxygen tension increases itaconic acid accumulation, glucose consumption, and the expression and activity of alternative oxidase in *Aspergillus terreus*. *Appl. Microbiol. Biotechnol.* **2018**, *102*, 8799–8808. [CrossRef]

44. Chenyu Du, A.A. Fermentative Itaconic Acid Production. *J. Biodivers. Bioprospecting Dev.* **2014**, *1*, 2. [CrossRef]

45. Bradford, M.M. A rapid and sensitive method for the quantitation of microgram quantities of protein utilizing the principle of protein-dye binding. *Anal. Biochem.* **1976**, *72*, 248–254. [CrossRef]

Sample Availability: Not available.

molecules

Article

Polish Varieties of Industrial Hemp and Their Utilisation in the Efficient Production of Lignocellulosic Ethanol

Aleksandra Wawro *, Jolanta Batog and Weronika Gieparda

Institute of Natural Fibres and Medicinal Plants, National Research Institute, Wojska Polskiego 71B, 60-630 Poznan, Poland; jolanta.batog@iwnirz.pl (J.B.); weronika.gieparda@iwnirz.pl (W.G.)
* Correspondence: aleksandra.wawro@iwnirz.pl; Tel.: +48-61-84-55-814

Abstract: Nowadays, more and more attention is paid to the development and the intensification of the use of renewable energy sources. Hemp might be an alternative plant for bioenergy production. In this paper, four varieties of Polish industrial hemp (Białobrzeskie, Tygra, Henola, and Rajan) were investigated in order to determine which of them are the most advantageous raw materials for the effective production of bioethanol. At the beginning, physical and chemical pretreatment of hemp biomass was carried out. It was found that the most effective is the alkaline treatment with 2% NaOH, and the biomasses of the two varieties were selected for next stages of research: Tygra and Rajan. Hemp biomass before and after pretreatment was analyzed by FTIR and SEM, which confirmed the effectiveness of the pretreatment. Next, an enzymatic hydrolysis process was carried out on the previously selected parameters using the response surface methodology. Subsequently, the two approaches were analyzed: separated hydrolysis and fermentation (SHF) and a simultaneous saccharification and fermentation (SSF) process. For Tygra biomass in the SHF process, the ethanol concentration was 10.5 g·L^{-1} (3.04 m^3·ha^{-1}), and for Rajan biomass at the SSF process, the ethanol concentration was 7.5 g·L^{-1} (2.23 m^3·ha^{-1}). In conclusion, the biomass of Polish varieties of hemp, i.e., Tygra and Rajan, was found to be an interesting and promising raw material for bioethanol production.

Keywords: hemp biomass; alkaline pretreatment; SEM; FTIR; response surface methodology; SHF; SSF; bioethanol

check for **updates**

Citation: Wawro, A.; Batog, J.; Gieparda, W. Polish Varieties of Industrial Hemp and Their Utilisation in the Efficient Production of Lignocellulosic Ethanol. *Molecules* **2021**, 26, 6467. https://doi.org/10.3390/molecules26216467

Academic Editors: Alejandro Rodriguez Pascual, Eduardo Espinosa Víctor and Carlos Martín

Received: 10 June 2021
Accepted: 21 October 2021
Published: 26 October 2021

Publisher's Note: MDPI stays neutral with regard to jurisdictional claims in published maps and institutional affiliations.

1. Introduction

The European Union countries have been obliged to achieve a certain share of biofuels in transport and to take measures to reduce greenhouse gas emissions. It is, therefore, necessary to replace diesel and gasoline with biofuels which are produced from lignocellulosic raw materials and represent an advantageous option for the fuels currently in use due to their renewable nature and the emission of an acceptable quality exhaust gases. Currently, mainly three biofuels are produced: bioethanol, biodiesel, and biogas. According to the EU RED II directive, the contributions of advanced biofuels and biogas produced from raw materials listed in Annex IX, part A to this directive, including lignocellulosic feedstocks as a share of final energy consumption in the transport sector are expected to be at least: 0.2% in 2022, 1% in 2025, and 3.5% in 2030 [1]. The production of biofuels from plant biomass is innovative and contributes to the solution of the key issue in the production of biofuels for transport fuels.

In Poland, high expectations are associated with plant biomass due to a significant amount of waste, including that from agri-food sector and the available acreage of agricultural land that can be used for the cultivation of energy crops. In recent years, there has been an increase in the acreage of cultivated industrial hemp (*Cannabis sativa* L.) in Poland (over 1000 ha). The cultivation of hemp for seed purposes is intensively developed, and there is unused hemp biomass in the field, which can be a suitable raw material for the production of lignocellulosic ethanol. Hemp is an environmentally friendly plant characterized by a short vegetation period (3–4 months), a rapid growth up to 4 m in height, and a dry matter yield up to 15 Mg·ha^{-1}. These plants improve soil quality and are useful for the remediation of degraded land (e.g., in the region of lignite mine). Hemp is also extremely resistant, perfectly adapts to various climatic conditions, is resistant to various pests, requires a slight number of pesticide treatments, and the cultivation of 1 ha of hemp in one season absorbs approximately 11 Mg of CO_2 from the atmosphere [2–4].

The dry matter yields of these hemp varieties described in this study were as follows: Białobrzeskie 8–10 Mg·ha^{-1}, Tygra 8–11 Mg·ha^{-1}, Henola 7–8 Mg·ha^{-1}, and Rajan 9–12 Mg·ha^{-1}. Based on the data from 2016 (the own research of the Institute of Natural Fibres and Medicinal Plants—National Research Institute), it was estimated that, in Poland, the acreage of devastated and degraded land requiring reclamation and constituting a potential area for hemp cultivation for energy purposes amounts to approximately 65,000 ha. In the INF&MP-NRI, the Hemp Program is carried out, the aim of which is to develop hemp cultivation for seed reproduction. As part of this program, cultivation acreage is significantly increasing every year: 420 ha in 2018, 1000 ha in 2019, and 2000 ha in 2020.

The use of hemp waste straw for the production of lignocellulosic ethanol is beneficial for the environment, as it results in a rational management of bio-waste. The straw remaining after the ginning of the hemp panicles is not suitable for textile purposes, but it can be used, e.g., as a raw material for the production of bioethanol. Additionally, the dynamic growth of the hemp cultivation acreage in Poland may significantly contribute to increasing the efficiency process of obtaining bioethanol.

The process of plant biomass conversion to lignocellulosic ethanol includes several stages, from the preparation of plant material (effective pretreatment), through enzymatic hydrolysis, i.e., the decomposition of polysaccharides into fermentable sugars (the selection of effective enzymatic preparations) and to ethanol fermentation (the selection of appropriate microorganisms). The production of bioethanol from lignocellulosic raw material consists of the deconstruction of cell walls into individual polymers and the hydrolysis of carbohydrates into simple sugars. Currently, one of the main challenges is to increase the efficiency of the fermentation of organic substrates, and alternative solutions that directly affect the quantity and the quality composition of the final product are still being researched.

Hemp biomass as a lignocellulosic raw material contains a polymer complex, i.e., lignocellulose, which is relatively resistant to biodegradation. It is found in cell walls and consists of cellulose, hemicelluloses, and lignin. Cellulose and hemicelluloses are potential substrates in the fermentation process, while lignin adversely affects the conversion of hemp biomass. This necessitates the use of the pretreatment of biomass, the purpose of which is to fragmentate the solid phase and loosen the compact structure of lignocellulose. Recent advances in biomass pretreatment can be found in the review article by Bing et al. [5]. The second important stage in the process of obtaining bioethanol from hemp biomass is enzymatic hydrolysis, which determines the amount of simple sugars metabolized by yeast in the fermentation process. The hydrolysis process can be carried out as the SHF process—separate hydrolysis and fermentation (enzymes operate at 50–60 °C)—or the SSF (process of simultaneous saccharification and fermentation), where enzymes must be adapted to the conditions of the fermentation process, i.e., 30–40 °C. The last stage in the process of hemp biomass conversion is the ethanol fermentation of the obtained hydrolysates. A method that combines cellulose hydrolysis with sugar fermentation in one bioreactor seems to be more effective and economical [6–11].

The aim of the presented study was to indicate which of the four Polish varieties of industrial hemp (Białobrzeskie, Tygra, Henola, and Rajan) are the most suitable raw materials for the effective production of lignocellulosic ethanol. Thus far, no literature has been published about the possibility of using these Polish varieties of hemp as a raw material in the process of obtaining bioethanol.

2. Results and Discussion

2.1. Hemp Biomass Preparation

The hemp biomass of the four varieties (Białobrzeskie, Tygra, Henola, and Rajan) was first cut into fragments up to 1 cm in size and then was comminuted by the knife mill for mesh sizes of 2 and 4 mm. In order to choose the most favorable fractions, the content of reducing sugars released during an enzymatic test was determined. It was found that the highest values of reducing sugars were obtained for fractions up to 2 mm, and two varieties, Tygra and Białobrzeskie, showed higher sugar values for the tested fractions compared to Rajan and Henola (Table 1).

Table 1. The amount of reducing sugars ($mg \cdot g^{-1}$) released after an enzymatic test depending on the fraction size.

Hemp Biomass	4 mm	2 mm
Białobrzeskie	65.1 ± 0.06	68.5 ± 0.26
Tygra	73.3 ± 0.14	76.3 ± 0.17
Rajan	51.3 ± 0.32	57.8 ± 0.04
Henola	50.4 ± 0.07	54.8 ± 0.16

2.2. Alkaline Pretreatment

The purpose of chemical pretreatment is removing lignin from materials with lignocelullose and increasing the accessibility of biomass structure. The type of a reagent used has a significant effect on the performance of the chemical pretreatment. Sodium hydroxide is one of the most popular alkaline reagents used in this process.

The optimization of the concentration of sodium hydroxide used in alkaline treatment for the four varieties of hemp biomass was carried out based on the amount of reducing sugars released after the enzymatic test. Concentrations ranging from 1.5% to 3% were tested (Table 2). It was found that, for Tygra and Białobrzeskie varieties, for 2% NaOH, the amount of released reducing sugars was about 13% higher than for 1.5%. In turn, for 3% sodium hydroxide, the content of reducing sugars was at a similar level. For Henola and Rajan varieties, a completely different correlation was observed; the lowest level of released reducing sugars was noted for the concentration of 3% NaOH, while, at the concentration of 2%, the content of reducing sugars was the highest. Moreover, it was noted that two of the varieties, i.e., Tygra and Rajan, were characterized by over 10% higher content of reducing sugars than Białobrzeskie and Henola, which proves that these are the varieties more susceptible to the alkaline pretreatment. Based on the obtained results, the concentration of sodium hydroxide at the level of 2% was selected for further research. Kumar et al. [12] conducted similar research and stated that the sodium hydroxide pretreatment of lignocellulosic biomass resulted in the highest level of delignification at 2% NaOH. In turn, Zhao et al. [10], in the research on the use of American industrial hemp for the production of bioethanol, used the alkaline treatment of 1% NaOH.

Table 2. The amount of reducing sugars ($mg \cdot g^{-1}$) released after an enzymatic test depending on the NaOH concentration.

Hemp Biomass	1.5%	2%	3%
Białobrzeskie	163 ± 0.78	178 ± 1.37	173 ± 0.23
Tygra	183 ± 1.43	206 ± 0.87	203 ± 1.70
Rajan	190 ± 2.54	180 ± 0.68	176 ± 0.55
Henola	159 ± 1.86	166 ± 1.15	147 ± 1.33

Efficient pretreatment should decrystallize cellulose, depolymerize hemicelluloses, reduce the formation of inhibitors that hinder carbohydrate hydrolysis, require low energy expenditure, and recover value-added products such as lignin.

To confirm the efficiency of the alkaline treatment, the determination of the chemical composition of hemp biomass after NaOH treatment was performed and compared to the chemical composition of the biomass before pretreatment. The results are presented in Table 3.

Table 3. The chemical composition of hemp biomass (percentage of dry matter); BP: before pretreatment; AP: after pretreatment.

Variety	Samples	Cellulose (%)	Hemicelluloses (%)	Lignin (%)
Białobrzeskie	BP	50.10 ± 0.18	32.10 ± 0.22	15.40 ± 0.03
	AP	61.46 ± 0.37	21.59 ± 0.06	15.12 ± 0.16
Tygra	BP	50.82 ± 0.12	27.79 ± 0.33	14.68 ± 0.46
	AP	62.70 ± 0.09	20.16 ± 0.16	15.12 ± 0.22
Henola	BP	46.82 ± 0.04	29.94 ± 0.45	15.48 ± 0.17
	AP	57.62 ± 0.08	20.33 ± 0.22	17.80 ± 0.06
Rajan	BP	48.69 ± 0.39	31.43 ± 0.04	16.72 ± 0.08
	AP	59.30 ± 0.33	19.91 ± 0.25	18.40 ± 0.18

The analysis of the chemical composition of the hemp biomass before and after treatment showed that, in all the four varieties, the alkaline treatment resulted in a visible increase in the cellulose content (by approximately 10%) and the partial degradation of hemicelluloses (as much as 12% for the Rajan variety). The highest content of cellulose after treatment was found in the following varieties: Tygra, Białobrzeskie, and Rajan, and its level was approximately 60%. In the case of the lignin content, only for the Białobrzeskie hemp biomass, a very slight reduction was observed after the alkaline pretreatment. In the case of the remaining varieties, the tendency was contrary. Similar observations were reported by Stevulova et al. [13], who examined the chemical composition of hemp biomass before and after the pretreatment with sodium hydroxide and proved that the content of lignin after the pretreatment was 7% higher than before.

At this stage, due to better properties, availability, and higher yield, only two types of biomass of the Tygra and the Rajan varieties were selected, for which further research on bioethanol production was continued.

The effect of the alkaline treatment on Tygra and Rajan hemp biomass was also confirmed by Fourier transform infrared spectrometer (FTIR) shown in Figure 1a,b and by scanning electron microscopy (SEM) shown in Figure 2.

An effective method for studying the structure of biomass after alkaline treatment is FT-IR [14]. Figure 1a,b show the changes in FTIR spectra after the alkali treatment of hemp biomass between 600 cm^{-1} and 4000 cm^{-1}. On the spectra of both varieties, typical vibration bands in the cellulose molecule were observed at 3300 cm^{-1}, 2900 cm^{-1}, and 1610 cm^{-1}. The broad band in the $3600–3100 \text{ cm}^{-1}$ region, which was due to the OH stretching vibration, gave considerable information concerning the hydrogen bonds. The peaks characteristic of hydrogen bonds from the spectra of the Rajan variety AP became a little sharper and more intense compared to the Rajan variety BP. However, in the case of the Tygra variety, this band was less intense. The 2900 cm^{-1} peak corresponding to the C–H stretching vibration in the case of the Tygra variety shifted to higher wavenumber values and slightly decreased in the intensity. These changes could have resulted from both the increased amount of cellulose after the treatment (2.2, Table 3) and the reduced cellulose crystallinity [15,16]. The band at 1610 cm^{-1} from the stretching vibrations of the O–H bonds, due to the adsorbed water in the sample, decreased, especially for the Rajan variety, which could be attributed to water loss due to drying the sample [17]. The intensities of peaks at $1160–1170 \text{ cm}^{-1}$ (asymmetric C–O–C stretching from cellulose) and $1110–1120 \text{ cm}^{-1}$ (C–OH skeletal vibration in cellulose) increased after pretreatment in both types of biomass. Furthermore, the intensity at $1050–1060 \text{ cm}^{-1}$ (C–O–C pyranose ring skeletal vibration ascribed to cellulose) also increased slightly after alkali pretreatment.

These changes were confirmed by the increase in the concentration of cellulose in the pretreated biomass in the chemical composition tests (2.2, Table 3) [18].

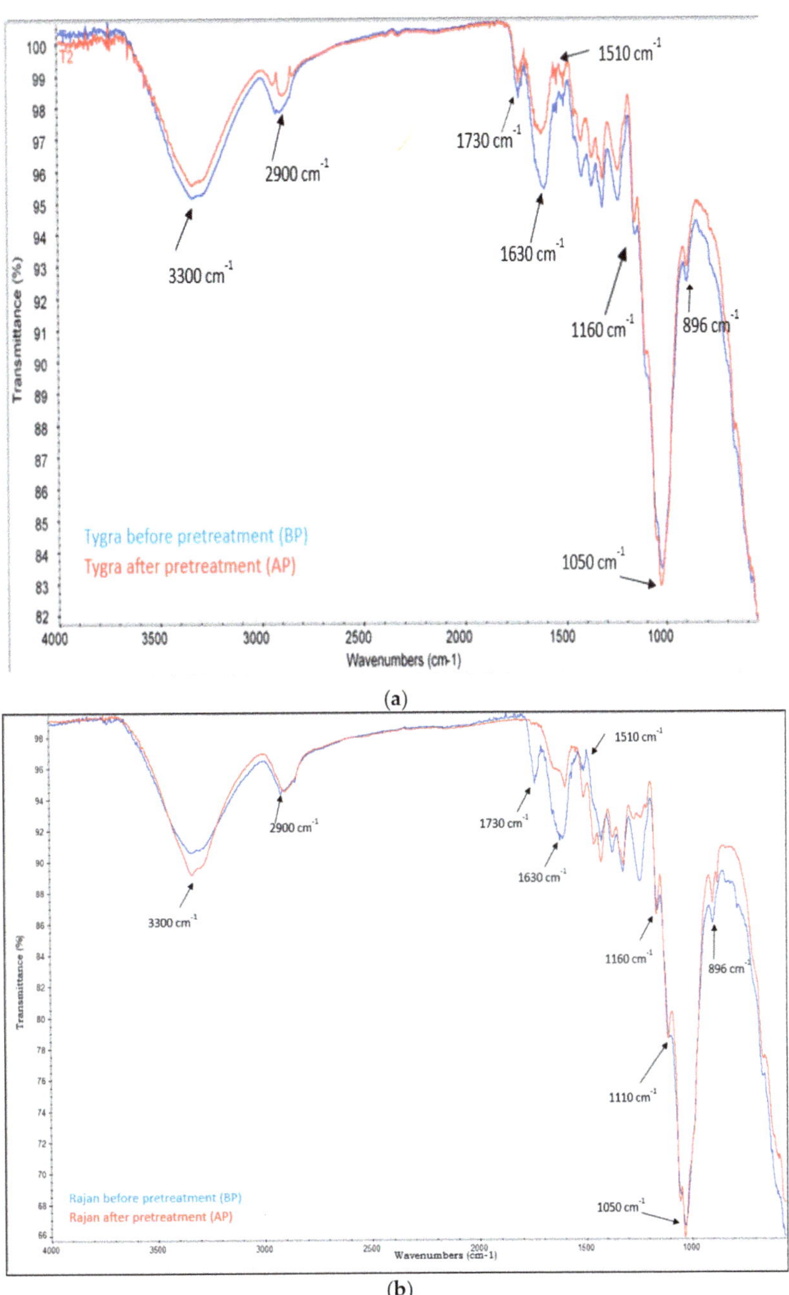

(**a**)

(**b**)

Figure 1. (**a**) The FTIR spectra of the Tygra biomass before and after the alkaline treatment. (**b**) The FTIR spectra of the Rajan biomass before and after the alkaline treatment.

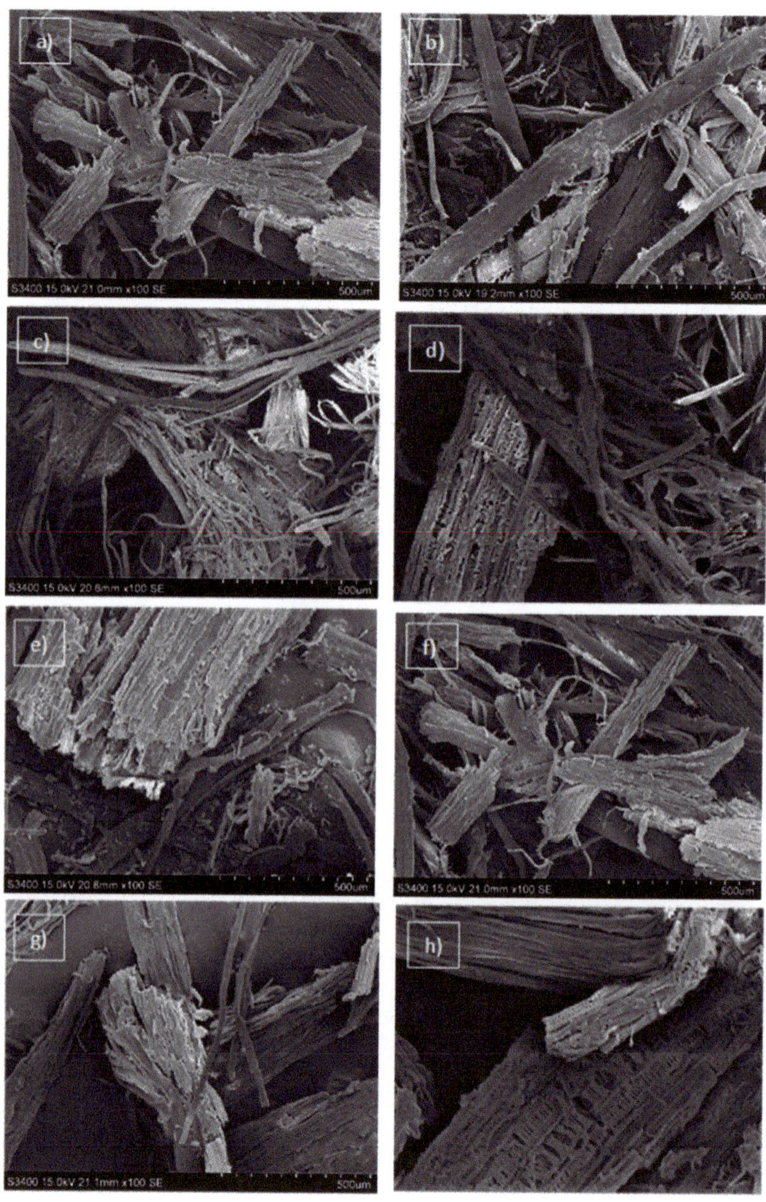

Figure 2. The SEM images of hemp biomass: (**a**) Tygra biomass before pretreatment, (**b**) Tygra biomass after pretreatment, (**c**) Tygra biomass after enzymatic hydrolysis, (**d**) Tygra biomass after enzymatic hydrolysis—selected fragment in high magnification, (**e**) Rajan biomass before pretreatment, (**f**) Rajan biomass after pretreatment, (**g**) Rajan biomass after enzymatic hydrolysis, (**h**) Rajan biomass after enzymatic hydrolysis—selected fragment in high magnification.

The vibration band visible at 1730 cm^{-1} (C=O stretching of acetyl groups in hemicelluloses and aldehydes in lignin) [19] was reduced in both hemp varieties after alkaline treatment, while the change was much more significant for the Rajan variety where the band almost disappeared. This occurred due to the decomposition of hemicelluloses and the solubilization of lignin during alkali pretreatment. This result correlates very well with the obtained results of the chemical composition of the biomass after pretreatment. According to these studies, the content of hemicelluloses after alkaline treatment in the Rajan variety decreased by as much as 12% (2.2, Table 3).

Moreover, the obtained FTIR spectra showed that the removal of lignin in the alkaline treatment process was problematic (band at 1510 cm^{-1}). This was confirmed by the results of the chemical composition of the lignin content presented in Table 2. This problem is widely described in the literature dealing with the studies of lignocellulosic biomass by infrared spectroscopy [20,21]. The process of the degradation or fragmentation of lignin is complicated due to the presence of strong C–C bonds and other functional groups, such as aromatic groups [13].

Significant changes on the surface of the biomass were observed and presented in the SEM images taken before and after the biomass pretreatment as well as after enzymatic hydrolysis (Figure 2). In the case of both varieties of hemp biomass (Tygra and Rajan), similar changes were observed in the biomass surface, which appeared as a result of subsequent processes. However, in the case of the Rajan variety, changes were more intensive, especially after the enzymatic hydrolysis process. The untreated hemp biomass was observed to have intact, rigid, and coarse structures with smooth surface and a well-ordered fiber skeleton (Figure 2a,e). This strongly blocked access to cellulose to limit enzymatic attack [22]. As a result of pretreatment, the biomass underwent various specific structural changes. The SEM images of hemp biomass after pretreatment showed that the surface area of the biomass was partially purified (Figure 2b,f). The morphological changes that indicated damage to the structure of biomass and that increased the surface area, making it more accessible to the cellulolytic enzymes [20,23,24], were observed. The enzymatic hydrolysis of the samples subjected to the previous alkaline treatment caused further significant changes in the structure of the biomass visible in the SEM pictures (Figure 2c,d,g,h). The appearance of micropores was very characteristic here.

Undoubtedly, the opening of the hemp biomass and creating the holes all over the biomass enhanced the enzyme accessibility of the structure and facilitated biomass digestibility [25]. Moreover, it was clearly visible that enzymatic hydrolysis of the Rajan hemp biomass made the fibrous structure fragile and was more successful.

2.3. Evaluation of Enzyme Preparations

The enzymatic hydrolysis process is the second main step in the process of obtaining bioethanol from plant biomass. Enzymatic hydrolysis determines the amounts of simple sugars that are metabolized by yeast in the fermentation process. The breakdown of cellulose into simple sugars requires the synergistic action of different enzymes: cellulases, endoglucanases, cellobiohydrolases, and ß-glucosidases. First, the commercial enzyme preparations of various compositions were gained (Flashzyme Plus 200, ACx8000L, Celluclast 1.5L, Cellobiase, Xylanase), and then their cellulolytic and xylanolytic activities were determined. Taking into account evaluated activity of the tested enzymes and their commercial availability, Flashzyme Plus 200 and Celluclast 1.5L preparations were selected for further research (Table 4).

In order to select the enzyme complex for the SHF and the SSF processes, enzymatic tests were performed using selected enzymes and their supplementation with glucosidase and xylanase and were partially described in our previous studies [4]. For the SHF process in the case of the Tygra biomass, an enzyme complex was selected with the composition of Flashzyme Plus 200, glucosidase, and xylanase, while composition for the Rajan was Flashzyme Plus 200:Celluclast 1.5L hemp biomass in the proportion of 70:30. For the SSF process, the enzyme complex for the Tygra biomass was selected as Flashzyme Plus 200:Celluclast 1.5L (70%:30%) and xylanase and for the Rajan was Flashzyme Plus 200:Celluclast 1.5L hemp biomass in the 50:50 ratio.

Table 4. The determination of cellulolytic and xylanolytic activities of commercial enzyme preparations (FPU·mL^{-1}).

Enzyme Preparations	Composition	Cellulolytic Activity	Xylanolytic Activity
Flashzyme Plus 200	endoglucanase, cellobiohydrolase, cellobiose, xylanase, mannase	90	2430
ACx3000L	endo-1,4-β-D-glucanase	63	487
Celluclast 1.5L	cellulase, xylanase	62	278
Accellerase 1500	exoglucanase, endoglucanase, cellobiose, β-glucosidase, hemicellulase	53	616
Alternafuel CMAX	cellulase, β-glucosidase, hemicellulase, arabinose	2.78	110
ACx8000L	cellulase	8	190
Cellobiase	glucosidase	0.18	325
Xylanase	xylanase	0.03	746

2.4. Separate Hydrolysis and Fermentation (SHF)

To determine the optimal conditions of the enzymatic hydrolysis method as a separate process of SHF, based on the literature data and the research experience, the following parameter ranges were selected for testing with the response surface methodology (RSM): dose of the enzyme 10–30 FPU·g^{-1} of solid, temperature 50–70 °C, pH 4.2–5.4, and time 24–72 h. The RSM method is an effective optimization tool consisting of mathematical and statistical techniques and used for the process of optimization [26–28].

Individual enzymatic tests of the hydrolysis process were performed for Tygra and Rajan biomasses, and the evaluation criterion was the amount of released glucose. Figure 3 presents various response surfaces and interaction effects of variables (temperature, time, enzymes' dose, and pH) on the glucose yield. The variable that had the most significant impact on the glucose content turned out to be the temperature. The lower the temperature was, the higher the glucose content was. The pH of the solution, the process time, and the enzymes' dose had lesser effects. However, slight differences were observed in the dependence of these variables on the glucose yield for the two biomass varieties. For the Tygra biomass, it was found that, in the SHF process, the optimal conditions for enzymatic hydrolysis were obtained for the substrate concentration of 5% using the following enzymes: Flashzyme Plus 200 30 FPU·g^{-1} of solid, glucosidase 20 CBU·g^{-1} of solid, and xylanase 500 XU·g^{-1} of solid. The process parameters were: temperature 50 °C, pH 4.2, and time 48 h. These parameters provided the opportunity to obtain a maximum glucose yield, which was 36.9 ± 0.64 (g·L^{-1}). In the SHF process for the Rajan biomass, optimal enzymatic hydrolysis conditions were obtained for a substrate concentration of 5% using the Flashzyme Plus 200:Celluclast 1.5L (70:30) enzyme complex with a dose of 10 FPU·g^{-1} of solid. The process parameters were: temperature 50 °C, pH 5.4, and time 72 h. The maximum glucose yield was 23.66 ± 0.16 (g·L^{-1}).

Similar research was conducted by Abraham [29]; during biomass hydrolysis at 50 °C and 18 FPU·g^{-1} of solid, the highest glucose yield was obtained. Salimi and others [30] optimized the enzymatic hydrolysis of lignocellulosic biomass using the RSM method. They applied the temperature range of 45–60 °C and the pH of 4.5—6.0. They obtained the highest content of monosaccharides at 45 °C and pH 6.0. Jambo et al. [31], in turn, optimized lignocellulosic biomass using similar parameters—temperature (30–60 °C), pH (3.8–5.8), and incubation time (12–72 h)—and obtained the glucose concentration at the level 24.24 g·L^{-1}.

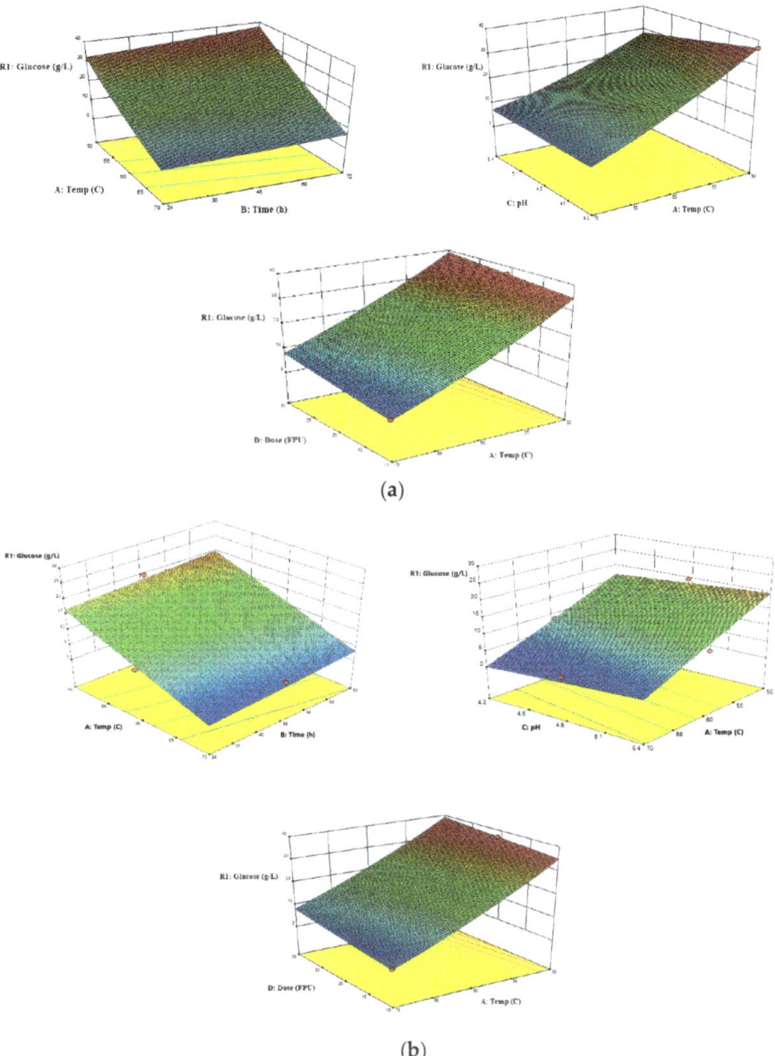

Figure 3. Enzymatic hydrolysis process of hemp biomass (RSM). Response surface representing the interaction effects of temperature, time, dose, and pH on glucose yield: (**a**) Tygra biomass, (**b**) Rajan biomass.

The next step in the conversion of hemp biomass to bioethanol in the SHF process was ethanol fermentation. In the SHF process for the Tygra biomass, the highest concentration of ethanol was observed at 48 hours and was 10.51 $g \cdot L^{-1}$. In the following hours of the process, no significant increase in ethanol concentration was observed. In turn, for the Rajan biomass, the highest ethanol concentration was noticed at 96 h, and it was only 2.76 $g \cdot L^{-1}$. Such a low concentration of ethanol in this case could be attributed to various reasons, which together may have had a significant negative impact on this parameter—for example, the chemical composition, which was characterized by a higher lignin content compared to the Tygra biomass. The reason could also have been the use of a low enzyme dose (10 FPU g^{-1} solid) in Rajan biomass as well as the low glucose concentration (23.66 g L^{-1}), which was determined immediately after the enzymatic hydrolysis step. Additionally, after optimizing this step, a pH of 5.4 was chosen based on the glucose concentration, which may have, to some extent, inhibited yeast activity in the initial phase of fermentation. Nevertheless, it was noticed that, with each passing day, the concentration of ethanol increased slightly, which ultimately indicates that the process itself was proceeding correctly. However, for Tygra biomass, no significant increase in ethanol concentration was observed with time extension of the SHF process (Figure 4). The average yield of bioethanol for the Tygra variety was equal to 253 L·Mg^{-1} (of hemp straw dry matter), i.e., 3.04 m^3·ha^{-1}. For the Rajan variety, the average yield of bioethanol was 69 L·Mg^{-1} (of hemp straw dry matter), i.e., 0.80 m^3·ha^{-1}. A similar study was presented by Kusmiyati et al. [32]. In their work, the conversion of the lignocelullosic biomass to bioethanol was carried out through pretreatment, saccharification, and fermentation processes. Their results showed that the SHF process gave a higher concentration of ethanol (8.11 g·L^{-1}). Fischer and others [33], in their research, dealt with lignocellulosic biomass and examined the SHF process, obtaining the ethanol concentration of 12.1 g·L^{-1}.

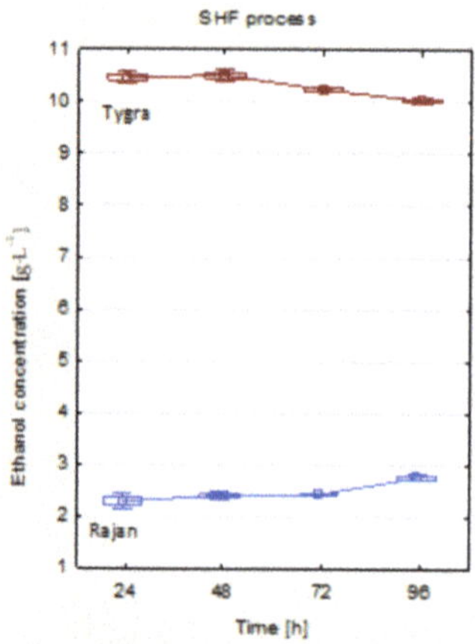

Figure 4. The ethanol concentration of hemp biomass in the SHF process. Optimized process conditions: Tygra biomass—substrate concentration 5%, enzymes: Flashzyme Plus 200 30 FPU·g^{-1} of solid, glucosidase 20 CBU·g^{-1} of solid and xylanase 500 XU·g^{-1} of solid, temperature 50 °C, pH 4.2, and time 48 h. Rajan biomass—substrate concentration 5%, Flashzyme Plus 200:Celluclast 1.5L (70:30) enzyme complex with a dose of 10 FPU·g^{-1} of solid, temperature 50 °C, pH 5.4, and time 72 h.

SHF as an alternative process in an industrial bioethanol plant manifests both potential and limitations. The main advantage of SHF is the possibility to optimize the process steps separately, especially to be able to run the enzymatic hydrolysis at an optimal temperature with respect to enzymes [34]. However, most literature reports confirm that SSF is a more promising and advantageous approach with respect to SHF because of a low production cost, less processing time, less reactor volume, higher ethanol productivity, a lower requirement of enzyme, the ability to overcome enzymatic inhibition by simultaneous end-product removal, and a lower requirement for sterile conditions, as bioethanol is produced immediately with glucose conversion [35,36].

2.5. Simultaneous Saccharification and Fermentation (SSF)

The simultaneous hydrolysis and fermentation must be carried out under conditions that ensure the optimal synergy of enzymes and distillery yeast. To optimize the SSF process according to the RSM, the following ranges of process parameters were selected: substrate content 5%–7% *w/v*, the dose of Flashzyme:Celluclast 1.5L enzymes (50:50)

10–30 FPU·g^{-1} of solid using *S. cerevisiae* yeast at 37 °C, pH 4.8, and 96 h. Then, the fermentation tests were carried out using the selected parameters, and the amount of ethanol (HPLC) was determined. The optimal conditions of the SSF process for the Tygra and the Rajan hemp biomasses were selected. The highest ethanol concentration for Tygra biomass was observed at a substrate content of 5% *w/v* and a dose of enzyme at 30 FPU·g^{-1} of solid, and it was 6.5 g·L^{-1}. In turn, for the Rajan biomass, the highest ethanol concentration equal to 7.5 g·L^{-1} was recorded for the substrate of 5% and for the enzyme 30 FPU·g^{-1} of solid. Higher substrate content above 5% *w/v* interfered with the effective mixing of the fermentation solution, which resulted in poorer access to biomass and thus less effective action of enzymes and yeast. It was also observed that enlarging the enzyme dose enhanced the conversion of cellulose to glucose and thus increased the concentration of ethanol. The enzyme dose of 10 FPU·g^{-1} of solid tested in the optimization process turned out to be too low to carry out efficient enzymatic hydrolysis.

For the Tygra biomass, there were no changes in the concentration of ethanol with the time lapse of the process. In turn, for the Rajan biomass, after 24 h, there was observed a decrease in ethanol concentration, and after 48 h of the SSF process, there was noticed a significant increase in ethanol concentration (Figure 5). It was observed that, for the Rajan biomass, in contrast to the Tygra biomass, a higher ethanol concentration was obtained in the SSF process than in the SHF process (Figures 4 and 5).

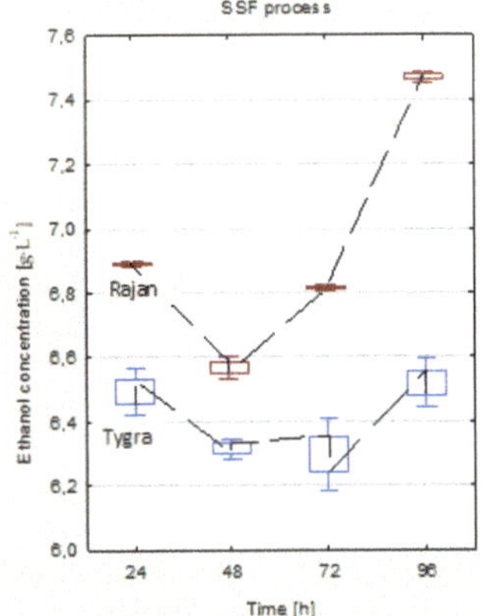

Figure 5. Ethanol concentration of hemp biomass in the SSF process. Optimized process conditions: Tygra and Rajan biomass-substrate concentration 5%, enzymes: Flashzyme:Celluclast 1.5L (50:50) 30 FPU·g^{-1} of solid, temperature 37 °C, pH 4.8, and time 96 h.

In their research, Fojas and Rosario [37] optimized the enzymatic saccharification of lignocellulosic biomass, and the SSF process was carried out with the following parameters: 3%–6% the amount of substrate, 20–25 FPU·g^{-1} of solid dose of enzyme, and temperature of 37 °C for 120 h. They achieved an ethanol content of about 9 g·L^{-1}.

Research on obtaining bioethanol from hemp biomass was also carried out by Orlygsson [38]; after the SSF process, the author obtained an ethanol concentration of approximately 1 g·L^{-1}.

On the basis of the average ethanol content, the hemp straw yield that could be obtained from 1 ha of hemp cultivation was specified. The highest average yield of bioethanol was estimated for the Rajan variety and was equal to 190 L·Mg^{-1} (of hemp straw dry matter), i.e., 2.23 m^3·ha^{-1}, while the average yield of bioethanol estimated for the Tygra variety was 165 L·Mg^{-1} (of hemp straw dry matter), i.e., 1.81 m^3·ha^{-1}.

Extensive research on industrial hemp as potential raw material for bioethanol production compared to other raw materials such as kenaf and sorghum was conducted by

Das et al. [39]. According to these studies, the ethanol yield from hemp was 250 L·Mg^{-1}, which turned out to be much higher than that of kenaf. Moreover, the cost analysis allowed the researchers to conclude that industrial hemp can generate higher gross profits per hectare than other crops. In conclusion, this scientific report emphasized that hemp has the potential to be a promising crop for the production of bioethanol.

3. Materials and Methods

3.1. Bioethanol Production Process

3.1.1. Hemp Biomass Preparation

The raw materials used in the study were Białobrzeskie, Tygra, Henola, and Rajan hemp (*Cannabis sativa* L.) biomasses from the Experimental Farm of the Institute of Natural Fibres and Medicinal Plants in Pętkowo. This material was subjected to preliminary crushing to the particles of size 20–40 mm and then dried in 50–55 °C for 24 h. Next, the material was disintegrated on a knife mill (Retsch SM-200, Germany) with the sieves of the mesh size of 2–4 mm. An enzymatic test for the crushed fractions was performed using the Celluclast 1.5L enzyme preparation, and the content of reducing sugars was determined by the Miller's method with 3,5-dinitrosalicylic acid (DNS) [40].

3.1.2. Alkaline Pretreatment

The evaluation of pretreatment conditions for hemp biomass was carried out at 5 h treatment with 1.5%–3% sodium hydroxide in 90 °C. NaOH:biomass weight ratio was 10:1. After the alkaline pretreatment was carried out, the biomass solution was filtered on a Büchner funnel, then washed with distilled water until neutralized and dried in a laboratory dryer at 50 °C for 24 h. The alkali effect on the biomass was evaluated in the enzymatic test, and content of the released reducing sugars was determined by the Miller's method. This test was performed with the use of Celluclast 1.5L (Novozymes, Bagsværd, Denmark) enzymatic preparation at the dose of 10 FPU·g^{-1} of solid. The raw material was incubated at 55 °C in 0.05 M citrate buffer of pH 4.8 for 24 h. Then, after the enzymatic test, the supernatant was diluted, a DNS reagent was added, and the mixture was incubated in a boiling water bath for 10 min. After cooling to room temperature, the absorbance of the supernatant was measured at 530 nm on UV–VIS Spectrophotometer V-630 (Jasco, Pfungstadt, Germany).

3.1.3. Enzyme Complex

The enzymes cellulolytic activity was determined according to NREL LAP Measurement of Cellulase Activities. In turn, enzymes xylanolytic activity was determined according to the Osaka University procedure (with changes) [41].

In order to select the enzyme complex for SHF and SSF processes, tests were performed using selected enzymes (Flashzyme Plus 200:Celluclast 1.5L) and their supplementation with glucosidase 20 CBU·g^{-1} of solid and xylanase 500 XU·g^{-1} of solid (Sigma-Aldrich, Darmstadt, Germany). Enzymatic tests were carried out for 5% of biomass with the enzyme in the amount of 10 FPU·g^{-1} of solid, at a pH of 4.8, and during 24 h at 55 °C for the SHF process and at 38 °C for the SSF process. The selection criterion was the content of reducing sugars determined by the Miller's method.

3.1.4. Separate Hydrolysis and Fermentation (SHF)

The optimization of the enzymatic hydrolysis of hemp biomass in the SHF process was carried out according to the response surface methodology (RSM) using the parameters: biomass content 5%–7% *w/v*, temperature 50–70 °C, time 24–72 h, pH 4.2–5.4, dose of enzyme 10–30 FPU·g^{-1} of solid. Then, tests of the hemp biomass hydrolysis process were performed, and the evaluation criterion was the amount of released glucose.

In the next stage, the obtained hydrolyzate was subjected to the ethanol fermentation process carried out in bioreactor Biostat B Plus (Sartorius, Göttingen, Germany) in 2 L vessel equipped with pH, temperature, stirring, and foaming controls. The temperature was maintained at 37 °C and stirring at 900 rpm, while pH was controlled at 4.2 for Tygra and 5.4 for Rajan by adding 1 M NaOH or 1 M HCl. Non-hydrated freeze-dried distillery yeast *S. cerevisiae* (Ethanol Red, Lesaffre, France) at a dose of 1 g·L^{-1} was used in the process, which corresponded to cell concentration after inoculation of about 1×10^7 cfu/mL. Inoculum grew for 24 h at 30 °C. After inoculation, a 96 h fermentation was carried out, and samples were taken every 24 h. All experiments were performed in triplicate.

3.1.5. Simultaneous Saccharification and Fermentation (SSF)

To optimize the SSF process according to the RSM, the ranges of process parameters were selected: substrate content 5%–7% *w/v*, dose of (Flashzyme:Celluclast 1.5L) enzymes 10–30 FPU·g^{-1} of solid. The SSF process was carried out in bioreactor Biostat B Plus (Sartorius, Göttingen, Germany) in 2 L vessel equipped with pH, temperature, stirring, and foaming controls. The temperature was maintained at 37 °C and stirring at 900 rpm, while pH was controlled at 4.8 by adding 1 M NaOH or 1 M HCl. In the fermentation process, non-hydrated freeze-dried distillery yeast *S. cerevisiae* (Ethanol Red, Lesaffre, France) at a dose of 1 g·L^{-1} was used, which corresponded to cell concentration after inoculation of about 1×10^7 cfu/mL. The duration of ethanol fermentation was 96 h. All experiments were performed in triplicate.

3.2. Analytical Methods

The chemical composition of hemp biomass before and after pretreatment was determined, i.e., cellulose according to TAPPI T17 m-55 [42], hemicelluloses as the difference of holocellulose and cellulose according to TAPPI T9 m-54 [43], and lignin according to TAPPI T13 m-54 [44].

In order to provide a more complete picture of the molecular structure of hemp biomass before and after the chemical pretreatment, the analysis of FTIR spectroscopy was performed using a Fourier Transform Infrared Spectrometer ISS 66v/S (Bruker, Bremen, Germany) at infrared wavenumbers of 400–4000 cm^{-1} [13].

The physical morphologies of hemp biomass before and after the chemical treatment and after enzymatic hydrolysis were performed using Scanning Electron Microscope S-3400N (Hitachi, Japan) in high vacuum conditions. The samples were covered with gold dust.

The contents of glucose and ethanol were determined by high performance liquid chromatography on Elite LaChrom (Hitachi, Tokio, Japan) using an RI L-2490 detector, Rezex ROA 300x7.80 mm column (Phenomenex, Torrance, CA, USA), as the mobile phase used 0.005 N H$_2$SO$_4$ at a flow rate of 0.6 mL/min at 40 °C.

3.3. Calculations

The ethanol yield from 100 g of raw material Y$_s$ (g/100 g of raw material) was calculated according to the Equation (1) [45]:

$$Y_s = \frac{Et \times 100}{M} \tag{1}$$

where: *Et*—amount of ethanol in 1000 mL of tested sample (g); *M*—mass of material weighed in 1000 mL fermentation sample (g).

Then, based on the ethanol yield from 100 g of raw material, the amount of ethanol in L per ton of straw dry matter (L·Mg^{-1}) was calculated, and on the basis of straw yield, the ethanol yield per hectare (m^3·ha^{-1}) was determined.

3.4. Statistical Analysis

The experiments of ethanol fermentation were carried out in triplicates. Standard deviations were calculated using the analysis of variance ANOVA, Statistica 13.0 software ($p < 0.05$).

4. Conclusions

To sum up, the Tygra and the Rajan varieties of hemp were selected, which proved to be proper sources of second generation bioethanol as alternatives to petroleum-oil based fossil fuels. Pretreatment, enzymatic hydrolysis, and ethanol fermentation were optimized. Alkaline pretreatment caused an increase in cellulose content and partial degradation of hemicelluloses. Enzymatic hydrolysis allowed us to achieve glucose yield at the level up to 36.9 g·L^{-1}. For the Tygra biomass in the SHF process, the ethanol concentration was 10.5 g·L^{-1} (3.04 m^3·ha^{-1}), and for the Rajan biomass in the SSF process, the ethanol concentration was 7.5 g·L^{-1} (2.23 m^3·ha^{-1}).

In the future, it will be important to conduct research on the mixtures of different varieties of hemp biomass in order to determine their potential for the production of ligno-cellulosic ethanol, which seems important in practical application, because the industrial production of biofuels occurs most often in large refineries which process the biomass of different varieties and species of plants.

Author Contributions: Conceptualization, A.W., J.B., W.G.; Methodology, A.W., J.B., W.G.; Software, A.W., W.G.; Validation, J.B.; Formal Analysis, J.B., A.W.; Investigation, A.W., W.G.; Resources, A.W., J.B. WG.; Data Curation, J.B., A.W., W.G.; Writing—Original Draft Preparation, A.W., W.G; Writing—Review & Editing, A.W., J.B., W.G.; Visualization, A.W.; Supervision, J.B.; Project Administration, J.B.; Funding Acquisition, J.B. All authors have read and agreed to the published version of the manuscript.

Funding: This research was funded by the Ministry of Agriculture and Rural Development, Poland, Multiannual Program (2017–2020).

Institutional Review Board Statement: Not applicable.

Informed Consent Statement: Not applicable.

Data Availability Statement: Data is contained within the article.

Acknowledgments: The study was conducted as a research project- Multiannual Program (2017–2020): Reconstruction and sustainable development of production and processing of natural fiber raw materials for the needs of agriculture and the economy was financed by the Ministry of Agriculture and Rural Development, Poland.

Conflicts of Interest: The authors declare no conflict of interest.

References

1. Directive 2009/28/EC of the European Parliament and of the Council of 23 April 2009 on the promotion of the use of energy from renewable sources and amending and subsequently repealing Directives 2001/77/EC and 2003/30/EC (Text with EEA relevance). Available online: https://eur-lex.europa.eu/legal-content/EN/ALL/?uri=CELEX:32009L0028 (accessed on 8 April 2021).
2. Cattaneo, C.; Givonetti, A.; Leoni, V.; Guerrieri, N.; Manfredi, M.; Giorgi, A.; Cavaletto, M. Biochemical aspects of seeds from *Cannabis sativa* L. plants grown in a mountain environment. *Sci. Rep.* **2021**, *11*, 3927. [CrossRef]
3. Zadrożniak, B.; Radwańska, K.; Baranowska, A.; Mystkowska, I. Possibility of industrial hemp cultivation in areas of high nature value. *Econ. Reg. Stud.* **2017**, *10*, 114–127. [CrossRef]
4. Wawro, A.; Batog, J.; Gieparda, W. Chemical and Enzymatic Treatment of Hemp Biomass for Bioethanol Production. *Appl. Sci.* **2019**, *9*, 5348. [CrossRef]
5. Bing, S.; Richen, L.; Chun Ho, L.; Hao, W.; To-Hung, T.; Yun, Y. Recent advances and challenges of inter-disciplinary biomass valorization by integrating hydrothermal and biological techniques. *Renew. Sustain. Energy Rev.* **2021**, *135*, 110370.
6. Bajpai, P. Pretreatment of Lignocelluloses Biomass for Bioethanol Production. In *Developments in Bioethanol, Green Energy and Technology*; Springer: Singapore, 2021. [CrossRef]
7. Vasić, K.; Knez, Ž.; Leitgeb, M. Bioethanol Production by Enzymatic Hydrolysis from Different Lignocellulosic Sources. *Molecules* **2021**, *26*, 753. [CrossRef]
8. Abo, B.O.; Gao, M.; Wang, Y.; Wu, C.; Ma, H.; Wang, Q. Lignocellulosic biomass for bioethanol: An overview on pretreatment, hydrolysis and fermentation processes. *Rev. Env. Health* **2019**, *34*, 57–68. [CrossRef] [PubMed]
9. Rahmati, S.; Doherty, W.; Dubal, D.; Atanda, L.; Moghaddam, L.; Sonar, P.; Hessel, V.; Ostrikov, K. Pretreatment and fermentation of lignocellulosic biomass: Reaction mechanisms and process engineering. *React. Chem. Eng.* **2020**, *5*, 2017–2047. [CrossRef]
10. Zhao, J.; Xu, Y.; Wang, W.; Griffin, J.; Roozeboom, K.; Wang, D. Bioconversion of industrial hemp biomass for bioethanol production: A review. *Fuel* **2020**, *281*, 118725. [CrossRef]

11. Rheay, H.T.; Omondi, E.C.; Brewer, C.E. Potential of hemp (*Cannabis sativa* L.) for paired phytoremediation and bioenergy production. *Bioenergy* **2021**, *13*, 525–536.

12. Kumar, A.K.; Sharma, S. Recent updates on different methods of pretreatment of lignocellulosic feedstocks: A review. *Bioresour. Bioprocess* **2017**, *4*, 7. [CrossRef]

13. Stevulova, N.; Cigasova, J.; Estokova, A.; Terpakova, E.; Geert, A.; Kacik, F.; Singovszka, E.; Holub, M. Properties characterization of chemically modified hemp hurds. *Materials* **2014**, *7*, 8131–8150. [CrossRef]

14. Kumar, A.; Rapoport, A.; Kunze, G.; Kumar, S.; Singh, D.; Singh, B. Multifarious pretreatment strategies for the lignocellulosic substrates for the generation of renewable and sustainable biofuels: A review. *Renew Energy* **2020**, *160*, 1228–1252.

15. Araújo, D.; Vilarinho, M.; Machado, A. Effect of combined dilute-alkaline and green pretreatments on corncob fractionation: Pretreated biomass characterization and regenerated cellulose film production. *Ind. Crop. Prod.* **2019**, *149*, 111785. [CrossRef]

16. Ciolacu, D.; Ciolacu, F.; Popa, V.I. Amorphous cellulose-structure and characterization. *Cellul. Chem. Technol.* **2011**, *45*, 13.

17. Gupta, A.D.; Pandey, S.; Kumar Jaiswal, V.; Bhadauria, V.; Singh, H. Simultaneous oxidation and esterification of cellulose for use in treatment of water containing Cu (II) ions. *Carbohydr. Polym.* **2019**, *222*, 114964. [CrossRef] [PubMed]

18. Zhao, J.; Xu, Y.; Wang, W.; Griffin, J.; Wang, D. High Ethanol Concentration (77 g/L) of Industrial Hemp Biomass Achieved Through Optimizing the Relationship between Ethanol Yield/Concentration and Solid Loading. *ACS Omega* **2020**, *5*, 21913–21921. [CrossRef]

19. Sun, X.F.; Xu, F.; Sun, R.C.; Fowler, P.; Baird, M.S. Characteristics of degraded cellulose obtained from steam-exploded wheat straw. *Carbohyd. Res.* **2005**, *340*, 97–106. [CrossRef]

20. Putnina, A.; Kukle, S.; Gravitis, J. STEX treated and untreated hemp fiber comparative structural analysis. *Mater. Science. Text. Cloth. Technol.* **2011**, *6*, 36–42.

21. Plácido, J.; Capareda, S. Analysis of alkali ultrasonication pretreatment in bioethanol production from cotton gin trash using FT-IR spectroscopy and principal component analysis. *Bioresour. Bioprocess* **2014**, *1*, 23. [CrossRef]

22. Zhao, J.; Xu, Y.; Zhang, M.; Wang, D. Integrating bran starch hydrolysates with alkaline pretreated soft wheat bran to boost sugar concentration. *Bioresour. Technol.* **2020**, *302*, 122826. [CrossRef] [PubMed]

23. Fang, G.; Chen, H.; Chen, A.; Mao, K.; Wang, Q. An efficient method of bio-chemical combined treatment for obtaining high-quality hemp fiber. *BioResources* **2017**, *12*, 1566–1578. [CrossRef]

24. Abraham, R.E.; Barrow, C.J.; Puri, M. Relationship to reducing sugar production and scanning electron microscope structure to pretreated hemp hurd biomass (*Cannabis sativa*). *Biomass Bioenerg.* **2013**, *58*, 180–187. [CrossRef]

25. Abraham, R.E.; Vongsvivut, J.; Barrow, C.J.; Puri, M. Understanding physicochemical changes in pretreated and enzyme hydrolysed hemp (Cannabis sativa) biomass for biorefinery development. *Biomass Conv. Bioref.* **2016**, *6*, 127–138. [CrossRef]

26. Lavudi, S.; Oberoi, H.S.; Mangamoori, L.N. Ethanol production from sweet sorghum bagasse through process optimization using response surface methodology. *Biotech* **2017**, *7*, 233. [CrossRef] [PubMed]

27. Campos, L.M.A.; Moura, H.O.M.A.; Cruz, A.J.G. Response surface methodology (RSM) for assessing the effects of pretreatment, feedstock, and enzyme complex association on cellulose hydrolysis. *Biomass Conv. Bioref.* **2020**, *12*, 124. [CrossRef]

28. Jaisamut, K.; Paulová, L.; Patáková, P.; Rychtera, M.; Melzoch, K. Optimization of alkali pretreatment of wheat straw to be used as substrate for biofuels production. *Plant. Soil. Environ.* **2013**, *59*, 537–542. [CrossRef]

29. Abraham, R.E. Bioprocessing of Hemp Hurd (*Cannabis sativa*) for Biofuel Production. Ph.D Thesis, Deakin University, Melbourne, Australia, 30 October 2014.

30. Salimi, M.N.; Lim, S.E.; Yusoff, A.H.M.; Jamlos, M.F. Conversion of rice husk into fermentable sugar by two stage hydrolysis. *J. Phys. Conf. Ser.* **2017**, *908*, 012056. [CrossRef]

31. Jambo, A.; Abdulla, R.; Marbawi, H.; Gansau, J.A. Response surface optimization of bioethanol production from third generation feedstock -Eucheuma cottonii. *Renew. Energy.* **2019**, *132*, 1–10. [CrossRef]

32. Maryanto, D.; Sonifa, R.; Kurniawan, S.A.; Hadiyanto, H. Pretreatment of Starch-Free Sugar Palm Trunk (*Arenga pinnata*) to Enhance Saccharification in Bioethanol Production. *MATEC Web. Conf.* **2018**, *156*, 6.

33. Fischer, J.; Lopes, V.S.; Cardoso, L.; Coutinho Filho, U.; Cardoso, V.L. Machine learning techniques applied to lignocellulosic ethanol in simultaneous hydrolysis and fermentation. *Braz. J. Chem. Eng.* **2017**, *34*, 53–63. [CrossRef]

34. Moreno, A.D.; Alvira, P.; Ibarra, D.; Tomas-Pejo, E. Production of ethanol from lignocellulosic biomass. In *Production of Platform Chemicals from Sustainable Resources*; Biofuels and Biorefineries; Feng, Z., Smith, R.L., Qi, X., Eds.; Springer: Singapore, 2017; pp. 375–410.

35. Robak, K.; Balcerek, M. Current state-of-the-art in ethanol production from lignocellulosic feedstocks. *Microbiol. Res.* **2020**, *240*, 126534. [CrossRef]

36. Dey, P.; Pal, P.; Kevin, J.D.; Das, D.B. Lignocellulosic bioethanol production: Prospects of emerging membrane technologies to improve the process—A critical review. *Rev. Chem. Eng.* **2020**, *36*, 333–336. [CrossRef]

37. Fojas, J.J.R.; Rosario, E.J.D. Optimization of pretreatment and enzymatic saccharification of cogon grass prior ethanol production. *In. Chem. Mol. Eng.* **2013**, *7*, 296–299.

38. Orlygsson, J. Ethanol production from biomass by a moderate thermophile, Clostridium AK1. *Icel. Agric. Sci.* **2012**, *25*, 25–35.

39. Das, L.; Liu, E.; Saeed, A.; Williams, D.W.; Hu, H.; Li, C.; Ray, A.E.; Shi, J. Industrial hemp as a potential bioenergy crop in comparison with kenaf, switchgrass and biomass sorghum. *Bioresour. Technol.* **2017**, *244*, 641–649. [CrossRef] [PubMed]

40. Miller, G.L. Use of dinitrosalicylic acid reagent for determination of reducing sugars. *Anal. Chem.* **1959**, *31*, 426–428. [CrossRef]

41. Measurement of Cellulase Activities-. Available online: https://www.nrel.gov/docs/gen/fy08/42628.pdf (accessed on 7 May 2021).
42. TAPPI 17 m-55. In *Cellulose in Wood*; TAPPI Press: Atlanta, GA, USA, 1955.
43. TAPPI T9 m-54. In *Holocellulose in Wood*; TAPPI Press: Atlanta, GA, USA, 1998.
44. Bagby, M.O.; Nelson, G.H.; Helman, E.G.; Clark, T.F. Determination of lignin non-wood plant fiber sources. *TAPPI J.* **1971**, *54*, 11.
45. Kawa-Rygielska, J.; Pietrzak, W. Zagospodarowanie odpadowe pieczywa do produkcji bioetanolu. *Żywność Nauka Technol. Jakość* **2011**, *79*, 105–118.

Article

A Direct Silanization Protocol for Dialdehyde Cellulose

Arianna Lucia [1,2], Markus Bacher [2], Hendrikus W. G. van Herwijnen [1] and
Thomas Rosenau [2,3,*]

1   Wood K Plus–Competence Center for Wood Composites and Wood Chemistry, Kompetenzzentrum Holz
    GmbH, Altenberger Straße 69, A-4040 Linz, Austria; a.lucia@wood-kplus.at (A.L.);
    e.herwijnen@wood-kplus.at (H.W.G.v.H.)
2   Institute for Chemistry of Renewable Resources, University of Natural Resources and Life Science
    Vienna (BOKU), Konrad-Lorenz-Straße 24, A-3430 Tulln an der Donau, Austria; markus.bacher@boku.ac.at
3   Johan Gadolin Process Chemistry Centre, Åbo Akademi University, Porthansgatan 3,
    FI-20500 Åbo/Turku, Finland
*   Correspondence: thomas.rosenau@boku.ac.at; Tel.: +43-1-47654-77411

Academic Editors: Alejandro Rodríguez and Eduardo Espinosa
Received: 26 April 2020; Accepted: 22 May 2020; Published: 25 May 2020

Abstract: Cellulose derivatives have many potential applications in the field of biomaterials and composites, in addition to several ways of modification leading to them. Silanization in aqueous media is one of the most promising routes to create multipurpose and organic–inorganic hybrid materials. Silanization has been widely used for cellulosic and nano-structured celluloses, but was a problem so far if to be applied to the common cellulose derivative "dialdehyde cellulose" (DAC), i.e., highly periodate-oxidized celluloses. In this work, a straightforward silanization protocol for dialdehyde cellulose is proposed, which can be readily modified with (3-aminopropyl)triethoxysilane. After thermal treatment and freeze-drying, the resulting product showed condensation and cross-linking, which was studied with infrared spectroscopy and ^{13}C and ^{29}Si solid-state nuclear magnetic resonance (NMR) spectroscopy. The cross-linking involves both links of the hydroxyl group of the oxidized cellulose with the silanol groups (Si-O-C) and imine-type bonds between the amino group and keto functions of the DAC (-HC=N-). The modification was achieved in aqueous medium under mild reaction conditions. Different treatments cause different levels of hydrolysis of the organosilane compound, which resulted in diverse condensed silica networks in the modified dialdehyde cellulose structure.

Keywords: biomaterials; cellulose; dialdehyde cellulose; organosilane chemistry; ^{29}Si NMR; solid state NMR; silanization

1. Introduction

The booming developments in the biopolymers field is evidently engaging cellulose and cellulose derivatives in crescent number of applications and new materials, such as fillers or matrices in polymer composites, aerogels, and separation media. Cellulosic components are simply central in the evolution of novel bio-materials [1]. One particular modification that is especially promising in the area of organic–inorganic hybrid materials is the silanization of cellulose—or polysaccharides in general—by this way generating polysaccharide–silane/silica interfaces with differing amounts of covalent bonds between the two bordering constituents. Silanization is widespread for enhancing the properties of cellulose composites: the coupling with silane agents involve an improved interfacial adhesion between fibers and matrix, better resistance to water leaching, hydrophobicity, thermostabilization, and improved fiber strength [2–4]. Quite many applications of novel cellulosic materials modified by silanes can be found in literature: hybrid substances formed by silica gel and dicarboxylic cellulose for dye absorption [5], films made with cellulose acetate and silane with isocyanate moieties [6]; aerogels with cross-linked cellulose, acrylamide polymers, and methyltrichlorosilane for oil/water separation [7]; or composite aerogels with silica and cellulosic fibers for thermal insulation [8] are just a few examples. The chemistry between cellulose derivatives and modified silica gels is used also in the field of chiral separation, producing packing materials for liquid chromatography [9]. Modifications of cellulosic materials are performed in wet state with different silanes [10,11].

A special cellulose derivative with great reverberation in research and applications is dialdehyde cellulose (DAC), usually produced by oxidation of cellulosics (to different degrees) with sodium periodate. Several applications of this type of oxidized cellulose are reported in the literature, such as films for packaging [12], as nanoparticles for drug delivery systems [13], the formation of nanocrystal aerogels with superabsorbent properties [14], or as self-healing nanocomposite hydrogels [15]. In this work, we describe the direct modification of DAC according to a straightforward silanization protocol, carried out in aqueous media and without severe thermal treatment. The chosen silane is (3-aminopropyl)triethoxysilane (APTES). This reagent is cheap and readily available, which is an important factor when it comes to up-scaling. APTES has been studied already in systems with cellulosic fibers [16]. Its utilization has already been reported for strengthening the interfaces of hybrid organic-inorganic coatings [17], in grafting reaction with tosylcellulose [18], for surface functionalization of cellulose nanocrystals [19], and also with DAC as the stationary phase in chromatography [20]. In the latter case, silica gel was modified by APTES in toluene, and then utilized for the modification of DAC in pyridine at high temperature during several hours which is considered neither practical and general nor compatible with green chemistry principles. In our approach, the silanization of DAC involves hydrolysis of APTES and self-condensation as well as condensation with the hydroxyl groups of the cellulose derivative, most notably using aqueous media and employing mild reaction conditions and short reaction times. The studies leading to the DAC modification protocol are described in the present account.

2. Results and Discussion

Periodate oxidation of cellulose, if reaching sufficiently high oxidation degrees of about 60% and above, provides a water-soluble material [21]. During the oxidation and solubilization process a great variety of masked aldehyde structures is formed, mainly aldehyde hydrates, hemiacetals, and hemialdals with intra- and intermolecular bonds [12,22,23]. The determination of the molecular weight of DAC, because of inter-chain crosslinking, is rather complicated and requires special approaches and precautions which have been addressed previously [23] and cannot be discussed here. The degree of oxidation was 59% for the thermally treated DAC and 62% for the freeze-dried DAC in our experiments, which corresponds to 7.4 mmol/g and 7.8 mmol/g of aldehyde groups, respectively. The acidic pH during the periodate oxidation promotes the formation and stability of hemiacetal structures [24], which are detected in the Fourier-transform infrared (FTIR) spectrum of the oxidized cellulose [25] (band at 876 cm^{-1} in Figure 1). Also, the influence of periodate oxidation on cellulose structure and morphology, according to the cellulose allomorphs, has been studied [26]. Since the acidic environment at the same time catalyzes hydrolysis of APTES [27], the organosilane was directly added to the DAC solution without a previous hydrolysis step, which is usually performed when this silane is used with cellulose. It has been reported that cellulose and other cellulose derivatives bind covalently with silanes after thermal treatment [3,18,28,29], although it was by no means clear that the behavior would be similar between DAC and APTES, since the chemistry of DAC—and its reactive moieties—is rather different from that of celluloses. Nevertheless, hydroxyl groups present in DAC in high number because of the solubilization process can be expected to be available for cross-linking and condensation with the silanol groups. A similar reactivity of DAC hydroxyl groups and the hydroxyl groups of polysaccharides with regard to hydrogen bond network formation has already been noted in the literature, e.g., for polyvinyl alcohol [30]. So far, modification of DAC with APTES took only imine formation [20] into account, but not a possible reaction between the hydrolyzed organosilane and DAC (hemiacetal/aldehyde hydrate/hemialdal) hydroxyl groups. While solubilized DAC is a film-like materials after thermal drying and foam-like after lyophilization (freeze-drying), the morphology changes upon derivatization and silanized DAC is a powder.

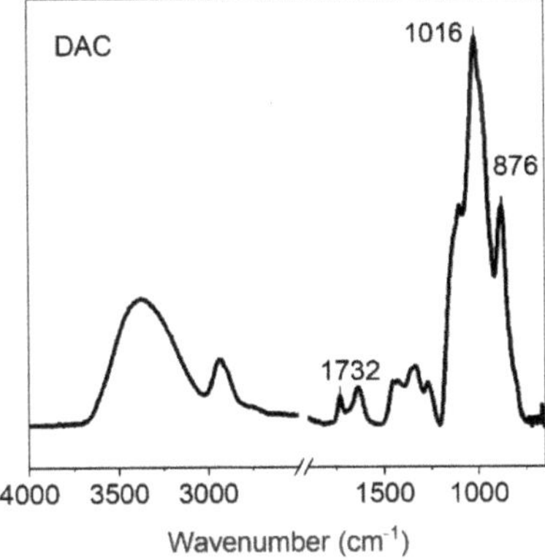
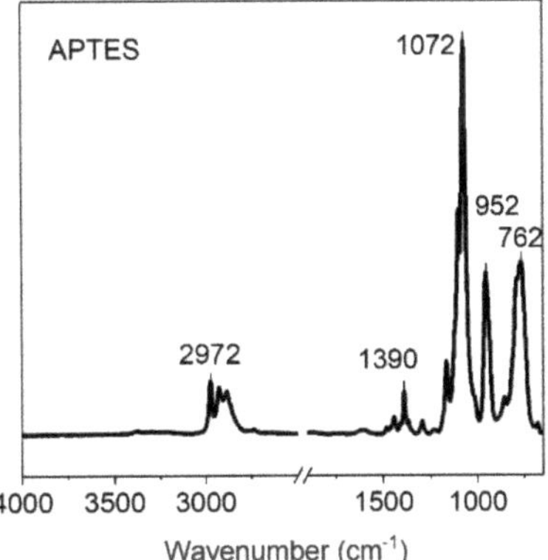

Figure 1. Fourier-transform infrared (FTIR) spectra of dialdehyde cellulose (DAC) (left) and (3-aminopropyl)triethoxysilane (APTES) (right). The labelled bands, from left to right, correspond to: carbonyl stretching (1732 cm^{-1}), C-O stretching (1016 cm^{-1}), and hemiacetal bond stretching (876 cm^{-1}) in the DAC spectrum; C-H methyl group asymmetrical stretching (2972 cm^{-1}), C-H methyl group deformation (1390 cm^{-1}), Si-O stretching (1072 cm^{-1}), skeletal vibration (952 cm^{-1}), and C-H bending (762 cm^{-1}) in the APTES spectrum.

Covalent bonding between cellulose and organosilane can be studied with ^{29}Si NMR: if the spectra do not change anymore over time after an initial period, further self-condensation of the silane structures is blocked by the covalent bonding with the cellulose molecules [28]. In this work we apply this approach to the reaction between APTES and DAC to study presence and type of covalent binding between the cellulose and the silane networks.

Figure 2 shows the cross-polarization/magic angle spinning (CP/MAS) ^{29}Si NMR spectra of the DAC-APTES condensation product after thermal treatment, revealing peaks of the two main condensation structures of the silane, namely T^2 and T^3 [6], with chemical shifts δ at −60.1 ppm and −68.5 ppm, respectively, in the one-day-old sample and δ −60.3 ppm and −68.4 ppm, respectively, for the same sample after two years. A minor contribution of the T^1 structure is visible at −50 ppm as a shoulder in both spectra. It was evident that the spectra did not change significantly within the two years of sample storage. Therefore, the silane structure formation is completed already in the first sample and stays constant afterwards, and the silane moieties are almost completely crosslinked in a tridimensional structure after the thermal treatment, involving the covalent bond with DAC. The ^{29}Si-CP/MAS NMR spectra did not change over time, because of the bonding between the silanol group and the hydroxyl group in DAC: if those bond had not occurred, hydrolyzed APTES would have continued to be engaged in self-condensation and have formed more highly condensed structure over time, which would have significantly changed the spectra (increase of T^3).

Given in Scheme 1, the condensation reaction of silanes and DAC occurred between the DAC's different hydroxyl groups and the silanol groups during the thermal treatment, which withdraws the water, this way shifting the equilibrium.

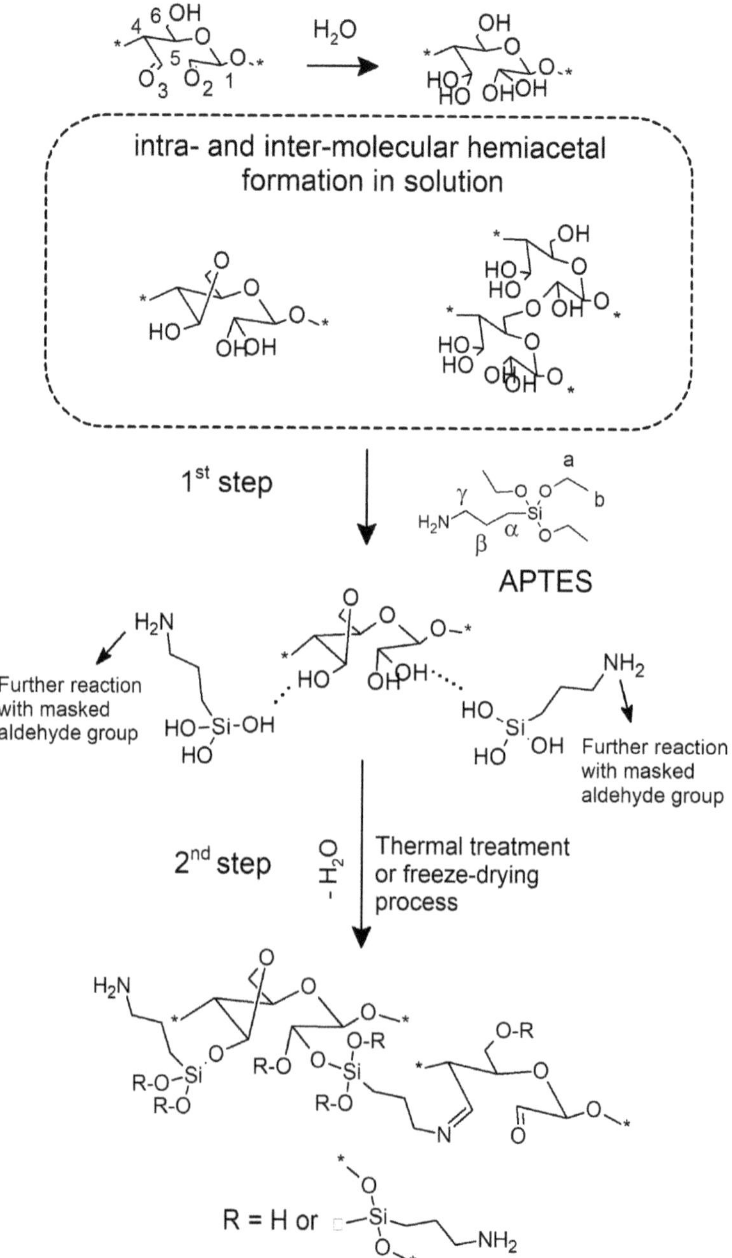

Scheme 1. Proposed reaction for the direct silanization of DAC with APTES, with the structures schematically presenting the binary cross-linking of DAC and APTES both by Si-O-C and imine bonds. In mildly acidic medium, the silanol groups of APTES bind covalently onto DAC's hydroxyl groups. After thermal treatment or simple freeze-drying treatment this bonding type is complemented by imine formation between the amino groups of APTES and the DAC's (masked) aldehyde functions.

Please, note once more that the structures in Scheme 1 are just examples of possible structures for the masked aldehyde groups [23].

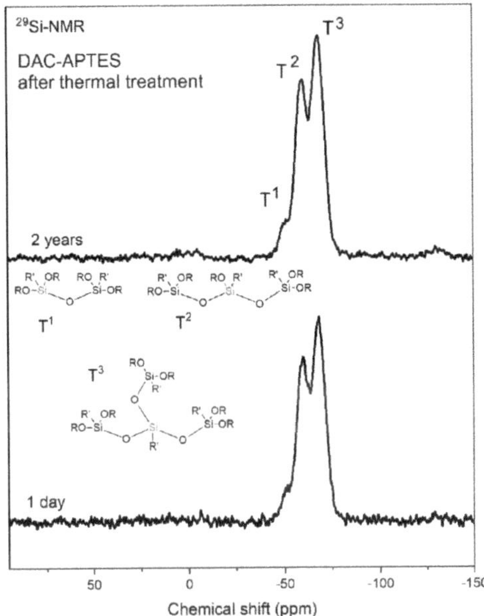

Figure 2. ^{29}Si cross-polarization/magic angle spinning (CP/MAS) nuclear magnetic resonance (NMR) spectra of DAC-APTES reaction product after thermal treatment: the spectrum after 1 day (bottom), and spectrum of the same sample after 2 years (top), with the schematic representations of the structures T^1, T^2, and T^3. R=H, C (from DAC); R′=C (from $(CH_2)_3$-NH_2 side chain in APTES).

In addition to the formation of Si-O-C structures by hydroxyl group condensation, a Schiff-base reaction of (masked) aldehydes with the amino group of APTES can occur. The aldehyde function can be present either in its free form or in its masked forms, which react the same way because of the underlying dynamic equilibria. We started from the hypothesis that the thermal treatment was mainly supporting the condensation of APTES' amino group with the DAC structures (besides further promoting condensation among the hydrolyzed silanes). The corresponding nucleophilic substitution reactions in the mild media present would initially produce hemiaminal structures, which would need more drastic media for the subsequent water elimination leading to the double bond structure of imines. We thus suspected elevated temperatures to be a suitable means for that, and indeed, imine bond formation was evident under these conditions. Surprisingly, the same happened also when we used the alternative approach of freeze-drying to remove water from the reaction mixture. Also, this much milder approach was obviously sufficient to move the equilibrium toward condensation product formation. Both thermal treatment and freeze-drying were effective enough in removing the reaction water, the second technique being clearly more energy-efficient and much milder. The final product (Scheme 1, bottom), involves both imine moieties as well as the Si-O-C structures among silanes and DAC.

The ^{29}Si-CP/MAS NMR of the freeze-dried DAC-APTES in Figure 3 shows the cross-linked structures of the silane networks, but with different proportions with respect to Figure 2. While in Figure 2 (thermally treated sample), T^3 was most prominent and T^1 present only in small amounts, Figure 3 (freeze-dried sample) shows a more equal contribution of the three condensed structures with an order of $T^2 > T^1 > T^3$, meaning that the condensation degree (Si-O-Si) was generally smaller here. In Figure 2, signals of the different silane structures were detected at δ = −49.8, −59.5, −68.6 ppm in the sample after one day, and at $\delta = -50.1, -59.5, -67.8$ ppm in the sample after two months' storage. The integral ratios of the three peaks did not change over time, showing the stability of the cross-linking even without thermal treatment. This result is different from cellulose reacting with silanization agents [16,28], where the condensation is slowly progressing and changing over much longer times.

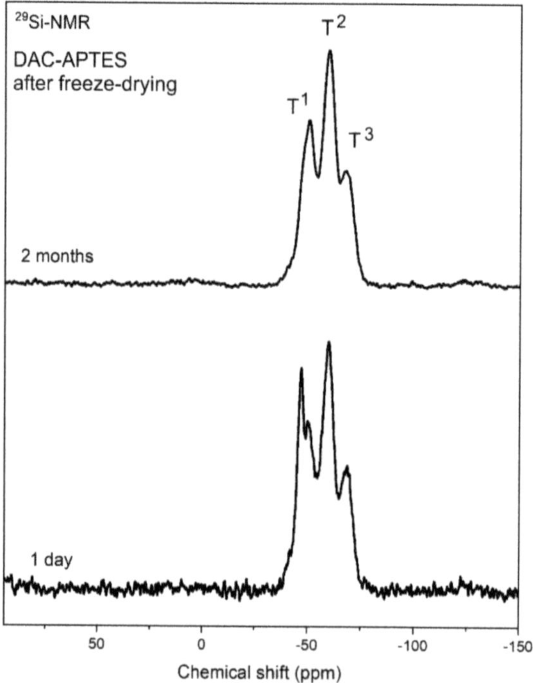

Figure 3. ^{29}Si CP/MAS NMR spectra of the DAC-APTES freeze-dried sample, after one day (lower spectrum) and after two months (upper spectrum), with the labels of the different condensed silica structures. The peak at −46.9 ppm in the one-day sample (lower spectrum) is attributable to residual, non-reacted starting material that was not removed by washing. The peak of the pure APTES (in the absence of the DAC matrix) appears at δ −45 ppm.

The presence of non-reacted APTES, which was concluded from ^{29}Si NMR (see caption of Figure 3), can also be seen in the FTIR spectra of the samples (Figure 4). In the freeze-dried sample after one day, three typical bands of neat APTES are visible at 2974 cm^{-1} (methyl group, asymmetrical stretching), at 1388 cm^{-1} (methyl group, deformation), and at 952 cm^{-1} (skeletal vibration), identical with those of pure APTES in Figure 1 [31]. It is known that APTES is moisture-sensitive and degrades nearly completely within 28 days [32]; this explains why the peak intensities decreased and eventually disappeared.

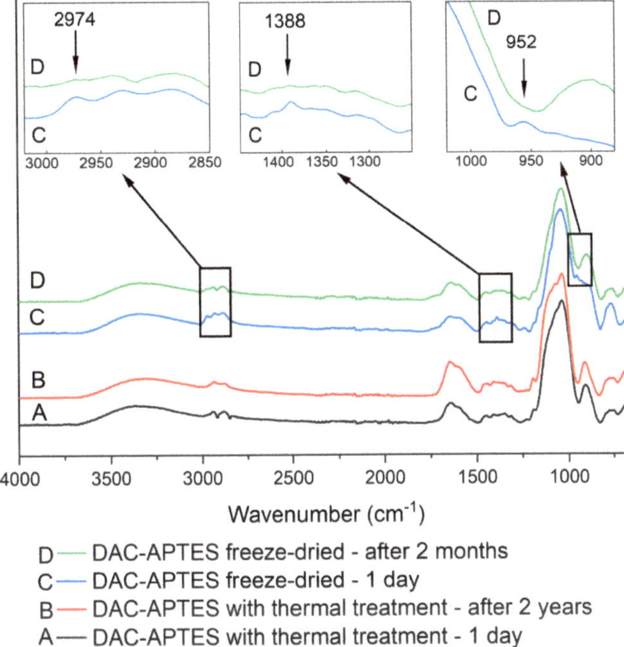

Figure 4. FTIR spectra of modified DAC with APTES for both processes, i.e., thermal treatment and freeze-drying treatment. The insets are showing the bands of unreacted APTES, which disappear in the freeze-dried samples after two months because of the instability of neat APTES.

The DAC-APTES structures show the same bands in the FTIR spectra both for thermal and freeze-drying after-treatment. The bands of the O-Si-O network are at 1644 and 1042 cm^{-1}: the latter is rather close to one of the C-O bands of DAC at 1016 cm^{-1}, so both partly superimpose. In the thermally treated samples a double peak was detected, with the second peak at 1070 cm^{-1}, ascribable to siloxane (Si-O-Si) bonds [33]. The strong intensity of this peak can be explained with the higher condensation degree of the silica network, which corresponds well with the results of the ^{29}Si NMR. The band at 898 cm^{-1} in the freeze-dried samples is in the range of the free silanol group stretching vibration and partly overlapped with the hemiacetal band in DAC. It shifted to 914 cm^{-1} in case of the thermal treated sample, with the hemiacetal contribution being reduced after thermal treatment. The bands at 762 and 684 cm^{-1} correspond to the C-H deformation in the aliphatic chain and Si-C vibration [34], whereas the band at 1590 cm^{-1} is attributed to the imine bond C=N vibration [35].

In the ^{13}C-CP/MAS NMR spectra of the freeze-dried samples, shown in Figure 5A, the trace of non-reacted APTES is clearly visible, the sharp signals at δ = 58.4 ppm and 18.5 ppm (Ca and Cb in the APTES structure of Scheme 1) being identical to those of pure APTES, shown in Figure 5B. Since there were no changes in the condensed structures over time as seen by ^{29}Si NMR, the decrease of these signals is caused by slow decomposition by atmospheric humidity. The peaks of the aliphatic chain in these APTES hydrolysates/condensates are slightly shifted relative to pure APTES. The hydrolysis—and the corresponding shift of the resonances—started already after about one hour of the reaction [16], and arrived at the ultimate shift values of δ 10.6–10.3 ppm (Cα) and at δ 21.5–21.2 (Cβ) ppm after one day (not changes up to two months). The Cγ peak is broadened and resulted in a shoulder at around 42 ppm. The peaks at around 15.2 ppm in the freeze-dried sample after one day and 14.9 ppm in the freeze-dried sample after two months storage arise from the APTES side chains in condensed oligomers. The ^{13}C resonances for the DAC carbons, except C2/C3, are still present in the spectra, with chemical shifts almost identical to non-modified DAC, shown in Figure 5C. Only the small peak of free (non-masked) aldehyde carbons C2/C3 (δ = 201.2 ppm) was not detected any longer, whereas a new signal at δ 170 ppm, characteristic for imine carbons, appeared [35]. The peak corresponding to the carbon in the C-O-Si bridge is located in the region between 60 and 50 ppm, but the overlap of resonances made an unambiguous assignment impossible.

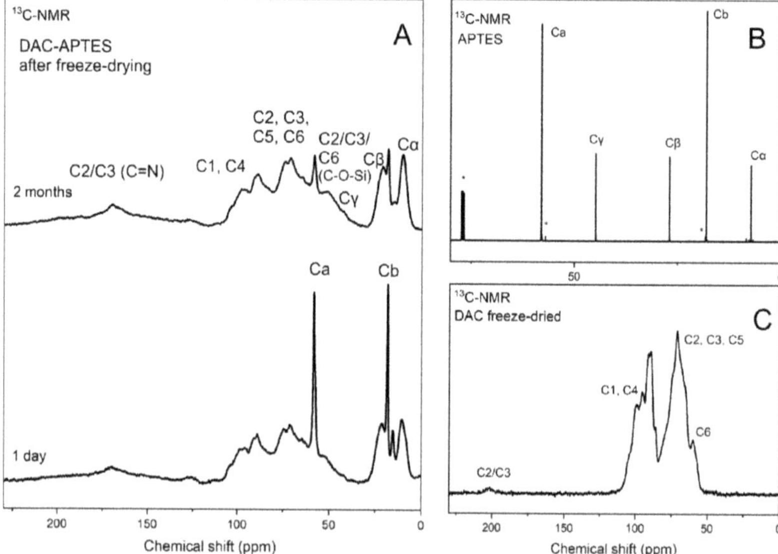

Figure 5. (**A**) ^{13}C CP/MAS NMR spectra of DAC-APTES freeze-dried after one day and after two months, the numbering referring to the structures indicated in Scheme 1: C1–6 from the DAC, a-b and α-β-γ from APTES structures. For the spectrum after two months, δ: 170.4, 98.3, 89.0, 74.1, 71.3, 64.9, 58.4, 51.6, 21.2, 18.4, 14.9, 10.3; for the spectrum after one day, δ: 170.1, 98.3, 88.8, 74.8, 71.2, 64.7, 58.4, 53.2, 21.5, 18.4, 15.3, 10.6. (**B**) ^{13}C NMR of APTES in CDCl$_3$, δ: 58.1, 44.9, 27.0, 18.1, 7.8 ppm; the stars are indicating chloroform and ethanol solvent impurities. (**C**) ^{13}C CP/MAS NMR spectrum of non-modified DAC, freeze-dried after solubilization, δ: 201.2, 99.1, 95.3, 91.0, 89.2, 71.1, 60.0 ppm.

The same characteristic signals and chemical shifts as for the freeze-dried samples can be seen in the spectrum of the thermally treated sample (Figure 6). The resonances from residual non-reacted APTES are absent, because it is fully condensed into the silanes network and with DAC. Furthermore, there is a decrease of the C1 and C4 intensities from DAC demonstrating a more pronounced condensation, and a significant increase in imine-type structures which point to the same conclusion. For both treatment options, thermally induced and freeze-drying, the NMR analyses confirmed the direct silanization of DAC with presence of both DAC-silanol Si-O-C interlinks and imine bonds. Differences in the degree of condensation arise from the different processes of drying, with the severity being higher in the case of the thermal option.

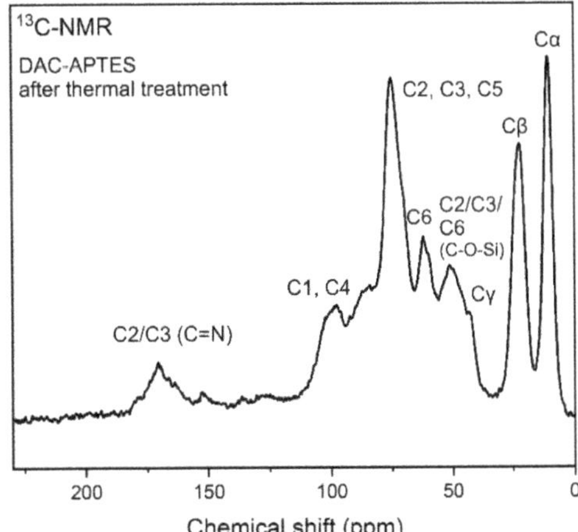

Figure 6. ^{13}C CP/MAS NMR spectrum of DAC-APTES after thermal treatment. The numbering refers to the structures indicated in Scheme 1: C1–6 from the DAC, a-b and α-β-γ from APTES structures, δ: 170.6, 97.7, 75.1, 62, 51.3, 42.9, 22.2, 10.6 ppm.

3. Materials and Methods

Microcrystalline cellulose (Avicel PH101), sodium metaperiodate (ACS reagent, ≥ 99.8%), acetic acid (glacial, ≥ 99%), and (3-aminopropyl)triethoxysilane (APTES, 99%) were purchased from Sigma-Aldrich (Schnelldorf, Germany).

3.1. Cellulose Oxidation and Solubilization

Microcrystalline cellulose was suspended in an aqueous solution (deionized water) of sodium metaperiodate with a 1.25 molar ratio between the oxidant and cellulose (anhydroglucose unit). The suspension was stirred for 24 h in the dark at 35 °C. The product was separated by centrifugation (5000 rpm for 20 min, Hettich Rotina 380, Westphalia, Germany) and washed. The never-dried DAC was suspended in water (5% of solid content) and heated to 100 °C for 90 min for solubilization [36]. The pH was adjusted to 3.5 with acetic acid. Determination of the aldehyde content was performed with the oxime titration method, as reported in the literature [37].

3.2. Silanization Protocol

An aliquot of APTES was added to the DAC solution under stirring at 250 rpm at r.t., with a molar ratio of aldehyde: organosilane of 2.5. After 1 h, the precipitated product was collected and washed by centrifugation (5000 rpm for 15 min, 700 mL of water in total per sample). For the freeze-drying treatment, the sample was frozen at −80 °C and then lyophilized (Christ Beta 1–8 LD Plus, Martin Christ Gefriertrocknungsanlagen GmbH, Osterode am Harz, Germany). For the thermal treatment, the sample was placed in an oven at 105 °C for 1 h (Memmert UNB 400, Schwabach, Germany). The samples were stored in vials in a ventilated cupboard.

3.3. Solid-State NMR and FTIR Measurements

All solid state NMR experiments were performed on a Bruker Avance III HD 400 spectrometer (Rheinstetten, Germany), resonance frequency of ^{1}H at 400.13 MHz, ^{13}C at 100.61 MHz, and ^{29}Si at 79.54 MHz, respectively, equipped with a 4 mm dual broadband CP/MAS probe. ^{13}C spectra were acquired by using the total sideband suppression (TOSS) sequence at ambient temperature with a spinning rate of 5 kHz, a cross-polarization (CP) contact time of 2 ms, a recycle delay of 2 s, SPINAL−64 ^{1}H decoupling and an acquisition time of 49 ms whereas the spectral width was set to 250 ppm. ^{13}C chemical shifts were referenced externally against the carbonyl signal of glycine at δ = 176.03 ppm. ^{29}Si NMR spectra were acquired with the normal CP pulse sequence using a spectral width of 300 ppm and a contact time of 2 ms. Chemical shifts were referenced externally against DSS with δ = 0 ppm.

The samples were analyzed with a PerkinElmer Frontier FTIR Single-Range spectrometer in ATR mode (PerkinElmer Frontier, Waltham, MA, United States).

4. Conclusions

In this work we presented a straightforward protocol, which meets common green chemistry principles, for the direct silanization of DAC, one of the most recently developed and studied cellulose derivatives. We describe the direct silanization of DAC with APTES, through thermal treatment and freeze-drying, both with imine formation and hydroxyl group condensation. The grafting and covalent binding between the organic and inorganic counterparts was proven and the structures characterized by ^{29}Si solid-state nuclear magnetic resonance, together with infrared spectroscopy (Fourier-transform Infrared, FTIR). Interestingly, condensation occurred not only after thermal treatment (as it does in the case of cellulose), but also after a simple freeze-drying process. The condensation of the hemiacetal/aldehyde hydrate/hemialdal groups from DAC and the silanol groups in the hydrolyzed APTES occurred simultaneously with imine link formation between the masked aldehyde structures of DAC in APTES' amino group. The resulting material was characterized by NMR techniques: the absence of changes in the spectra over time confirmed the proposed cross-linking and its stability. The constancy of the spectra indicated that formation of the crosslinked network was completed after one day and did not proceed further or change afterwards. Apparently, the reaction centers were consumed or became inaccessible because of the decreased internal mobility of the structure. In any case, this relatively fast process in the DAC case is different from the slow, continuously changing process in the case of celluloses as the co-reactant of silanes.

The modification was achieved in aqueous media and with mild reaction conditions, even by an energy-saving and byproduct-reducing freeze-drying process instead of a thermal treatment. ^{13}C CP/MAS NMR confirms different grades of hydrolysis and condensation severity, depending on the drying process. In all cases, the imine bond is confirmed. The product showed a larger silane network for the specimens after thermal treatment. FTIR spectra confirmed all conclusions derived from NMR. The resulting condensed hybrid product, combining an organic and inorganic phase, represent a class of important biomaterials with diverse applications reaching from materials science over separation science and chromatography to medicine.

Author Contributions: A.L.: study concept, design of experiments, data collection and drafting of manuscript, M.B.: data collection, NMR experiments and evaluation, H.W.G.v.H.: study concept, drafting and revision of manuscript, T.R.: study concept, design of experiments, drafting and revision of manuscript. All authors have read and agreed to the published version of the manuscript.

Funding: This research was funded by Austrian Research Promotion Agency (FFG-Forschungsförderungsgesellschaft, Project Nr. 855592) and Metadynea Austria GmbH, the Austrian Biorefinery Center Tulln (ABCT) and the doctoral school "Advanced biorefineries – chemistry and materials" (ABCM).

Acknowledgments: The authors are grateful to Julien Jaxel and Hubert Hettegger for the fruitful discussions.

Conflicts of Interest: The authors declare no conflict of interest.

References

1. Shaghaleh, H.; Xu, X.; Wang, S. Current progress in production of biopolymeric materials based on cellulose, cellulose nanofibers, and cellulose derivatives. *RSC Adv.* **2018**, *8*, 825–842. [CrossRef]
2. Hubbe, M.A.; Rojas, O.J.; Lucia, L.A. Green Modification of Surface Characteristics of Cellulosic Materials at the Molecular or Nano Scale: A Review. *Bioresources* **2015**, *10*, 6095–6206. [CrossRef]
3. Xie, Y.; Hill, C.A.S.; Xiao, Z.; Militz, H.; Mai, C. Silane coupling agents used for natural fiber/polymer composites: A review. *Compos. Part A-Appl. S.* **2010**, *41*, 806–819. [CrossRef]

4. Abdelmouleh, M.; Boufi, S.; ben Salah, A.; Belgacem, M.N.; Gandini, A. Interaction of Silane Coupling Agents with Cellulose. *Langmuir* **2002**, *18*, 3203–3208. [CrossRef]

5. Salama, A.; Aljohani, H.A.; Shoueir, K.R. Oxidized cellulose reinforced silica gel: New hybrid for dye adsorption. *Mater. Lett.* **2018**, *230*, 293–296. [CrossRef]

6. Achtel, C.; Härling, S.M.; Hering, W.; Westerhausen, M.; Heinze, T. Synthesis of Biopolymer-Based Precursors for the Formation of Organic–Inorganic Hybrid Materials. *Macromol. Rapid Commun.* **2018**, *39*, 1800199. [CrossRef]

7. Liao, Q.; Su, X.; Zhu, W.; Hua, W.; Qian, Z.; Liu, L.; Yao, J. Flexible and durable cellulose aerogels for highly effective oil/water separation. *RSC Adv.* **2016**, *6*, 63773–63781. [CrossRef]

8. Jaxel, J.; Markevicius, G.; Rigacci, A.; Budtova, T. Thermal superinsulating silica aerogels reinforced with short man-made cellulose fibers. *Compos. Part A- Appl. S.* **2017**, *103*, 113–121. [CrossRef]

9. Okamoto, Y.; Ikai, T.; Shen, J. Controlled Immobilization of Polysaccharide Derivatives for Efficient Chiral Separation. *Isr. J. Chem.* **2011**, *51*, 1096–1106. [CrossRef]

10. Hettegger, H.; Sumerskii, I.; Sortino, S.; Potthast, A.; Rosenau, T. Silane Meets Click Chemistry: Towards the Functionalization of Wet Bacterial Cellulose Sheets. *ChemSusChem* **2015**, *8*, 680–687. [CrossRef]

11. Beaumont, M.; Bacher, M.; Opietnik, M.; Gindl-Altmutter, W.; Potthast, A.; Rosenau, T. A General Aqueous Silanization Protocol to Introduce Vinyl, Mercapto or Azido Functionalities onto Cellulose Fibers and Nanocelluloses. *Molecules* **2018**, *23*, 1427. [CrossRef] [PubMed]

12. Plappert, S.F.; Quraishi, S.; Pircher, N.; Mikkonen, K.S.; Veigel, S.; Klinger, K.M.; Potthast, A.; Rosenau, T.; Liebner, F.W. Transparent, Flexible, and Strong 2,3-Dialdehyde Cellulose Films with High Oxygen Barrier Properties. *Biomacromolecules* **2018**, *19*, 2969–2978. [CrossRef] [PubMed]

13. Peng, X.; Liu, P.; Pang, B.; Yao, Y.; Wang, J.; Zhang, K. Facile fabrication of pH-responsive nanoparticles from cellulose derivatives via Schiff base formation for controlled release. *Carbohyd. Polym.* **2019**, *216*, 113–118. [CrossRef] [PubMed]

14. Yang, X.; Cranston, E.D. Chemically Cross-Linked Cellulose Nanocrystal Aerogels with Shape Recovery and Superabsorbent Properties. *Chem. Mater.* **2014**, *26*, 6016–6025. [CrossRef]

15. Xiao, G.; Wang, Y.; Zhang, H.; Chen, L.; Fu, S. Facile strategy to construct a self-healing and biocompatible cellulose nanocomposite hydrogel via reversible acylhydrazone. *Carbohyd. Polym.* **2019**, *218*, 68–77. [CrossRef]

16. Brochier Salon, M.-C.; Abdelmouleh, M.; Boufi, S.; Belgacem, M.N.; Gandini, A. Silane adsorption onto cellulose fibers: Hydrolysis and condensation reactions. *J. Colloid Interface Sci.* **2005**, *289*, 249–261. [CrossRef]

17. Li, C.; Wilkes, G.L. The mechanism for 3-aminopropyltriethoxysilane to strengthen the interface of polycarbonate substrates with hybrid organic–inorganic sol-gel coatings. *J. Inorg. Organomet. Polym.* **1998**, *8*, 33–45. [CrossRef]

18. El-Sayed, N.S.; El-Sakhawy, M.; Brun, N.; Hesemann, P.; Kamel, S. New approach for immobilization of 3-aminopropyltrimethoxysilane and TiO2 nanoparticles into cellulose for BJ1 skin cells proliferation. *Carbohyd. Polym.* **2018**, *199*, 193–204. [CrossRef]

19. Khanjanzadeh, H.; Behrooz, R.; Bahramifar, N.; Gindl-Altmutter, W.; Bacher, M.; Edler, M.; Griesser, T. Surface chemical functionalization of cellulose nanocrystals by 3-aminopropyltriethoxysilane. *Int. J. Biol. Macromol.* **2018**, *106*, 1288–1296. [CrossRef]

20. Gao, J.; Chen, L.; Wu, Q.; Li, H.; Dong, S.; Qin, P.; Yang, F.; Zhao, L. Preparation and chromatographic performance of a multifunctional immobilized chiral stationary phase based on dialdehyde microcrystalline cellulose derivatives. *Chirality* **2019**, *31*, 669–681. [CrossRef]

21. Hell, S.; Ohkawa, K.; Amer, H.; Potthast, A.; Rosenau, T. "Dialdehyde Cellulose" Nanofibers by Electrospinning as Polyvinyl Alcohol Blends: Manufacture and Product Characterization. *J. Wood Chem. Technol.* **2018**, *38*, 96–110. [CrossRef]

22. Potthast, A.; Rosenau, T.; Kosma, P.; Saariaho, A.-M.; Vuorinen, T. On the Nature of Carbonyl Groups in Cellulosic Pulps. *Cellulose* **2005**, *12*, 43–50. [CrossRef]

23. Sulaeva, I.; Klinger, K.M.; Amer, H.; Henniges, U.; Rosenau, T.; Potthast, A. Determination of molar mass distributions of highly oxidized dialdehyde cellulose by size exclusion chromatography and asymmetric flow field-flow fractionation. *Cellulose* **2015**, *22*, 3569–3581. [CrossRef]

24. Münster, L.; Vícha, J.; Klofáč, J.; Masař, M.; Kucharczyk, P.; Kuřitka, I. Stability and aging of solubilized dialdehyde cellulose. *Cellulose* **2017**, *24*, 2753–2766. [CrossRef]

25. Calvini, P.; Gorassini, A.; Luciano, G.; Franceschi, E. FTIR and WAXS analysis of periodate oxycellulose: evidence for a cluster mechanism of oxidation. *Vib. Spectrosc.* **2006**, *40*, 177–183. [CrossRef]

26. Siller, M.; Amer, H.; Bacher, M.; Rosenau, T.; Potthast, A. Effects of periodate oxidation on cellulose polymorphs. *Cellulose* **2015**, *22*, 2245–2261. [CrossRef]

27. Bel-Hassen, R.; Boufi, S.; Salon, M.-C.B.; Abdelmouleh, M.; Belgacem, M.N. Adsorption of silane onto cellulose fibers. II. The effect of pH on silane hydrolysis, condensation, and adsorption behavior. *J. Appl. Polym. Sci.* **2008**, *108*, 1958–1968. [CrossRef]

28. Salon, M.-C.B.; Gerbaud, G.; Abdelmouleh, M.; Bruzzese, C.; Boufi, S.; Belgacem, M.N. Studies of interactions between silane coupling agents and cellulose fibers with liquid and solid-state NMR. *Magn. Reson. Chem.* **2007**, *45*, 473–483. [CrossRef]

29. de Morais Zanata, D.; Battirola, L.C.; Gonçalves, M.d.C. Chemically cross-linked aerogels based on cellulose nanocrystals and polysilsesquioxane. *Cellulose* **2018**, *25*, 7225–7238. [CrossRef]

30. Münster, L.; Vícha, J.; Klofáč, J.; Masař, M.; Hurajová, A.; Kuřitka, I. Dialdehyde cellulose crosslinked poly(vinyl alcohol) hydrogels: Influence of catalyst and crosslinker shelf life. *Carbohyd. Polym.* **2018**, *198*, 181–190. [CrossRef]

31. Putz, A.-M.; Putz, M. Spectral Inverse Quantum (Spectral-IQ) Method for Modeling Mesoporous Systems: Application on Silica Films by FTIR. *Int. J. Mol. Sci.* **2012**, *13*, 15925–15941. [CrossRef]

32. Morrill, A.R.; Duong, D.T.; Lee, S.J.; Moskovits, M. Imaging 3-aminopropyltriethoxysilane self-assembled monolayers on nanostructured titania and tin (IV) oxide nanowires using colloidal silver nanoparticles. *Chem. Phys. Lett.* **2009**, *473*, 116–119. [CrossRef]

33. Lenza, R.F.S.; Vasconcelos, W.L. Preparation of silica by sol-gel method using formamide. *Mater. Res.-Ibero-Am. J.* **2001**, *4*, 189–194. [CrossRef]

34. Oh, T.; Choi, C. Comparison between SiOC thin films fabricated by using plasma enhanced chemical vapor deposition and SiO2 thin films by using Fourier transform infrared spectroscopy. *J. Korean Phys. Soc.* **2010**, *56*, 1150–1155. [CrossRef]

35. Keshk, S.M.; Ramadan, A.M.; Bondock, S. Physicochemical characterization of novel Schiff bases derived from developed bacterial cellulose 2, 3-dialdehyde. *Carbohyd. polym.* **2015**, *127*, 246–251. [CrossRef]

36. Kim, U.-J.; Wada, M.; Kuga, S. Solubilization of dialdehyde cellulose by hot water. *Carbohyd. polym.* **2004**, *56*, 7–10. [CrossRef]

37. Lucia, A.; van Herwijnen, H.W.G.; Oberlerchner, J.T.; Rosenau, T.; Beaumont, M. Resource-Saving Production of Dialdehyde Cellulose: Optimization of the Process at High Pulp Consistency. *ChemSusChem* **2019**, *12*, 4679–4684. [CrossRef]

Sample Availability: Samples of the compounds are available from the authors.

MDPI
St. Alban-Anlage 66
4052 Basel
Switzerland
Tel. +41 61 683 77 34
Fax +41 61 302 89 18
www.mdpi.com

Molecules Editorial Office
E-mail: molecules@mdpi.com
www.mdpi.com/journal/molecules